MINERAL NITROGEN IN THE PLANT–SOIL SYSTEM

Physiological Ecology

A Series of Monographs, Texts, and Treatises

Edited by

T. T. KOZLOWSKI

University of Wisconsin
Madison, Wisconsin

Recently published volumes

MINERAL NITROGEN IN THE PLANT–SOIL SYSTEM

R. J. HAYNES

Agricultural Research Division
Ministry of Agriculture and Fisheries
Lincoln, Canterbury, New Zealand

With Contributions by

K. C. CAMERON
K. M. GOH
R. R. SHERLOCK
Department of Soil Science
Lincoln College
Canterbury, New Zealand

1986

ACADEMIC PRESS, INC.

Harcourt Brace Jovanovich, Publishers
Orlando   San Diego   New York   Austin
Boston   London   Sydney   Tokyo   Toronto

ACADEMIC PRESS, INC.
Orlando, Florida 32887

United Kingdom Edition published by
ACADEMIC PRESS INC. (LONDON) LTD.
24–28 Oval Road, London NW1 7DX

Library of Congress Cataloging in Publication Data

Haynes, R. J.
    Mineral nitrogen in the plant-soil system.

    (Physiological ecology)
    Includes index.
    1. Plants, Effect of nitrogen on. 2. Soils–
Nitrogen content. 3. Nitrogen cycle. 4. Nitrification.
5. Plant-soil relationships. I. Title. II. Series.
QK753.N54H39 1986   581.1'335   85-28719
ISBN 0–12–334910–9 (alk. paper)

PRINTED IN THE UNITED STATES OF AMERICA

86 87 88 89   9 8 7 6 5 4 3 2 1

Contents

Chapter 2   **The Decomposition Process:**
            Mineralization, Immobilization,
            Humus Formation, and Degradation
            R. J. Haynes

Chapter 3   **Nitrification**
            R. J. Haynes

Chapter 4 Retention and Movement of Nitrogen in Soils
K. C. Cameron and R. J. Haynes

Chapter 5 Gaseous Losses of Nitrogen
R. J. Haynes and R. R. Sherlock

Chapter 6   **Uptake and Assimilation of Mineral Nitrogen by Plants**
R. J. Haynes

Chapter 7   **Nitrogen and Agronomic Practice**
K. M. Goh and R. J. Haynes

Preface

Commercial synthesis of nitrogenous fertilizers from atmospheric nitrogen has probably been the single most important factor resulting in dramatically increased crop yields over the past forty years. Indeed, nitrogen is required by plants in large quantities, and it is the most common key limiting factor to crop production when soil water supply is adequate. It is therefore not surprising that on an overall basis considerably more nitrogen than any other element is supplied to crops as fertilizer. Losses of fertilizer nitrogen from agricultural systems are, however, of considerable concern to both agriculturists and environmentalists since they not only represent an economic loss but may also result in pollution of ground or surface waters and the atmosphere. An understanding of the processes by which mineral nitrogen is formed and transformed in soils, absorbed and used by plants, and lost from the plant–soil system is therefore of particular importance from both agricultural and ecological viewpoints.

This comprehensive monograph is planned as an advanced text and reference for graduate students and researchers in the broad area of agriculture. The subject matter overlaps into a variety of disciplines and will be of interest to agronomists, soil scientists, plant physiologists, horticulturists, and foresters. This monograph fills a gap in the literature by providing an integrated account of the transformations and fate of mineral nitrogen in the plant–soil system. Throughout the text, emphasis is placed on a broad understanding of the processes being discussed and, in particular, on the major factors which influence each process. Physiological and biochemical aspects of biological nitrogen fixation are discussed in detail in a large number of reviews, so only its role in nitrogen cycling in natural and agricultural systems is discussed in this book.

The introductory chapter outlines the origin, distribution, and cycling of nitrogen in both natural and agricultural terrestrial ecosystems, and presents a broad perspective of the role and importance of mineral nitrogen in the plant–soil system. The processes of decomposition and mineralization–immobilization turnover are discussed in Chapter 2, while Chapter 3 outlines the processes of nitrification. Separate chapters follow on the adsorption of mineral nitrogen by soil components and leaching losses of nitrate, gaseous losses of nitrogen, plant uptake, translocation and use of nitrogen, and, finally, the use of nitrogen in agronomic practice.

I am indebted to my colleagues who accepted invitations to coauthor chapters in their particular fields of expertise. I also express my sincere thanks to F. E. Broadbent, W. B. Bowden, C. A. Campbell, J. R. Freney, W. A. Jackson, D. R. Keeney, H. M. Reisenauer, T. Rosswall, E. L. Schmidt, F. J. Stevenson, and A. Wild, who kindly reviewed various chapters of the book.

R. J. Haynes

Chapter 1

Origin, Distribution, and Cycling of Nitrogen in Terrestrial Ecosystems

R. J. HAYNES

I. INTRODUCTION

During ecosystem development, from unproductive rocks devoid of soil and vegetation to an ecosystem with a deep soil profile and abundant vegetation, both total biomass N and soil N increase (Stevens and Walker, 1970; Jenny, 1980). Such an increase is achieved by wet and dry deposition of atmospheric N and through the actions of a specialized group of microorganisms that fix atmospheric N_2. An equilibrium level of N is obtained within the mature ecosystem.

Indeed, when a natural ecosystem is in a steady state, the rates of both N input and loss are characteristically very small and equal. In contrast, large quantities of N are cycled within the system. On a global scale, 90 to 97% of the N content of the net primary production of plant biomass is derived from recycling of N within the biosphere (Rosswall, 1976), leaving approximately 3 to 10% as annually fixed.

Mineral N in the soil represents a very small and usually transitory pool of N in terms of the total N stock of any ecosystem. Indeed, the major forms of mineral N (NH_4^+- and NO_3^--N) usually account for less than 2% of the total N content of soils (Melillo, 1981; Woodmansee et al., 1981). It is, nevertheless, this N that is available for direct uptake by plants.

Nitrate N is easily lost from soils through leaching to groundwater and through denitrification. Maintenance of a low rate of nitrification (which itself results in gaseous losses of N) is therefore essential in N conservation in most natural ecosystems (Verstraete, 1981); in general, the rate of nitrification appears to be regulated by the supply of NH_4^+-N within the soil (Robertson and Vitousek, 1981; Adams and Attiwill, 1982). Thus, the internal cycling processes within terrestrial ecosystems that prevent accumulation of NH_4^+-N and therefore nitrification in the soil are particularly important in N conservation.

In contrast to mature natural ecosystems, many agricultural ecosystems sustain large inputs (via fertilizer and symbiotic N_2 fixation) and outputs (via leaching losses, gaseous losses, and product removal) of N. In industrial areas the N cycle has also been modified since combustion processes can lead to emission of significant quantities of N oxides to the atmosphere.

In this introductory chapter, the origin, distribution, and cycling of N within terrestrial ecosystems are discussed to present a broad perspective of the role and importance of mineral N in the plant–soil system. Where appropriate, man's influence on the N cycle is highlighted.

II. THE NITROGEN CYCLE

A. Geobiological Distribution of Nitrogen

The distribution of N on earth and within the biosphere is outlined below, followed by a discussion of the processes by which N is cycled within terrestrial ecosystems.

1. Lithosphere and Atmosphere

The bulk of the earth's N (98%) is held in rocks and minerals (Table I). In general this N exists as nitrides of iron, titanium, and other metals, or as NH_4^+ ions held in the lattice structure of primary silicate minerals (Stevenson, 1965). The primary (igneous) rocks of the earth's crust hold approximately 97.8% of the global N. These rocks contribute very little to the cycling of N; they may perhaps give up in the region of 5 Tg (10^{12} gm) N yr^{-1} via outgassing through the earth's crust and a considerably lesser quantity through volcanic action (Burns and Hardy, 1975). Only about 0.2% of the global N is in sedimentary rocks, but it has been suggested (Chalk and Keeney, 1971) that the solution leached from such rocks could make significant contributions to the N content of groundwater. Although the lithosphere contains the bulk of the earth's N, mineral weathering does not represent an important source of N to the biosphere.

Gaseous N in the atmosphere represents only 1.9% of the earth's total

Table I

Biogeochemical Distribution of N on Earth[a]

Pool of nitrogen	Total mass (Tg N)	Percentage of total N mass (%)
Atmosphere N_2	3.9×10^9	1.9
N_2 dissolved in oceans	2.2×10^7	0.01
Biosphere	2.4×10^7	0.01
Lithosphere		
Igneous rocks	1.9×10^{11}	97.8
Sedimentary rocks	4.0×10^8	0.2
Total N mass	1.94×10^{11}	

[a] Data from Stevenson (1965) and Burns and Hardy (1975).

N mass (Table I). The bulk of the atmospheric N exists as molecular N_2; the triple bond $N \equiv N$ of this diatomic molecule is very stable. Over millions of years the earth's atmosphere (78.1% N on a molar basis) has apparently constituted virtually the sole source of N for nutrition of all forms of life. The atmosphere also contains small but significant quantities of other nitrogenous compounds (Table II) such as N_2O, NO, NO_2, NH_3, HNO_3^-, NO_3^-, NH_4^+, and organic N (the latter three are present as aerosols).

Table II

Distribution of Atmospheric N

Form	Atmospheric mass (Tg N)	
	Söderlund and Svensson (1976)	Galbally and Roy (1983)
N_2	3.9×10^9	—
N_2O	1.3×10^3	1.5×10^3
NH_3	0.9	1.7
NH_4^+	1.8	0.4
NO_x	2	0.6
NO_3^-	0.5	0.1
HNO_3	—	0.2
Organic N	1	—

2. Biosphere

In comparison with N contained in the atmosphere and lithosphere, the quantities present in the biosphere are very small (approximately 0.01%, Table I). The predominant form of N in the biosphere is, in fact, N_2 gas dissolved in the oceans. The distribution of biomass N between land and oceans is illustrated in Table III. Organic N constitutes a sizable proportion of the N contained by oceans (50%) and land (73%). On land the microbial conversion of this dead organic N into mineral forms (NH_4^+ and NO_3^-) supports the active growing biomass (plants and animals).

Nevertheless, mineral N makes up only a very small proportion of the calculated land biomass N (Table III). Rosswall (1976) calculated that apportionment of N for the global terrestrial distribution was as follows: plants 4%, plant litter 1%, microorganisms 0.2%, and soil organic matter 94%. Less than 1% of terrestrial N was stored as the mineral (plant available) forms in the soil.

3. Ecosystems

The range of variation in the distribution of N among the major pools in terrestrial ecosystems is very wide (Table IV). The Amazonian caatinga forest, situated on low fertility soils, represents an extreme case. The partitioning of N within the different components of this ecosystem in fact favors the living component by over 55%. The trees of the Amazonian

Table III

Global Distribution of N Within the Biosphere[a]

Pool of nitrogen	Tg N	
	Oceans	Land
Soluble N		1,000
NH_4^+	5,000	
NO_3^-	650,000	
NO_2^-	5,000	
Insoluble N		100,000
Coal N		100,000
Organic N	650,000	550,000
Animals	3,000	1,000
Plants	1,000	10,000
	1,300,000	750,000
N_2 dissolved in sea	22,000,000	

[a] Data from Burns and Hardy (1975).

forest have a very extensive fine root system that is presumably an adaptation to low fertility conditions. This explains why the root mass constitutes the most important N pool.

In both the deciduous oak–hickory forest and the shortgrass prairie of the United States most of the N is present in the soil (Table IV). The relatively greater proportion of N in the total and aboveground vegetation in the deciduous forest compared with the grassland means that removal of biota has a much greater effect on the N cycle of forests than that of grasslands. For example, severe fires may have little effect on the N cycle in grasslands but may have catastrophic effects on N cycling in forests (Woodmansee and Wallach, 1981).

In situations where environmental conditions do not favor vigorous plant growth (e.g., desert or tundra ecosystems), an overwhelming proportion of the organic N is present in the soil (Skujins, 1981; Van Cleve and Alexander, 1981). For example, in an Alaskan wet meadow tundra (Table IV) over 98% of the total organic N was in the soil.

B. Cycling Processes

1. Additions, Losses, and Transformations

A generalized N cycle is shown in Fig. 1. The magnitude of the global flows of N due to some of the processes illustrated in Fig. 1 is shown in Table V. The major additions of N to the soil occur through the processes of wet and dry deposition and by the action of microorganisms that fix atmospheric N_2. Fixation of N_2 can also occur through the action of lightning. Man is increasingly active in fixing N_2, both intentionally by industrial processes and unintentionally by use of internal combustion engines. In many areas, most nitrates in precipitation originate from sources related to man. The quantity of N fixed industrially and applied to agricultural lands is of the same order as that fixed by microorganisms (Table V).

Losses of N occur through leaching of NO_3^-, erosion and surface runoff, volatilization of ammonia, gaseous losses of N_2 and N_2O, and, in agricultural ecosystems, plant or animal removal. The estimates in Table V indicate that the greatest losses of N from terrestrial ecosystems originate from denitrification. In the context of Table V, however, denitrification simply refers to gaseous losses of N_2O and N_2. In fact, as illustrated in Fig. 1, N_2O emission can occur as the result of two separate processes: (1) microbial reduction of NO_3^- to yield N_2O and N_2 (dentrification) and (2) microbial oxidation of NH_4^+ to NO_2^- (the first step of nitrification). The relative importance of the two processes of N_2O production is, as yet,

Table IV

Distribution of N in a Nutrient-Poor Amazonian Rain Forest, an Oak–Hickory Forest, a Shortgrass Prairie, and a Wet Meadow Tundra Ecosystem

Nitrogen pool	Amazonian forest[a] kg N ha^{-1}	%	Oak–hickory forest[b] kg N ha^{-1}	%	Shortgrass prairie[c] kg N ha^{-1}	%	Meadow tundra[d] kg N ha^{-1}	%
Total vegetation	1179	56.2	492	8.4	66	1.9	0.92	1.0
Above ground	336	16.2	357	6.1	15	0.4	0.24	0.3
Roots	843	40.0	135	2.3	51	1.5	0.68	0.7
Litter	132	6.3	274	4.7	76	2.1	0.10	0.1
Soil	785	37.7	5080	86.9	3374	96.0	91	98.9
(sampling depth in cm)	(75 cm)		(60 cm)		(30 cm)		(20 cm)	
Total	2096		5846		3516		92	

[a] From Herrera and Jordan (1981).
[b] From Henderson and Harris (1975).
[c] From Woodmansee et al. (1981).
[d] From Van Cleve and Alexander (1981).

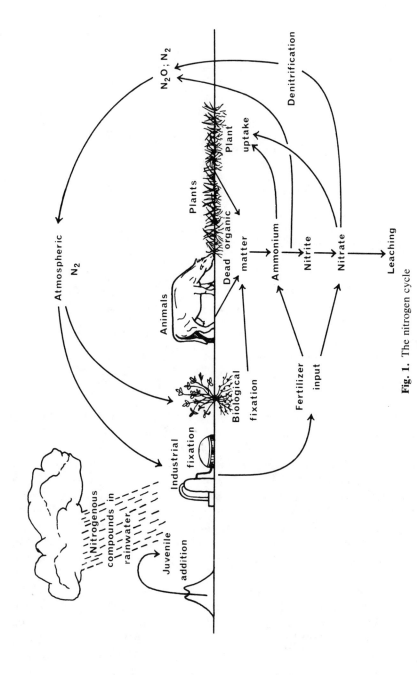

Fig. 1. The nitrogen cycle

Table V

Global Fluxes of Nitrogen into and out of the Terrestrial Biosphere[a]

	Process rate (Tg N yr^{-1})
Inputs	
Wet and dry deposition (NH_3/NH_4^+)	90–200
Wet and dry deposition (NO_x)	30–80
Wet and dry deposition (organic N)	10–100
Atmospheric fixation (lightning)	0.5–30
Biological fixation	100–200
Industrial fixation (fertilizers)	60
Outputs	
Ammonia volatilization	36–250
Denitrification ($N_2 + N_2O$)	40–350
Biogenic NO_x production	1–15
Fossil fuel burning (NO_x)	10–20
Fires (NO_x)	10–20
Leaching and runoff (inorganic)	5–20
Leaching and runoff (organic)	5–20

[a] Data compiled from Delwiche (1977), Söderlund and Rosswall (1982), Crutzen (1983), Galbally and Roy (1983), and Rosswall (1983).

unknown. Nonetheless, in terms of total losses of N, denitrification is by far the most important process since losses of N_2 during denitrification are an order of magnitude greater than total losses of N_2O (Rosswall, 1983).

An internal N cycle operates within the plant–soil system. Nitrogenous organic residues are microbially decomposed with the release of NH_4^+-N (mineralization), which can then be oxidized by microorganisms to NO_3^--N (nitrification). These mineral forms of N (NH_4^+ and NO_3^-) are utilized by microorganisms (immobilized), particularly during decomposition of organic residues with a low N content. The positively charged NH_4^+ ion can be held by negatively charged soil colloids or fixed by clay minerals but the NO_3^- anion is highly mobile in soils.

The major portion of the pool of mineral N that is not immobilized by soil microorganisms is absorbed and assimilated by plants during their growth. The organic N in the plant material may then be either consumed by animals or returned directly to the soil following death of plants. In either event, the eventual repository is the soil, where microbial decomposition again returns N to an inorganic form for repetition of the cycle.

The residence time of N in soil is highly variable since the longevity of an N-containing molecule in the soil is a function of its solubility, the degree to which it is bound to the soil colloids, and the ease with which it is broken down by microorganisms. Thus a given N atom may be cycled rapidly or may persist in the soil for years or even centuries.

It is evident from the preceding discussion that the distribution of N in its various pools and transfer of N between these pools is dominated by biological processes. The cycle is maintained by input of solar energy to the plant biomass, which allows it to incorporate N from the inorganic pool of mineral N in the soil. This pool is maintained by microorganisms, which decompose litter and soil organic matter.

2. Cycling within Ecosystems

Thus far, the distribution of N within the biosphere and the processes involved in the terrestrial N cycle have been outlined. In this section, the relatively closed N cycle of mature natural ecosystems is contrasted with the relatively open cycle in agricultural ecosystems.

When a natural ecosystem is in a steady state (e.g., a mature forest) the rate of N input by precipitation and biological N_2 fixation balances outputs by denitrification, volatilization, and groundwater and stream loss. Furthermore, in such ecosystems little N enters or leaves the system in comparison with the quantity that is cycled annually by vegetation (Rosswall, 1976). Indeed, within the global plant–soil system the quantity of N cycled annually by vegetation is approximately 2500 Tg N yr^{-1} compared with losses of 194 Tg N yr^{-1} and gains of 194 Tg N yr^{-1} (Rosswall, 1976).

In a compilation of nutrient budgets from 14 temperate deciduous and coniferous forests of the world, Likens *et al.* (1977) reported an average streamwater loss of N of 2.0 kg N ha^{-1} yr^{-1} (range 0.4 to 5.6 kg N ha^{-1} yr^{-1}) and an average precipitation input of 8.0 kg N ha^{-1} yr^{-1} (range 1.1 to 20.7 kg N ha^{-1} yr^{-1}). Although such data do not account for gaseous losses of N, nor inputs through biological N_2 fixation, they do demonstrate that in temperate forests inputs and outputs of N are small (only a few kg N ha^{-1} yr^{-1}) compared to the amounts of N annually cycled by the vegetation (30 to 90 kg N ha^{-1} yr^{-1}).

When natural ecosystems are disturbed by natural or anthropogenic events (e.g., following an intense forest fire or harvesting of a forest) the internal N cycle is interrupted because of the death and/or removal of vegetation. Thus losses of N as gases and through leaching may then increase markedly (Bormann and Likens, 1979; Khanna, 1981).

There is a great contrast between nutrient cycles in natural ecosystems and those in agricultural ecosystems. In the latter the cycle can be inter-

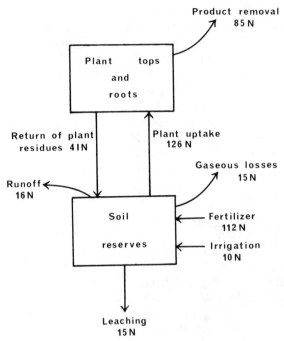

Fig. 2. The nitrogen cycle in a corn crop in northern Indiana, USA. Values in kg N ha⁻¹
yr⁻¹. [Data from Thomas and Gilliam (1978).]

rupted by product removal. The nitrogen cycle for a corn crop grown for
grain is shown in Fig. 2. The corn crop is harvested and removed from the
field and 85 kg N ha⁻¹ yr⁻¹ is lost. When losses of N through runoff,
leaching, and gaseous losses are included, total losses amount to 131 kg N
ha⁻¹ yr⁻¹. Such losses are large when compared with the amount of N
cycled by the crop plant (41 kg N ha⁻¹ yr⁻¹) and enormous in comparison
with estimated losses of N from natural ecosystems, which are only a few
kg N ha⁻¹ yr⁻¹. In order to balance such losses, fertilizer inputs of 112 kg
N ha⁻¹ yr⁻¹ are required while some N inputs also occur in irrigation
water.

III. ADDITIONS OF NITROGEN TO ECOSYSTEMS

A. Wet and Dry Deposition

Wet and dry deposition processes are important in the redistribution of
N in terrestrial ecosystems. In wet deposition, gaseous and particulate

matter is removed from the atmosphere by precipitation including both rain and snow. In dry deposition, removal takes place by gravitational settling, turbulent transport in eddies, molecular diffusion, and impaction by inertia.

In Section IV, emissions of ammonia (NH_3), nitric oxide (NO), and nitrogen dioxide (NO_2) to the atmosphere from the plant–soil system are discussed. The gases NO and NO_2 are often referred to collectively as NO_x. The atmospheric lifetime of both NH_3 and NO_x is approximately 1–2 weeks (Galbally and Roy, 1983). The bulk of these gases is returned to the earth's surface by wet and dry deposition.

Ammonia readily dissolves and ionizes to NH_4^+ in an atmospheric water vapor with the formation of aerosols (suspensions of liquid or solid material in the atmosphere). These aerosols, containing NH_4^+ salts such as $(NH_4)_2SO_4$, NH_4HSO_4, and NH_4NO_3, are then returned to the earth's surface by wet and dry deposition (see Chapter 5).

In the atmosphere, emitted NO is quickly oxidized to NO_2 by atmospheric ozone. A substantial fraction of the NO_2 undergoes hydrolysis with the formation of HNO_3. Nitrate is incorporated into atmospheric aerosols by condensation of HNO_3 vapor into existing aerosols. Such aerosols are then removed from the atmosphere by wet and dry deposition. The dry deposition process also involves gaseous deposition of both NH_3 and NO_x back to the plant–soil system (Chapter 5).

Estimates of total quantities of N reaching the global land masses by wet and dry deposition are approximately 100–280 Tg N yr^{-1} with NH_3/NH_4^+ inputs being 90–200 Tg N yr^{-1} and NO_3^- inputs being 30–80 Tg N yr^{-1} (Table V). Of these inputs, approximately half is deposited by dry deposition and half by wet deposition (Söderlund and Svensson, 1976; Galbally and Roy, 1983). Such values do not represent total global values for wet and dry deposition since significant quantities of atmospheric N are deposited over the oceans.

1. Wet Deposition

The amounts of N deposited during wet deposition are reasonably well known. Measurements for different geographical regions and localities are available in the literature (Eriksson, 1952; Steinhart, 1973; Likens *et al.*, 1977; Söderlund, 1977; Böttger *et al.*, 1978). However, due to large variations in the chemical composition of rainfall over large areas as well as small-scale variations in time and space it is difficult to make generalizations.

For example, maximum deposition of NO_3^- occurs in industrialized areas since nitrogen oxides, which are released during the combustion of fossil fuels, constitute the major source of nitrates in precipitation. Söderlund (1977) estimated that wet deposition of NO_3^--N in preindustrial times in northwestern Europe was approximately 0.2 Tg N yr^{-1} in comparison with a 1977 estimate of 1.4 Tg N yr^{-1}.

High deposition rates of ammonium occur in eastern Europe and southeast Asia (Söderlund, 1981). This may be due to agricultural activity, especially the great number of livestock in these areas since the excreta of livestock is thought to be a large source of atmospheric ammonia (Söderlund and Svensson, 1976).

2. Dry Deposition

The deposition of N in gaseous form or in particulate matter has not received a great deal of study. With the exception of NH_3, little reliable quantitative or qualitative data is available regarding deposition rates of nitrogenous gases and much research into absorption of such gases by plants and other surfaces has been carried out in polluted or laboratory atmospheres at concentrations well above tropospheric background levels.

It is, however, known that NH_3 is extremely soluble in water and it can be readily absorbed by water bodies, vegetation, and soils (see Chapter 5). The effect of growing plants acting as a natural sink for gaseous NH_3 may have a profound effect in reducing the magnitude of NH_3 volatilization from soils with plant cover (Denmead et al., 1976, 1978).

Soils have been shown to be capable of sorbing both NO_2 (Abeles et al., 1971) and N_2O (Freney et al., 1978) from the atmosphere while vegetation has been shown capable of removing significant quantities of NO_2 and NO from artificial atmospheres (Tingey, 1968; Hill and Chamberlain, 1976).

B. Biological Nitrogen Fixation

Biological fixation of atmospheric N_2 is accomplished solely by prokaryotic organisms living freely or in association with certain plants. Almost one-quarter of the estimated global biological N_2 fixation is carried out by the root nodule bacterium, *Rhizobium,* in association with agricultural legumes (Table VI). The remainder is fixed by various bacteria and actinomycetes either living freely or in association with vegetation such as ferns, grasses, shrubs, or trees.

The process of biological N_2 fixation is estimated to add approximately 100 to 200 Tg N yr^{-1} to terrestrial ecosystems on a global basis (Table V). The N_2 fixed by microorganisms is released into the soil upon microbial decomposition; hence soil factors affecting mineralization also affect the quantity of fixed N_2 that is released as mineral N. Similarly, below-ground transfer of the N_2 fixed symbiotically by legumes and nonlegumes is by decay of roots and root nodules.

Table VI

Estimates of Nitrogen Fixation in Various Ecosystems[a]

Ecosystem	Area (10^6 ha)	Dinitrogen fixed (kg N ha^{-1} yr^{-1})	Nitrogen input (Tg N yr^{-1})
Natural			
Wasteland	4900	2	10
Forest	4100	10	40
Agricultural			
Legumes	250	140	35
Nonlegumes	1150	35	9
Grassland	3000	15	45

[a] Source: Burns and Hardy (1975).

1. Nitrogen-Fixing Organisms

Organisms that can fix atmospheric N_2 can be divided into five broad groupings: (1) free-living bacteria; (2) symbiotic cyanobacteria; (3) rhizocoenoses; (4) actinomycete nodule symbiosis; and (5) *Rhizobium* symbiosis.

Three groups of free-living bacteria fix N_2: the heterotrophs (Jensen, 1981), chemoautotrophs (Dalton, 1981), and photoautotrophs (Gallon, 1981). Free-living heterotrophs capable of fixing N_2 (e.g., *Azotobacter* and *Beijerinckia*) are common in most soil habitats. These organisms require organic matter as a source of carbon and energy substrate and their capacity to fix N_2 is generally low (Knowles, 1977). Both chemoautotrophs and photoautotrophs are able to fix CO_2 and N_2 but the latter utilize light energy rather than energy derived from chemical reactions to supply their metabolic needs. The three major groups of photoautotrophs are the photosynthetic green bacteria (Chlorobiaceae), the purple bacteria (Rhodospirillaceae), and the cyanobacteria (also known as blue-green algae).

There is increasing evidence of the role of cyanobacteria in the N economy of rice paddies (Ito and Watanabe, 1981; Wetselaar, 1981) although there is uncertainty in regard to both the levels fixed and the dominant N_2-fixing species. Values of N_2 fixation of the order of 30 kg N ha^{-1} yr^{-1} have been reported (Balandreau *et al.*, 1975).

The cyanobacteria may also be found growing in association with algae, fungi (e.g., as lichens), bryophytes, ferns, gymnosperms, and angiosperms (Stewart *et al.*, 1981).

Bacteria associated with the roots of a large number of plants, most of them Gramineae, are known to fix N_2 (Döbereiner and De-Polli, 1981). As yet, many of these associations (known as rhizocoenoses) have not been adequately defined. The most well known is the *Azospirillum* rhizocoenosi, which is common in many parts of the world (Dobereiner and De-Polli, 1981). The amounts of N_2 fixed are variable and usually small (often in the range of 10 to 50 kg N ha^{-1} yr^{-1}).

Actinomycete root nodule symbiosis of angiosperms has been confirmed in at least 140 plant species belonging to 17 genera (Akkermans and Roelofsen, 1981). Many nodulated plants such as *Alnus, Hippophae,* and *Casuarina* spp. are typical colonizers of barren, nitrogen-poor soils.

By far the greatest contribution to soil N comes from the symbiotic association between bacteria of the genus *Rhizobium* and members of the plant family Leguminosae. There are approximately 700 different genera of legumes and approximately 14,000 individual species, although only 200 or so are exploited in agriculture. On the basis of their restricted host range the *Rhizobium* bacteria are classified into six major species groups: the slow-growing *R. japonicum* and *R. lupini* species and the faster-growing *R. leguminosarum, R. meliloti, R. phaseoli,* and *R. trifolii* species.

Burns and Hardy (1975) estimated that the quantity of N fixed by *Rhizobium* symbiosis in temperate agricultural soils was in the range of 50 to 300 kg ha^{-1} yr^{-1} while that for free-living soil bacteria was estimated at 0.4 to 0.8 kg N ha^{-1} yr^{-1}. The high fixation rates recorded for symbiotically associated bacteria is apparently related to the direct coupling of the energy-producing photosynthetic system of higher plants with the dinitrogen-fixing process.

2. Processes of Nitrogen Fixation

The various aspects of the physiology and biochemistry of biological N_2 fixation have been reviewed extensively elsewhere (e.g., Bergersen, 1981; Burris *et al.,* 1981; Eady *et al.,* 1981; Evans *et al.,* 1981; Stewart, 1980; Robson and Postgate, 1980). The major features are briefly outlined below since the process is not dealt with in later chapters.

The nitrogenase enzyme complex is responsible for reduction of dinitrogen to ammonia in all the major groups of N_2-fixing bacteria (Child, 1981). Nitrogenase consists of two iron–sulfur proteins. The operation of the nitrogenase system is shown in Fig. 3; its major requirements include a source of reductant and a supply of energy. During the fixation process electrons flow from the reducing agent (e.g., ferredoxin) to the MgATP–iron protein complex. The reduced MgATP–iron protein then effects the reduction of the Mo–iron protein, which is, in turn, implicated in reduction of N_2.

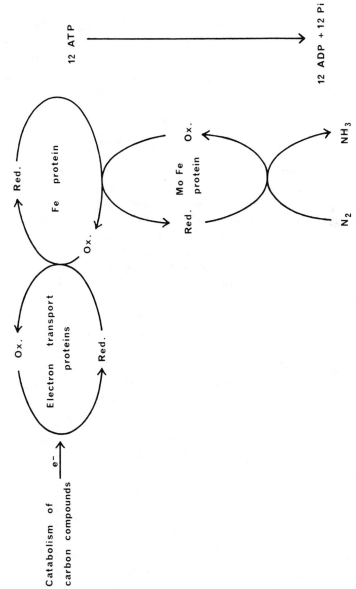

Fig. 3. Operation of the nitrogenase system.

C. Industrial Nitrogen Fixation

1. Usage of Fertilizer Nitrogen

The high-producing agricultural systems of the western world that are characterized by intensive production methods rely on large fertilizer inputs to sustain productivity (Bollin and Arrhenius, 1977). This is because little, if any, of the organic wastes from domestic and farm sources are returned to the soil since the crop is fed to animals or people located elsewhere. Indeed, agricultural ecosystems often have a greater N requirement than natural ecosystems (see Section V) while, as illustrated in Fig. 2, much of the N used by crop plants is harvested and removed from the ecosystem. On a global basis Child (1981) estimated that 140 Tg N yr^{-1} are removed as agricultural produce from the world's arable land.

Although a small quantity of natural products such as sodium nitrate is used, by far the major portion (over 80%) of fertilizer N that is applied to agricultural lands originates from the commercial synthesis of ammonia.

The magnitude and intensity of fertilizer N usage on a world scale are shown in Table VII. There was a large increase in the use of fertilizer N between 1976 and 1980 that amounted to approximately 10 Tg N yr^{-1}. Current estimates are that global inputs of fertilizer N are of the order of 30 to 60% of those supplied by biological N$_2$ fixation (Table V). Fertilizer N is primarily applied to arable lands (areas listed in Table VII) although a

Table VII

Consumption and Intensity of Fertilizer Nitrogen Usage in Regions of the World[a]

Region	Arable land (×1000 ha) 1976	N consumption (Tg N yr^{-1}) 1976	N consumption (Tg N yr^{-1}) 1980	Average intensity of use (kg N ha^{-1} yr^{-1}) 1976	Average intensity of use (kg N ha^{-1} yr^{-1}) 1980
North America	230,129	10.0	11.2	43.5	48.6
Latin America	116,732	2.0	2.9	17.1	24.8
Western Europe	76,130	7.7	8.8	101.1	115.6
Eastern Europe	51,303	4.5	5.4	87.7	105.3
USSR	227,400	7.3	9.1	32.1	40.0
Asia	454,908	10.2	14.2	22.4	31.2
Africa	194,910	1.3	1.7	6.7	8.7
Oceania	46,212	0.19	0.28	4.1	6.1
Total		43.19	53.58		

[a] From Hauck (1981).

minor portion of the total is applied to plantation crops, forests, and permanent pastures. The intensity of fertilizer use (calculated in terms of arable land area) gives an indication of the rate of fertilizer N application in the various regions.

2. Manufacture of Fertilizer Nitrogen

Ammonia is synthesized commercially by the Haber process. Hydrogen and nitrogen are combined in a 3 : 1 ratio at elevated temperature (300 to 500°C) and pressure (400 to 1000 atmospheres) in the presence of a catalyst (e.g., reduced iron):

$$N_2 + 3H_2 \rightleftharpoons 2NH_3$$

Natural gas is now the commonest source of hydrogen for the process, although naphtha, water, coal, or oil are sometimes used. Much of a modern ammonia plant is involved in the manufacture of a pure hydrogen/nitrogen mixture ready to pass over the iron catalyst.

Detailed accounts of the methods and reactions involved in the commercial synthesis of ammonia have been presented elsewhere (Snyder and Burnett, 1966; Pesek et al., 1971; Rankin, 1978).

IV. LOSSES OF NITROGEN FROM ECOSYSTEMS

A. Gaseous Losses

Numerous nitrogenous compounds are emitted or escape from land or oceans into the atmosphere. The most abundant oxide of N in the lower atmosphere is N_2O (Table II), although NO and NO_2 are more important in terms of air pollution since they play a prominent role in the generation of photochemical smog. Recently, however, N_2O emissions have received considerable attention because of concern that an increase in the atmospheric concentration of N_2O might contribute to the degradation of the stratospheric ozone layer, which helps absorb harmful UV radiation from the sun (Crutzen, 1974; see Chapter 5).

Typical concentrations of gaseous nitrogenous compounds in the rural atmosphere are N_2O, 330 ppb; NH_3, 1–5 ppb; NO + NO_2, 1–10 ppb; and HNO_3, 0.1–5.0 ppb (Rasmussen et al., 1975; Galbally and Roy, 1983). Concentrations of NO, NO_2, and thus HNO_3 in many urban atmospheres are high due to the dominating influence of combustion processes. Typical concentrations of N oxides in urban atmospheres are NO + NO_3, 20–500 ppb, and HNO_3, 1–20 ppb (Galbally and Roy, 1983).

1. Dinitrogen

The loss of N_2 from the terrestrial biosphere amounts to 25–280 Tg N yr^{-1} (Rosswall, 1983). Although such losses are relatively large in relation to other losses of N (Table V), they are rather small in relation to the atmospheric mass of N_2 (Table II). The major source of the loss of N_2 is biological denitrification, which occurs under anaerobic soil conditions (Chapter 5).

2. Nitrogen Oxides

a. NO_x. The major sources of NO_x in the atmosphere are formation of NO and NO_2 during combustion and emissions from soils (Table V). Small quantities of NO_x are also produced by oxidation of atmospheric NH_3 and by meteoroid impact on the upper atmosphere.

Anthropogenic sources of NO_x are primarily combustion processes in which the temperature is high enough to oxidize atmospheric N_2 (Knelson and Lee, 1977). NO_x is also produced during combustion by oxidation of N compounds contained in fossil fuels such as coal and oil. Indeed, combustion of gasoline and diesel oils can be a major source of atmospheric NO_x (Table VIII) and can constitute 20 to 50% of the anthropogenic sources (Söderlund, 1977). Estimates of global NO_x emissions from combustion of fossil fuels are often in the range of 10 to 20 Tg N yr^{-1} (Robinson and Robbins, 1970; Crutzen, 1983).

Burning of forests and other vegetation is also a source of NO_x. The NO_x resulting from this process is mainly as a product of oxidation of fixed N already bound in the biomass. Few estimates of the magnitude of this source of NO_x have been reported though Söderlund and Rosswall (1982) estimated 17 Tg N yr^{-1} and Crutzen (1983) suggested 20 Tg N yr^{-1}.

Emissions of NO_x directly from the soil are thought to occur through the process of chemodenitrification (see Chapter 5), which involves several chemical reactions of NO_2^- with soil constituents with the release of NO_x. Some laboratory studies have also linked NO emission to the actions of autotrophic nitrifying bacteria, which oxidize soil NH_4^+ to NO_2^- (Lipschultz et al., 1981). The proposed range for NO_x release from soils is 1–15 Tg N yr^{-1} (Söderlund and Svensson, 1976; Crutzen, 1983; Galbally and Roy, 1983).

b. N_2O. The major sources of N_2O emission are thought to be through the actions of soil microorganisms. Anaerobic denitrifying bacteria reduce soil NO_3^- and N_2O and N_2 while N_2O is also released by aerobic autotrophic nitrifying bacteria, which oxidize soil NH_4^+ to NO_2^- (see Chapter 5). While global estimates of total denitrification losses ($N_2O + N_2$) are about 40–350 Tg N yr^{-1} (Table V), losses of N_2O are estimated at about

Table VIII

Sources of Anthropogenic NO$_x$ over Northwestern Europe[a]

Source	Emission (Tg N yr^{-1})
Coal	
Lignite-brown coal	1–3
Hard coal	0.1–0.5
Other combustion	0.1–1.5
Oil	
Electric power generation	0.3–0.6
Gasoline	0.8–1.2
Diesel fuels	1.2
Other combustion	0.1–0.8
Natural Gas	
Electric power generation	0.03–0.1
Industrial combustion	0.1–0.4
Domestic combustion	0.02–0.03
Industrial processes	
Petroleum refinery	0.06
Steel manufacturing	0.01–0.1

[a] Data from Söderlund (1977).

15–70 Tg N yr^{-1} (Rosswall, 1983). No reliable estimates of N$_2$O losses during nitrification have been made on a global scale. Although concern has been expressed that the increasing use of fertilizer N may greatly increase losses of N$_2$O through denitrification, it has been calculated that the global source of N$_2$O from N fertilization is only a few Tg N yr^{-1} (Crutzen, 1981).

An increasing amount of N$_2$O is being produced by combustion of fossil fuels (Weiss and Craig, 1976) and this source may be increasing at a rate of 3.5% per year. Indeed, Weiss (1981) found that stratospheric N$_2$O levels were increasing at a rate of 0.2% annually and that this increase could be satisfied by the increases in N$_2$O emissions from fossil fuel burning. Estimates of losses of N$_2$O by combustion of fossil fuels are about 1–3 Tg N yr^{-1} and those for burning of vegetation 1–2 Tg N yr^{-1} (Crutzen, 1983).

3. Ammonia

The volatilization of NH$_3$ from the soil surface requires a supply of free ammonia (NH$_3$(aq) and NH$_3$(g)) near the soil surface (see Chapter 5). The origin of this free NH_3 in the soil is the NH_4^+ ion. The decomposition of amino acids, amides, and protein in dead plants, animals, and microorganisms is evidently the major global source of soil NH_4^+ and thence atmospheric NH_3 (Freney *et al.*, 1981). Substantial amounts can also come from animal excreta and ammoniacal fertilizers.

Losses of NH_3 due to decomposition processes have been estimated at 27 Tg N yr^{-1} (Dawson, 1977), those due to wild animal excreta 2–6 Tg N yr^{-1} (Söderlund and Svensson, 1976), and those due to domestic animals 20–30 Tg N yr^{-1} (Söderlund and Svensson, 1976; Crutzen, 1983). Losses of NH_3 caused by applications of fertilizers are estimated at only 3 Tg N yr^{-1} (Crutzen, 1983).

No estimates of losses of NH_3 from plants are available since plants can both absorb and emit NH_3 (Chapter 5). Ammonia emitted from the soil surface can be reabsorbed within the plant canopy above (Denmead *et al.*, 1976) thus reducing potential losses of N. Ammonia can, however, be emitted from the top of the canopy.

Anthropogenic sources of NH_3 (combustion of fossil fuels) are small in comparison with biological sources amounting to 4–12 Tg N yr^{-1} (Söderlund and Svensson, 1976). Losses of NH_3 due to burning of vegetation are also likely to be small.

B. Leaching Losses

The mobility of the two major forms of mineral N in soils (NH_4^+ and NO_3^-) differs markedly. Ammonium is unlikely to be leached because (1) NH_4^+ is held in soil by cation exchange, fixation by clay lattices, and microbial immobilization, and (2) under many conditions, NH_4^+-N is quickly nitrified to NO_3^-. In contrast to NH_4^+, there is little tendency for the NO_3^- anion to be absorbed by soil colloids, which commonly possess a net negative charge. Nitrate N is thus susceptible to diffusion and transport in soil water.

1. Processes of Leaching

Two of the major factors controlling leaching losses of NO_3^- are the quantity of water passing through the soil profile and the concentration of nitrate in the soil profile at the time of leaching.

Most of the soil NO_3^- in terrestrial ecosystems is derived from mineralization and subsequent nitrification of organic N. Inputs of N in rainfall are usually very low, sometimes higher and sometimes lower than the corresponding quantities exported in percolating soil water (Likens *et al.*, 1977; Feller and Kimmins, 1979). Moisture and temperature are the major environmental factors controlling mineralization and nitrification in eco-systems. These processes are discussed in detail in Chapters 2 and 3, respectively.

The processes involved in leaching of NO_3^- and the factors influencing the quantities leached are discussed in detail in Chapter 4.

2. Ecosystem Disturbance

The amount of nitrate leached from natural and agricultural ecosystems often increases following natural or anthropogenic disturbances (Khanna, 1981) such as fire, harvesting, fallowing, cultivating, and fertilizing.

Figure 4 demonstrates the effect of clear-cutting an experimental forest watershed on levels of NO_3^- in streamwater. Massive losses of NO_3^- occurred immediately after disturbance. This apparently was due to uncoupling of forest floor mineralization and nitrification from nitrogen uptake

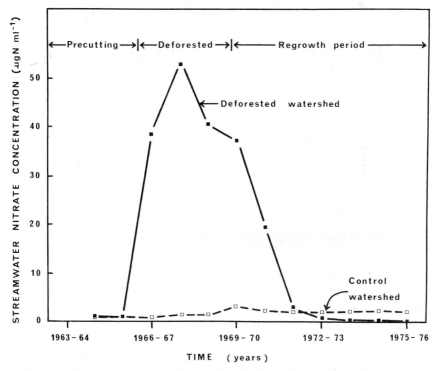

Fig. 4. Annual weighted concentrations of nitrate in streamwater from a forested reference watershed and an experimentally devegetated watershed. Phases of the devegetated watershed are indicated. [Data from Bormann and Likens (1979).]

by growing vegetation (Bormann and Likens, 1979). Plant uptake of N may be extremely important in minimizing leaching losses of NO_3^- from natural ecosystems. Indeed, Vitousek and Reiners (1975) argued that losses of NO_3^- from ecosystems are at a minimum in the intermediate stages of ecological succession (see Section VI,B), when plant biomass accumulation (and N-uptake rate) is greatest.

In agricultural ecosystems, which often are continually disturbed, leaching losses of NO_3^- can be large (Wild and Cameron, 1980). Studies on agricultural lands have indicated that leaching of applied fertilizer N can be substantial, and that NO_3^--N can move rapidly, especially in light sandy soil under intensive irrigation (Endelman et al., 1974; Kissel et al., 1974; Chichester and Smith, 1978). Leaching losses of NO_3^- from fertilized irrigated corn (Zea mays) crops in the United States can be in the range of 10 to 100 kg N ha^{-1} (Smika et al., 1977; Gast et al., 1978; Timmons and Dylla, 1981) with fertilizer inputs of 100–300 kg N ha^{-1}.

However, a major source of groundwater NO_3^- from agricultural land can originate from mineralization of organic N following cultivation, rather than from the fertilizer itself (Kolenbrander, 1975).

C. Soil Erosion

In terms of ecosystem stability, soil erosion may be the most important potential destabilizing force for terrestrial ecosystems (Bormann and Likens, 1979). This is because in its most severe form (mass soil movement) a significant proportion of the soil profile can be removed so that part of the ecosystem is returned to a more primitive level of development with lower production. On the other hand, erosion can often represent a transfer of soil, and thus N, from one part of an ecosystem to another, rather than a loss of N from the ecosystem.

1. Processes of Erosion

Losses from wind or water erosion depend mainly on plant cover, topography, inherent soil stability, and intensity of wind or runoff events (Troeh et al., 1980). In many natural grassland and forest ecosystems, the vegetation cover is adequate to overcome the adverse effects of steep topography, unstable soils, and high intensities of rain or wind. Thus erosion losses from such ecosystems usually are negligible (Kilmer, 1974; Likens et al., 1977) and normally are not included in their N budgets (Gosz, 1981; Melillo, 1981; Woodmansee et al., 1981). However, significant losses of N through erosion mechanisms can occur in natural arid (Skujins, 1981) and semiarid (Bate, 1981) ecosystems, although the magnitude of such losses varies widely (see Table IX).

Table IX

Losses of Nitrogen through Erosion from Two Arid Ecosystems[a]

	kg N ha^{-1} yr^{-1}			
	Great Basin Desert		Sonoran Desert	
Erosion type	Mean	Range	Mean	Range
Wind erosion	1.6	(0.2–2.5)	1.2	(0.2–3)
Water erosion	3.4	(0.3–5.1)	1.7	(1.0–5.6)

[a] Data from Fletcher et al. (1978).

2. Ecosystem Disturbance

Erosion (particularly runoff) can be severe when natural or anthropogenic disturbances destroy vegetation (Troeh et al., 1980). Such disturbances include fire, cultivation, logging, overgrazing, mining, and construction. For example, clear-cutting of a forest can affect many erosional processes, causing substantial losses of particulate N in susceptible sites (Vitousek, 1981; see Chapter 4).

In cultivated agricultural ecosystems, erosion by wind and water has resulted in reduced productivity of many soils (Moldenhauer, 1980; Young, 1980). The major causes of such erosion are lack of a complete vegetation cover during much of the year (Allison, 1973) and disturbance of the topsoil by tillage practices such as plowing (Romkens et al., 1973). Losses of N by runoff from cultivated agricultural lands can be quite large. For example, Alberts et al. (1978) found the mean losses of N contained in eroded sediments in runoff from two fertilized contour-farmed, corn-cropped watersheds to be 50.8 and 62.9 kg N ha^{-1} yr^{-1}, respectively.

V. TRANSFERS OF NITROGEN WITHIN ECOSYSTEMS

Most ecosystems rely chiefly on the flow of N from soil to plant and back to the soil and on its conversion to forms available for the next cycle of plant growth. Return to the soil occurs primarily as detritus (dead organic matter). This consists mainly of plant litter, but also feces and carcasses of herbivores and their predators. The detritus is broken down by the combined action of the decomposer community, which is composed of bacteria, fungi, protozoa, and invertebrate animals.

The storages and transfers of N within the global plant–soil system are shown in Fig. 5. It is evident that in comparison with the amount of N resident in the soil, the quantity of N cycled is small. Calculated turnover times for the various compartments of the global system (Fig. 5) along with data from two selected ecosystems are shown in Table X. The large buffering pool of soil N has a slow turnover time while the N in microorganisms and in inorganic form has a very short residence time. The rate of turnover of N in litter and plants is about 10 times that of N in microorganisms and in inorganic form but approximately one-hundredth of that in soil organic matter.

A. Uptake of Nitrogen by Plants

Nitrogen plays a central role in plant productivity because it is a major component of amino acids, proteins, nucleic acids, and chlorophyll. Organic N commonly constitutes 1.5 to 5% of the dry weight of plants, although there is some variation with age, species, and plant organ. In leaves and stems approximately 60% of the N is present as enzyme or membrane protein and most of the remainder is in the form of free amino acid nitrogen (Parsons and Tinsley, 1975). In seeds over 90% of the N is in the form of storage proteins.

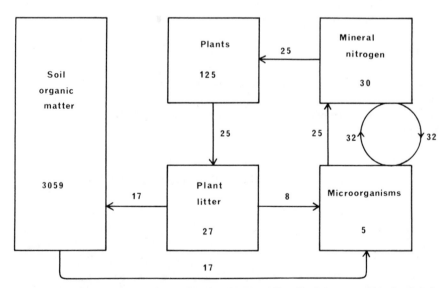

Fig. 5. Storages (Tg N × 10^2) and transfers (Tg N × 10^2 yr^{-1}) of nitrogen within the global plant–soil system. [After Rosswall (1976).]

Table X

Turnover Time of Nitrogen in Various Parts of the Global
Terrestrial Nitrogen Cycle[a]

| | Ecosystem | | |
	"World"	Oak–hickory forest	Tundra mire
Primary producers	4.9	4.1	5.6
Plant litter	1.1	2.9	1.7
Soil organic matter	177	150	372
Microorganisms	0.09	0.15	0.32
Inorganic soil N	0.53	0.19	0.30

[a] Data from Rosswall (1976).

1. Forms of Nitrogen

In both fertilized and unfertilized soils ammonium and nitrate are the only major ionic forms of N actively absorbed by plants (Haynes and Goh, 1978). Uptake of nutrient elements in ionic form by roots is an active physiological process. Absorption of nutrients from soil solution is thus affected by many soil and environmental factors that can affect both the uptake process as well as the availability of the nutrients within the soil. The processes involved in the uptake of NH_4^+ and NO_3^- by plants are discussed in detail in Chapter 6.

2. Rates of Nitrogen Uptake

The uptake rates of N by vegetation are specific to a given ecosystem and ecosystem condition. In general, plants with a low production rate usually have a low N-uptake demand; nevertheless it is not always clear whether a low production rate limits N uptake or whether a low uptake limits production (Cole, 1981). It does, however, seem that plants in N-deficient ecosystems make more efficient use of the N they have already taken up either by retaining foliage for longer periods or by translocating N back to living tissues at the time of tissue senescence (Cole, 1981; Staaf and Berg, 1981). The significance of such translocation processes in terms of litter input of N to the soil is discussed in the next section.

The rate of N uptake by plants per unit area of ground in climates that are not conducive to vigorous plant growth, for example, the rate in tundra and desert ecosystems, is considerably lower than that of temperate forests and grasslands (Table XI). The uptake rate in intensively man-

Table XI

Nitrogen Uptake by Vegetation from Some Selected Ecosystems

Ecosystem	Nitrogen uptake (kg N ha^{-1} yr^{-1})	Reference
Natural systems		
Deciduous forest	70	Cole (1981)
Coniferous forest	39	Cole (1981)
Natural grassland	65	Woodmansee et al. (1981)
Tundra vegetation	11	Rosswall et al. (1975)
Desert vegetation	13	West and Klemmedson (1978)
Agricultural systems		
Maize crop	160	Date (1973)
Wheat crop	190	Date (1973)
Sugar beet crop	290	Date (1973)
Cabbage crop	280	Date (1973)
Potato crop	145	Thomas and Gilliam (1977)
Sorghum crop	380	Date (1973)

aged agricultural ecosystems is, in general, very much higher than that in most natural forest and grassland ecosystems (Table XI).

B. Input of Detritus

Plant litter formation represents a substantial coupling between two major parts of terrestrial ecosystems: the primary producers and the soil. Litter movement is the major pathway that supplies energy and nitrogen to the soil community in most ecosystems. Voluminous qualitative and quantitative data are available on aboveground litter production and recently there has been an increasing awareness of the quantitative importance of root litter formation (Staaf and Berg, 1981).

1. Aboveground Litter

Genetic control of vegetative production rate, life span of organs, and redistribution of N within plants are the major factors influencing the flow of N to the soil via plant litter (Staaf and Berg, 1981). The internal cycling of N within plants is an important consideration in the overall N cycle of ecosystems. Indeed, before short-lived physiologically active organs, such as flowers and leaves, are shed much of the N is usually withdrawn. In grasslands, for example, translocation of N from dying tissues to perennial or actively growing tissues results in conservation of approximately one-third to two-thirds of the total N in living tissue (Clark, 1977; Woodmansee *et al.*, 1978). Such a process also gives deciduous trees a degree of nutritional independence from the soil N pool during the critical early spring growth period (Kramer and Kozlowski, 1979; Melillo, 1981). Changes in dry weight and N content of *Quercus coccinea* leaves during the growing season are shown in Fig. 6. The decline in N content during leaf senescence prior to abscission in late October is clear.

Environmental factors may influence litter deposition by either triggering physiological processes that initiate plant senescence or directly causing premature shedding of plant parts. The latter function may tend to increase the rate of N input since deposition occurs before N redistribution can occur.

2. Belowground Litter

In forest, shrub, and herbaceous ecosystems most of the litter input comes from root production and decomposition (Coleman, 1976). In forests, several workers (Harris *et al.*, 1973; Cox *et al.*, 1978; Persson, 1978;

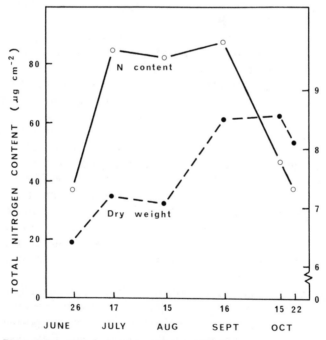

Fig. 6. Dry weight and nitrogen content of leaves of *Quercus coccinea* during the growing season at Upton, New York. [Data from Woodwell (1974).]

McClaugherty *et al.*, 1982; Vogt *et al.*, 1982) have found that fine root litter (roots less than 0.5 mm in diameter) and associated mycorrhizal biomass is the most important pathway of N transfer between plants and the soil. In a mature oak–hickory forest Harris *et al.* (1973), for example, estimated that 68 kg N ha^{-1} yr^{-1} enters the soil in root litter in comparison with 33 kg N ha^{-1} yr^{-1} in aboveground litter fall.

In perennial grasslands, where the mass of roots is often two to four times as large as that of tops (Woodmansee *et al.*, 1981), it is estimated that 0.2 to 0.25% of the living root mass may die each year (Dahlman and Kucera, 1968; Clark, 1977).

In addition to root death, exudation and sloughing of root cells may also contribute to transfer of nitrogen from root to soil (Coleman, 1976).

3. Role of Herbivores

The energy flow through the herbivore subsystem is usually only a minor fraction of that transferred to the detritus pool. In forests it usually amounts to not more than 1–2% but in grazed grasslands it can be in the region of 10–20% or even higher (Swift *et al.*, 1979; Staaf and Berg, 1981). However, by consumption of nutritive plant parts, animals may transfer larger amounts of N than is indicated by the energy flow. For example, Perkins *et al.* (1978) found that the quantity of dry matter consumed by sheep on a temperate grassland was less than half of that directly entering the aboveground detritus pool. In contrast, the proportion of N entering the detrital pool via the herbivore was 81%.

In most ecosystems, rodents and insects are probably not significant grazers (Woodmansee *et al.*, 1978) except during years of population outbreaks when their impact may be significant.

Herbivores can influence litter deposition and content in three major ways: (1) By consumption of herbage, N is transferred to the grazer food chain. The composition of animals and animal excreta is markedly different from that of the tissues they consume; thus they have different decomposition characteristics. (2) Premature shedding of plant parts can be caused by trampling, spoilage, and nibbling of herbage. Such processes tend to increase the rate of N input to the soil because litter deposition often occurs before N redistribution within the plant can occur. (3) Grazers affect plant growth and mortality by decreasing the general condition of the plant.

Most of the N entering the herbivore subsystem eventually passes in the form of excreta and carcasses to the decomposition subsystem. Some losses of N can occur through emigration of herbivore populations or harvesting of herbivores as in agricultural ecosystems. Gaseous losses of N from urine patches also are common.

C. Decomposition of Litter

As outlined above, in mature ecosystems the bulk of the internally cycled N is shed in plant litter while a small portion is cycled by herbivorous animals. The detritus is then broken down by decomposer organisms. Eventually the decomposers die and their carcasses enter the detritus compartment and in turn are acted on by other decomposer organisms.

1. Processes of Decomposition

Decomposition can be viewed as consisting of three interrelated processes. (Swift *et al.,* 1979). These are (1) leaching, (2) catabolism, and (3) comminution.

Leaching is a physical process that occurs very soon after litter fall and involves removal of soluble matter from detritus by the action of water. The percentage of total N content of litter leached can be as high as 10–25% (Berg and Staaf, 1981).

Catabolism comprises energy-yielding enzymatic reactions (or chains of reactions) that involve transformation of complex organic compounds to smaller and simpler ones. Some products of catabolism are inorganic (e.g., NH_4^+), others are intermediates that enter the metabolic pool of the decomposer organisms and are resynthesized into complex compounds (e.g., proteins), and still others may be incorporated into noncellular organic matter (e.g., humus). Catabolism is mediated by extracellular enzymes secreted by saprotrophic bacteria and fungi as well as by enzymes in the digestive system of saprotrophic invertebrates such as protozoans, nematodes, annelids (earthworms), and arthropods (e.g., mites and collembolans).

Comminution is the reduction in particle size of detritus. It is a physical process that is largely brought about by the feeding activity of decomposer animals (Anderson *et al.,* 1981). Comminution is an important component of the decomposition process since it results in fragmentation of detritus, thus exposing a greater surface area for microbial colonization and attack.

2. Mineralization–Immobilization

Decomposition performs two major functions within ecosystems: (1) mineralization of nutrient elements and (2) formation of soil organic matter.

Mineralization occurs when inorganic forms of an element (e.g., NH_4^+-N) are released during catabolism. The last step in mineralization of N, in which simple organic nitrogenous substances are metabolized with the release of NH_4^+-N, is known as ammonification. The outcome of catabolism is the release of energy for anabolic activity (reactions in which cell components are built up from organic and/or inorganic precursors). Anabolic activity requires uptake and use (immobilization) of mineral N by decomposer organisms; thus, immobilization inevitably accompanies mineralization. Rosswall (1976) estimated that in the global plant–soil system, immobilization was responsible for the fate of slightly more than half of the annual gross mineralization of N. In the deciduous forest at Hubbard Brook, Aber *et al.* (1978) estimated that one-third of the total amount of N mineralized during one year was immobilized during that year. Microorganisms are apparently more successful than higher plants in competing for mineral N (Jansson, 1958). Hence, availability of N to plants depends primarily on the magnitude of net mineralization—the extent to which mineralization exceeds immobilization.

Major factors that influence the rate of decomposition of litter, and thus the mineralization/immobilization balance of N, are environmental parameters (moisture regime, aeration, pH, and temperature) and litter "quality" (carbon : nitrogen ratio and lignin and polyphenol content) (Swift *et al.*, 1979). The influences of these factors on decomposition are discussed in Chapter 2.

3. Formation of Soil Organic Matter

The complete decomposition of any piece of detritus occurs over a long time span (hundreds or thousands of years). The residues of the decomposition process contribute to the formation of soil organic matter. The soil organic matter can be divided into the cellular and humic components. The cellular fraction consists of particulate matter formed by the action of decomposer organisms (e.g., partially digested plant material, animal feces, carcasses, and microbial cells). The humic component is less readily identifiable and consists of a mixture of complex polymeric organic molecules with amorphous character (see chapter 2). Humic substances arise from the chemical and biological degradation of plant and animal residues and from the synthetic activity of microorganisms. The products so formed tend to associate into complex chemical structures that are more stable than the starting materials.

D. Nitrification

Nitrification is the biological oxidation of ammonium to nitrite and thence to nitrate. The processes involved in this oxidation are discussed in Chapter 3.

It is generally carried out by autotrophic bacteria, which derive their energy solely from these oxidations of NH_4^+ and NO_2^- and not from oxidations of carbonaceous compounds. Several genera of autotrophic bacteria are able to oxidize ammonium to nitrite, including *Nitrosomonas, Nitrosolobus,* and *Nitrospira,* while *Nitrobacter* appears to be the dominant or only nitrite oxidizer in terrestrial ecosystems (Belser, 1979). Particularly in some acid soils, a slow form of nitrification appears to occur that is carried out by heterotrophic fungi and bacteria (Focht and Verstraete, 1977; Verstraete, 1981).

Since only a few groups of bacteria are capable of nitrification, environmental factors can have rather marked influences on the process (see Chapter 3). However, many natural terrestrial ecosystems fail to produce nitrate to any significant extent despite apparently suitable environmental conditions for nitrification (Borman and Likens, 1979; Vitousek *et al.,* 1979; Skujins, 1981; Woodmansee *et al.,* 1981). Indeed, it is now generally believed that nitrification plays a minor role in the N cycle of undisturbed ecosystems (see Chapter 3).

Nonetheless, following ecosystem disturbance, nitrification can make a significant contribution to the N cycle as shown by increased leaching of NO_3^- following disturbance (Vitousek, 1981). Nitrification is also an important transformation of N in many agricultural soils (Hauck, 1981; Verstraete, 1981).

VI. NITROGEN CONTENT OF SOILS

A. Soil-Forming Factors

Since, on average, about 99% of the N in terrestrial ecosystems is organically bound (Rosswall, 1976) accumulation of soil N closely follows that of soil organic matter. Thus the N content of soils is determined by an equilibrium between the input of products of plant litter decomposition and their losses from the soil. In natural ecosystems, the N content of the soil approaches an equilibrium value (Stevenson, 1965; Jenny, 1980). However, since the soil N system is dynamic, any change to the environment (e.g., a change of climate) may lead to a new equilibrium level of soil N. Therefore, the N content of soils is very diverse, ranging from less than 0.1% in desert soils to over 2% in highly organic soils.

Jenny (1961) described the factors that influence soil development in the form of a general equation.

$$S = f(cl, o, r, p, t \ ...)$$

That is, the soil state (S) (for example, organic matter or N content) is a function of climate (cl), a biotic factor (o), which generally refers to vegetation type in the area, topography (r), parent material (p), and time (t). The dots indicate that other factors may be involved. The concept that each factor can be treated as an independent variable can be criticized since this is seldom, if ever, the case in nature. Despite this, Jenny's ideas on soil formation within an ecosystem framework have contributed substantially to our understanding of factors that influence the N content of soils.

To some extent, all the factors that influence retention of N by soils are interdependent. Indeed, it is the interactions of environmental factors affecting microbial activity and vegetation type that largely determine the N content of the soil.

Nitrogen can be maintained at high levels in the soil only when microbial activity is inhibited during at least some period of the year. The activities of the soil microflora may be restricted by low temperature, poor drainage, low pH, presence of toxic inhibitory substances (e.g., allochemicals), and the formation of metal–clay–organic matter complexes.

1. Climate

Climate is the most important factor influencing the types of plant species growing in a given area, the quantity of plant material produced, and the intensity of microbial activity in the soil. As such, it is the main factor influencing soil organic matter and N levels in soils.

In general, the organic matter content of soils declines from the tundra to the tropics and generally is lower in drier parts of similar latitudes (Swift *et al.,* 1979; Stevenson, 1982). Soils formed under restricted drainage do not follow a climatic pattern since O_2 deficiency inhibits microbial decomposition of organic residues over a wide temperature range.

2. Vegetation

The type and quantity of vegetation directly influence the nature and quantity of organic matter that is formed during decomposition of plant litter. For example, on desert soils where vegetation is sparse, the amounts of organic residues added each year are extremely small and the organic matter content of the soils is correspondingly low.

It is widely known that, other factors being equal, the organic matter content of grassland soils is substantially higher than that of forest soils. Incorporation of carbon and N into soil organic matter is favored by a continuous carbon (energy) supply in combination with a deficiency of available N as found in rhizosphere soil under grass (Huntjens and Albers, 1978).

In forest soils, there is a general distinction between organic matter accumulation under coniferous and deciduous trees. The slowly decomposing coniferous residues characteristically accumulate at the soil surface, while the more readily decomposable residues from deciduous trees are generally rapidly mixed with the soil to a depth of 30 cm or more. The complex nature and interactions of the many factors that result in the two types of organic matter accumulation outlined above are discussed in more detail in Section VI,A,6 dealing with interactions among soil-forming factors.

3. Topography

Topography can affect the N content of soil through its influence on climate, runoff, evaporation, and transpiration. For example, soils on slopes or knolls generally have lower N contents than those in depressions. Naturally wet sites in depressions are usually high in N since the anaerobic conditions, which occur in wet periods of the year, restrict microbial activity and, therefore, organic matter destruction.

4. Parent Material

Parent material exerts its influence on soil organic matter accumulation primarily through its effect on soil texture (e.g., clay content). The formation of complexes between organic compounds and inorganic colloids in soil (Theng, 1974, 1979) has long been recognized as a factor responsible for the stability and resistance of humus to biodegradation. Thus, other factors remaining constant, the organic matter content of heavy soils is higher than that of loamy soils, which is, in turn, higher than that of sandy soils.

Retention of organic matter is also affected by the type of clay minerals present. Clays that have high absorption capacities for organic molecules (e.g., montmorillonite) are particularly effective in protecting nitrogenous substances from microbial attack.

5. Time

The rate and pattern of organic matter and N accumulation in soils with time are discussed later in this chapter. Following a variable period of time (100 to 10,000 years) an equilibrium level of organic matter is attained in soils. This equilibrium is controlled by the soil-forming factors of climate, vegetation, topography, and parent material. The great variability in the organic matter content of soils is the result of the numerous combinations under which the soil-forming factors can act over time.

6. Interactions of Factors

A consideration of two extreme types of organic layers, mull and mor, illustrates the complex interrelations that influence the N content and rate of cycling of N in soils. Some characteristics of mull and mor humus are shown in Table XII. In general, conditions that promote rapid decomposition of detritus favor mull formation. In mull, the main accumulation of material is soil humus, whereas that of mor is cellular material. Although climate is a major determinant of these factors, intervention of other factors can produce local variations.

In many areas, for example, succession results in deciduous hardwoods replacing conifers, causing a change from mor to mull humus formation (Gosz, 1981). Planting an area that was formerly in hardwood vegetation with conifer species can cause a change from mull to mor humus formation (Nihlgärd, 1971). Furthermore, differences in the supply and cycling rate of N can contribute to formation of mull or mor humus under the same species. A change from mor to mull humus formation can be induced by increasing the quantity of N (e.g., fertilizing) as well as by practices that simply increase the cycling rate of N by increasing litter decomposition, such as liming (raising the pH) or disking (Gosz, 1981).

Table XII

Some Characteristics of Mull and Mor Humus Types[a]

Characteristic	Mull	Mor
Vegetation	Deciduous trees, grasses	Conifers, heathland
Moisture status	Leaching and flushing—gentle topography and/or warm conditions	Strongly leached—high precipitation and/or free drainage due to topography or high sand content
Soil horizons[b]	O rarely recognizable, other horizons show good mixing of organic matter and mineral components	O well defined, A_1 well defined, deep and largely organic
Organic matter	Low cellular component, C : N ratio 15	High cellular component, C : N ratio 20
Organisms	Bacterial counts high, fungal mycelium not obvious, earthworms present	Bacterial counts low, fungal mycelium abundant, earthworms absent

[a] Compiled from Swift et al. (1979).

[b] O = Organic material above the surface of the mineral soil. A_1 = Surface mineral horizon with high organic matter content.

B. Plant Succession and Nitrogen Accumulation

From the preceding discussion it is evident that small variations in climate, topography, vegetation type, and parent material can result in markedly different rates of accumulation of organic matter and N turnover. In the same way that organic matter accumulation is affected by variations in space, it is also strongly influenced by time. The most fundamental aspect of such changes occurs during ecosystem development through the process of succession.

Succession is a central concept of ecology (Odum, 1969; Whittaker, 1975; Grime, 1979). During succession there is a progressive alteration in the structure and species composition of the vegetation. Two types of succession can be defined: primary succession involves the colonization of a previously unoccupied habitat devoid of soil and vegetation (e.g., the moraine of a receding glacier) while secondary succession is the process of recolonization of a disturbed habitat (e.g., following a forest fire). The mature community that ends a succession, the climax ecosystem, is one in which a steady-state system in equilibrium with the environment has developed. Its population and species composition remains relatively constant.

The widely accepted model of primary and secondary succession (Odum, 1969; Whittaker, 1975) is characterized by an asymptotic curve of biomass accumulation culminating in a steady-state climax condition (Fig. 7). Nevertheless biomass accumulation during ecosystem development often follows a more complex pattern (Major, 1974a; Borman and Likens, 1979) that can be broadly divided into four phases (Fig. 7), including (1) a slow initial phase, (2) an aggradation phase in which biomass steadily increases to a maximum, (3) a transition phase in which a decline in biomass occurs, and, finally, (4) a steady-state phase in which biomass stabilizes with somewhat irregular oscillations about a mean.

1. Pattern of Nitrogen Accumulation

Since almost all the input of detritus to the soil is derived either directly or indirectly from living plant biomass, it is not surprising that levels of soil organic matter and N during succession follow trends similar to those of total ecosystem biomass (Figs. 7 and 8). The pattern and rate of soil N accumulation during succession have been reviewed by several workers (Stevens and Walker, 1970; Major, 1974b; Jenny, 1980).

The time required to reach equilibrium steady-state N levels varies greatly. For example, about 110 years was required to reach equilibrium on the recessional moraines of Alaskan glaciers (Crocker and Major, 1955), while on the Manawatu sand dune systems of New Zealand accu-

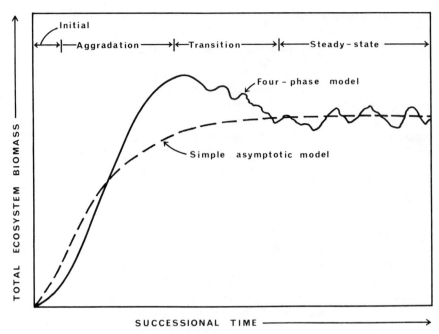

Fig. 7. Two hypothetical models of ecosystem development during primary succession. Developmental phases of the four-phase model are indicated.

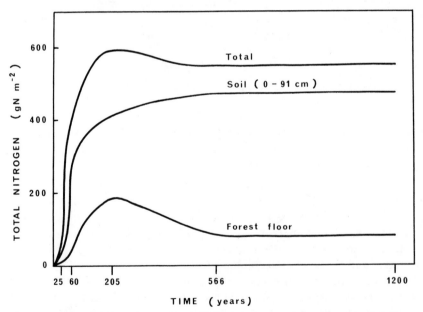

Fig. 8. Accumulation of nitrogen in the forest floor litter layer, the upper soil horizons, and in the soil plus litter (total) during primary succession of a conifer forest at Mount Shasta, California. [From Dickson and Crocker (1953).]

mulation of N approached equilibrium after 10,000 years (Syers *et al.*, 1970). Intermediate times have been reported by other workers, such as 2000 years for spodosols developed on the Rhone glacier moraines (Jenny, 1980) and in excess of 1000 years for the Lake Michigan sand dunes (Olson, 1958).

2. Basis of Nitrogen Accumulation

The major source of soil N during early development of ecosystems is biological N_2 fixation by both free-living and symbiotic organisms. Wet and dry deposition of N can also be important sources (Söderlund, 1981).

During early stages of ecosystem development unproductive rocks, stones, sands, and minerals represent a harsh environment for plant growth. Cyanobacteria are particularly hardy and are able to develop in environments where the supply of available nutrients is low (Fogg *et al.*, 1973). Thus the cyanobacteria in both free-living forms and lichen symbioses often are primary colonizers and characteristic of marginal areas.

When soil-forming processes have created the necessary base for growth of higher plants, the nitrogen demand becomes limiting and many types of nitrogen-fixing plants characteristically become dominant (Fogg *et al.*, 1973; Burns and Hardy, 1975; Granhall, 1981). Thus, in many primary successions the development of a stable ecosystem depends on colonization by N_2-fixing species, which accumulate N in the system, subsequently enabling nonfixing species to grow and eventually dominate (Crocker and Major, 1955; Crocker and Dickson, 1957; Lawrence *et al.*, 1967; Stevens and Walker, 1970).

The general patterns of N_2 fixation with ecosystem development as visualized by Gorham *et al.*, (1979) are illustrated in Fig. 9. The large peak early in primary succession is well documented whereas the early peak in secondary succession depends on N reserves remaining following ecosystem disturbance and presence or absence of N_2-fixing populations. The late minor rise in N_2 fixation represents fixation activity in woody litter with a high C/N ratio.

C. Ecosystem Disturbance and Recovery

In the previous discussion it was shown that the N content of the soil approaches an equilibrium value that is dependent on soil-forming factors. However, any alteration in the environment may lead to a change in N content of the soil. Indeed, when a stable ecosystem structure is disturbed the cycling of nutrients, including N, is also disturbed (Gorham *et al.*, 1979). For example, ecosystems can be disturbed by clear-cutting a forest or by conversion of a mature forest or grassland to arable farming. These will be discussed separately.

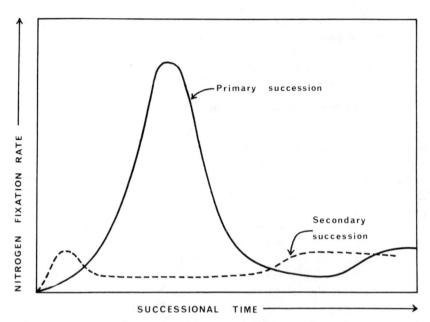

Fig. 9. General patterns of biological nitrogen fixation during succession. [After Gorham *et al.* (1979). Reproduced, with permission, from the Annual Review of Ecology and Systematics, Volume 10, © 1979 by Annual Reviews Inc.]

1. Clear-cutting

Uptake of N and water by trees is substantially reduced following clear-cutting of a forest. This period may be short lived where rapidly growing pioneer vegetation can quickly reoccupy the site and restore plant growth (Marks and Bormann, 1972). Following clear-cutting, both mineralization and immobilization of N may be increased; on fertile sites mineralization generally exceeds immobilization since the C/N ratio is relatively low (Vitousek, 1981). In the Hubbard Brook forest, for example, Aber *et al.* (1978) estimated that in the first five years following forest harvesting 580 kg N ha^{-1} yr^{-1} were mineralized while only 215 kg N ha^{-1} yr^{-1} were immobilized, mainly in decaying wood.

The potential for excess N mineralization over mineral N utilization by a forest system following clear-cutting of a fertile forest is illustrated in Fig. 10. Where nitrification is rapid, much of the excess mineral N can be lost from the ecosystem by leaching of NO_3^--N (Tamm *et al.*, 1974; Vitousek and Melillo, 1979) or by denitrification. Where nitrification is slow, NH_4^+-N may accumulate in the soil or be lost through NH_3 volatilization (Vitousek, 1981).

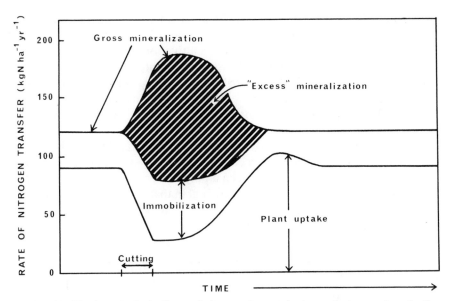

Fig. 10. The hypothetical effects of clear-cutting on the internal nitrogen in a fertile deciduous forest. [After Vitousek (1981).]

Where significant quantities of N are lost from the system directly following clear-cutting, similar quantities can be accumulated during secondary succession although this may take decades or even centuries. In the Hubbard Brook forest Bormann and Likens (1979) calculated that 70% of the N added to the ecosystem during secondary succession was through N_2 fixation and 30% by precipitation.

2. Conversion to Arable Farmland

The conversion of a natural forest or grassland ecosystem to arable farmland is tantamount to reversing the successional process and reintroducing an immature seral vegetative stage. It is, therefore, not surprising that one of the major changes that occurs is a decline in soil organic matter content. Indeed, the conditions responsible for the steady state of accumulation of soil organic matter over centuries are suddenly and drastically changed.

Cultivation is generally considered an oxidative process since it promotes soil aeration and exposes soil surfaces to the atmosphere (Allison, 1973; Power, 1981). Consequently, cultivation enhances processes such as oxidation of organic matter, mineralization of organic N, and nitrification of NH_4^+-N. Under cultivation erosion by water and wind is also

tremendously accelerated (Allison, 1973). In addition, much of the vegetation produced under many cropping situations is removed so the N
cycle is interrupted. Thus, in general, cultivation of mature grasslands
results in soil organic matter and organic N contents decreasing rather
markedly for the first 25–50 years (Stevenson, 1965; Allison, 1973; Campbell, 1978; Fig. 11). Steady-state conditions often appear to be reached
within 50–100 years after commencement of cultivation (Stevenson,
1965).

If cultivated land is allowed to revert to a more natural vegetative state
(i.e., secondary succession proceeds) the organic matter and N content of
the soil increase again. This has been amply illustrated at the Rothamsted
Agricultural Station in England, where a field that had been cropped with
wheat for centuries was allowed to revert to natural woodland (Jenkin-

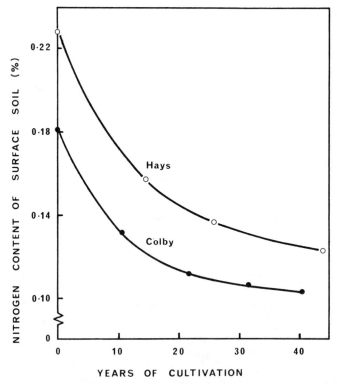

Fig. 11. Decline in nitrogen content in the surface soil (0–18 cm) of two virgin prairie soils
of Kansas during cultivation and cropping for 40 years. [Data from Hobbs and Brown (1957).
Reproduced from *Agron. J.* **49,** p. 259 by permission of the American Society of Agronomy.]

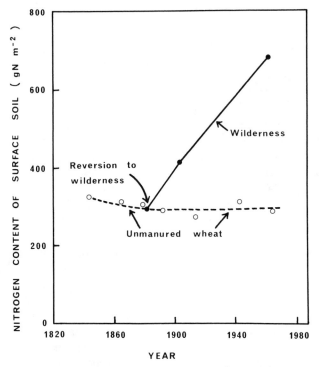

Fig. 12. Total nitrogen content of the surface soil (0–23 cm) of the Broadbalk Wilderness and the unmanured plots under continuous wheat. [Data from Jenkinson (1977).]

son, 1971, 1977). Legumes were evident in the early stages, but within 15 years trees and shrubs became dominant and the area, known as Broadbalk Wilderness, is now a well established deciduous woodland. The dramatic rise in total N content of the surface soil on the Broadbalk Wilderness is shown in Fig. 12. Between 1883 and 1964 the wilderness accumulated 3930 kg N ha^{-1} in the top 23 cm, a mean accumulation rate of 49 kg N ha^{-1} yr^{-1} (Jenkinson, 1971). The net gain attributed to biological N$_2$ fixation is 34 kg N ha^{-1} yr^{-1}.

In many tropical countries the buildup of organic matter during secondary succession is the basis of the maintenance of soil fertility. The system of shifting cultivation used in these localities involves an alteration between cropping for a few years on selected clearings and a lengthy period when soil fertility is restored by regeneration of natural vegetation (Nye and Greenland, 1960; Aweto, 1981). Cultivation consequently shifts within an area otherwise covered by natural vegetation.

VII. CONCLUSIONS

In natural ecosystems, the major inputs of N occur through fixation of atmospheric N_2 by N_2-fixing microorganisms, either free living or in association with higher plants, and through wet and dry deposition of nitrogenous substances from the atmosphere. Major losses of N from ecosystems occur through leaching of NO_3^-, loss of soluble and particulate N during wind and water erosion, volatilization of NH_3, and loss of N_2O during nitrification and N_2O and N_2 during denitrification.

Within ecosystems an internal N cycle operates in which organic N in detritus is decomposed, with the release of mineral N, which is then utilized by the plant biomass. Eventually, this organic N is returned to the soil as detritus, which again decomposes. Detritus is deposited to the soil through shedding of plant parts and/or death of plants. In many ecosystems most litter input to the soil occurs below ground due to the rapid turnover of fine root material. The role of herbivores in cycling of N and the input of detritus is not generally significant except in grazed grasslands.

The deposited detritus undergoes decomposition; this process performs two major functions: (1) mineralization of nutrient elements and (2) formation of soil organic matter. Ammonification, the last step in mineralization of N, involves microbial conversion of simple organic nitrogenous compounds to NH_4^+-N. The NH_4^+-N may then be taken up by the microbial biomass (immobilized) or used by growing plants. Net mineralization (gross mineralization minus immobilization) represents the quantity of mineral N available to plants since microorganisms are generally better competitors than plants for NH_4^+-N.

In general, nitrification (the microbial conversion of NH_4^+-N to NO_3^--N) plays a minor role in the N cycle of undisturbed natural ecosystems. Nevertheless, in agricultural ecosystems and following the disturbance of natural ecosystems nitrification can be an important transformation of mineral N.

The formation of soil organic matter is a complex process. Soil organic matter can be divided into (1) the cellular fraction, consisting of particulate matter formed by the action of decomposer organisms, and (2) the humic fraction, which consists of complex polymeric organic molecules originating from chemical and biological degradation of plant and animal residues and from synthetic activities of microorganisms.

Since almost all the input of detritus to the soil is derived directly or indirectly from the living plant biomass, the pattern of soil organic matter and N accumulation during colonization of a previously unoccupied site and subsequent ecosystem development (primary succession) follows trends similar to that of total ecosystem biomass. In general, there is rapid accumulation of soil N in the first few years, which diminishes slowly and eventually reaches equilibrium in 100 to 10,000 years. The interaction of environmental factors affecting microbial activity and the quantity of plant material produced largely determines the N content of the soil at equilibrium.

When an ecosystem reaches an equilibrium condition, the gains of N through biological N_2 fixation and precipitation more or less balance losses of N as gases, by leaching of NO_3^--N, and through erosion. Furthermore, the magnitude of such gains and losses is very small in comparison with the large amounts of N that are conserved through internal cycling within the ecosystem.

When the internal N cycle within the ecosystem is interrupted (e.g., by removal of vegetation during clear-cutting a forest or cultivating natural ecosystems for agricultural use) significant losses of N from the system can occur, which may result in a decrease in N content of the soil. Such losses of N can, in the long term, be balanced by similar gains that take place if secondary succession is allowed to proceed.

Man has significantly influenced the N cycle principally on a regional and local basis. The development of industry has, for example, led to marked increases in the rates of emission of NO_x to the atmosphere; concomitant increases in the rate of wet deposition of N have also occurred in industrialized areas. Similarly, intensive livestock production in agricultural areas can lead to increased emissions of NH_3 with concomitant increases in wet deposition of N.

Within agricultural ecosystems the cycling of N often contrasts with that in natural ecosystems. There is generally a large loss of N from the high-producing open agricultural systems of the western world since the crop is usually fed to people or animals located elsewhere. Such systems rely on large inputs of fertilizer N to sustain productivity; by far the major portion of fertilizer N originates from industrial synthesis of ammonia from atmospheric N_2. The use of legumes also represents a significant input of N to some agricultural ecosystems.

Introduction of agricultural activity has increased the turnover rate of N in the biosphere via leaching of NO_3^--N to groundwater and gaseous losses of N to the atmosphere as N_2 and N_2O. The major factor underlying this increased turnover rate appears to be that nitrification is often a predominant transformation of N in agricultural soils. Lack of vegetation cover and soil disturbance through tillage practices can also contribute to losses of N from agricultural lands through erosion.

REFERENCES

Abeles, F. C., Cracker, L. E., Forrence, L. E., and Leather, G. R. (1971). Fate of air pollutants: Removal of ethylene, sulfur dioxide and nitrogen dioxide by soil. *Science* **173**, 914–916.

Aber, J. D., Botkin, D. B., and Melillo, J. M. (1978). Predicting the effects of different harvest regimes on forest floor dynamics in northern hardwoods. *Can. J. For. Res.* **8**, 306–315.

Adams, M. A., and Attiwill, P. M. (1982). Nitrogen mineralisation and nitrate reduction in forests. *Soil Biol. Biochem.* **14**, 197–202.

Akkermans, A. D. L., and Roelofsen, W. (1981). Symbiotic nitrogen fixation by actinomycetes in *Alnus*-type root nodules. *In* "Nitrogen Fixation" (W. D. P. Stewart and J. R. Gallon, eds.), pp. 279–299. Academic Press, New York.

Alberts, E. E., Schuman, G. E., and Burwell, R. D. (1978). Seasonal runoff losses of nitrogen and phosphorus from Missouri Valley loess watersheds. *J. Environ. Qual.* **7**, 203–208.

Allison, F. E. (1973). "Soil Organic Matter and Its Role in Crop Production." Am. Elsevier, New York.

Anderson, R. V., Coleman, D. D., and Cole, C. V. (1981). Effects of saprotrophic grazing on net mineralisation. *In* "Terrestrial Nitrogen Cycles: Processes, Ecosystem Strategies and Management Impacts" (F. E. Clark and T. Rosswall, eds.), pp. 201–216. Ecological Bulletins, Stockholm.

Aweto, A. E. (1981). Secondary succession and soil fertility restoration in south-western Nigeria. II. Soil fertility restoration. *J. Ecol.* **69**, 609–614.

Balandreau, J., Rinaudo, G., Fares-Hamad, I., and Dommergues, Y. (1975). Nitrogen fixation in the rhizosphere of rice plants. *In* "Nitrogen Fixation by Free-living Microorganisms" (W. D. P. Stewart, ed.), pp. 57–70. Cambridge Univ. Press, London and New York.

Bate, G. C. (1981). Nitrogen cycling in savanna ecosystems. *In* "Terrestrial Nitrogen Cycles: Processes, Ecosystem Strategies and Management Impacts" (F. E. Clark and T. Rosswall, eds.), pp. 463–475. Ecological Bulletins, Stockholm.

Belser, L. W. (1979). Population ecology of nitrifying bacteria. *Annu. Rev. Microbiol.* **33**, 309–333.

Berg, B., and Staaf, H. (1981). Leaching, accumulation and release of nitrogen in decomposing forest litter. *In* "Terrestrial Nitrogen Cycles: Processes, Ecosystem Strategies and Management Impacts" (F. E. Clark and T. Rosswall, eds.), pp. 163–178 Ecological Bulletins, Stockholm.

Bergersen, F. J. (1981). Leghaemoglobin, oxygen supply and nitrogen fixation: Studies with soybean nodules. *In* "Nitrogen Fixation" (W. D. P. Stewart and J. R. Gallon, eds.), pp. 139–160. Academic Press, New York.

Bollin, B., and Arrhenius, E. (1977). Nitrogen—an essential life factor and a growing environmental hazard. *Ambio* **6**, 96–105.

Bormann, F. H., and Likens, G. E. (1979). "Pattern and Process in a Forested Ecosystem." Springer-Verlag, Berlin and New York.

Böttger, A., Ehhalt, D. H., and Gravenhorst, G. (1978). Atmospharische Kreislaufe von Stickoxiden und Ammoniak," Rep. No. 1558. Kernforschungsanlage, Julich, F.R.G.

Burns, R. C., and Hardy, R. W. F. (1975). "Nitrogen Fixation in Bacteria and Higher Plants." Springer Verlag, Berlin and New York.

Burris, R. H., Arp, D. J., Benson, D. R., Emerich, D. W., Hageman, R. V., Jones, T.,

Ludden, P. W., and Sweet, W. J. (1981). The biochemistry of nitrogenase. *In* "Nitrogen Fixation" (W. D. P. Stewart and J. R. Gallon, eds.), pp. 37–54. Academic Press, New York.

Campbell, C. A. (1978). Soil organic carbon, nitrogen and fertility. *In* "Soil Organic Matter" (M. Schnitzer and S. U. Khan, eds.), pp. 173–271. Am. Elsevier, New York.

Chalk, P. M., and Keeney, D. R. (1971). Nitrate and ammonium contents of Wisconsin limestones. *Nature (London)* **229**, 42.

Chichester, F. W., and Smith, S. J. (1978). Disposition of N-labelled fertiliser nitrate applied during corn culture in field lysimeters. *J. Environ. Qual.* **7**, 227–233.

Child, J. J. (1981). Biological nitrogen fixation. *In* "Soil Biochemistry" (E. A. Paul and J. N. Ladd, eds.), Vol. 5, pp. 297–322. Dekker, New York.

Clark, F. E. (1977). Internal cycling of ^{15}nitrogen in shortgrass prairie. *Ecology* **38**, 1322–1333.

Cole, D. W. (1981). Nitrogen uptake and translocation by forest ecosystems. *In* "Terrestrial Nitrogen Cycles: Processes, Ecosystem Strategies and Management Impacts" (F. E. Clark and T. Rosswall, eds.), pp. 219–232. Ecological Bulletins, Stockholm.

Coleman, D. C. (1976). A review of root production processes and their influence on soil biota in terrestrial ecosystems. *In* "The Role of Terrestrial and Aquatic Organisms in Decomposition Processes" (J. M. Andersson and A. Macfadyen, eds.), pp. 417–434. Blackwell, Oxford.

Cox, T. L., Harris, W. F., Ausmus, B. S., and Edwards, N. T. (1978). The role of roots in biogeochemical cycles in an eastern deciduous forest. *Pedobiologia* **18**, 264–271.

Crocker, R. L., and Dickson, B. A. (1957). Soil development on the recessional moraines of the Herbert and Mendenhall glaciers, southeastern Alaska. *J. Ecol.* **45**, 169–185.

Crocker, R. L., and Major, J. (1955). Soil development in relation to vegetation and surface age at Glacier Bay, Alaska. *J. Ecol.* **43**, 427–448.

Crutzen, P. J. (1974). Estimation of possible variations in total ozone due to natural causes and human activities. *Ambio* **3**, 201–210.

Crutzen, P. J. (1983). Atmospheric interactions—homogeneous gas reactions of C, N, and S containing compounds. *In* "The Major Biogeochemical Cycles and Their Interactions" (B. Bolin and R. B. Cook, eds.), pp. 67–114. Wiley, New York.

Dahlman, R. G., and Kucera, C. L. (1968). Tagging native grassland vegetation with carbon-14. *Ecology* **49**, 1199–1203.

Dalton, H. (1981). Chemoautotrophic nitrogen fixation. *In* "Nitrogen Fixation" (W. D. P. Stewart and J. R. Gallon, eds.), pp. 177–195. Academic Press, New York.

Date, R. A. (1973). Nitrogen, a major limitation in the productivity of natural communities, crops and pastures in the Pacific Area. *Soil Biol. Biochem.* **5**, 5–18.

Dawson, G. A. (1977). Atmospheric ammonia from undisturbed land. *JGR, J. Geophys. Res.* **82**, 3125–3133.

Delwiche, C. C. (1977). Energy relations in the global nitrogen cycle. *Ambio* **6**, 106–111.

Denmead, O. T., Freney, J. R., and Simpson, J. R. (1976). A closed ammonia cycle within a plant canopy. *Soil Biol. Biochem.* **8**, 161–164.

Denmead, O. T., Nulsen, R., and Thurtell, G. W. (1978). Ammonia exchange over a corn crop. *Soil Sci. Soc. Am. J.* **42**, 840–842.

Dickson, B. A., and Crocker, R. L. (1953). A chronosequence of soil's and vegetation near Mt. Shasta, California. II. The development of the forest floors and the carbon and nitrogen profiles of the soils. *J. Soil Sci.* **4**, 142–154.

Döbereiner, J., and De-Polli, H. (1981). Diazotrophic rhizocoenoses. *In* "Nitrogen Fixation" (W. D. P. Stewart and J. R. Gallon, eds.), pp. 301–333. Academic Press, New York.

Eady, R. R., Imad, S., Lowe, D. J., Miller, R. W., and Smith, B. E., and Thorneley, R. N. F. (1981). The molecular enzymology of nitrogenase. *In* "Nitrogen Fixation" (W. D. P. Stewart and J. R. Gallon, eds.), pp. 19–35. Academic Press, New York.

Endelman, F. J., Keeney, D. R., Gilmore, J. T., and Saffigna, P. G. (1974). Nitrate and chloride movement in the Plainfield loamy sand under intensive irrigation. *J. Environ. Qual.* **3**, 295–298.

Eriksson, E. (1952). Composition of atmospheric precipitation. I. Nitrogen compounds. *Tellus* **4**, 215–232.

Evans, H. J., Emerich, D. W., Lepo, J. E. E., Maier, R. J., Carter, K. R., Hanus, F. H., and Russell, S. A. (1981). The role of hydrogenase in nodule bacteroids and free-living rhizobia. *In* "Nitrogen Fixation" (W. D. P. Stewart and J. R. Gallon, eds.), pp. 55–81. Academic Press, New York.

Feller, M. C., and Kimmins, J. P. (1979). Chemical characteristics of small streams near Haney in south-western British Columbia. *Water Resour. Res.* **15**, 257–258.

Fletcher, J. E., Sorensen, D. L., and Porcella, D. B. (1978). Erosional transfer of nitrogen in desert ecosystems. *In* "Nitrogen in Desert Ecosystems" (N. E. West and K. Skujins, eds.), pp. 171–181. Dowden, Hutchinson & Ross, Stroudsburg, Pennsylvania.

Focht, D. D., and Verstraete, W. (1977). Biochemical ecology of nitrification and denitrification. *Adv. Microb. Ecol.* **1**, 135–214.

Fogg, G. I., Stewart, W. D. P., and Walsby, A. E. (1973). "The Blue-Green Algae." Academic Press, New York.

Freney, J. R., Denmead, O. T., and Simpson, J. R. (1978). Soil as a source or sink for atmospheric nitrous oxide. *Nature (London)* **273**, 530–532.

Freney, J. R., Simpson, J. R., and Denmead, O. T. (1981). Ammonia volatilisation. *In* "Terrestrial Nitrogen Cycles: Processes, Ecosystem Strategies and Management Impacts" (F. E. Clark and T. Rosswall, eds.), pp. 291–302. Ecological Bulletins, Stockholm.

Galbally, I. E., and Roy, C. R. (1983). The fate of nitrogen compounds in the atmosphere. *In* "Gaseous Loss of Nitrogen from Plant–Soil Systems" (J. R. Freney and J. R. Simpson, eds.), pp. 265–284. Martinus Nijhoff/Dr. W. Junk, The Hague.

Gallon, J. R. (1981). Nitrogen fixation by photoautotrophs. *In* "Nitrogen Fixation" (W. D. P. Stewart and J. R. Gallon, eds.), pp. 197–238. Academic Press, New York.

Gast, R. G., Nelson, W. W., and Randall, G. W. (1978). Nitrate accumulation in soils and loss in tile drainage following nitrogen applications to continuous corn. *J. Environ. Qual.* **7**, 258–261.

Gorham, E., Vitousek, P. M., and Reiners, W. A. (1979). The regulation of chemical budgets over the course of terrestrial ecosystem succession. *Annu. Rev. Ecol. Syst.* **10**, 53–84.

Gosz, J. R. (1981). Nitrogen cycling in coniferous ecosystems. *In* "Terrestrial Nitrogen Cycles: Processes, Ecosystem Strategies and Management Impacts." (F. E. Clark and T. Rosswall, eds.), pp. 405–426. Ecological Bulletins, Stockholm.

Granhall, U. (1981). Biological nitrogen fixation in relation to environmental factors and functioning of natural ecosystems. *In* "Terrestrial Nitrogen Cycles: Processes, Ecosystem Strategies and Management Impacts" (F. E. Clark and T. Rosswall, eds.), pp. 131–144. Ecological Bulletins, Stockholm.

Grime, J. P. (1979). "Plant Strategies and Vegetational Processes." Wiley, New York.

Harris, W. F., Goldstein, R. A., and Henderson, G. S. (1973). Analysis of forest biomass pools: Annual primary production and turnover of biomass for a mixed deciduous watershed. *In* "IUPRO Biomass Studies 84, 01" (H. E. Young, eds.), pp. 43–66. IUFRO, Nancy, France.

Hauck, R. D. (1981). Nitrogen fertiliser effects on nitrogen cycle processes. *In* "Terrestrial Nitrogen Cycles: Processes, Ecosystem Strategies and Management Impacts" (F. E. Clark and T. Rosswall, eds.), pp. 551–562. Ecological Bulletins, Stockholm.

Haynes, R. J., and Goh, K. M. (1978). Ammonium and nitrate nutrition of plants. *Bio. Rev. Cambridge Philos. Soc.* **53**, 465–510.

Henderson, G. S., and Harris, W. F. (1975). An ecosystem approach to characterisation of the nitrogen cycle in a deciduous forest watershed. *In* "Forest Soils and Forest Land Management" (B. Bernier and C. F. Winget, eds.), pp. 179–193. Les Presses de l'Université Laval, Quebec.

Herrera, R., and Jordan, C. F. (1981). Nitrogen cycle in a tropical Amazonian rain forest: The caatinga of low mineral nutrient status. *In* "Terrestrial Nitrogen Cycles: Processes, Ecosystem Strategies and Management Impacts" (F. E. Clark and T. Rosswall, eds.), pp. 493–505. Ecological Bulletins, Stockholm.

Hill, C. A., and Chamberlain, E. M. (1976). The removal of water soluble gases from the atmosphere by vegetation. *In* "Atmosphere–Surface Exchange of Particulate and Gaseous Pollutants" (G. A. Sehmel, ed.), ERDA Symp. Ser. 38, pp. 153–170. U.S. Energy Res. Dev. Admin., Oak Ridge, Tennessee.

Hobbs, J. A., and Brown, P. L. (1957). Nitrogen changes in cultivated dryland soils. *Agron. J.* **49**, 257–260.

Huntjens, J. L. M., and Albers, R. A. J. M. (1978). A model experiment to study the influence of living plants on the accumulation of soil organic matter in pastures. *Plant Soil* **50**, 411–418.

Ito, O., and Watanabe, I. (1981). Immobilization, mineralization and availability to rice plants of nitrogen derived from heterotrophic nitrogen fixation in flooded soil. *Soil Sci. Plant Nutr.* **27**, 169–176.

Jansson, S. L. (1958). Tracer studies on nitrogen transformations in soil with special attention to mineralization–immobilization relationships. *Lantbrukshoegsk. Ann.* **24**, 101–361.

Jenkinson, D. S. (1971). The accumulation of organic matter in soil left uncultivated. *Rep. Rothamsted Exp. Stn. 1970*, Pt 2, pp. 113–137.

Jenkinson, D. S. (1977). The nitrogen economy of the Broadbalk experiments. I. Nitrogen balance in the experiments. *Rep. Rothamsted Exp. Stn. 1976*, Pt. 2, pp. 103–109.

Jenny, H. (1961). Derivation of state factor equations of soils and ecosystems. *Soil Sci. Soc. Am. Proc.* **25**, 385–388.

Jenny, H. (1980). "The Soil Resource Origin and Behaviour." Springer-Verlag, Berlin and New York.

Jensen, V. (1981). Heterotrophic microorganisms. *In* "Nitrogen Fixation" (W. J. Broughton, ed.), Vol. 1, pp. 30–56. Oxford Univ Press (Clarendon) London and New York.

Khanna, P. K. (1981). Leaching of nitrogen from terrestrial ecosystems—patterns, mechanisms and ecosystem responses. *In* "Terrestrial Nitrogen Cycles: Processes, Ecosystem Strategies and Management Impacts" (F. E. Clark and T. Rosswall, eds.), pp. 343–352. Ecological Bulletins, Stockholm.

Kilmer, V. J. (1974). Nutrient losses from grasslands through leaching and runoff. *In* "Forage Fertilization" (D. A. Mays, ed.), pp. 341–362. Am. Soc. Agron., Madison, Wisconsin.

Kissel, D. E., Ritchie, J. T., and Burnett, E. (1974). Nitrate and chloride leaching in a swelling clay soil. *J. Environ. Qual.* **3**, 401–404.

Knelson, J. H., and Lee, R. (1977). Oxides of nitrogen in the atmosphere: Origin, fate and public health implications. *Ambio* **6**, 126–130.

Knowles, R. (1977). The significance of asymbiotic dinitrogen fixation by bacteria. *In* "A Treatise on Dinitrogen Fixation. Section IV. Agronomy and Ecology" (R. W. F. Hardy and A. H. Gibson, eds.), pp. 33–83. Wiley, New York.

Kolenbrander, G. J. (1975). Nitrogen in organic matter and fertiliser as a source of pollution. *Int. Assoc. Water Pollut. Res.* **1**, 93–114.

Kramer, P. J., and Kozlowski, T. T. (1979). "Physiology of Woody Plants." Academic Press, New York.

Lawrence, D. B., Schoenike, R. E., Quispl, A., and Bond, G. (1967). The role of *Dryas drummondii* in vegetation development following ice recession at Glacier Bay, Alaska, with special reference to its nitrogen fixation by root nodules. *J. Ecol.* **55**, 793–813.

Likens, G. E., Bormann, R. H., Pierce, R. S., Eaton, J. S., and Johnson, M. J. (1977). "Biogeochemistry of a Forested Ecosystem." Springer-Verlag, Berlin and New York.

Lipschultz, F., Zafirou, O. C., Wofsy, S. C., McElroy, M. B., Valois, F. W., and Watson, S. W. (1981). Production of NO and N_2O by nitrifying bacteria. *Nature (London)* **294**, 641–643.

McClaugherty, C. A., Aber, J. D., and Melillo, J. M. (1982). The role of fine roots in the organic matter and nitrogen budgets of two forested ecosystems. *Ecol. Monogr.* **63**, 1481–1490.

Major, J. (1974a). Kinds and rates of changes in vegetation and chronofunctions. *In* "Vegetation Dynamics" (R. Knapp, ed.), pp. 9–18. Dr. W. Junk, The Hague.

Major, J. (1974b). Nitrogen accumulation in succession. *In* "Vegetation Dynamics" (R. Knapp, ed.), pp. 207–213. Dr. W. Junk, The Hague.

Marks, P.L., and Bormann, F. H. (1972). Revegetation following forest cutting: Mechanisms for return to steady state cycling. *Science* **176**, 914–915.

Melillo, J. M. (1981). Nitrogen cycling in deciduous forests. *In* "Terrestrial Nitrogen Cycles: Processes, Ecosystem Strategies and Management Impacts" (F. E. Clark and T. Rosswall, eds.), pp. 427–442. Ecological Bulletins, Stockholm.

Moldenhauer, W. C. (1980). Soil erosion—a global problem. *In* "Assessment of Erosion" (M. DeBoodt and D. Gabriels, eds.), pp. 3–8. Wiley, New York.

Nihlgärd, B. (1971). Pedological influence of spruce planted on former beech forest in Scania, South Sweden. *Oikos* **22**, 302–314.

Nye, P. H., and Greenland, D. J. (1960). "The Soil Under Shifting Cultivation, Tech. Commun. No. 15. Commonw. Agric. Bur., Harpenden, England.

Odum, E. P. (1969). The strategy of ecosystem development. *Science* **164**, 262–270.

Olson, J. S. (1958). Rates of succession and soil changes on southern Lake Michigan sand dunes. *Bot. Gaz. (Chicago)* **119**, 125–170.

Parsons, J. W., and Tinsley, J. (1975). Nitrogenous substances. *In* "Soil Components" (J. E. Gieseking, ed.), pp. 263–304. Springer-Verlag, Berlin and New York.

Perkins, D. F., Jones, V., Millar, R. O., and Neep, R. (1978). Primary production, mineral nutrients and litter decomposition in the grassland ecosystem. *In* "Production Ecology of British Moors and Montane Grasslands" (O. W. Heal and D. F. Perkins, eds.), pp. 304–331. Springer-Verlag, Berlin and New York.

Persson, H. (1978). Root dynamics in a Scots pine stand in central Sweden. *Oikos* **30**, 508–519.

Pesek, J., Stanford, G., and Case, N. L. (1971). Nitrogen production and use. *In* "Fertilizer Technology and Use" (R. A. Olson, T. J. Army, J. J. Hanway, and V. J. Kilmer, eds.), pp. 217–269. Soil Sci. Soc. Am., Madison, Wisconsin.

Power, J. F. (1981). Nitrogen in the cultivated ecosystem. *In* "Terrestrial Nitrogen Cycles' Processes, Ecosystem Strategies and Management Impacts" (F. E. Clark and T. Rosswall, eds.), pp. 529–546. Ecological Bulletins, Stockholm.

Rankin, J. D. (1978). Catalysts in ammonia production. *Proc.—Fert. Soc.* **168.**

Rasmussen, K. H., Taheri, M., and Kabel, R. L. (1975). Global emissions and natural processes for removal of gaseous pollutants. *Water, Air, Soil Pollut.* **4,** 33–64.

Robertson, G. P., and Vitousek, P. M. (1981). Nitrification potentials in primary and secondary succession. *Ecol. Monogr.* **62,** 376–387.

Robinson, E., and Robbins, R. C. (1970). Gaseous nitrogen compound pollutants from urban and natural sources. *J. Air Pollut. Control Assoc.* **20,** 303–306.

Robson, R. L., and Postgate, J. R. (1980). Oxygen and hydrogen in biological nitrogen fixation. *Annu. Rev. Microbiol.* **34,** 183–207.

Romkens, M. J. M., Nelson, D. W., and Mannering, J. V. (1973). Nitrogen and phosphorus composition of surface runoff as affected by tillage method. *J. Environ. Qual.* **2,** 292–295.

Rosswall, T. (1976). The internal nitrogen cycle between micro-organisms, vegetation and soil. *In* "Nitrogen, Phosphorus and Sulfur—Global Cycles" (B. H. Svensson and R. Söderlund, eds.), pp. 157–167 Ecological Bulletins, Stockholm.

Rosswall, T. (1983). The nitrogen cycle. *In* "The Major Biogeochemical Cycles and Their Interactions" (B. Bolin and R. B. Cook, eds.), pp. 46–50. Wiley, New York.

Rosswall, T., Flower-Ellis, J. G. K., Johansson, J. G., Jonsson, S., Rydén, B. E., and Sonesson, M. (1975). Stordalen, Abisko, Sweden. *In* "Structure and Function of Tundra Ecosystems" (T. Rosswall and O. W. Heal, eds.), pp. 265–294. Ecological Bulletins, Stockholm.

Skujins, J. (1981). Nitrogen cycling in arid ecosystems. *In* "Terrestrial Nitrogen Cycles: Processes, Ecosystem Strategies and Management Impacts" (F. E. Clark and T. Rosswall, eds.), pp. 477–491. Ecological Bulletins, Stockholm.

Smika, D. E., Heermann, D. F., Duke, H. R., and Batchelder, A. R. (1977). Nitrate-N percolation through irrigated sandy soil affected by water management. *Agron. J.* **69,** 623–626.

Snyder, J. L., and Burnett, J. A. (1966). Manufacturing processes for ammonia. *In* "Agricultural Anhydrous Ammonia" (M. H. McVickar, W. P. Martin, I. E. Miles, and H. H. Tucker, eds.), pp. 1–20. Am. Soc. Agron., Madison, Wisconsin.

Söderlund, R. (1977). NO pollutants and ammonia emissions—a mass balance for the atmosphere over N.W. Europe. *Ambio* **6,** 118–122.

Söderlund, R. (1981). Dry and wet deposition of nitrogen compounds. *In* "Terrestrial Nitrogen Cycles: Processes, Ecosystem Strategies and Management Impacts" (F. E. Clark and T. Rosswall, eds.), pp. 123–130. Ecological Bulletins, Stockholm.

Söderlund, R., and Rosswall, T. (1982). The nitrogen cycles. *In* "The Natural Environment and the Biogeochemical Cycles" (O. Hutzinger, ed.), pp. 61–81. Springer-Verlag, Berlin and New York.

Söderlund, R., and Svensson, B. H. (1976). The global nitrogen cycle. *In* "Nitrogen, Phosphorus and Sulphur—Global Cycles" (B. H. Svensson and R. Söderlund, eds.), pp. 23–73. Ecological Bulletins, Stockholm.

Staaf, H., and Berg, B. (1981). Plant litter input to soil. *In* "Terrestrial Nitrogen Cycles: Processes, Ecosystem Strategies and Management Impacts" (F. E. Clark and T. Rosswall, eds.), pp. 147–162. Ecological Bulletins, Stockholm.

Steinhart, U. (1973). Input of chemical elements from the atmosphere. A tabular review of literature. *Göttinger Bodenkd. Ber.* **29,** 93–132.

Stevens, P. R., and Walker, T. W. (1970). The chronosequence concept and soil formation. *Q. Rev. Biol.* **45,** 333–350.

Stevenson, F. J. (1965). Origin and distribution of nitrogen in soil. *In* "Soil Nitrogen" (W. V. Bartholomew and F. E. Clark, eds.), pp. 1–42. Am. Soc. Agron., Madison, Wisconsin.

Stevenson, F. J. (1982). "Humus Chemistry. Genesis, Composition, Reactions." Wiley, New York.

Stewart, W. D. P. (1980). Some aspects of structure and function in N_2-fixing cyanobacteria. *Annu. Rev. Microbiol.* **34,** 497–536.

Stewart, W. D. P., Rowell, P., and Rai, A. N. (1981). Symbiotic nitrogen fixing cyanobacteria. *In* "Nitrogen Fixation" (W. D. P. Stewart and J. R. Gallon, eds.), pp. 239–277. Academic Press, New York.

Swift, M. J., Heal, O. W., and Anderson, J. M. (1979). "Decomposition in Terrestrial Ecosystems." Blackwell, Oxford.

Syers, J. K., Adams, J. A., and Walker, T. W. (1970). Accumulation of organic matter in a chronosequence of soils developed on wind-blown sand in New Zealand. *J. Soil. Sci.* **21,** 46–153.

Tamm, C. O., Holmen, H., Popovic, B., and Wiklander, G. (1974). Leaching of plant nutrients from soils as a consequence of forestry operations. *Ambio* **3,** 211–221.

Theng, B. K. G. (1974). "The Chemistry of Clay–Organic Reactions." Hilger, London.

Theng, B. K. G. (1979). "Formation of Clay Polymer Complexes." Elsevier, Amsterdam.

Thomas, G. W., and Gilliam, J. W. (1978). Agro-ecosystems in the U.S.A. *In* "Cycling of Mineral Nutrients in Agricultural Ecosystems" (M. J. Frissel, ed.), pp. 182–243. Elsevier, Amsterdam.

Timmons, D. R., and Dylla, A. S. (1981). Nitrogen leaching as influenced by nitrogen management and supplemental irrigation level. *J. Environ. Qual.* **10,** 421–426.

Tingey, D. T. (1968). Foliar absorption of nitrogen dioxide. M.A. Thesis, University of Utah, Salt Lake City.

Troeh, F. R., Hobbs, H. A., and Donahue, R. L. (1980). "Soil and Water Conservation for Productivity and Environmental Protection." Prentice-Hall, Englewood Cliffs, New Jersey.

Van Cleve, K., and Alexander, V. (1981). Nitrogen cycling in tundra and boreal ecosystems. *In* "Terrestrial Nitrogen Cycles: Processes, Ecosystem Strategies and Management Impacts" (F. E. Clark and T. Rosswall, eds.), pp. 375–404. Ecological Bulletins, Stockholm.

Verstraete, W. (1981). Nitrification. *In* "Terrestrial Nitrogen Cycles: Processes, Ecosystem Strategies and Management Impacts" (F. E. Clark and T. Rosswall, eds.), pp. 303–314. Ecological Bulletins, Stockholm.

Vitousek, P. M. (1981). Clear-cutting and the nitrogen cycle. *In* "Terrestrial Nitrogen Cycles: Processes, Ecosystem Strategies and Management Impacts" (F. E. Clark and T. Rosswall, eds.), pp. 631–642. Ecological Bulletins, Stockholm.

Vitousek, P. M., and Melillo, J. M. (1979). Nitrate losses from disturbed forests: Patterns and mechanisms. *For. Sci.* **25,** 605–619.

Vitousek, P. M., and Reiners, W. A. (1975). Ecosystem succession and nutrient retention: A hypothesis. *BioScience* **25,** 376–381.

Vitousek, P. M., Gosz, J. R., Grier, C. C., Melillo, J. M., Reiners, J. M., and Todd, R. L. (1979). Nitrate losses from disturbed ecosystems. *Science* **204,** 469–474.

Vogt, K. A., Grier, C. C., Meir, C. E., and Edmonds, R. L. (1982). Mycorrhizal role in net primary production and nutrient cycling in *Abies amabilis* ecosystems in western Washington. *Ecol Monogr.* **63,** 370–380.

Weiss, R. F. (1981). The temporal and spatial distribution of tropospheric nitrous oxide. *JGR, J. Geophys. Res.* **86,** 7185–7195.

Weiss, R. F., and Craig, H. (1976). Production of atmospheric nitrous oxide by combustion. *Geophys. Res. Lett.* **3,** 751–753.

West, N. E., and Klemmedson, T. O. (1978). Structural distribution of nitrogen in desert ecosystems. *In* "Nitrogen in Desert Ecosystems" (N. E. West and J. Skujins, eds.), pp. 1–16. Dowden, Hutchinson and Ross, Stroudsberg, Pennsylvania.

Wetselaar, R. (1981). Nitrogen inputs and outputs of an unfertilised paddy field. *In* "Terrestrial Nitrogen Cycles: Processes, Ecosystem Strategies and Management Impacts" (F. E. Clark and T. Rosswall, eds.), pp. 573–583. Ecological Bulletins, Stockholm.

Whittaker, R. H. (1975). "Communities and Ecosystems." Macmillan, New York.

Wild, A., and Cameron, K. C. (1980). Soil nitrogen and nitrate leaching. *In* "Soils in Agriculture" (P. B. Tinker, ed.), pp. 35–70. Blackwell, Oxford.

Woodmansee, R. G., and Wallach, L. S. (1981). Effects of fire regimes on biogeochemical cycles. *In* "Terrestrial Nitrogen Cycles: Processes, Ecosystem Strategies and Management Impacts" (F. E. Clark and T. Rosswall, eds.), pp. 649–669. Ecological Bulletins Stockholm.

Woodmansee, R. G., Dodd, J. L., Bowman, R. A., Clark, F. E., and Dickinson, C. E. (1978). Nitrogen budget of a shortgrass prairie ecosystem. *Oecologia* **34,** 363–376.

Woodmansee, R. G., Vallis, I., and Mott, J. J. (1981). Grassland nitrogen. *In* "Terrestrial Nitrogen Cycles: Processes, Ecosystem Strategies and Management Impacts" (F. E. Clark and T. Rosswall, eds.), pp. 443–462. Ecological Bulletins, Stockholm.

Woodwell, G. M. (1974). Variation in the nutrient content of leaves of *Quercus alba, Quercus coccinea* and *Pinus rigida* in the Brookhaven Forest from budbreak to abscission. *Am. J. Bot.* **61,** 749–753.

Young, K. K. (1980). The impacts of erosion on the productivity of soils in the United States. *In* "Assessment of Erosion" (M. DeBoodt and D. Gabriels, eds.), pp. 295–303. Wiley, New York.

Chapter 2

The Decomposition Process: Mineralization, Immobilization, Humus Formation, and Degradation

R. J. HAYNES

I. INTRODUCTION

Organic N-containing compounds, the products of microbial decomposition of plant and animal remains, account for over 90% of the total N in most soils. Litter, originating from both above- and belowground plant parts, is the major pathway of supply of energy and N to the soil in most terrestrial ecosystems (Staaf and Berg, 1981) and decomposition constitutes the means by which N held in the structure of plant tissues is released into the soil for reuse by plants. Indeed, the process of litter decomposition represents a very important link in the N cycle of most natural ecosystems (Swift *et al.*, 1979) and many agricultural ecosystems (Floate, 1981).

As a result of decomposition of plant and animal residues, C is recirculated to the atmosphere as CO_2 and organic N is made available (mineralized) as NH_4^+- and NO_3^--N while other essential nutrients appear in plant-available forms. During decomposition, some of the C and N is assimilated (immobilized) into microbial tissue and part is microbially converted into resistant humic substances (humus), which constitute the bulk of the soil organic matter. Nevertheless, some of the native soil humus is mineralized concurrently so that the total soil organic matter and N content of a soil may remain at a steady-state level.

Indeed, although there is a progressive increase in soil organic N levels during ecosystem development, when an ecosystem reaches a steady-state equilibrium situation annual decomposition of organic matter balances annual input so that the soil organic matter content remains constant. Thus organic matter turnover, the process in which losses and gains proceed simultaneously, is a process central to ecosystem stability.

In agriculture, the decomposition of organic detritus is very important in the N cycle of grassland farming ecosystems (Floate, 1981; Woodmansee et al., 1981) and in cultivated cropping systems where significant quantities of crop residues remain after harvest or where organic residues are added (Allison, 1973; Power, 1981). The decomposition of soil organic matter N is also important in terms of N availability to crop plants in cultivated ecosystems, particularly where the rapid microbial immobilization of added fertilizer N is followed by a slow net remineralization over a period of years.

The process of detrital decay is complex and is facilitated by the activities of a wide range of macro- and microorganisms. These activities are influenced by numerous factors such as the chemical composition and physical structure of the detritus and environmental factors such as temperature, moisture, aeration, and pH (Berg and Staaf, 1981; Jenkinson, 1981). The decomposition of soil organic matter is less well understood but it is known to be strongly influenced by environmental factors (Allison, 1973).

In this chapter, the process of decomposition is discussed in terms of the mineralization/immobilization balance of N in the soil and thus the availability of mineral N to growing plants.

II. PROCESSES OF DECOMPOSITION

The decomposition of different plant materials in different soils under different environmental conditions has been studied extensively (see Dickinson and Pugh, 1974; Swift et al., 1979). Much research has centered on the initial stages of decomposition (i.e., the first year or so) when the decomposition rate can be simply estimated by periodic measurements of weight loss of the litter. However, in recent times, the use of isotopically labeled plant materials has allowed the process of decomposition to be followed over a period of many years, long after the initial litter input has become unrecognizable.

A. Breakdown of Organic Residues

Since the C content of plants is in the region of 40 to 50% by weight, the decomposition of ^{14}C from labeled plant materials gives a reasonable measure of the overall decomposition process. This provides a basis from which to compare the decomposition of ^{15}N from plant residues. The general patterns of decomposition of ^{14}C- and ^{15}N-labeled plant materials are therefore discussed below.

1. Plant Carbon

A number of long-term field studies have been carried out to estimate the rates of decomposition of ^{14}C-labeled plant tissues (Jenkinson, 1965, 1971, 1977a,b; Führ and Sauerbeck, 1968; Smith and Douglas, 1971; Sauerbeck et al., 1972; Oberlander, 1973; Shields and Paul, 1973; Jenkinson and Ayanaba, 1977; Sauerbeck and Gonzalez, 1977; Ladd et al., 1981). In general, the concentration of organic ^{14}C in soils (residual ^{14}C-labeled plant material and derived microbial cells and products) decreases rapidly in the first few months, after which the net rate of decomposition slows considerably. A range of crop residues, when incorporated with the soil, has been shown to lose approximately 60 to 70% of their C as CO_2 during the first year in the field and another 20% or so in the following 6 to 9 years (Jenkinson, 1965; Führ and Sauerbeck, 1968; Martin et al., 1980; Ladd et al., 1981). The residual C is incorporated into the microbial biomass and the soil humus.

2. Plant Nitrogen

In the context of this chapter, it is the decomposition of N from plant materials that is of particular importance. The release of C and N from residues differs in that C is generally volatilized as CO_2 or methane while N tends to be conserved. Part of this conservation is caused by the demand for N by the decomposer microorganisms since nitrogen can be a major factor limiting microbial growth during decomposition. Indeed in some cases (as discussed in the following sections) the microbial biomass may even incorporate mineral N from the surrounding soil or litter during the decomposition of organic residues with a wide C : N ratio.

The relative rates of decomposition of labeled C and N from plant residues reflect such conservation of N (Paul, 1976; Broadbent and Nakashima, 1974; Ladd et al., 1981). Ladd et al. (1981), for example, found that in four South Australian soils 60 to 65% of ^{15}N-labeled medic (C/N ratio 8.7 : 1) remained as residues after 32 weeks of decomposition; the percentage decreased to 45 to 50% after 4 years. In contrast, more than 50% of medic ^{14}C had disappeared from all soils after 4 weeks and only 15 to 20% of the ^{14}C remained in the residues after 4 years. Amato and Ladd (1980) demonstrated a similar pattern of decomposition of ^{14}C-, ^{15}N-labeled medic in a laboratory study (Fig. 1).

It is clearly evident from Fig. 1 that the decomposition of ^{14}C and ^{15}N from labeled plant residues followed very similar patterns with respect to time. That is, there was an initial rapid loss of both C and N followed by a slow decomposition phase. This second slow phase is indicative of the formation of recalcitrant N-containing humic substances during decomposition.

3. Plant Constituents

Except for a small proportion of mineral constituents (comprising between 1 and 8% by weight) plant tissues are largely composed of complex organic compounds. These include ether-soluble fats, oils, waxes, and resins (0.5 to 5%), water-soluble compounds (5 to 20%), cellulose (15 to 60%), hemicelluloses (10 to 30%), crude protein (5 to 15%), and lignin (5 to 30%). The proportion of each different organic constituent varies according to plant species and tissue. Detailed reviews of the synthesis, structure, and function of such compounds are presented elsewhere (e.g., Miller, 1973).

The N content of plant material varies from 0.1 to 6% depending on plant part, age, and species. The majority of mature materials contain less

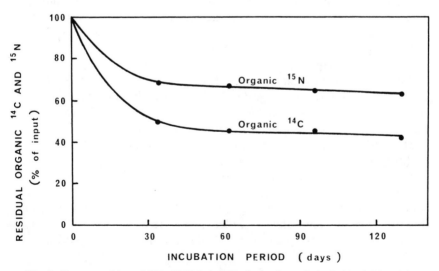

Fig. 1. Decomposition of ^{14}C-, ^{15}N-labeled *Medicago littoralis* leaf material in a laboratory study. [Redrawn from Amato and Ladd (1980). Reprinted by permission from Pergamon Press.]

than 1% N. In leaves and stems approximately 60% of the N is present as enzyme or membrane protein while most of the remainder is accounted for as free (water-soluble) amino acid N (Parsons and Tinsley, 1975). In seeds, over 90% of N can be in the form of storage protein. The biochemistry of the degradation of proteins and amino acids with the release of NH_4^+ is discussed in Section III,E.

Individual organic components of plant material decompose at different rates when they are added to soils. For example, lignin, the most resistant of the major plant components, loses only about 20 to 30% of its C over a 6-month to 1-year period (Martin and Haider, 1977, 1979). On the other hand, sugars, many polysaccharides, proteins, amino acids, aliphatic acids, and other highly degradable organic substances may lose 80 to 90% of their C during the first 3 to 6 months (Martin and Focht, 1977). Decomposition rates of litter tend to vary according to the proportion of each component that is present and are generally slow where lignins, fats, and waxes constitute a large proportion.

4. Animal Constituents

Most of the biochemical constituents discussed above (e.g., proteins and amino acids) are also present in animal residues and their turnover rates are similar to those from dead plant material. Constituents of animals that are not present in plant residues are largely components of specialized structures. For example, the polymeric amino sugar chitin is a component of the exoskeleton of arthropods and eggs of nematodes (Parsons and Tinsley, 1975) and is generally rather stable in soils (Sørensen, 1977).

As noted in Chapter 1, in some ecosystems (e.g., grazed pastures) the herbivore can form a significant pathway in the cycling of N. This is because the proportion of ingested N that is used by the animal is rather small (e.g., 15 to 20% for cattle) in comparison to the quantity returned to the plant–soil system via the urine and feces (Floate, 1981; Woodmansee *et al.*, 1981). Urine usually contains 50 to 80% of the excreted N.

Nitrogen in the urine is predominantly in the form of urea (NH_2CONH_2), which undergoes rapid hydrolysis in the soil to form NH_4^+ (see Section III,E). The feces (principally undigested forage) generally contain mineral elements bound in relatively resistant organic fractions that are released only slowly. Indeed, N is released from feces more slowly than from fresh plant material (Barrow, 1961; Floate, 1970b).

B. Phases of Nitrogen Release and Accumulation

In general, the N dynamics during litter decomposition at the soil surface can be separated into three phases: leaching, accumulation, and final release (Berg and Staaf, 1981). Not all three phases are always seen in practice. In Fig. 2 three different cases are illustrated: (a) all three phases, (b) accumulation and release, and (c) leaching followed by release or simply release. When organic material is incorporated into the soil the leaching phase is unlikely to occur.

1. Leaching

During the very early period following litter fall, weight loss and nutrient release are not caused by microbial action but rather by leaching of soluble substances from the litter. The presence of water-soluble nitrogenous substances in plant tissues has been demonstrated by short-term leaching of plant tissue with distilled water (Nykvist, 1959, 1963).

Several field studies of litter decomposition have clearly illustrated the presence of a leaching phase of N release (Gosz *et al.*, 1973; see Fig. 3, beech leaves; Howard and Howard, 1974; Hodkinson, 1975; Staaf and Berg, 1977; Berg and Söderström, 1979). The quantity of N leached ex-

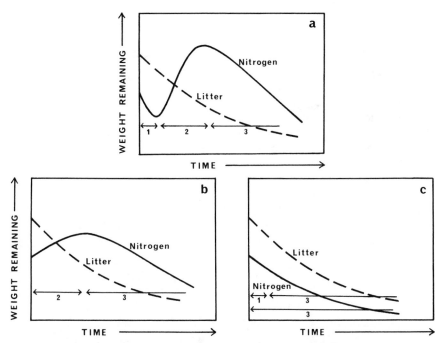

Fig. 2. Typical trends in total litter weight and weight of N in litter during the decomposition process. Phases of N release and accumulation shown are: (1) leaching, (2) accumulation, and (3) release. Three different cases are shown: (a) leaching, accumulation, and release; (b) accumulation and release; and (c) leaching and release or release only. [Redrawn from Berg and Staaf (1981).]

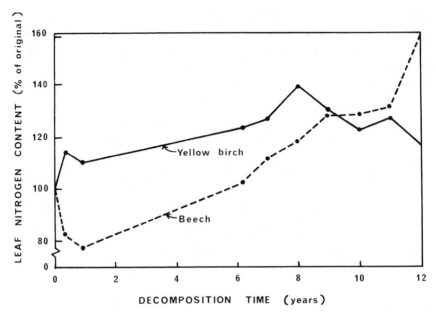

Fig. 3. Nitrogen content (expressed as a percentage of the original weight of leaves of yellow birch (*Betula allegheniensis* Britt.) and beech (*Fagus grandifolia* Ehrh.) during decomposition. [Data from Gosz *et al.* (1973).]

pressed as a percentage of the total initial litter N content appears to be in the region of 10% (Berg and Staaf, 1981). The extent, and more particularly the rate, of leaching is largely determined by the quantity of water percolating through the litter; a high rainfall would tend to result in high leaching losses.

2. Accumulation

a. Occurrence. An increase in the relative amount of N during decomposition of litter (increasing N content relative to remaining litter weight) is a generally occurring and well-known phenomenon (Aber and Melillo, 1980; Berg and Staaf, 1981; Melillo *et al.,* 1982). Such an increase occurs whether or not there is an absolute increase (increase in N content relative to initial litter weight) or loss of N from the litter (e.g., Fig. 4). It occurs initially because N in organic debris is generally in short supply; thus the decomposer microorganisms utilize and retain most of the N through its incorporation into microbial cells. Later, some of this N is converted into recalcitrant humic substances. Most of the N is therefore retained while the amount of C is progressively reduced (through evolu-

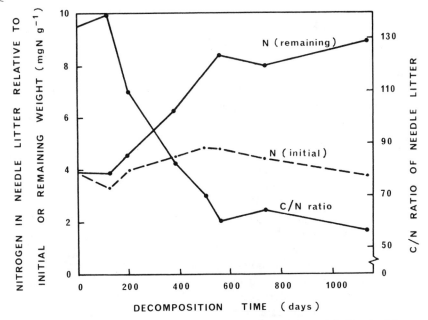

Fig. 4. Nitrogen content (relative to initial or remaining weight) and C : N ratio of decomposing Scots pine (*Pinus sylvestris* L.) needle litter. [Data from Berg and Söderström (1979). Reprinted with permission from Pergamon Press.]

tion of CO_2) so that the C : N ratio of the litter narrows as decomposition proceeds (Fig. 4).

Under some conditions there is also an absolute increase in the quantity of N in the litter as it decomposes (Hayes, 1965; Will, 1967; Bocock, 1964; Anderson, 1973; Gosz *et al.*, 1973; Howard and Howard, 1974; Kaarik, 1974; Dowding, 1976; Berg and Söderström, 1979; Melillo *et al.*, 1982). Such a phase can be difficult to distinguish in field studies (Berg and Staaf, 1981) or indeed it may be absent. In some cases there is no change in the absolute N content of litters (Anderson, 1973; Staaf and Berg, 1977; Berg and Staaf, 1980) during decomposition while in others there is a consistent absolute release of N (Bocock, 1964; Hayes, 1965; Berg and Staaf, 1980; Ladd *et al.*, 1981). When organic residues (e.g., straw) are incorporated into agricultural soils an absolute increase in their N content is a common phenomenon (Bartholomew, 1965; Allison, 1973).

b. Explanation. The actual origin of the absolute increases in the N mass of forest litter during decomposition has been explained in a number of ways, including biological N_2 fixation, absorption of atmospheric NH_3,

rainfall, throughfall, dust, insect frass, green litter, and fungal translocation (Bocock, 1964; Gosz *et al.,* 1973; Staaf and Berg, 1977; Berg and
Söderström, 1979; Bormann and Likens, 1979).

It seems likely that the relative importance of the various proposed
pathways of N input will differ in different situations. Nevertheless, the
underlying reason for such an accumulation is clearly the demand for N
by the microbial biomass decomposing the carbonaceous litter material.
Indeed, it is thought that when the C : N ratio of litter is high (>25 to 30 : 1;
N content < 1.4–1.8%) net immobilization of N will generally occur,
resulting in the net importation of N from outside the litter system. In
such a situation, the heterotrophic decomposer biomass invariably out-
competes the nitrifier organisms and plants for NH_4^+-N and this can cause
a deficiency of mineral N in the surrounding soil (Allison, 1973).

As already noted, when organic residues with a high C : N ratio (>25 to
30 : 1) are incorporated into agricultural soils, net immobilization of N
commonly occurs (Allison, 1973). The mineral N for decomposition is
supplied through mineralization of soil organic N. Thus, N fertilizers are
often applied along with the residues to overcome a depression of plant-
available N that occurs during decomposition (Bartholomew, 1965). The
immobilization, and subsequent release, of N during the decomposition of
wheat straw (C : N ratio $\approx$ 100) incorporated into a soil, amended with
$NaNO_3$, is illustrated in Fig. 5. Immobilization was very rapid in the first 7
days and closely paralleled microbial activity as estimated by CO_2 evolution. Fertilizer N additions can stimulate the rate of decomposition if the
supply of mineral N is low. However, in fertile agricultural soils the
supply of mineral N is often adequate and fertilizer additions have little or
no effect on the decomposition rate (e.g., Smith and Douglas, 1968, 1971).
In contrast, when residues are left at the soil surface N supply can severely limit decomposition and incorporation of residues into the soil can
greatly increase the decomposition rate (Brown and Dickey, 1970). Thus,
decomposition of surface litter in natural systems is generally very slow in
comparison with that of crop residues that are incorporated into agricultural soils.

3. Net Release

As illustrated in Figs. 4 and 5, the accumulation (net immobilization) of
N in the litter is followed by a slow net release (net mineralization) of N.
This is because during decomposition the C : N ratio progressively decreases and at some point N becomes no longer limiting to microbial
growth and activity. At that critical point there is a switch from net
immobilization to net mineralization. In litter with a high N content (e.g.,
legume residues) N may not limit microbial growth at any time and in such

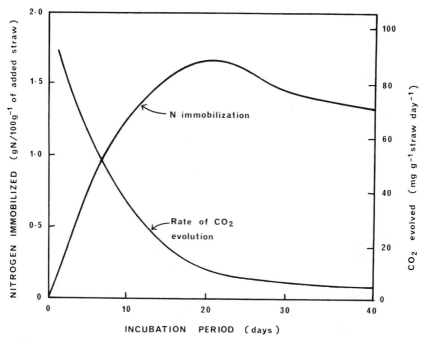

Fig. 5. Immobilization and release of nitrogen and rate of CO_2 evolution from a soil receiving wheat straw and nitrate-nitrogen. [Redrawn from Allison and Klein (1962). Reprinted with permission from the Williams and Wilkins Co., Baltimore.]

cases there is no accumulation phase and net release occurs immediately following litter deposition (e.g., Fig. 1).

As already noted, the critical N content above which net mineralization will occur is generally considered to be in the range >1.4 to 1.8% N and C:N ratio < 25 to 30. Nonetheless, in natural and agricultural ecosystems, the N level in litter at which net release of N occurs varies enormously (see Section IV,A).

Once the release phase has started the pattern of N loss appears to follow that of weight loss (Wood, 1974); Staaf and Berg (1977), for example, found that the release of N from Scots pine needles was linearly related to weight loss ($r = 0.93$). Nevertheless, the N concentration of the litter typically continues to increase (e.g., Fig. 4). This is indicative of the fact that relative to C, N is retained in the decomposing litter. Much of this retained N is likely to be incorporated into humic polymers, which are synthesized through the actions of the decomposer microflora (Section III).

C. Role of Decomposer Organisms

The major groups of litter-decomposing organisms are bacteria, actino-mycetes, fungi, protozoa, nematodes, microarthropods, enchytraeid worms, and lumbricid worms (Edwards, 1974; Harding and Stuttard, 1974; Lofty, 1974; Pugh, 1974; Swift *et al.*, 1979). The role of these organisms in the decomposition process, with particular emphasis on the mineralization of organic N, is discussed below.

1. Microflora

Litter decomposition is characterized by the accumulation of a huge biomass of decomposer microorganisms. In general, the microfloral population associated with dead plant and animal materials is extremely diverse (Swift, 1976; Swift *et al.*, 1979). Novack and Whittingham (1968), for example, found 161 species of fungi in deciduous leaf litter from a single forest. When bacteria and actinomycetes, as well as fungi, are included, a comprehensive list of the microflora of decaying detritus would yield many hundreds of species.

Microbial attack of plant parts, in fact, begins long before senescence occurs. Studies have shown that the surfaces of leaves become colonized by phylloplane bacteria and fungi as soon as they unfold or even before bud-burst (Bell, 1974; Jensen, 1974). Roots are also invaded by bacteria and fungi well before they slough (Martin, 1977).

In most environments fungi constitute the bulk of the primary decomposer population of plant materials (Swift *et al.*, 1979). Fungi are well adapted to their role as primary decomposers since they possess a fila-mentous, mycelial network that can permeate the relatively massive structure of the recently deposited litter. The bacteria often occur as the secondary population since their unicellular mode of life means they are adapted to the occupation of particulate detritus with a high surface to volume ratio. Furthermore, many of the bacterial decomposer organisms are, in fact, mycolytic and exert considerable activity in the breakdown of fungal mycelia. The actinomycetes generally appear to have a lower competitive ability than the common, rapidly growing species of soil fungi and bacteria (Goodfellow and Cross, 1974; Alexander, 1977) yet they form a characteristic although minor component of the decomposer microflora.

2. Roles of Microflora

Through actions of a number of secreted extracellular enzymes the fungi and some bacteria are active decomposers of plant proteins, amino acids, and other nitrogenous organic compounds, resulting in the eventual release of NH_4^+ (see Section III,C). Among the genera of fungi known to possess proteolytic ability are *Alternaria, Aspergillus, Mucor, Penicillium,* and *Rhizopus* (Wainwright, 1981). While this overall process is termed mineralization, the final step in which NH_4^+ is liberated is known as ammonification.

For synthesis of their own proteins the decomposer microflora can utilize either mineral N (NH_4^+ or NO_3^-) or simple nitrogenous compounds (e.g., amino acids). The ability of fungi to absorb simple organic nitrogenous substances, which are deaminated within the cells, means that the ammonification process is not always necessary in terms of fungal nutrition. Even so, a large proportion of the mineral N released by fungal enzymes is almost immediately absorbed (immobilized) by the fungi themselves; thus mineralization can, in the short term, be offset by subsequent N immobilization (Wainwright, 1981). In a shortgrass prairie ecosystem Woodmansee *et al.* (1981), for example, estimated that 50% of the N annually ammonified was reimmobilized by the microbial biomass.

The preference of the microbial biomass for either NH_4^+ or NO_3^- is of some importance in relation to the efficient use of fertilizer N. Although Jansson (1958) emphasized the preferential use of NH_4^+ by soil microorganisms, Alexander (1977) states that many bacteria and filamentous fungi develop readily on media containing either NH_4^+ or NO_3^-. Many workers (e.g., Allison and Klein, 1962; Broadbent and Tyler, 1962) have shown that either NH_4^+ or NO_3^- are readily immobilized by microorganisms. Nevertheless, when both forms of mineral N are present at the same time, the soil biomass apparently assimilates NH_4^+ considerably more rapidly than NO_3^- (Broadbent, 1968), perhaps because less energy is required for assimilation of the former (Ahmad *et al.*, 1972).

Regardless of the form in which the N is assimilated by fungi, it is eventually released since the fungal biomass produced during litter decomposition is itself, in turn, decomposed. This is accomplished through autolysis or by the actions of mycolytic bacteria and/or soil microfauna that feed on both bacteria and fungi. Bacterial N may also be released by other bacteriolytic bacteria or simply by autolysis. Thus the ability of fungi to immobilize N in the biomass is offset by recycling of N through the eventual lysis of microbial tissue and/or faunal feeding, resulting in turnover of extracellular pools of N even when N contents of litter are low (Heal *et al.*, 1982).

3. Fauna

Evidence is accumulating that terrestrial decomposition processes are greatly influenced by animals, particularly invertebrates (Coleman *et al.*, 1977; Crossley, 1977; Anderson *et al.*, 1981; Hole, 1981). Studies have generally indicated that when fauna are present, the loss of plant litter mass per annum is 25 to 80% greater than when fauna are excluded (Jensen, 1974). The four major groupings of decomposer fauna are the annelids (earthworms belonging to the families Lumbricidae and Enchytraeidae), arthropods (collembolans, mites, insects, termites, and ants), nematodes, and protozoa.

The major effects of earthworms originate from the large amounts of dead plant material they ingest and distribute throughout the soil (Wallwork, 1976; Syers *et al.*, 1979). The collembola are widely distributed in soil and they appear to feed principally on fungi (Parkinson *et al.*, 1979). Mites are also abundant in many soils and can be divided into those that feed on decomposing litter and those that feed on fungi and/or bacteria (Wallwork, 1976). Adult insects can play an important role in the decomposition of woody litter and animal dung (Gillard, 1967; Anderson *et al.*, 1981). The termites are herbivores, while the ants can be herbivores or carnivores.

Soil-inhabiting nematodes show great diversity in feeding habits and they may be broadly grouped into plant feeders, microbial feeders, predators, and omnivores (Wallwork, 1976). The microscopic soil protozoans feed principally on bacteria (Habte and Alexander, 1977; Charholm, 1981).

4. Roles of Fauna

Soil fauna have three major effects on decomposition processes: (1) a physical effect of redistributing organic materials, (2) a chemical effect of concentrating certain elements in their bodies and accelerating nutrient cycling, and (3) a biological effect of regulating microbial activity.

a. Physical. During this process, large litter components are fragmented thereby exposing greater surface areas for colonization and attack by microorganisms. The organic material may also be incorporated into deeper soil horizons. Such actions are carried out primarily by the macrofauna such as earthworms, ants, mites, and termites.

b. Chemical. Soil fauna may make little quantitative difference to the flow of energy through the ecosystem but they have a definite effect on accelerating nutrient cycling (Edwards *et al.*, 1970; Reichle, 1977). Soil fauna generally excrete N as NH_4^+, urea, or amino acids and because of a respiratory loss of C the C/N ratio of excretory products is generally lower than that of the ingested material. Thus, the action of soil fauna tends to promote net mineralization of N rather than immobilization (Anderson *et al.*, 1981). For example, earthworms are known to produce worm casts with a lower C/N ratio and higher levels of NH_4^+ and NO_3^- than the litter they ingest (Barley and Jennings, 1959; Syers *et al.*, 1979).

The soil fauna that feed on primary decomposers (fungi and bacteria), or on one another, increase decomposition rates and increase nutrient mineralization since they have relatively short generation times, and although most of the ingested C-containing compounds are used to sustain metabolic activity, the bulk of the N that they consume is excreted (Anderson et al., 1981). Ineson et al. (1982), for example, showed that the effect of collembolan grazing on decomposing litter was to increase the leaching of NH_4^+ and NO_3^- from that litter. Anderson et al. (1981) calculated that the bacteria-feeding nematodes in forest litter could consume approximately 800 kg of bacteria ha^{-1} yr^{-1}, resulting in 20 to 130 kg N ha^{-1} yr^{-1} being mineralized.

c. *Biological.* The activities of soil fauna can influence the microbial activity of the litter and soil and thus the rate of organic matter decomposition. For example, there is evidence that earthworm casts are enriched with microflora, which may contribute to an increased rate of decomposition (Edwards and Lofty, 1977). By their movements, fauna can disperse fungal spores, resulting in concentrations of hyphae that may, in turn, attract soil fauna (Hole, 1981).

The fauna that feed on microorganisms (e.g., protozoa, nematodes, and collembola) can increase primary decomposer actions, as measured by CO_2 evolution, even though numbers of some microbial types may be reduced (Habte and Alexander, 1978; Hanlon and Anderson, 1979; Ineson et al., 1982). Such fauna thus tend to regulate microbial population sizes, composition, and activities (Anderson et al., 1981).

D. Degradation of the Microbial Biomass

In the preceding discussions it was shown that during the decomposition of litter net immobilization of N by the decomposer biomass can occur. Even where net mineralization of N does occur, a significant proportion of the total mineral N released by fungal enzymes is almost immediately absorbed by the microbial biomass. Much of this N is, however, eventually released to the soil through lysis of microbial tissues or faunal feeding. Thus the decomposition of microbial tissues is of great importance in terms of the final release to the soil of N originally bound in plant litter.

The results of Ladd et al. (1981) (Fig. 6) demonstrate the pattern of buildup and then decay of biomass ^{14}C and ^{15}N during the decomposition of ^{14}C-, ^{15}N-labeled plant residues. In samples taken at 8 weeks, biomass ^{14}C and ^{15}N accounted for 14 and 22%, respectively, of the total ^{14}C and ^{15}N residues. Thereafter, concentrations of biomass ^{14}C and ^{15}N de-

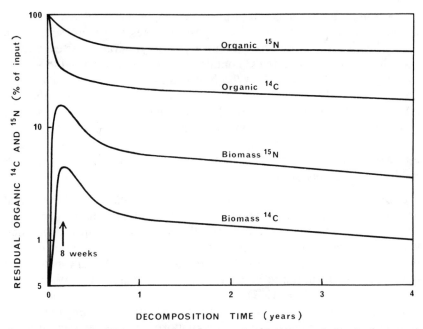

Fig. 6. Formation and decay of isotope-labeled microbial biomass during the decomposition of ¹⁴C-, ¹⁵N-labeled *Medicago littoralis* leaf material in a soil under field conditions. [Redrawn from Ladd *et al.* (1981). Reprinted with permission from Pergamon Press.]

creased rapidly and then more slowly until at 208 weeks they accounted for 5 and 9%, respectively, of the total ¹⁴C and ¹⁵N residues.

1. Forms and Microbial Nitrogen

The N content of microbial cells is somewhat higher than that of plants, amounting to 5–10% of dry matter. Much of this is in the form of storage protein (Parsons and Tinsley, 1975) and most of the remainder is present in cell walls as highly polymerized heteropolymers, the components of which are amino sugars and amino acids.

Amino sugars occur in nature largely in the form of amino polysaccharide polymers. Amino sugars found in soils are derived principally from the polymers chitin, peptidoglycans, and teichoic acids. Chitin is a major component of fungal cell walls as well as of invertebrate exoskeletons. The peptidoglycans occur in the cell walls of all bacteria except extreme halophiles, while teichoic acids are major components of the cell walls of gram-positive bacteria. Many amino sugars occur in such polysaccharides and the three predominant ones are glucosamine, galactos-

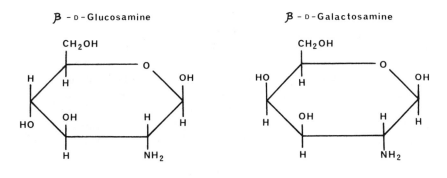

β - D - Glucosamine   β - D - Galactosamine

β - Muramic acid

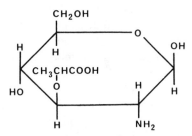

Fig. 7. The structure of β-D-glucosamine, β-D-galactosamine, and β-muramic acid.

amine, and muramic acid (Fig. 7), which often occur as their N-acetyl derivatives.

The pathways of decomposition of amino sugars and amino acids, with the eventual release of NH_4^+-N, are discussed in detail in Section III,E.

2. Microbial Nitrogen as a Source of Mineralizable Soil Nitrogen

The soil microbial biomass is in a constant state of turnover (Jenkinson and Ladd, 1981) and dead microbial cells are readily mineralized by the remaining microflora (e.g., Nelson *et al.*, 1979). Thus the biomass can contribute substantial amounts of nutrients to the pool of mobile, plant-available nutrients in the soil (Anderson and Domsch, 1980; Marumoto *et al.*, 1982a; Paul, 1984).

The quantity of N held in the biomass is relatively large; Anderson and Domsch (1980) calculated that microbial biomass N in the surface 12.5 cm of 26 agricultural soils accounted for between 0.5 and 15.3% of the total soil N and amounted to approximately 108 kg N ha^{-1}, while Jenkinson and Ladd (1981) estimated the N in the microbial biomass of an unmanured wheat field to be 95 kg N ha^{-1} (to 23 cm depth). In the same field, Jenkinson and Ladd (1981) calculated the N flux through the microbial biomass to be 34 kg N ha^{-1} yr^{-1}, which was greater than the annual offtake in wheat grain and straw of 24 kg N ha^{-1} yr^{-1}. Field studies utilizing ^{15}N fertilizers have emphasized the role of the microbial biomass N as both a source and sink of mineral N and have demonstrated the rapid remineralization of immobilized N (Carter and Rennie, 1984).

Sudden changes in environmental conditions that cause death of a large proportion of the microbial biomass and a subsequent large flush of N mineralization include drying and rewetting (Ahmad *et al.*, 1973; Kai *et al.*, 1973; Marumoto *et al.*, 1974, 1977a,b,c, 1982a,b), freezing and thawing (Witkamp, 1969; Shields *et al.*, 1974), and fluctuating soil temperatures (Biederbeck and Campbell, 1973). Marumoto *et al.* (1982a) calculated that during the first 4 weeks following a drying and rewetting cycle about 40 kg of mineral N ha^{-1} in the upper 12.5 cm of soil was derived from microbial cells after death. Furthermore, the rate of decomposition of dead microbial ^{15}N in soil after drying and rewetting has been shown to be almost five times as great as that of mobile soil organic N (Ladd *et al.*, 1976; Amato and Ladd, 1980; Marumoto *et al.*, 1982a). Marumoto *et al.* (1982b) estimated that following drying and rewetting of a soil, about 76% of the flush of N mineralization was derived from the dead microbial biomass in that soil and the remaining 24% was derived from native soil organic matter.

It is evident that the microbial biomass plays a dual role in the soil: as an agent of decomposition and release of N from fresh organic residues and soil organic matter and second as a labile pool of soil N.

III. HUMUS FORMATION, COMPOSITION, AND DEGRADATION

As outlined in Chapter 1, the decomposition of any piece of plant detritus and its transformation product (humus) is completed over a time span that can take from a few days to hundreds or thousands of years. The residues of the decomposition process contribute to the formation of soil organic matter. In the short term such residues constitute a cellular fraction consisting of partially digested plant material, animal carcasses, feces, and microbial cells. A second fraction of organic residues that is not readily available as an energy source for the heterotrophic soil microflora also accumulates in the soil. These latter residues constitute humus.

An understanding of the formation, structure, and degradation of humic substances is important since a very large proportion of soil nitrogen (i.e., over 90% in most soils) is associated with such substances, which are formed during the decomposition process. Despite this, our knowledge regarding the nature of humic substances is incomplete due principally to their extremely complex and heterogeneous polymeric structure.

A. Origin and Formation

1. Polyphenol Theory

Although several pathways have been proposed for the formation of humic substances, most soil chemists now favor the polyphenol theory, which is shown schematically in Fig. 8. Numerous studies have demonstrated that phenolic compounds can be chemically or enzymatically polymerized to humic-like structures (e.g., Bondietti *et al.*, 1971; Flaig *et al.*, 1975; Liu *et al.*, 1981). The three major sources of phenolic structural units during decomposition are (1) polyphenols derived from the micro-

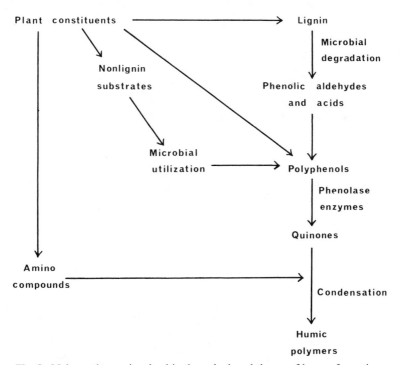

Fig. 8. Major pathways involved in the polyphenol theory of humus formation.

bial degradation of plant lignins (Cheshire *et al.*, 1967; Hurst and Burges, 1967); (2) phenolic polymers (melanins) synthesized by many fungi, actinomycetes, and bacteria from simple aliphatic compounds that arise from the degradation of nonlignin C sources (Martin and Haider, 1971; Flaig *et al.*, 1975; Saiz-Jimenez *et al.*, 1975); and (3) polyphenols of plant origin, which can be of some importance in forest litter with a high polyphenol content (see Section IV,A).

The polyphenols arising from these various sources are not stable but are subject to decomposition by the soil biomass. Alternatively, they may undergo recombination, either alone or with other organic molecules, often after conversion to quinones. Although phenols can be spontaneously oxidized to quinones in alkali media, the phenoloxidase and peroxidase enzymes that are synthesized by many microorganisms and are present in soils (Ladd, 1978) are thought to be the primary factors responsible for the conversion in neutral and acid soils (Martin *et al.*, 1979; Martin and Haider, 1980; Sjoblad and Bollag, 1981).

The quinones may then undergo self-condensation or combine with amino compounds to form N-containing polymers. Self-condensation of quinones is greatly enhanced in the presence of amino compounds such as amino acids, peptides, and proteins (Flaig *et al.*, 1975). Examples of the types of reactions postulated to occur between amino acids and quinones are illustrated in Fig. 9.

2. Perspective

Although the polyphenol theory provides a mechanism for the synthesis of humic-like polymers, it nevertheless seems unlikely that any newly

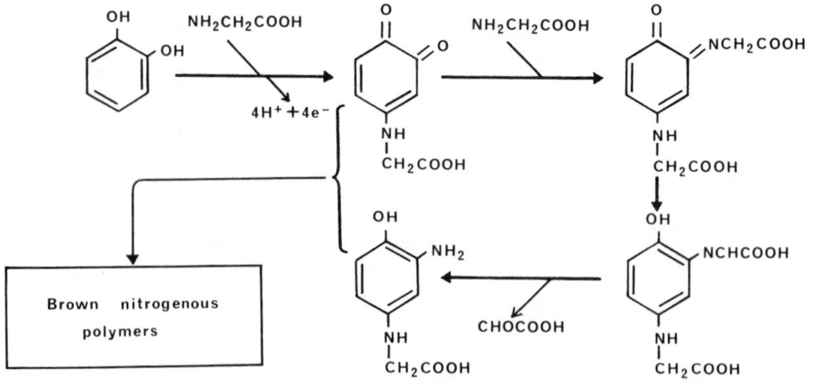

Fig. 9. Formation of humic substances from quinones and amino acids as illustrated by the reaction between catechol and glycine. [Redrawn from Stevenson (1982a).]

microbially synthesized humic-type structure would have a composition comparable with any soil humic substance. In the soil environment, compounds and components of plant, animal, and microbial origin are likely to become bound to the phenolic compounds. Nelson *et al.* (1979), for instance, showed that cell wall and cytoplasmic fractions of bacteria, yeasts, and filamentous fungi were stabilized against microbial degradation by linkage into phenolic polymers. Thus, in reality, soil humic materials represent highly heterogeneous mixtures formed as a result of many chemical and biological synthesis reactions (Hayes and Swift, 1978). In addition, such mixtures consist of old and newly synthesized polymers. With residence times in the range of 200 to 2000 years (Stevenson, 1982a), structural changes are likely to take place that could involve further biological transformations, chemical rearrangements, and condensation reactions within the polymers.

B. Structure

A detailed discussion of the structure of humic substances is well beyond the scope of this chapter. Nevertheless, for completeness their basic molecular structure is outlined below. Methods involved in extracting humic substances from soils and the characteristics and proposed structures of such extracted substances have been reviewed in detail elsewhere (Hayes and Swift, 1978; Kowalenko, 1978; Stout *et al.*, 1981; Stevenson, 1982a,b). As discussed above, the formation of humic substances is thought to involve enzymatic conversion of polyphenols to quinones, which undergo self-condensation or combine with amino compounds to form N-containing polymers. The products so formed tend to associate into complex chemical structures.

The humic fraction of soil has indeed been found to consist of a complex system of molecules that have a wide range of generally high molecular weights. For example, Stevenson (1982a) observed that the average range for humic acids (alkali-soluble, acid-insoluble humic fractions) and fulvic acids (alkali- and acid-soluble fractions) is on the order of 50,000 to 100,000 and 500 to 2000, respectively. A few humic acids have molecular weights exceeding 250,000.

The basic structure of humic acid is believed to be an aromatic ring, of di- or trihydroxyphenol type, bridged by —O—, —CH$_2$—, —NH—, —N=, —S—, and other groups and containing both free hydroxyl groups and the double linkages of quinones (Stevenson, 1982a,b). In their natural state, such molecules are thought to contain attached proteinaceous and carbohydrate residues. A hypothetical model of humic acid is shown in Fig. 10. Nitrogen is shown incorporated into the humic acid in

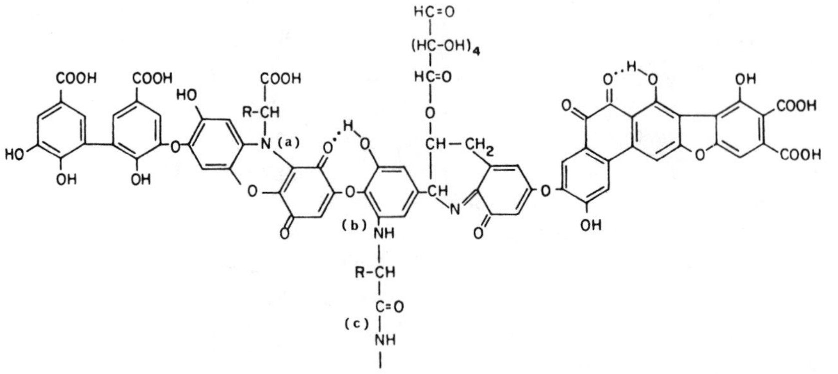

Fig. 10. Hypothetical structure of humic acid according to Stevenson (1982a). Nitrogen is incorporated into the humic acid in three ways: (a) as a bridge unit; (b) in the form of *N*-phenylamino acid; and (c) in the form of a peptide bond.

three different ways: as a bridge unit, in the form of *N*-phenylamino acid, and in the form of a peptide bond.

Humic substances are thought to exist in soils as heterogeneous, complex, three-dimensional amorphous structures. Their heterogeneity is a function of the disorderly condensation of phenolic units, copolymerization reactions, extensive cross-linkages between the randomly arranged phenolic units, and their large molecular size.

C. Chemical Forms of Soil Organic Nitrogen

Even though over 90% of the N in the surface layer of most soils occurs in organic forms, the exact nature of this N is only partially understood. Several workers have reviewed our present knowledge of the nature of soil organic N (Stevenson and Wagner, 1970; Parsons and Tinsley, 1975; Kowalenko, 1978; Stevenson, 1982a,b).

1. Extraction Procedure

To identify organic nitrogenous compounds in soils, first they must be extracted from the soils. The generally accepted method of extraction involves acid hydrolysis (Bremner, 1965a, 1967; Kowalenko, 1978; Stevenson, 1982a).

The results of a partial fractionation of soil N following acid hydrolysis are shown in Table I. Acid hydrolysis does not dissolve all the soil N and the nature of the nonhydrolyzable fraction (which usually accounts for on the order of 15 to 35% of the total soil N content) is not completely

Table I

Classical Fractionation of Soil Nitrogen from 32 Canadian Soils (Surface Horizons)[a]

Fraction after 6 N HCl reflux	Quantity (% of total soil N)
Nonhydrolyzable	15 ± 6
Hydrolyzable	
Total	85
Ammonium	21 ± 5
Amino acid	40 ± 7
Amino sugar	7 ± 2
Unidentified	19

[a] Data from Kowalenko (1978).

known. Approximately 20 to 35% of the soil N is normally recovered in the acid hydrolysate as NH_4^+, 30 to 45% as amino acid N, 5 to 10% as amino sugar (hexosamine) N, and another 10 to 20% has not yet been identified (Stevenson, 1982b).

Some of the 10 to 20% of soil N that is recovered in acid hydrolysates as NH_4^+ is derived from indigenous clay-fixed NH_4^+ and another part comes from the degradation of amino acid amines (asparagine and glutamine), amino sugars, and some amino acids (e.g., tryptophan is completely lost during hydrolysis). However, the origin of approximately 50% of the NH_4^+ in acid hydrolysates is still unknown (Stevenson, 1982a,b). Some may be derived from complexes formed by fixation reactions (e.g., the fixation of NH_3 by soil organic matter, see Chapter 5).

The unidentified portion of the acid-hydrolyzable N may occur mainly as non-α-amino N in arginine, tryptophan, lysine, and proline (Greenfield, 1972; Goh and Edmeades, 1979). The non-amino N in these amino acids is not included in amino N values as determined by conventional methods of analysis (Stevenson, 1982a).

2. Major Forms

a. Amino acids. The amino acid composition of soils is extremely variable (e.g., Sowden *et al.,* 1977; Singh *et al.,* 1978, 1981; Goh and Edmeades, 1979). For example, in some soils over one-third of the amino acid N has been reported to be in the form of basic amino acids (lysine, histidine, arginine, and ornithine) while in others less than one-tenth has been reported in these compounds (Stevenson, 1982b). Equally divergent results have been reported for individual amino acids within each group.

Much of the amino acid material that accumulates in soils appears to be derived from peptides, mucoproteins, and teichoic acids of microbial cells (Kowalenko, 1978; Stevenson, 1982b). Glycine, alanine, aspartic acid, and glutamic acid, which are often the prevalent amino acids in soils (Kowalenko, 1978), are also the dominant amino acids in bacterial cells (Stevenson, 1982a). Indeed, many of the amino acids found in soils are not normal constituents of proteins but represent products synthesized by microorganisms; these include ornithine, taurine-3,4-dihydroxyphenyla-lanine, β-alanine, α-amino-n-butyric acid, and γ-amino-n-butyric acid.

Amino acids are thought to exist in soils in several forms, including those free in soil solution and those bound to humic polymers and clay minerals. Concentrations in soil solution are low ($<$2 μg gm^{-1} soil), but in the rhizosphere they may be sevenfold higher (Stevenson, 1982a) due to exudation by plant roots.

Approximately half the total N content of extracted soil humic acids occurs as amino acid N (Bremner, 1965a,b; Sequi *et al.*, 1975; Carter and Mitterer, 1978; Tsutsuki and Kuwatsuka, 1978). Some possible types of bonding of amino acids to humic polymers are illustrated in Fig. 11. These are (a) amino acids bonded by peptide bonds, such as those in proteins, (b) those linked by quinone rings, and (c) those bonded directly to pheno-lic rings. It is thought that bonding types (a) and (b) may represent acid-hydrolyzable humic N while type (c) represents N that is not released from humic acids without subsequent alkaline hydrolysis (Piper and Posner, 1972).

Amino acids and other nitrogenous compounds can be absorbed on both the external and internal (interlayer) surfaces of clay minerals

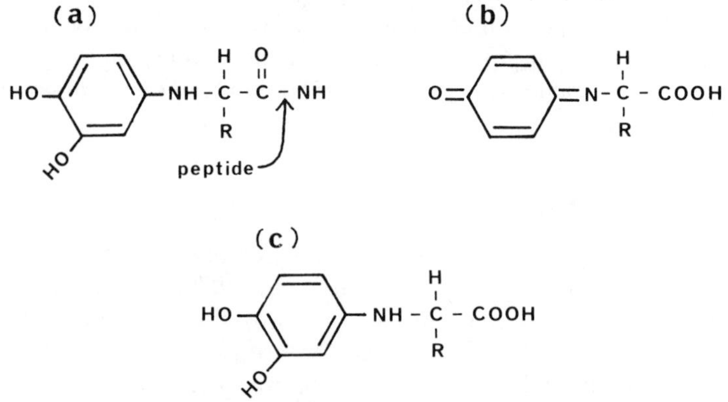

Fig. 11. Three possible bonding structures of amino acids with phenolic polymers: (a) as a peptide bond; (b) linked by a quinone ring; and (c) bonded directly to a phenolic ring.

(Theng, 1974, 1979). Such adsorption reactions may well be important factors influencing amino acid distribution patterns in soils (Stevenson, 1982a) although their significance in terms of the magnitude and composition of the soil amino acid component is, as yet, relatively unknown.

b. Amino sugars. The dominant amino sugar present in soils is D-glucosamine, while significant amounts of D-galactosamine have also been found (Kowalenko, 1978; Parsons, 1981; Stevenson, 1982b). Small amounts of N-acetylglucosamine, D-mannosamine, 2-deoxy-2-amino-D-talose, and muramic acid have also been detected in soils (Parsons and Tinsley, 1975; Stevenson, 1982b). The presence of significant quantities of amino sugars in soils provides further evidence that much of the soil N is of microbial origin since amino sugars are present in only trace amounts in plant components but they are major structural components of the cell walls of bacteria and fungi (see Section II,D).

Fractionation of soil extracts has failed to provide clarification of the nature of amino sugar-containing polymers present in soils. There is some evidence for the presence of chitin-like polymers but in general it appears that amino sugars occur in polymers along with neutral sugars, amino acids, and other compounds (Parsons, 1981).

Much of the amino sugar N present in soils is bound to humic components; approximately 2–8% of humic acid N can be accounted for as amino sugars. Indeed, amino monosaccharides and amino polysaccharides, possessing free amino groups, can be complexed with phenolic polymers in ways similar to those for amino acids (e.g., Fig. 11).

D. Degradation

Since during the process of decomposition humus is being continuously formed, it follows that humus is also continuously being degraded because in steady-state ecosystems the organic matter and N content of soils remains at a constant level (see Chapter 1). The breakdown of soil humus, and the consequent mineralization of soil N, thus represents an important source of the pool of mineral N in soils. Nevertheless, there is little information available concerning the organisms and biochemical pathways involved in the degradation of humic substances. To a large extent this reflects our incomplete knowledge of the chemistry and structure of the humic materials themselves.

1. Recalcitrance of Humic Nitrogen

As already noted, humic substances are characterized by their inherent stability and slow decomposition rate. Indeed, the stability of organic nitrogenous substances against microbial degradation can be greatly increased by linkage into humic-like phenolic polymers. The degradation of proteins in soils, for instance, is significantly decreased by intimate mixing with soil humic acid or model humic-like polymers (Mayaudon, 1969; Verma *et al.*, 1975). Other research (Haider *et al.*, 1965; Martin and Haider, 1969; Bondietti *et al.*, 1972) has shown that amino acids and amino sugars are stabilized against microbial degradation when these compounds are oxidatively polymerized with phenol mixtures. Such stabilization is generally attributed to reactions between amino groups and quinones (such as those depicted in Fig. 9), resulting in the formation of relatively stable *N*-phenolamino compounds.

As will be discussed in following sections of this chapter, in the soil *in situ,* several other important factors contribute to the stability of soil organic N. These include the formation of biologically stable clay–organic matter complexes and the physical inaccessibility of organic matter present within soil aggregates.

2. Processes of Degradation

It is thought that during humus degradation high molecular weight units of humic acids are broken down by microbial attack and oxidation to form smaller molecules, with a preferential loss of nitrogenous materials (Swift and Posner, 1972). During this process the amino acids incorporated by peptide linkages would be attacked and rapidly removed by hydrolytic enzymes, but amino acids bound to phenolic nuclei would be more stable and less susceptible to attack. Nevertheless, the oxidation of aromatic polymers and cleavage of aromatic rings during humus degradation would eventually result in the release of N present as bridging units or as *N*-phenolamino acids.

In relation to the mineralization of relatively accessible humic N, important hydrolytic enzymes are likely to be the proteases and peptidases, which catalyze the hydrolysis of peptide bonds, and the enzymes involved in the hydrolysis of amino sugars (aminoglycan hydrolases, amino sugar kinases, and deaminases). For example, amino acids are known to be released from humic acids through the action of proteolytic enzymes (Ladd and Butler, 1969). A detailed summary of the biochemical pathways of the decomposition of proteins, amino acids, amino sugars, and other nitrogenous compounds commonly present in the soil environment is presented in Section III,E.

Although it is generally considered that the phenolic "core" of humic acids is resistant to decomposition, the capability to oxidize complex aromatic humic polymers appears to be widespread in the microbial world (Tate, 1980). It seems that it is the complex structure of soil humic substances that is largely responsible for their stability. Several bacterial species, especially pseudomonads (Huntjens, 1972; Taha *et al.*, 1973) as well as actinomycetes (Steinbrenner and Mundstock, 1975; Monib *et al.*, 1981), and several fungal species (Biederbeck and Paul, 1971; Ruocco and Barton, 1978; Khandelwal and Gauer, 1980) have been reported to catabolize humic acid in laboratory cultures. An etherase enzyme system is thought to catalyze the cleavage of aromatic subunits (Paul and Mathur, 1967).

3. Chemical Fractionation in Relation to Degradation

Chemical fractionation of soil N, following acid hydrolysis, has been used with limited success to follow the decomposition of soil organic N. For example, several incubation experiments (Isirimah and Keeney, 1973; Singh *et al.*, 1978, 1981) have indicated that much of the readily mineralizable soil N is derived from the acid-hydrolyzable amino acid, amino sugar, and unidentified fractions. Much of the unidentified N fraction of soils can be degraded by soil microorganisms with the release of NH_4^+-N (Ivarson and Schnitzer, 1979). Similarly, following net immobilization of mineral N, most of the immobilized N is found in the acid-hydrolyzable fraction (amino acid, amino sugar, and unidentified N) (Broadbent, 1968; Stewart *et al.*, 1963; McGill, 1971; Allen *et al.*, 1973; Ladd and Paul, 1973; Smith *et al.*, 1978). Subsequent remineralization leads to decreases in the amounts of all forms of organic N with the acid-hydrolyzable amino acid and amino sugar N again being major contributors (McGill, 1971; Ladd and Paul, 1973).

Such results may, however, reflect a turnover of the microbial biomass more than the breakdown of humic N. In other words, since the amino acid and amino sugar fractions are primarily of microbial origin and are major constituents of the biomass, it is not surprising that in the short term they represent a significant portion of newly immobilized and readily mineralizable pools of soil N.

In long-term field studies of soil organic matter decomposition, the chemical fractionation of soil N has not yielded conclusive results (Keeney and Bremner, 1964; Fleige and Baeumer, 1974; Meints and Peterson, 1977; Rao and Ghosh, 1981). Overall, cultivation has only a small effect on the relative sizes of the chemical N fractions although it results in decreases in the total N content of the soil. This indicates that in the long term all forms of chemically extracted N are biodegradable. Indeed, acid hydrolysis appears to be of little or no practical value as a means of soil testing for plant-available N or predicting crop response when N is limiting (Kadirgamathaiyah and MacKenzie, 1970; Moore and Russell, 1970; Osborne, 1977).

E. Biochemistry of Nitrogen Mineralization

The major biochemical pathways by which the predominant nitrogenous compounds present in soils are degraded are outlined below. The
discussion is relevant to the decomposition of plant and animal tissues
(containing proteins, peptides, amides, amino acids, nucleic acids, and
purines and pyrimidines) as well as the soil biomass and soil humic components, both of which contain significant quantities of amino sugars as
well as the above-mentioned compounds. The hydrolysis of urea is also
outlined although it represents only a transient compound in soils and
does not constitute a component of humic substances. A more detailed
account of the biochemistry of N mineralization has been presented by
Ladd and Jackson (1982).

1. Proteins and Peptides

The formation of NH_4^+ from the degradation of proteins and peptides
requires that they are initially hydrolyzed by a sequence of reactions to
form amino acids. These reactions involve the hydrolysis of peptide
bonds and are catalyzed by the proteinase and peptidase enzymes. Peptide bond hydrolysis is shown schematically in Fig. 12. The hydrolysis
involves nucleophilic substitution in which the active (basic) site of the
enzyme becomes bonded to the electrophilic C atom of the CO group in
the peptide. At the same time the N atom is displaced and it receives a
donated proton from water or the enzyme. The compound containing the
displaced N atom may be an amino acid or a peptide. The peptide will
undergo further sequential hydrolysis in the presence of appropriate proteinases and peptidases until only amino acids are formed.

The proteinases and peptidases are widespread in soils and are thought
to be of diverse origin (Ladd and Jackson, 1982). Although a wide range of

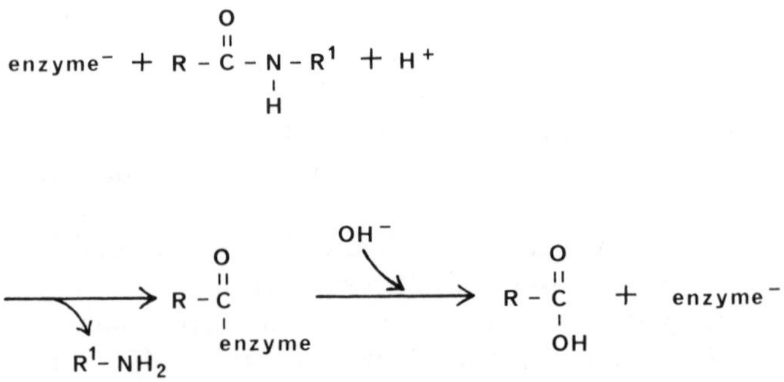

Fig. 12. Enzymatic peptide bond hydrolysis.

proteolytic microorganisms can be readily isolated from soils (Ladd and Paul, 1973; Mayaudon et al., 1975), seasonal changes in proteinase activity in field soils do not appear to be closely related with changes in microbial populations (Ladd et al., 1976). Indeed soil proteinases and peptidases are probably also derived from plant and animal sources (Ladd and Jackson, 1982) although the relative contributions of microbial, plant, and animal enzymes to the total activities in soils are unknown. Furthermore, such enzymes may become stabilized in the soil and hence may persist for long periods after the original sources have been extensively decomposed (Burns, 1978).

2. Amides and Amidines

The amidohydrolases and amidinohydrolases hydrolyze a wide variety of linear and cyclic amides and amidines often with the release of amino acids and NH_4^+ or urea. The hydrolysis of amides is a nucleophilic substitution in which the N atom leaves as an NH_4^+ group.

3. Amino Acids

Oxidative deamination of amino acids is catalyzed by amino acid dehydrogenases and amino acid oxidases. The respective reaction mechanisms are shown in Figs. 13a and b. Both reactions involve initial oxidation of amino acids and the formation of amino acid intermediate and finally yield α-oxo acids and NH_4^+. However, while the dehydrogenases utilize nicotinamide adenine dinucleotide (NAD^+) as an H-accepting coenzyme, the oxidases are flavoproteins and the flavin adenine dinucleotide (FAD) is reduced initially and then reoxidized by O_2 with the formation of H_2O_2.

Most available evidence indicates that the activity of these enzymes in soils is dependent on simultaneous microbial growth and activity.

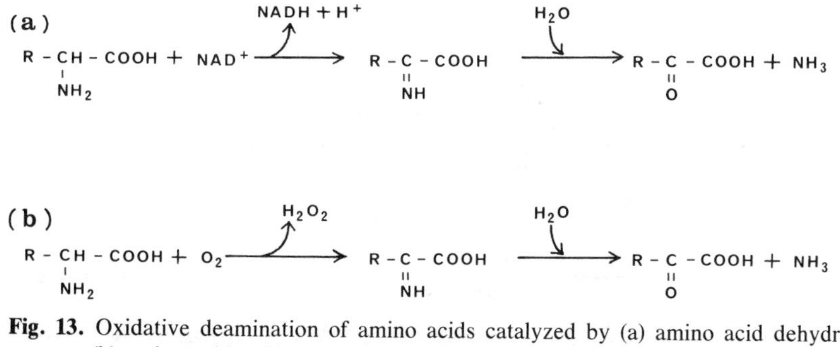

Fig. 13. Oxidative deamination of amino acids catalyzed by (a) amino acid dehydrogenases or (b) amino acid oxidases to yield α-oxo acids.

4. Amino Sugars

Amino sugars occur in soils primarily in the form of polymers (e.g., chitin, peptidoglycans, and teichoic acids) and the formation of NH_4^+ from sugar polymers requires prior hydrolysis to amino monosaccharides. The enzymes that catalyze the hydrolysis of amino polysaccharides to amino monosaccharides are known as the aminoglycanhydrolases.

The amino monosaccharides so formed (predominantly glucosamine, galactosamine, and muramic acid) are then degraded through a series of reactions with the production of NH_4^+. The pathways by which galactosamine and muramic acid are degraded appear to be unknown (Ladd and Jackson, 1982) but that for glucosamine is shown in Fig. 14.

The first step is the formation of glucosamine 6-phosphate by transfer of phosphate groups of ATP and is catalyzed by glucosamine kinase. The glucose 6-phosphate is then deaminated through the action of the enzyme glucosamine-6-phosphate isomerase. The fructose 6-phosphate thus formed undergoes glycolysis with the formation of lactic acid and ATP.

5. Urea

In the soil, urea (originating primarily from animal urine or applied fertilizer) is hydrolyzed to CO_2 and NH_4^+ through a reaction catalyzed by the enzyme urease:

$$NH_2CONH_2 + H_2O \longrightarrow 2NH_3 + CO_2$$

The reaction is thought to involve two steps in which carbamate (NH_2COOH) is the obligatory intermediate compound (Ladd and Jackson, 1982).

Urease in soils is thought to originate from both soil microorganisms and plant roots although the relative importance of these two sources probably differs in differing situations (Bremner and Mulvaney, 1978).

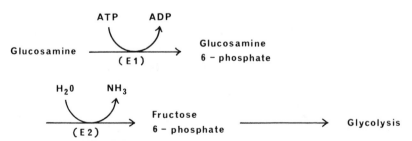

Fig. 14. Enzymatic hydrolysis of glucosamine. Enzymes involved: E_1, glucosamine kinase; E_2, glucosamine-6-phosphate isomerase.

Urease is known to be an extracellular enzyme of bacteria, fungi, and actinomycetes (Bremner and Mulvaney, 1978) while plants are known to be rich sources of the enzyme (Reithel, 1971) and exocellular ureases from plant roots have been demonstrated (Mahaptra et al., 1977).

The activity of urease in soils is a function of urease present in various states of biological and physicochemical stability (Ladd and Jackson, 1982) since it is generally agreed that bonding of the enzyme to inorganic and organic soil constituents confers on it varying degrees of stability (Burns, 1978; Ceccanti et al., 1978). Many factors are known to influence soil urease activity and these have been reviewed elsewhere (Bremner and Mulvaney, 1978; Mulvaney and Bremner, 1981).

6. Other Nitrogenous Compounds

Other nitrogenous compounds found in soils include nucleic acids, purines, and pyrimidines.

A large number of enzymes are required to convert nucleic acid N to NH_4^+. The nucleases catalyze the depolymerization of nucleic acids to mononucleotides, which are then dephosphorylated by nucleotidases to N-glycosides of purines and pyrimidines. These nucleosides are hydrolyzed to purines, pyrimidines, and pentoses by nucleosidases.

The amidohydrolases and amidinohydrolases catalyze the conversion of purine and pyrimidine N to NH_4^+.

IV. FACTORS AFFECTING DECOMPOSITION

Since the decomposition of organic materials and the release of mineral N, from either native soil organic matter or decaying litter, is the result of complex interactions between microbial populations and activities it is affected by many factors. These include the composition of the decomposing litter (substrate quality), environmental factors (particularly moisture and temperature), and other soil factors. The major factors affecting the decomposition process are discussed below, with particular reference to the mineralization of organic N.

A. Substrate Quality

Substrate quality, as defined by chemical composition of the decomposing material, has long been recognized as a critical factor determining the rate of litter decay (e.g., Waksman and Tenney, 1927). Chemical indices of substrate quality include elemental concentrations and concentrations of various classes of organic compounds.

1. Nitrogen Content and the C : N Ratio

a. Concept. Nitrogen content of plant material has been shown to be an important factor controlling the rate of decomposition in many studies (Cowling and Merrill, 1966; Aber and Melillo, 1980). Several studies have demonstrated that the addition of supplementary N to natural litter materials (Mahendrappa, 1978) and incorporated crop residues (Allison and Cover, 1960; Bartholomew, 1965) can enhance their rate of decomposition.

The concept of the regulatory effect of the C : N ratio on the release of N from decomposing litter was discussed in Section II,B. A low C : N ratio (high N content) in litter facilitates N mineralization by encouraging a high rate of decomposition and ensuring that N mineralization exceeds immobilization by a considerable extent.

The C : N ratios of some organic materials are shown in Table II. The low C : N ratio of microbial tissue in comparison with other organic residues is obvious. The generally low C : N ratios of leguminous crop residues and the very high C : N ratios of woody materials are also notable.

Table II

The Carbon : Nitrogen Ratios of Some Organic Materials[a]

Material	C : N ratio[b]
Mirobial tissue	8 : 1
Soil humus	10 : 1
Alfalfa hay	13 : 1
Clover residues	23 : 1
Corn stalks	60 : 1
Oak leaves	65 : 1
Oat straw	80 : 1
Timothy	80 : 1
Pine needles	225 : 1
Sawdust	400 : 1

[a] Reprinted with permission from Volk, B. G. and Loeppert, R. H. (1982). Soil organic matter. *In* "Handbook of Soils and Climate in Agriculture" (V. J. Kilmer, ed.), pp. 211–268. Copyright CRC Press, Boca Raton, Florida.

[b] Values are only approximate.

When organic residues with a high C : N ratio (e.g., straw or sawdust) are added to agricultural soils it is advisable to add fertilizer N concomitantly in order to lower the C : N ratio below 20 to 25 and thus avoid net immobilization and consequent N deficiency (Allison, 1973).

   b. Validity. In the field situation, it appears that the concept of a fixed critical C : N ratio that controls the mineralization–immobilization balance is not of general validity. Two types of experimental data can be utilized to illustrate the variability in critical N levels. First, Berg and Staaf (1981) compiled an extensive list of decomposition studies in which they tabulated the initial N level versus whether or not net mineralization or immobilization of N occurred. The data they compiled showed that the N level for which net immobilization of N occurred varied greatly, ranging from about 0.3 to 1.4%, while the N level for which net mineralization occurred varied from 0.58 to 3.06%. Second, in studies in which net immobilization of N has occurred during decomposition, the threshold N level at which net immobilization gives way to net mineralization can be estimated. Some such estimates, shown in Table III, indicate that such threshold N levels range from 0.3 to 1.8% (C : N ratios of 167 : 1 to 27 : 1). Similarly, when crop residues are incorporated into soils, the threshold N levels vary from 0.8 to 1.5% (Brown and Dickey, 1970; Smith and Douglas, 1971).

Table III

Critical Nitrogen Concentration and Carbon : Nitrogen Ratio of Some Decomposing Forest Litters that Accumulate Nitrogen Before Its Release

Litter type	Initial N level (%)	Critical N level (%)	Critical C : N ratio	Reference
Betula allegheniensis leaves	0.85	1.83	27 : 1	Gosz *et al.* (1973)
Betula verrucosa leaves	0.76	1.30	39 : 1	Berg and Staaf (1981)
Quercus petraea leaves	0.77	1.61	31 : 1	Bocock (1964)
Pinus radiata needles	0.56	1.40	36 : 1	Will (1967)
Pinus silvestris needles	0.38	0.74	68 : 1	Staaf and Berg (1977)
Pinus silvestris cones	0.20	0.30	167 : 1	Berg and Staaf (1981)

Thus there is a large range of variability in critical N and C : N values. Within this range factors other than N content may be expected to regulate the mineralization–immobilization balance. These include other parameters of substrate quality such as lignin and polyphenol content as well as environmental factors.

2. Lignin Content

Several studies have indicated that the initial lignin content of the litter exerts more control over the rate of decomposition than does N (Bollen, 1953; Fogel and Cromack, 1977; Melillo *et al.*, 1982). Indeed, the higher the lignin content the lesser is the influence of initial N content on decomposition rate (Berg and Staaf, 1980). Moreover, the higher the initial lignin content the greater the quantity of N retained in the litter during decomposition (Coldwell and Delong, 1950; Toth *et al.*, 1974; Melillo *et al.*, 1982).

The reason that a high lignin content slows decomposition and favors N accumulation is, as already discussed, that its degradation products (phenolic compounds) constitute an important source of structural units for the synthesis of N-containing humic polymers. In fact, under some circumstances the amount of humus formed from decomposing plant litter is positively correlated with the litter's initial lignin content (De Haan, 1977).

The importance of lignin as a source of structural units for humus was demonstrated by Martin *et al.* (1980), who utilized specifically ^{14}C-labeled organic substrates to study humus formation. They found that the majority of lignin carbons were incorporated into the more resistant or aromatic portions of soil humus while added polysaccharide carbons were metabolized and utilized as energy sources for the decomposer microflora and for synthesis of cellular proteins, polysaccharides, and some phenolic compounds.

3. Polyphenol Content

Polyphenolic compounds of plant origin are well known as modifiers of the rate of decomposition (Williams and Gray, 1974; Swift *et al.*, 1979). In general, the higher the polyphenol content of the litter the lower is the rate of litter decomposition and N release.

Since polyphenols constitute the major structural units from which N-containing humic polymers are formed (e.g., Fig. 9) it is not surprising that their presence generally serves to decrease the rate of decomposition and N release. The formation of relatively insoluble recalcitrant nitrogenous compounds originating from the reactions of plant polyphenols with proteins has been observed by several workers (Basaraba and Starkey,

1966; Benoit and Starkey, 1968; Benoit *et al.*, 1968; Lewis and Starkey, 1968).

Polyphenols may also decrease the rate of decomposition by direct inhibition of fungal and/or faunal activity. Harrison (1971) found that the high polyphenolic content of oak and bench leaves appeared to inhibit the growth of many fungi. There also appears to be an inverse relationship between polyphenol content and the rate at which leaves are broken down due to the feeding activities of soil fauna (King and Heath, 1967; Satchell and Lowe, 1967).

Plants appear to be able to regulate the polyphenol content in their foliage in response to the external N supply. Indeed, trees often produce higher amounts of polyphenolic substances in their leaves when provided with a low supply of N (Davies *et al.*, 1964; Lamb, 1975). Thus, when N is in short supply, the rate of litter decomposition is decreased and more N is combined into humic substances. In this way the rate of N mineralization is decreased and N is conserved (Gosz, 1981).

The low N status of some mor humus sites (Chapter 1, Table XIV) may be the major factor determining high levels of polyphenols in vegetation, which, in turn, contributes to the mor humus formation. Studies have shown that the addition of inorganic N to such sites results in a higher N content and lower polyphenol content of plant tissue and an increased rate of litter decomposition and N mineralization (Gosz, 1981).

B. Moisture

Soil moisture can influence the mineralization of N in three major ways: (1) moisture stress inhibits microbial growth directly; (2) as moisture content increases, aeration decreases and microbial growth is inhibited; and (3) cycles of wetting and drying tend to increase the amount of available substrate.

1. Moisture Content

Decomposer organisms differ in their response to the moisture content of their environment. In general fungi and actinomycetes are relatively tolerant of low moisture potentials. Indeed, an active mycoflora is maintained down to a soil moisture potential of approximately -1500 kPa while bacteria become inactive below -800 to -1500 kPa (Chen and Griffin, 1966; Wilson and Griffin, 1975). It is generally believed that the catabolic activity of the total microbial biomass may be limited at moisture potentials below -1000 to -5000 kPa.

At very high soil moisture contents the rates of biological activity and decomposition are decreased through lack of oxygen. The majority of soil fungi and actinomycetes are aerobes as are many of the bacteria. Under anaerobic conditions, decomposition is dependent on anaerobic bacteria (e.g., clostridia strains), which operate at a much lower energy level and are less efficient than aerobic organisms (Yoshida, 1975; Campbell, 1978; Patrick, 1982). Thus the metabolic processes of decomposition and synthesis are slowed under anaerobic conditions and poorly drained soils are therefore characteristically high in organic matter.

A characteristic feature of anaerobic bacterial degradation is its low N requirement, which leads to a more rapid release of NH_4^+ ions than would ordinarily be expected on the basis of a wide C/N ratio of the decomposing material and its slow rate of decomposition (Patrick, 1982). Williams *et al.* (1968), for example, concluded that the N requirement for the decomposition of rice straw in submerged soils was one-third (0.5 vs. 1.5% N) the average N content required for aerobic decomposition, while Waring and Bremner (1964) observed a more rapid release of inorganic N under waterlogged than aerobic conditions in a number of soils.

There are few data available to predict the effects of moisture content on the decomposition of litter. However, in general, high (>100 to 150%) and low (<30 to 50%) moisture contents (dry weight basis) tend to slow the decomposition of litter at the soil surface (Henningsson, 1967; Pechmann *et al.*, 1967; Van Cleve and Sprague, 1971; de Boois, 1974) and thus the release of mineral N.

The breakdown of native soil organic matter with the release of NH_4^+ generally increases with increasing moisture content between $-15,000$ and -10 to -50 kPa (permanent wilting point and field capacity), while above and below these limits the rate of ammonification decreases (Miller and Johnson, 1964; Reichman *et al.*, 1966; Stanford and Epstein, 1974). The influence of soil moisture content on N mineralization is illustrated in Fig. 15. The optimum soil moisture potential for ammonification is between 10 and 50 kPa (Miller and Johnson, 1964; Reichman *et al.*, 1966; Sabey, 1969; Stanford and Epstein, 1974; Myers *et al.*, 1982). The lower limit at which no net ammonification occurred was shown by Myers *et al.* (1982) to be close to -4000 kPa, although Wetselaar (1968) reported significant ammonification at a soil moisture potential of -5000 kPa while Robinson (1957) found little ammonification at less than -1500 kPa.

2. Drying and Rewetting

The effect of drying and rewetting on the decomposition of plant residues is unclear. Van Schreven (1968), for example, found that although drying stimulated the subsequent mineralization of C and N from soil humus it retarded mineralization of fresh plant materials. Haider and Martin (1981) found that drying and rewetting had no effect on the decom-

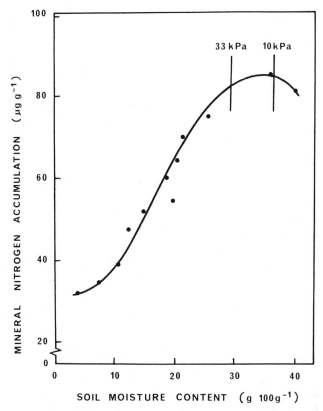

Fig. 15. Mineral nitrogen accumulation in a soil after a 2-week incubation at different soil moisture contents. [Data from Stanford and Epstein (1974). Reproduced from *Soil Sci. Soc. Am. Proc.* **38,** p. 105 by permission from the Soil Science Society of America.]

position of ^{14}C-labeled lignins when they were incorporated into the soil but Sørensen (1974) found that drying and rewetting cycles increased the decomposition of added cellulose in soils. Birch (1964) showed that drying and rewetting released N from grass tissues added to an acid soil compared to no release of N under constant moisture conditions.

Drying and rewetting cycles are known to be important factors influencing the evolution of CO_2 from soils and the mineralization of soil organic N (Birch, 1960; Broadbent *et al.,* 1964; Van Schreven, 1968; Agarwal *et al.,* 1971a; Campbell *et al.,* 1975). The stimulatory effects of drying and rewetting cycles on the mineralization of soil organic N are illustrated in Fig. 16. Cycles of drying and rewetting generally cause flushes in N mineralization; each successive cycle causes a slightly smaller flush and the

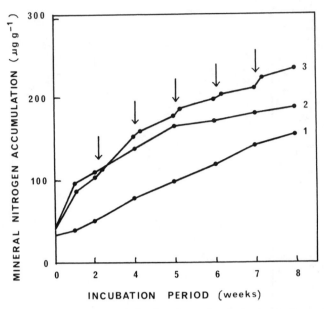

Fig. 16. Mineral nitrogen accumulation in soil samples during incubation and treatment as follows: (1) fresh moist soil incubated continuously; (2) soil initially dried at 35°C, then rewetted and incubated continuously; and (3) soil periodically dried at 35°C and rewetted during incubation. Drying–rewetting is denoted by arrows. [Data from Van Schreven (1968).]

size of the flush is positively related to humus content, the dryness of the soil, and the length of time the soil has remained dry (Birch, 1960). Heating the soil to high temperatures enhances the subsequent effect of rewetting on mineralization (Agarwal *et al.*, 1971a).

Intensive Canadian research into the effects of environmental factors on soil microbial populations and activities (Biederbeck and Campbell, 1971; Campbell *et al.*, 1971, 1973, 1975; Biederbeck *et al.*, 1977; Campbell and Biederbeck, 1982) has revealed that drying and rewetting is very important in influencing microbial numbers and activity; even dew formation on very dry soils can enhance microbial growth. Campbell and Biederbeck (1976, 1982) found that very small quantities of precipitation falling on dry summer fallow soil sometimes caused flushes in microbial growth and N mineralization as great as did larger amounts of precipitation. In rice paddy soils Ventura and Watanabe (1978) showed that the prolonged drying prior to the wet season was one of the major factors stimulating the mineralization of soil N.

The fact that a burst of biological activity occurs immediately following remoistening a dried soil indicates that the drying process results in the release of organic compounds since biological activity is, in general, directly proportional to the available energy supply. There are two obvious new sources of substrate following drying.

First, swelling and shrinking of the soil upon wetting and drying probably physically disrupts the soil aggregates thus exposing organic matter not previously accessible to microbial attack (see Section IV,J). This is much like the explanation forwarded by many workers to explain the stimulatory effect of cultivation on the mineralization of soil N (e.g., Rovira and Graecen, 1957).

Second, during the dry period there would be an accumulation of dead microbial cells; these would serve as an N-rich substrate for the surviving microbial population following rewetting (Campbell and Biederbeck, 1982). The evidence for this explanation has already been discussed in detail in Section II,D.

It seems probable that the new pool of readily mineralizable N that becomes available following drying and rewetting a soil consists of both previously inaccessible soil organic matter and dead N-rich microbial cells.

C. Temperature

1. General Effect

Temperature is a major factor influencing the decomposition of organic materials. Decomposer organisms have different temperature optima and growth ranges. The mesophilic bacteria actinomycetes and fungi (temperature optima in the range 0 to 45°C) are common soil inhabitants while the thermophilic bacteria and actinomycetes (range 45 to 60°C) are minor inhabitants (Alexander, 1977).

Progress in the decomposition process is similar at various temperatures except the rate of tissue breakdown and CO_2 evolution is greater at higher temperatures. In a laboratory experiment with grass litter Floate (1970a) found that the amounts of CO_2 evolved over a 12-week period were reduced from an average of 40% of the original C content at 30°C to 25% at 10°C and 12% at 5°C. Over the same period, the net production of NH_4^+-N at 30, 10, and 5°C was 61.0, 16.5, and 0.10 mg N/100 gm of added plant material, respectively.

The combined effect of high temperature and moisture is more prominent than that of temperature alone. Indeed, periods of dry, warm conditions during summer have a most deleterious effect on the rate of decomposition of surface litter (Karenlampi, 1971). Conditions of high moisture and temperature generally favor microbial growth and thus decomposition. Hence, the rate of litter decomposition in tropical climates is considerably higher than that in cool temperate climates (Olson, 1963; Jenkinson and Ayanaba, 1977).

The rate of mineralization of native soil N is profoundly influenced by temperature within the range normally encountered in soils under field conditions. The lower temperature limit for ammonification is generally around freezing (Sabey et al., 1956; Stanford et al., 1973a). In contrast to most microbial transformations in soils (e.g., nitrification), the optimum temperature for ammonification is in the thermophilic (45 to 60°C) rather than the mesophilic temperature range (Alexander, 1977). Thus in a tropical Australian soil, Myers (1975) showed that ammonification had an optimal temperature of around 50°C (Table IV).

2. Fluctuating Temperatures

Under field conditions, marked diurnal and seasonal fluctuations in surface soil temperature are common (e.g., Biederbeck and Campbell, 1973). In general, pure culture studies have demonstrated that microbial growth is greater at constant rather than fluctuating temperatures, particularly if the amplitude of the fluctuations is greater than 10°C (Biederbeck and Campbell, 1971).

Although it has generally been found that microbial growth is inhibited by fluctuating temperatures, some research has indicated that minerali-

Table IV

Effect of Temperature on Mineral Nitrogen Accumulation (μg N gm^{-1}) in an Unamended Tropical Australian Soil[a]

Temperature (°C)	Time of incubation (days)	
	7	28
20	7.5 (1.6)[b]	11.8 (1.5)
30	9.4 (1.6)	14.8 (2.6)
40	14.1 (1.7)	23.5 (2.9)
50	20.0 (1.9)	35.7 (2.4)
60	16.4 (2.3)	24.8 (2.1)

[a] Source: Myers (1975). Reprinted with permission from Pergamon Press.

[b] Standard errors are shown in parentheses.

zation of N remains virtually unaffected in the mesophilic temperature range (Stanford et al., 1973b, 1975). For example, Stanford et al. (1975) found that different sequences of fluctuating temperatures between 5 and 35°C imposed on three soils during incubations of 52 days had no effect on the amount of N mineralized.

In contrast with the above findings Biederbeck and Campbell (1973) emphasized that the rate of mineralization of N at a given temperature is a function of the preceding temperature regime. In an incubation study, they showed that at optimum temperatures the microbial biomass was high but with the onset of unfavorable temperatures (e.g., a decrease in temperature) there was initially a large kill of microorganisms. This provided considerable amounts of readily available nitrogenous substrate for subsequent mineralization as the surviving organisms adapted to the new temperature regime. The validity of these laboratory data was supported by four years of field data that showed that the onset of the first cold spell each autumn and late frosts in spring resulted in sudden flushes of nitrate production.

3. Freezing and Thawing

The thawing of previously frozen surface detritus may result in the immediate release of large amounts of soluble material (Witkamp, 1969; Bunnell et al., 1975). This is thought to represent the release of materials previously immobilized in microbial tissue (Witkamp, 1969). Such a release of soluble materials contributes significantly to the burst of decomposer activity that occurs at the onset of snow melt in tundra ecosystems (Bunnell et al., 1975).

In general, freezing and thawing also stimulates decomposition of native soil organic matter and thence the mineralization of soil N (Gasser, 1958). Studies of freezing and thawing have generally shown that it has a similar but lesser effect than does drying and rewetting (Soulides and Allison, 1961; Mack, 1962, 1963), although Shields et al. (1974) observed that under Canadian conditions freezing and thawing was more effective in influencing the decomposition of soil organic matter than was drying and rewetting.

Explanations for freezing and thawing effects are analogous to those for drying and rewetting effects. Shields et al. (1974) postulated that much of the soil biomass is killed by the freeze–thaw cycles. This was supported by the laboratory data of Biederbeck and Campbell (1971) and field data of Campbell et al. (1971) and Biederbeck and Campbell (1973). The readily decomposable dead microbial tissue may then be decomposed by the surviving organisms, resulting in a flush of N mineralization. Alternatively it can be argued (Shields *et al.*, 1974) that freezing and thawing physically disrupts soil aggregates thus exposing previously inaccessible organic matter to microbial attack.

D. Soil pH

The pH is one of the most important factors influencing decomposition. Decomposition typically proceeds more readily in neutral than in acid soils. Consequently, the treatment of acid soil with lime accelerates the decay of plant tissues, simple carbonaceous compounds, and soil organic matter (Alexander, 1977).

Many alterations are known to occur to soil microbial populations and activities as soil pH changes. Characteristically, the population shifts from bacteria to actinomycetes to fungi as soil pH declines although acid tolerance of individual species varies widely (Alexander, 1980). The effect of soil pH on the abundance of most soil animals is not usually very pronounced (Swift *et al.*, 1979; Abrahamsen *et al.*, 1980) although different species have different pH optima.

The pH of plant materials is normally acidic. The leaves of temperate deciduous trees are often in the range of pH 5.0 to 6.5 while those of coniferous needles are more acid (pH 3.5 to 4.2) (Swift *et al.*, 1979). Such differences largely reflect the nutrient status and pH of the soils on which deciduous and coniferous trees naturally grow (Gosz, 1981). It is interesting to note that the rate of litter decomposition and N mineralization on the floor of coniferous forests is characteristically lower than that of deciduous hardwood forests (Gosz, 1981). Applications of lime to the floor of coniferous forests are known to increase the rates of litter decomposition and N mineralization (Adams *et al.*, 1978; Nommik, 1978).

Since mineralization of native soil organic N is carried out by a diverse range of microflora, the process does not show a marked sensitivity to pH (Alexander, 1980). Nonetheless, liming acid soils often causes an increase in the N mineralization rates (White, 1959; Ayres, 1961; Bornemisza *et al.*, 1967; Nyborg and Hoyt, 1978; Edmeades *et al.*, 1981), as illustrated by the results shown in Table V, although this may only be a temporary effect (Nyborg and Hoyt, 1978). Increased uptake of N by plants has been suggested as a reason for a positive response to lime in several studies (Awad and Edwards, 1977; Nyborg and Hoyt, 1978; Edmeades *et al.*, 1981). The greater tolerance of mineralization than nitrification to low pH is reflected in the finding that ammonium is generally the dominant form of N in acidic soils while nitrate predominates in nonacidic soils (Haynes and Goh, 1978; Rorison, 1980).

Table V

The Mineralization of Organic Nitrogen in 40 Soils Incubated with or without Lime[a,b]

		Organic N mineralized in 120 days	
		Concentration (μg N gm^{-1})	Percentage of total soil N
Treatment[c]			
No lime	Average	34	1.6
	Range	-1 to 136	-0.1 to 3.8
Lime	Average	72	3.5
	Range	3 to 212	0.4 to 5.6

[a] Source: Nyborg and Hoyt (1978).

[b] Soil samples ranged in texture from sandy loam to clay, pH (0.1 M CaCl) from 4.0 to 5.6 (average 5.0), and in total N content from 0.076 to 0.458% (average 0.21%).

[c] Lime added to raise soil pH to 6.7.

E. Inorganic Nutrients

1. Deficiencies

When organic materials are added to a soil, the microorganisms decomposing them obtain the nutrients necessary for their nutrition (e.g., N, P, K, Ca, Mg) from the organic materials themselves or from the pool of available nutrients in the surrounding soil or decomposing litter. Since the nutrient required by microorganisms in the greatest amounts is N, it is not surprising that N is most often the nutrient limiting microbial activity in surface litter. This aspect of decomposition has already been discussed in Section IV,A.

Nevertheless, nutrients other than N can limit decomposition. For example, Ausmus et al. (1976) estimated the immobilization of nutrients by microbial populations in the litter layers of a hardwood forest at Oak Ridge, Tennessee, and concluded that P was probably limiting microbial populations from March through July and K from April through July. Gosz et al. (1973) observed immobilization of N, S, P, and Zn in sugar maple and beech litter, while during the decomposition of woody tissues, immobilization of N, P (Swift, 1978), and K (Swift, 1973) has been recorded. Immobilization of P has also been noted during the decomposition of pasture grass residues (Floate, 1970b) and soil-incorporated wheat straw (Brown and Dickey, 1970).

Absolute deficiencies of nutrient elements in soils can limit microbial activity in soils and thus the mineralization of soil organic N. For instance, on P-deficient soils, applications of phosphate can increase the quantity of soil N that is mineralized (Munevar and Wallum, 1977). Bertrand (1971) showed that additions of small quantities of Cu increased the rate of N mineralization in some soils while addition of Mo-containing industrial wastes to a soil increased proteolytic activity (Kanatchinova, 1969).

2. Toxicities

Contamination of the environment by trace elements (defined here as elements that are, when present in sufficient quantities, toxic to living systems) has generated concern over the effects of such elements on the biogeochemical reactions in the biosphere. The trace element content of soils can be substantially increased by the application of waste materials such as sewage sludge and industrial mining wastes (Cast, 1976; Jones and Jarvis, 1981).

Several studies have demonstrated the inhibitory effect of high rates of trace element application on mineralization of N (Premi and Cornfield, 1969; Quarishi and Cornfield, 1973; Tyler, 1975; Liang and Tabatabai, 1977; Chang and Broadbent, 1982). Such inhibitory effects are clearly illustrated in Table VI.

Liang and Tabatabai (1977) showed that the addition of any of 19 trace elements (added at 5 μmol gm^{-1} soil) inhibited N mineralization in soils; their degrees of effectiveness varied in the four soils studied. Chang and

Table VI

Effect of Metal Additions on the Concentrations of Mineral Nitrogen in a Soil (μg N gm^{-1}) after 4 Weeks[a]

Concentration of added metal (μg gm^{-1})	Metal added[b]					
	Cd	Cr	Cu	Pb	Zn	Mn
0	82.5	82.5	82.5	82.5	82.5	82.5
100	40.5	45.6	38.4	52.7	42.5	36.4
200	34.9	30.4	19.5	45.4	30.5	48.4
400	25.5	29.3	20.4	55.6	28.6	54.3

[a] Source: Chang and Broadbent (1982). Reproduced from the *J. Environ. Qual.* **11**, p. 3 by permission of the American Society of Agronomy.

[b] Standard error of means is within one concentration of added metal ($= 1.43$).

Broadbent (1982) investigated the influence of trace metals on N immobilization–mineralization following additions of organic residues to incubated soils. At low levels of metal addition (100 and 200 μg gm^{-1}) Mn(II) and Pb(II) stimulated immobilization of added NH$_4^+$-N but at 400 μg gm^{-1} all metals tested were inhibitory to both immobilization and mineralization.

Nevertheless, several soil properties influence the toxic effects of trace elements. With increasing soil pH, the availability and toxicity of most metal ions (e.g., Zn, Cu, Fe, Mn, Co, Cr, Ni, and Pb) decrease while clay minerals and particularly soil organic matter adsorb such metals and thus reduce their toxicity to the microbial biomass (Gadd and Griffiths, 1978; Jones and Jarvis, 1981). Variations in soil properties (e.g., pH, clay, and organic matter content) may help explain why several workers have observed small or inconsistent effects (Bhuiya and Cornfield, 1974; Rother et al., 1982) or even stimulation of mineralization (Premi and Cornfield, 1969; Chang and Broadbent, 1982) following applications of high rates of trace elements to soils.

It is noted here that short-term studies involving the additions of high concentrations of trace elements to uncontaminated soils may not give a true indication of the long-term effects of such additions. This is because the sensitivity of microorganisms to toxic elements can vary and some species are known to develop tolerance. There are numerous studies in which tolerance to heavy metals has been induced under laboratory conditions for a wide range of bacteria (Gadd and Griffiths, 1978; Sterrit and Lester, 1980). Metal-tolerant organisms have also been isolated from soil where high concentrations of metals occur naturally or from soils that have been polluted by heavy metals (Hartman, 1974; Doelman and Haanstra, 1979). Nevertheless, Rother et al. (1982) observed that the effects of adding heavy metals (Cd, Zn, and Pb) to soils on the N mineralization were not consistently correlated with the extent of the soils' previous heavy metal contamination.

3. General Salt Effect

The addition of salts and fertilizers to soils can cause net mineralization of soil organic N (i.e., have a "priming effect") (Singh et al., 1969; Fig. 17; Broadbent and Nakashima, 1971; Agarwal et al., 1971b; Westerman and Tucker, 1974; Heilman, 1975; Laura, 1977) although in some cases decreases or no change in mineralization have been observed (Broadbent, 1970; Laura, 1974).

In general, the ability of cations to stimulate mineralization of N follows the same order as their replacing power on cation exchange sites in soils (e.g., Al^{3+} > Fe^{3+} > Ca^{2+} > Mg^{2+} > K$^+$ > Na$^+$) (Singh et al., 1969;

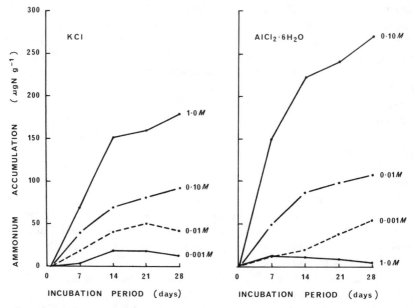

Fig. 17. Effect of increasing concentrations of KCl and AlCl₃ · 6H₂O on ammonium accumulation in a Hawaiian Typic Hydrandept. [Data from Singh *et al.* (1969). Reproduced from *Soil Sci. Soc. Am. Proc.* **33,** p. 558 by permission of the Soil Science Society of America.]

Agarwal *et al.,* 1971b; Heilman, 1975). The greater stimulatory effect of additions of AlCl₂ · 6H₂O rather than KCl (at concentrations at or below 0.10 *M*) is obvious from Fig. 17. The effect also depends on the anion species associated with the particular cation (Agarwal *et al.,* 1971b; Heilman, 1975); for salts of K the order of effectiveness is KCl > K₂SO₄ > K₂CO₃ > KHPO₄ (Heilman, 1975). Such a salt effect may explain the sometimes stimulatory effect of trace element additions on N mineralization (e.g., Premi and Cornfield, 1969). The mechanisms by which salts stimulate mineralization are the subject of much speculation and controversy.

Broadbent (1970) and Broadbent and Nakashima (1971) postulated that osmotic effects contributed to the salt-stimulated mineralization of soil organic N, the effect being at least partially due to extraction of organic N by the salt solution. The organic N thus rendered soluble would constitute a pool of easily mineralizable N.

Since different salts at equivalent osmotic pressure result in different magnitudes of NH₄⁺ release (Broadbent and Nakashima, 1971), osmotic pressure cannot be the sole factor operating. Hence other workers (e.g.,

Singh *et al.,* 1969; Agarwal *et al.,* 1971b) have suggested that in addition to the osmotic effect certain chemical reactions and/or processes are involved that cause the splitting of NH_4^+-N from organoinorganic complexes in a soil. The NH_4^+ ions thus released could subsequently be exchanged with cations in the surrounding soil solution, which would explain why the magnitude of the stimulatory effect of cations occurs in the order of their replacing power (Agarwal *et al.,* 1971b).

Another possible contributing factor could be that high salt concentrations result in the death and breakdown of microbial cells with the release of readily available N for subsequent mineralization by the remaining microbial population. High salt concentrations are known to be toxic to the microbial soil population (Alexander, 1977).

4. Fertilizer Nitrogen

In addition to a general salt effect stimulating the mineralization of soil N, there appears to be a specific effect of added inorganic N (Westerman and Tucker, 1974). Nonetheless, applications of inorganic fertilizer N have been reported to stimulate, depress, or have no effect on the mineralization of native soil organic N (Gadet and Soubies, 1965; Broadbent, 1970). An apparent increase in mineralization of native soil N following the addition of ^{15}N-labeled fertilizer N has been observed under laboratory, greenhouse, and field conditions (Legg and Stanford, 1967; Saphozhnikov *et al.,* 1969; Broadbent and Nakashima, 1971; Westerman and Kurtz, 1973; Westerman and Tucker, 1974). With applications of fertilizer N in the range of 56 to 168 kg N ha^{-1}, Westerman and Kurtz (1973) found that the uptake of native soil N by *Sorghum sudanenses* was increased from 17 to 45% and 8 to 27%, respectively, at two separate field sites. The positive priming effect on soil organic N generally appears to be greater when NH_4^+ fertilizer rather than NO_3^- fertilizer is applied (Broadbent, 1965).

Controversy surrounds the origin and nature of the positive priming effect of applied fertilizer N. Several workers have attributed the effect to mineralization–immobilization turnover (Fig. 18) (Stewart *et al.,* 1963; Aleksic *et al.,* 1968; Nommik, 1969; Huntjens, 1971; Jansson, 1971; Jansson and Persson, 1982). As discussed previously, mineralization and immobilization occur simultaneously. The addition of labeled inorganic fertilizer N to the soil will result in some immobilization of labeled N and some mineralization of nonlabeled native soil N. Thus more nonlabeled mineral N will occur in the fertilized soil than in the smaller pool of mineral N in unfertilized controls. This does not necessarily mean that net mineralization of N has occurred, merely that labeled N is immobilized and at the same time unlabeled native N is mineralized. Huntjens (1971)

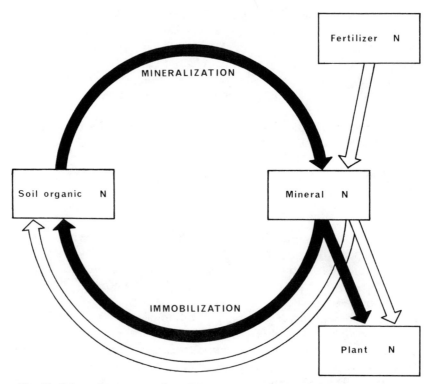

Fig. 18. Schematic representation of the turnover of nitrogen in relation to the priming effect of fertilizer nitrogen.

observed no priming effect when labeled fertilizer N was added to a dense turf grass. He attributed this to a rapid uptake of fertilizer N by the turf so that the turnover effect was of negligible significance.

Aleksic *et al.* (1968) and Sapozhnikov *et al.* (1969) suggested that the more rapid shoot and root development caused by N fertilization results in increased uptake of native inorganic N. The results of Sørensen (1982), who used a soil containing an organic fraction labeled with ^{15}N, tend to support this assertion. He found that fertilizing with 106 or 424 kg N ha^{-1} (as KNO_3) significantly increased barley yields and the plant uptake of labeled soil N at the first harvest (Table VII) but this was compensated for in subsequent crops and harvests. There was no indication that fertilizing with KNO_3 accelerated mineralization of organically bound ^{15}N. The observed priming effect appeared to arise from a more thorough initial search of the soil volume for native ^{15}N-labeled N by the better developed root system (Table VII) of the fertilized plants. The above suggestions

Table VII

Dry Matter Production and Nitrogen Uptake (from Nonlabeled N and Labeled Soil Organic N)[a] by a Barley Crop at the First Harvest of a Pot Experiment[b]

Barley crop	Nonlabeled addition (mg N/pot)			
	None	90	360	LSD ($P \leq 0.05$)
Total dry matter (gm/pot)	14.2	25.2	35.4	3.5
Root dry matter (gm/pot)	1.2	2.2	2.9	0.8
Nonlabeled N uptake (mg/pot)	91.5	178.3	429.5	14.1
Labeled N uptake (mg/pot)	31.8	36.5	37.9	2.2

[a] Plants were grown in a loam soil containing an organic fraction labeled with ^{15}N and were fertilized with 90 or 360 mg N/pot of unlabeled N in the form of KNO_3.

[b] Data from Sørensen (1982).

can, however, only be partial explanations since the priming effect of added inorganic N has also been demonstrated in field and laboratory incubation studies in the absence of plants.

As already noted Broadbent (1965) and Broadbent and Nakashima (1971) observed increased mineralization of soil organic N following solubilization of soil organic matter by added salts, including those of NH_4^+, which were attributed to changes in pH and osmotic concentration induced by salt additions. Other workers (Westerman and Kurtz, 1973; Westerman and Tucker, 1974) simply believe that additions of NH_4^+ salts stimulate microbial activity and hence cause an increase in net mineralization.

In summary, the effect of fertilizer N on the mineralization of soil organic N appears to be a partially real and partially apparent priming effect. No one explanation explains all the effects that have been reported. It is, however, likely that additions of N to the soil may influence the activities and population diversity of the microbial biomass through changes to the microbial environment (e.g., changes in pH). Such changes could conceivably alter the rate of mineralization. Nonetheless, mineralization–immobilization turnover will often explain the priming effect since it is a ubiquitous feature of the soil system.

F. Additions of Organic Residues

The addition of fresh organic materials to soils can apparently either stimulate or retard the decomposition of indigenous soil organic matter although the effects may be short-lived and small in comparison with the quantities of native soil organic matter present (Jenkinson, 1971). Several workers have observed a small positive priming effect on soil organic matter decomposition following additions of organic materials (Sauerbeck, 1966; Jansson, 1971; Jenkinson, 1971; Broadbent and Nakashima, 1974; Shields *et al.,* 1974; Nyhan, 1975; Sørensen, 1977; Dalenberg and Jager, 1981).

In a five-year laboratory study Broadbent and Nakashima (1974), for example, observed the priming effect of the addition of ^{15}N-labeled barley tops to a soil on the release of native soil organic N (Fig. 19). The net release of soil organic N was greater in the barley-amended than in the

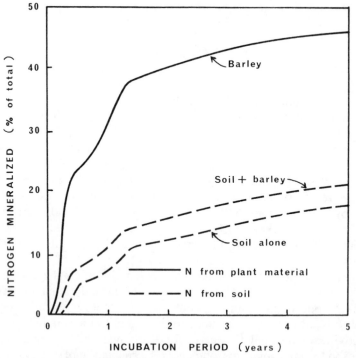

Fig. 19. Mineralization of barley ^{15}N and unlabeled soil organic N following the addition of barley residues to a soil as compared to mineralization of soil organic N from an unamended soil. [Data from Broadbent and Nakashima (1974). Reproduced from *Soil Sci. Soc. Am. Proc.* **38,** p. 314 by permission from the Soil Science Society of America.]

-unamended soil, particularly in the first half year when the decomposition of the barley residue itself was also greatest.

The exact mechanisms by which such a priming effect occurs are unknown (Clark, 1968; Jansson, 1971; Jenkinson, 1981). However, in the same way that additions of fertilizer N stimulate mineralization–immobilization turnover, additions of organic materials will also do so. The native microbial biomass in the soil is likely to respond massively immediately following additions of energy-rich materials. Additions of fresh residues provide a source of energy for microbial proliferation, resulting in a larger and more active microbial biomass; a consequent increase in enzyme activity could speed the breakdown of native soil organic matter as well as that of the added materials. A change in the species composition of the microbial biomass could also occur and this might result in increased or decreased attacks on native soil organic matter. The outcompeted species may die and their dead biomass would become liable to mineralization by the successful species.

Indirect effects may also occur through ecological modifications to the microbial environment caused by additions of fresh actively decomposing organic matter. Such modifications could involve localized changes in pH, aeration, or nutrient availability, all of which could have a transient effect on mineralization–immobilization turnover and thus the decomposition of native soil organic matter.

G. Pesticides

The use of pesticides (fumigants, herbicides, insecticides, and fungicides) is an integral part of modern agricultural practice. However, applications of such chemicals can have undesirable effects on nontarget organisms.

The activity of a pesticide in soils, and thus its effects on the soil biomass, depends on many factors, including its tendency to volatilize and the intrinsic capacity of the pesticide to resist degradation as influenced by temperature, moisture, and its degree of sorption onto mineral and particularly organic soil constituents (Calvet, 1981; Morill et al., 1982). The sorption of pesticides by soils is generally governed by a balance between their water solubility and the organic matter content of the soil (Hamaker and Thompson, 1972). If a pesticide is to have any significant long-term effect on the mineralization of N, it must accumulate in soil solution at concentrations high enough to affect microbial populations.

Increased mineralization of soil N frequently occurs at recommended rates of application of soil-applied fumigants, insecticides, and fungicides and at higher-than-recommended rates for herbicides (Anderson, 1978; Goring and Laskowski, 1982). Stimulation of mineralization by application of pesticides apparently occurs because a part of the soil microbial biomass is killed and this serves as an N-rich substrate for the surviving microbial population so that mineral N is released from the dead biomass (Jenkinson and Powlson, 1970). Repeated treatment of soils with pesticides does not continue to increase N mineralization once the susceptible microbial population is destroyed (Jenkinson and Powlson, 1970).

Although such results may be regarded as a general indication of a lack of harmful effects by pesticides, this is not necessarily the case. Dubey (1970), for example, reported that repeated applications of dithiocarbamates to a soil inhibited N mineralization and caused N deficiency in sugarcane. Furthermore, many pesticides applied at higher-than-normal rates tend to inhibit nitrification and soil fumigants are characteristically toxic to nitrifying organisms (see Chapter 3). Thus, although ammonification may be stimulated by pesticide applications, nitrification can be inhibited (e.g., Marsh and Greaves, 1979). This can result in an accumulation of NH_4^+-N in the soil that could be potentially toxic to many crop plants.

H. Growing Plants

The volume of soil immediately surrounding the root is termed the rhizosphere. The rhizosphere is a region highly favorable to microbial growth since the plant contributes excretory products (such as root exudates) and sloughed off tissues that act as an energy source for microbial growth (Warembourg and Billes, 1979; Sarkar and Wyn Jones, 1980).

The production of a ready supply of available C in the rhizosphere means that N often becomes the most limiting nutrient for microbial growth. Most of the mineral N released by decomposition in the rhizosphere is, therefore, first incorporated into the microbial biomass. The importance of living roots (supplying carbonaceous materials) was demonstrated by Huntjens (1971), who found net mineralization of soil N in soils containing dead grass roots but net immobilization of N in soils where living plant roots were present. In a model perfusion experiment Huntjens and Albers (1978) showed that a combination of a deficiency of available N and a continuous supply of carbonaceous material (conditions that occur in the extensive rhizosphere soils under permanent pasture) is most favorable for immobilization of N and the formation of soil organic matter.

Such results help explain why there often tends to be an accumulation of soil organic matter and soil N under permanent pasture (Barrow, 1957; Clement and Williams, 1967). Nonetheless, results obtained in studies of

N mineralization under crop plants compared with under fallow land have been rather contradictory. Plants have been reported to retard (Bartholomew and McDonald, 1966), increase (Cornish and Raison, 1977), or have no effect (Bartholomew and Hitbold, 1952) on N mineralization.

The effects of plants on decomposition processes are not all caused by rhizosphere effects. Several researchers have, for instance, shown that the presence of plants slows the rate of decomposition of added ^{14}C-labeled residues (Führ and Sauerbeck, 1968; Shields and Paul, 1973; Jenkinson, 1977b). Such an effect has been attributed to the desiccating effect of vegetation on soils during the summer (Jenkinson, 1977b).

I. Cultivation

As discussed in the final section of Chapter 1, cultivation of soils from under mature natural ecosystems results in a rapid decline in their organic matter and N content. Cultivation is an oxidative process since it invariably promotes good aeration and the rapid decomposition of soil organic matter, and consequently the mineralization of N (Campbell, 1978). Plowing also kills existing vegetation and the dead plants decompose rapidly. Often, nutrients so released are lost before the newly planted crop can use them.

As noted in Chapter 1, decomposition of soil organic matter is generally rather rapid in the first 25 to 50 years of cultivation and steady-state conditions are often reached within 50 to 100 years after conversion to arable cropping (Stevenson, 1965; Allison, 1973). The most extensive decomposition and mineralization occurs under irrigated conditions. Kononova (1966), for example, reported that in the USSR irrigated serozems supporting a row crop of cotton lost 53% of their total organic matter in 3 to 5 years while in the drier chernozem and chestnut soils losses were only 1% per year. A discussion of mineralization and N availability under herbicide management (minimum or zero tillage) versus conventional tillage (cultivation) is present in Chapter 7.

The increased mineralization following cultivation is thought to be at least partially attributable to the physical disruption of soil aggregates, resulting in the exposure of microsites where organic matter was previously physically inaccessible to microbes or their enzymes (Rovira and Greacen, 1957; Adu and Oades, 1978).

J. Clay Content

In general, soil organic matter contents tend to increase with increasing clay contents of soils (Jenny, 1941). Thus, for a given climate, provided vegetation and topography are constant, fine-textured soils have higher organic matter and N contents than their coarse-textured counterparts. The mechanisms by which clays increase soil humus formation and/or decrease its decomposition are not clear although three major ways have been suggested.

1. Microbial Activity

In general, clay minerals appear to exert a marked stimulatory effect on microbial growth and activity and may also increase the efficiency of C utilization by the microorganisms (Stotzky, 1967; Bondietti et al., 1971, 1972; Filip et al., 1972a,b; Martin et al., 1976). Bondietti et al. (1971) showed that additions of montmorillonite and vermiculite to aerobic cultures of *Hendersonula tortuloidea, Stachybotrys* spp., and *Aspergillus sydowi* greatly accelerated growth, glucose utilization, CO_2 evolution, phenol synthesis, and phenolic polymer formation. In some tests total microbial biomass was also increased by additions of clay.

The reason for this stimulatory effect of clays on microbial growth and activity is unclear. Nevertheless, it does seem that, to some extent, clays may encourage humus formation indirectly by stimulating microbial growth and activity and thus the microbial production and synthesis of phenols and N-containing phenolic polymers (i.e., humic substances). Nonetheless, some workers have suggested that clays have little influence on microbial growth and Marshman and Marshall (1981) showed no effect of clays on microbial growth efficiency.

2. Complexation with Organic Compounds

Clay and organic substances are known to interact in the soil to form complexes that often result in the substance being less susceptible to biodegradation (McLaren and Peterson, 1965; Theng, 1979; Stevenson, 1982a). Substances such as amino acids, peptides, proteins, purines, pyrimidines, nucleic acids, and nucleosides have been shown to be sorbed by clay minerals.

Because of their greater surface area the montmorillonite (2 : 1) clays are generally more effective than the kaolinite-type clays in stabilizing organic materials (Sørensen, 1972). In expanding layer silicate clays, interlayer complexes can be formed (Theng, 1974). Amorphous aluminosilicates (e.g., allophane) also form very stable bonds with organic materials (Wada and Inoue, 1967) and thus appreciably decrease their rate of decomposition (Zunino et al., 1982).

In addition to slowing the microbial attack of substrates that become sorbed by the clay surfaces, clays may also decrease rates of decomposition by complexing with newly synthesized materials, extracellular metabolites, or extracellular enzymes themselves (Burns, 1978). Sørensen

(1967, 1972), for instance, showed that additions of clay minerals to a sandy soil did not influence the rate of decomposition of ^{14}C-labeled carbohydrate but the contents of amino acid metabolites derived from the added carbohydrates were two to three times greater in the soils amended with clays in comparison with controls. The accumulation of ^{14}C-labeled amino acid metabolites in soils as a function of their clay content is illustrated in Fig. 20.

3. Complexation with Humic Substances

The complexing of humic substances to clays has long been recognized as a factor responsible for the stability and resistance of humus to degradation (e.g., Mattson, 1932). Indeed, the organic and mineral components of soils are known to be intimately associated. Greenland (1971), for example, summarized the work of five groups of workers and showed that the total C that was combined with the clay component in 11 soils studied ranged from 51.6 to 97.8% with a mean of 73%. The treatment of mineral soils with hydrofluoric acid to break down clay minerals is known to solubilize considerable quantities of organic N (Stevenson, 1982a); this also indicates the close association between the mineral and organic soil components.

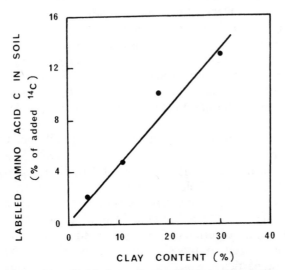

Fig. 20. Labeled amino acid C in four soils after 900 days of incubation with ^{14}C-labeled cellulose plotted against content of clay in soils (particles < 2 μm). Amino acids were synthesized during the decomposition of ^{14}C-labeled cellulose added to the soils. [Data from Sørensen (1981). Reprinted with permission from Pergamon Press.]

The mechanisms involved in the binding of soil humic substances to clay minerals have been reviewed extensively (Greenland, 1971; Theng, 1979; Burchill *et al.*, 1981). Edwards and Bremner (1967) proposed that soil microaggregate (<250 μm diameter) formation involves linkage of electrically neutral clay mineral and organic matter particles by polyvalent cations (e.g., Ca^{2+}, Fe^{2+}, Al^{3+}, Mn^{2+}) on exchange sites. These microaggregates could contribute, through polymerization, to the formation of larger aggregates. Mortland (1970) subsequently suggested that functional groups of the organic matter are actually linked to the polyvalent cations through a water bridge. In view of the structural complexity of humic substances and the many factors that determine their association with layer silicate clays (Theng, 1979) such models are somewhat tentative.

The formation of soil aggregates through interactions between clay minerals and humic substances may be of considerable importance in terms of the mineralization of soil organic N because it may render some soil organic matter inaccessible to the decomposer microflora. This aspect is discussed in the following section.

K. Physical Inaccessibility

As pointed out by Swift *et al.* (1979), the surface area and volume of detritus particles influence their decomposition in a number of ways. They determine the pattern of colonization by microorganisms (increasing surface-to-volume ratio selects for surface-growing unicellular forms as against penetrative mycelial forms) and the ability of animals to ingest food and may also be an important feature affecting the accessibility of substrates to enzymes. In nature, soil animals play a major role in comminuting organic matter and thus making it more accessible to microbial attack (Section II,C).

In general, finely divided or ground plant material decomposes more quickly than coarse material (Allison and Cover, 1960; Sims and Frederick, 1970; Cheshire *et al.*, 1974; Moore, 1974). Grinding exposes more surface area for microbial colonization and the action of enzymes, thus it accelerates the decomposition process. For material with a wide C : N ratio this results in the rapid immobilization of inorganic soil N. For example, Sims and Frederick (1970) observed that fine particles of maize stalk pith immobilized six times as much inorganic N in the first month than did coarse particles.

In previous sections, the effects of drying and rewetting, freezing and thawing, and cultivation in stimulating the mineralization of soil organic nitrogen were partially attributed to physical disruption of aggregates resulting in the exposure of organic matter not previously accessible to microbial attack. More detailed discussion of the accessibility of soil organic matter to organisms and enzymes is, therefore, warranted. The concepts of soil aggregate structure (Edwards and Bremner, 1967; Mortland, 1970) and the role of clay–polyvalent cation–organic matter bonding in such structures have already been outlined.

Much of the evidence regarding the inaccessibility of soil organic matter to microbial attack comes from studies of the effects of grinding on N mineralization. Generally, soils with a high clay content show large increases in mineralization due to grinding. Such increases are usually attributed to the release of organic N previously inaccessible to attack (Edwards and Bremner, 1967; Craswell and Waring, 1972a,b).

Calculations by Adu and Oades (1978) showed that only a very small fraction of organic matter in soils is likely to be in close proximity to soil organisms at any one time. They calculated that in a clay loam with a microbial population of 10^8 bacterial gm^{-1} soil, the organisms would occupy about 0.1% of the soil surfaces. Assuming that enzyme diffusion in soils is limited by adsorption, Adu and Oades (1978) concluded that at least 90% of surface in soils are not accessible to microorganisms or their enzymes. In a laboratory study Adu and Oades (1978) found that the artificial distribution of ^{14}C-labeled substrate into macro- and micropores of soil aggregates rendered a portion of them inaccessible to microbial attack. Disruption of aggregates, either by mechanical disturbance or a drying and rewetting cycle, during incubation resulted in a flush of $^{14}CO_2$ evolution.

Thus in the field situation it seems likely that, due to the inherent structure of aggregates, much of the organic matter in soils may be inaccessible to the soil microflora and extracellular enzymes.

V. CONCLUSIONS

Organic materials that undergo decomposition in soil originate from several sources. In many ecosystems vast quantities of plant remains and tree litter decompose at the soil surface. Death of subterranean plant parts, plus aboveground tissues that are mechanically incorporated into the soil by the action of decomposer fauna, results in decomposition processes occurring below the soil surface. Animal tissues and excretory products are also subject to decomposition.

The detritus is broken down by the combined action of the decomposer community, which is composed predominantly of microorganisms (bacteria and fungi) and invertebrate animals. Such organisms degrade the detritus and utilize some of the energy and nutrients released for their own growth. Eventually, the decomposer organisms die and their carcasses enter the detritus pool and are, in turn, acted upon by succeeding generations of decomposers.

The residues of decomposition contribute to the formation of soil organic matter. In the short term, such residues consist of partially digested plant matter, animal carcasses, feces, and microbial cells. Soil organic matter formation is not, however, wholly a degradative process. Indeed, the highly resistant humic component of soil organic matter consists of a mixture of large, complex polymeric molecules that are synthesized by the decomposer microflora during the decomposition process.

The major pathway of humic polymer formation appears to be through condensation reactions involving the polymerization quinones in the presence or absence of amino compounds. The two major sources of quinones are thought to be phenolic compounds released from lignin during microbial attack and polyphenols synthesized by microorganisms from non-lignin C sources (e.g., cellulose).

The release of C and N from decaying residues during decomposition differs in that C is generally volatilized as CO_2 or methane while N tends to be conserved within the decomposing residues and the remainder is released to the soil as NH_4^+-N. The conservation of N is caused by the demand for N by the decomposer biomass. During the decomposition of residues with a wide C : N ratio the microbial biomass may even incorporate mineral N from the surrounding soil or litter, resulting in an absolute increase in the quantity of N in the litter as it decomposes, that is, net immobilization of N occurs.

During the course of decomposition, the C : N ratio progressively decreases and at some point there is a switch from net immobilization to net mineralization of N. The critical C : N ratio below which net mineralization of N will occur is commonly quoted as being in the range 25 to 30 : 1 (N content 1.7 to 2.5%), although the N content is by no means the only factor influencing the mineralization–immobilization balance.

Substrate quality is, however, a critical factor determining the rate of litter decomposition and the release of mineral N. Apart from N content, other important indices of substrate quality include the concentrations of various classes of organic compounds such as polyphenols and particularly lignin in the litter. Environmental parameters, including moisture, temperature, pH, and supply of nutrients, can also strongly influence the rate of decomposition and mineralization of N.

The mineralization of soil organic N is also greatly affected by the above environmental parameters. Variations in environmental parameters, such as drying and rewetting, freezing and thawing, or fluctuating temperatures, appear to be particularly important. These phenomena often cause a flush in microbial activity and N mineralization. Two complementary explanations have been forwarded. First, the microbial biomass is thought to constitute a major pool of readily mineralizable N in soils. Fluctuations in environmental conditions can cause the death of a significant proportion of the biomass and the dead biomass is readily mineralized by the surviving microflora. Second, in the case of drying and rewetting or freezing and thawing, it is thought that such phenomena cause disruption of soil aggregates and the exposure of organic matter previously inaccessible to microbial attack.

The addition of organic matter or inorganic fertilizer materials (both nitrogenous and nonnitrogenous) can cause net mineralization of soil organic N (i.e., have a "priming effect"). No one explanation can explain all the effects that have been reported. Nevertheless, it seems possible that additions of such substances to the soil may influence the activity and population diversity of the microbial biomass through changes to the microbial environment; this could conceivably alter the rate of mineralization.

Man's activities can also significantly influence the rate of mineralization of soil N. Cultivation of agricultural lands, for example, can greatly stimulate mineralization through the disruption of soil aggregates and the exposure of previously inaccessible organic N to microbial attack. The use of soil-applied pesticides (herbicides, insecticides, fungicides, and fumigants) at high rates can result in a flush of N mineralization because part of the microbial biomass is killed and then decomposed by the surviving population. Applications of waste materials (e.g., sewage sludge) to lands can cause the buildup of high levels of trace elements (e.g., heavy metals) in soils, which can have an inhibitory effect on the mineralization of N.

REFERENCES

Aber, J. D., and Melillo, J. M. (1980). Litter decomposition: Measuring relative contributions of organic matter and nitrogen to forest soils. Can. J. Bot. **58**, 416–421.

Abrahamsen, G., Hovland, J., and Hagvar, S. (1980). Effects of artificial acid rain and liming on soil organisms and the decomposition of organic matter. In "Effects of Acid Precipitation on Terrestrial Ecosystems" (T. C. Hutchinson and M. Havas, eds.), pp. 341–362. Plenum, New York.

Adams, S. M., Cooper, J. E., Dickson, D. A., Dickson, E. L., and Seaby, D. A. (1978). Some effects of lime and fertiliser on a Sitka spruce plantation. Forestry **5**, 57–65.

Adu, J. K., and Oades, J. M. (1978). Utilisation of organic materials in soil aggregates by bacteria and fungi. Soil Biol. Biochem. **10**, 117–122.

Agarwal, A. S., Singh, B. R., and Kanehiro, Y. (1971a). Soil nitrogen and carbon mineralisation as affected by drying–re-wetting cycles. *Soil Sci. Soc. Am. Proc.* **35**, 96–100.

Agarwal, A. S., Singh, B. R., and Kanehiro, Y. (1971b). Ionic effect of salts on mineral nitrogen release in an allophanic soil. *Soil Sci. Soc. Am. Proc.* **35**, 454–457.

Ahmad, Z., Kai, H., and Harada, T. (1972). Effecs of nitrogenous forms on immobilisation and release of nitrogen in soil. *J. Fac. Agric., Kyushu Univ.* **17**, 49–65.

Ahmad, Z., Yahiro, Y., Kai, H., and Harada, T. (1973). Factors affecting immobilisation and release of nitrogen in soil and chemical characteristics of the nitrogen newly immobilised. *Soil Sci. Plant Nutr.* **19**, 287–298.

Aleksic, Z., Broeshart, H., and Midleboe, V. (1968). The effect of nitrogen fertilisation on the release of soil nitrogen. *Plant Soil* **29**, 474–478.

Alexander, M. (1977). "Introduction to Soil Microbiology." Wiley, New York.

Alexander, M. (1980). Effects of acidity on micro-organisms and microbial processes in soil. *In* "Effects of Acid Precipitation on Terrestrial Ecosystems" (T. C. Hutchinson and M. Havas, eds.), pp. 363–380. Plenum, New York.

Allen, A. L., Stevenson, F. J., and Kurtz, L. T. (1973). Chemical distribution of residual fertilizer nitrogen in soil as revealed by nitrogen-15 studies. *J. Environ. Qual.* **2**, 120–124.

Allison, F. E. (1973). "Soil Organic Matter and Its Role in Crop Production." Elsevier, Amsterdam.

Allison, F. E., and Cover, R. G. (1960). Rates of decomposition of short-leaf pine sawdust in soil at various levels of nitrogen and lime. *Soil Sci.* **89**, 194–201.

Allison, F. E., and Klein, C. J. (1962). Rates of immobilisation and release of nitrogen following additions of carbonaceous materials and nitrogen to soils. *Soil Sci.* **93**, 383–386.

Amato, M., and Ladd, J. M. (1980). Studies of nitrogen immobilisation and mineralisation in calcareous soils. V. Formation and distribution of isotope-labelled biomass during decomposition of ^{14}C- and ^{15}N-labelled plant material. *Soil Biol. Biochem.* **12**, 405–411.

Anderson, J. M. (1973). The breakdown and decomposition of sweet chestnut (*Castanea sativa* Mill.) and beech (*Fagus sylvatica* L.) leaf litter in two deciduous woodland soils. II. Changes in the carbon, hydrogen, nitrogen and polyphenol content. *Oecologia* **12**, 275–288.

Anderson, J. P. E., and Domsch, K. H. (1980). Quantities of plant nutrients in the microbial biomass of selected soils. *Soil Sci.* **130**, 211–216.

Anderson, J. R. (1978). Pesticide effects on non-target soil microorganisms. *In* "Pesticide Microbiology" (I. R. Hill and S. J. L. Wright, eds.), pp. 313–533. Academic Press, New York.

Anderson, R. V., Coleman, D. C., and Cole, C. V. (1981). Effects of saprotrophic grazing on net mineralisation. *In* "Terrestrial Nitrogen Cycles: Processes, Ecosystem Strategies and Management Impacts" (F. E. Clark and T. Rosswall, eds.), pp. 201–216. Ecological Bulletins, Stockholm.

Ausmus, B. S., Edwards, N. T., and Witkamp, M. (1976). Microbial immobilisation of carbon, nitrogen, phosphorus and potassium: Implications for forest ecosystem processes. *In* "The Role of Terrestrial and Aquatic Organisms in Decomposition Processes" (J. M. Andersson and A. Macfadyen, eds.), pp. 397–416. Blackwell, Oxford.

Awad, A. S., and Edwards, D. G. (1977). Reversal of adverse effects of heavy ammonium sulphate application on growth and nutrient status of kikuyu pasture. *Plant Soil* **48**, 169–183.

Ayres, A. S. (1961). Liming Hawaiian sugarcane soils. *Hawaii. Plant. Rec.* **56**, 227–244.

Barley, K. P., and Jennings, A. C. (1959). Earthworms and soil fertility. III. The influence of earthworms on the availability of nitrogen. *Aust. J. Agric. Res.* **10**, 364–370.

Barrow, N. J. (1957). Renovation of phalaris pastures with special reference to nitrogen and sulphur relationships. *Aust. J. Agric. Res.* **8**, 617–634.

Barrow, N. J. (1961). Mineralisation of N and S from sheep faeces. *Aust. J. Agric. Res.* **12**, 644–650.

Bartholomew, W. V. (1965). Mineralization and immobilization of nitrogen in the decomposition of plant and animal residues. *In* "Soil Nitrogen" (W. V. Bartholomew and F. E. Clark, eds.), pp. 285–306. Am. Soc. Agron., Madison, Wisconsin.

Bartholomew, M. V., and Hitbold, A. E. (1952). Recovery of fertiliser nitrogen by oats in the greenhouse. *Soil Sci.* **73**, 193–201.

Bartholomew, W. V., and McDonald, I. (1966). Measurement of organic material deposited in soil during the growth of some crop plants. *In* "The Use of Isotopes in Soil Organic Matter Studies," pp. 235–242. FAO/IAEA, Brunswick.

Basaraba, J., and Starkey, R. L. (1966). Effect of plant tannins on decomposition of organic substances. *Soil Sci.* **101**, 17–23.

Bell, M. K. (1974). Decomposition of herbaceous litter. In "Biology of Litter Decomposition" (C. H. Dickinson and G. J. F. Pugh, eds.), pp. 37–67. Academic Press, New York.

Benoit, R. E., and Starkey, R. L. (1968). Inhibition of decomposition of cellulose and some other carbohydrates by tannin. *Soil Sci.* **105**, 291–296.

Benoit, R. E., Starkey, R. L., and Basaraba, J. (1968). Effect of purified plant tannin on decomposition of some organic compounds and plant materials. *Soil Sci.* **105**, 153–158.

Berg, B., and Söderström, B. (1979). Fungal biomass and nitrogen in decomposing Scots pine needle litter. *Soil Biol. Biochem.* **11**, 339–341.

Berg, B., and Staaf, H. (1980). Decomposition rate and chemical changes of Scots pine needle litter. II. Influence of chemical composition. *In* "Structure and Function of Northern Coniferous Forests: An Ecosystem Study" (T. Persson, ed.), pp. 375–39. Ecological Bulletins, Stockholm."

Berg, B., and Staaf, H. (1981). Leaching, accumulation and release of nitrogen in decomposing forest litter. *In* "Terrestrial Nitrogen Cycles: Processes, Ecosystem Strategies and Management Impacts" (F. E. Clark and T. Rosswall, eds.), pp. 163–178. Ecological Bulletins, Stockholm.

Bertrand, D. (1971). Influence de taux de cuivre soluble du sol sur l'ammonification. *C.R. Seances Acad. Agric. Fr.* **57**, 1556–1561.

Bhuiya, M. R. H., and Cornfield, A. H. (1974). Incubation study on effect of pH on nitrogen mineralisation and nitrification in soils treated with 1000 ppm lead and zinc as oxides. *Environ. Pollut.* **7**, 161–164.

Biederbeck, V. O., and Campbell, C. A. (1971). Influence of simulated fall and spring conditions on the soil system. I. Effect on soil microflora. *Soil Sci. Soc. Am. Proc.* **35**, 474–479.

Biederbeck, V. O., and Campbell, C. A. (1973). Soil microbial activity as influenced by temperature trends and fluctuations. *Can. J. Soil Sci.* **53**, 363–376.

Biederbeck, V. O., and Paul, E. A. (1971). Fungal degradation of soil humic nitrogen. *Agron. Abstr.*, p. 80.

Biederbeck, V. O., Campbell, C. A., and Nicholaichuk, W. (1977). Simulated dew formation and microbial growth in soil of a semi-arid region of Western Canada. *Can. J. Soil Sci.* **57**, 93–102.

Birch, H. F. (1960). Nitrification in soils after different periods of dryness. *Plant Soil* **12**, 81–96.

Birch, H. F. (1964). Mineralization of plant nitrogen following alternate wet and dry conditions. *Plant Soil* **20**, 43–49.

Bocock, K. L. (1964). Changes in the amount of dry matter, nitrogen, carbon and energy in decomposing woodland leaf litter in relation to the activities of the soil fauna. *J. Ecol.* **52**, 273–284.

Bollen, W. B. (1953). Mulches and soil conditioners. Carbon and nitrogen in farm and forest products. *J. Agric. Food Chem.* **7**, 379–381.

Bondietti, E., Martin, J. P., and Haider, K. (1971). Influence of nitrogen source and clay on growth and phenolic polymer production by *Stachybotrys* spp., *Hendersonula toruloidea* and *Aspergillus sydowi*. *Soil Sci. Soc. Am. Proc.* **35**, 917–922.

Bondietti, E., Martin, J. P., and Haider, K. (1972). Stabilisation of amino sugar units in humic-type polymers. *Soil Sci. Soc. Am. Proc.* **36**, 597–602.

Bormann, F. H., and Likens, G. E. (1979). "Patterns and Process in a Forested Ecosystem." Springer-Verlag, Berlin and New York.

Bornemisza, E., La Roche, F. A., and Fassbender, H. W. (1967). Effects of liming on some chemical characteristics of a Costa Rican Latosol. Proc.—*Soil Crop Sci. Soc. Fla.* **27**, 219–226.

Bremner, J. M. (1965a). Organic forms of nitrogen. In "Methods of Soil Analysis" (C. A. Black, ed.), pp. 1148–1178. Am. Soc. Agron., Madison, Wisconsin.

Bremner, J. M. (1965b). Organic nitrogen in soils. *In* "Soil Nitrogen" (W. V. Bartholomew and F. E. Clark, eds.), pp. 93–149. Am. Soc. Agron., Madison, Wisconsin.

Bremner, J. M. (1967). Nitrogenous compounds. *In* "Soil Biochemistry" (A. D. McLaren and G. H. Peterson, eds.), Vol. 1, pp. 19–66. Dekker, New York.

Bremner, J. M., and Mulvaney, R. L. (1978). Urease activity in soils. *In* "Soil Enzymes" (R. G. Burns, ed.), pp. 149–196. Academic Press, New York.

Broadbent, F. E. (1965). Effect of fertiliser nitrogen on the release of soil nitrogen. *Soil Sci. Soc. Am. Proc.* **29**, 692–696.

Broadbent, F. E. (1968). Nitrogen immobilisation in relation to N-containing fractions of soil organic matter. *In* "Isotopes and Radiation in Soil Organic Matter Studies," pp. 131–140. IAEA/FAO, Vienna.

Broadbent, F. E. (1970). Variables affecting A values as a measure of soil nitrogen availability. *Soil Sci.* **110**, 19–23.

Broadbent, F. E., and Nakashima, T. (1971). Effect of added salts on nitrogen mineralization in three California soils. *Soil Sci. Soc. Am. Proc.* **35**, 457–460.

Broadbent, F. E., and Nakashima, T. (1974). Mineralization of carbon and nitrogen in soil amended with carbon-13 and nitrogen-15 labelled plant material. *Soil Sci. Soc. Am. Proc.* **38**, 313–315.

Broadbent, F. E., and Tyler (1962). Laboratory and greenhouse investigations of nitrogen immobilization. *Soil Sci. Soc. Am. Proc.* **26**, 459–469.

Broadbent, F. E., Jackman, R. H., and McNicoll, J. (1964). Mineralization of carbon and nitrogen in some New Zealand allophanic soils. *Soil Sci.* **98**, 118–128.

Brown, P., and Dickey, D. D. (1970). Losses of wheat straw residue under stimulated field conditions. *Soil Sci. Soc. Am. Proc.* **34**, 118–121.

Bunnell, F. L., MacLean, S. F., and Brown, J. (1975). Barrow Alaska, USA. *In* "Structure and Function of Tundra Ecosystems" (T. Rosswall and O. W. Heal, eds.), pp. 73–124. Ecological Bulletins, Stockholm.

Burchill, S., Hayes, M. H. B., and Greenland, D. J. (1981). Adsorption. *In* "The Chemistry of Soil Processes" (D. J. Greenland and M. H. B. Hayes, eds.), pp. 221–400. Wiley, New York.

Burns, R. G. (1978). Enzyme activity in soil. Some theoretical and practical considerations. *In* "Soil Enzymes" (R. G. Burns, ed.), pp. 295–340. Academic Press, New York.

Calvet, R. (1981). Adsorption–desorption phenomena. *In* "Interactions between Herbicides and the Soil" (R. J. Hance, ed.), pp. 1–30. Academic Press, New York.

Campbell, C. A. (1978). Soil organic carbon, nitrogen and fertility. *In* "Soil Organic Matter" (M. Schnitzer and S. U. Khan, eds.), pp. 173–271. Am. Elsevier, New York.

Campbell, C. A., and Biederbeck, V. O. (1976). Soil bacterial changes as affected by growing season weather conditions: a field and laboratory study. *Can. J. Soil Sci.* **56**, 293–310.

Campbell, C. A., and Biederbeck, V. O. (1982). Changes in mineral N and numbers of bacteria and actinomycetes during two years under wheat-fallow in Southwestern Saskatchewan. *Can. J. Soil Sci.* **62**, 125–137.

Campbell, C. A., Biederbeck, V. O., and Warder, F. G. (1971). Influence of simulated fall and spring conditions on the soil system. II. Effect on soil nitrogen. *Soil Sci. Soc. Am. Proc.* **35**, 480–483.

Campbell, C. A., Biederbeck, V. O., Warder, F. G., and Robertson, G. W. (1973). Effect of rainfall and subsequent drying on nitrogen and phosphorus changes in a dryland fallow loam. *Soil Sci. Soc. Am. Proc.* **37**, 909–915.

Campbell, C. A., Biederbeck, V. O., and Hinman, W. C. (1975). Relationships between nitrate in summer-fallowed surface soil and some environmental variables. *Can. J. Soil Sci.* **55**, 213–223.

Carter, M. R., and Rennie, D. A. (1984). Dynamics of soil microbial N under zero and shallow tillage for spring wheat, using ^{15}N urea. *Plant Soil* **76**, 157–164.

Carter, P. W., and Mitterer, R. M. (1978). Amino acid composition of organic matter associated with carbonate and non-carbonate sediments. *Geochim. Cosmochim. Acta* **42**, 1231–1238.

Ceccanti, B., Nannipieri, P., Cervelli, S., and Sequi, P. (1978). Fractionation of humus–urease complexes. *Soil Biol. Biochem.* **10**, 39–45.

Chang, F. H., and Broadbent, F. E. (1982). Influence of trace metals on some soil nitrogen transformations. *J. Environ. Qual.* **11**, 1–4.

Charholm, M. (1981). Protozoan grazing of bacteria in soil-impact and importance. *Microb. Ecol.* **7**, 343–350.

Chen, A. W., and Griffin, D. M. (1966). Soil physical factors and the ecology of fungi. V. Further studies in relatively dry soils. *Trans. Br. Mycol. Soc.* **49**, 419–426.

Cheshire, M. V., Falshaw, C. P., Floyd, A. J., and Haworth, R. D. (1967). Humic acid. II. Structure of humic acids. *Tetrahedron* **23**, 1669–1682.

Cheshire, M. V., Mundie, C. M., and Shepherd, H. (1974). Transformation of sugars when rye hemicellulose labelled with ^{14}C decomposes in soil. *J. Soil Sci.* **25**, 90–98.

Clark, F. E. (1968). The growth of bacteria in soil. *In* "The Ecology of Soil Bacteria" (T. R. G. Gray and D. Parkinson, eds.), pp. 441–457. Liverpool Univ. Press, Liverpool.

Clement, C. R., and Williams, T. E. (1967). Leys and soil organic matter. II. The accumulation of nitrogen in soils under different leys. *J. Agric. Sci.* **69**, 133–138.

Coldwell, B. B., and Delong, W. A. (1950). Studies of the composition of deciduous forest tree leaves before and after partial decomposition. *Sci. Agric.* **30**, 456–466.

Coleman, D. C., Cole, C. V., Anderson, R. V., Blaha, M., Campion, M. K., Clarholm, M., Elliot, E. T., Hunt, H. W., Shaefer, B., and Sinclair, J. (1977). An analysis of rhizosphere–saprophage interactions in terrestrial ecosystems. *In* "Soil Organisms as Components of Ecosystems" (U. Lohm and T. Persson, eds.), pp. 299–309. Ecological Bulletins, Stockholm.

Cornish, P. S., and Raison, R. J. (1977). Effects of phosphorus and plants on nitrogen mineralisation in three grassland soils. *Plant Soil* **47**, 289–295.

Council for Agricultural Science and Technology (CAST). (1976). "Application of sewage sludge to cropland: appaisal of potential hazards of heavy metals to plants and animals." CAST Rep. No 64, Ames, Iowa.

Cowling, E. B., and Merrill, W. (1966). Nitrogen in wood and its role in wood deterioration. *Can. J. Bot.* **44**, 1539–1554.

Craswell, E. T., and Waring, S. A. (1972a). Effect of grinding on the decomposition of soil organic matter. I. The mineralisation of organic nitrogen in relation to soil type. *Soil Biol. Biochem.* **4**, 427–433.

Craswell, E. T., and Waring, S. A. (1972b). Effect of grinding on the decomposition of soil organic matter. II. Oxygen uptake and nitrogen mineralisation in virgin and cultivated cracking clay soils. *Soil Biol. Biochem.* **4**, 435–442.

Crossley, D. A. (1977). The roles of terrestrial saprophagous arthropods in forest soils: Current status of concepts. *In* "The Role of Arthropods in Forest Ecosystems" (M. J. Mattson, ed.), pp. 49–56. Springer-Verlag, Berlin and New York.

Dalenberg, J. W., and Jager, G. (1981). Priming effect of small glucose additions to ^{14}C-labelled soil. *Soil Biol. Biochem.* **13**, 219–223.

Davies, R. I., Coulson, C. B., and Lewis, D. A. (1964). Polyphenols in plant, humus and soil. IV. Factors leading to increased biosynthesis of polyphenol in leaves and their relationship to mull and mor formation. *J. Soil Sci.* **15**, 310–318.

de Boois, H. M. (1974). Measurement of seasonal variations in the oxygen uptake of various litters of an oak forest. *Plant Soil* **40**, 545–555.

De Haan, S. (1977). Humus, its formation, its relation with the mineral part of the soil and its significance for soil productivity. *In* "Soil Organic Matter Studies," Vol. 1, pp. 21–30. IAEA/FAO, Vienna.

Dickinson, C. H., and Pugh, G. J. F., eds. (1974). "Biology of Plant Litter Decomposition," Vols. 1 and 2. Academic Press, New York.

Doelman, P., and Haanstra, L. (1979). Effect of lead on the soil microflora. *Soil Biol. Biochem.* **11**, 487–491.

Dowding, P. (1976). Allocation of resources, nutrient uptake and release by decomposer organisms. *In* "The Role of Terrestrial and Aquatic Organisms in Decomposition Processes" (J. M. Anderson and A. Macfadyen, eds.), pp. 169–183. Blackwell, Oxford.

Dubey, H. (1970). A nitrogen deficiency disease of sugar cane probably caused by repeated pesticide applications. *Phytophathology* **60**, 485–487.

Edmeades, D. C., Judd, M., and Sarathchandra, S. U. (1981). The effect of lime on nitrogen mineralization as measured by grass growth. *Plant Soil* **60**, 177–186.

Edwards, A. P., and Bremner, J. M. (1967). Micro-aggregates in soils. *J. Soil Sci.* **18**, 64–73.

Edwards, C. A. (1974). Macroarthropods. *In* "Biology of Plant Litter Decomposition" (C. H. Dickinson and G. J. F. Pugh, eds.), Vol. 2, pp. 533–554. Academic Press, New York.

Edwards, C. A., and Lofty, J. F. (1977). "Biology of Earthworms." John Wiley, New York.

Edwards, C. A., Reichle, D. E., and Crossley, D. A. (1970). The role of invertebrates in turnover of organic matter and nutrients. *In* "Analysis of Temperate Forest Ecosystems" (D. E. Riechle, ed.), pp. 147–172. Springer-Verlag, Berlin and New York.

Filip, Z., Haider, K., and Martin, J. P. (1972a). Influence of clay minerals on growth and metabolic activity of *Epicoccum nigrum* and *Stachybotrys chartarum*. *Soil Biol. Biochem.* **4**, 135–145.

Filip, Z., Haider, K., and Martin, J. P. (1972b). Influence of clay minerals on the formation of humic substances by *Epicoccum nigrum* and *Stachybotrys chartarum*. *Soil Biol. Biochem.* **4**, 147–154.

Flaig, W., Beutelspacher, H., and Rietz, E. (1975). Chemical composition and physical properties of humic substances. *In* "Soil Components" (J. E. Gieseking, ed.), Vol. 1, pp. 1–211. Springer-Verlag, Berlin and New York.

Fleige, H., and Baeumer, K. (1974). Effect of zero-tillage on organic carbon and total nitrogen content, and their distribution in different N-fractions in loessial soils. *Agro-Ecosystems*, **1**, 19–29.

Floate, M. J. S. (1970a). Decomposition of organic materials from hill soils and pastures. II. Comparative studies of the mineralisation of carbon, nitrogen and phosphorus from plant materials and sheep faeces. *Soil Biol. Biochem.* **2**, 173–185.

Floate, M. J. S. (1970b). Decomposition of organic materials from hill soils and pastures. III. The effect of temperature on mineralisation of carbon, nitrogen and phosphorus from plant materials and sheep faeces. *Soil Biol. Biochem.* **2**, 187–196.

Floate, M. J. S. (1981). Effects of grazing by large herbivores on nitrogen cycling in agricultural ecosystems. *In* "Terrestrial Nitrogen Cycles: Processes, Ecosystem Strategies and Management Impacts" (F. E. Clark and T. Rosswall, eds.), pp. 585–601. Ecological Bulletins, Stockholm.

Fogel, R., and Cromack, K. (1977). Effect of habitat and substrate quality on Douglas Fir litter decomposition in Western Oregon. *Can. J. Bot.* **55**, 1632–1640.

Führ, F., and Sauerbeck, D. (1968). Decomposition of wheat straw in the field as influenced by cropping and rotation. *In* "Isotopes and Radiation in Soil Organic Matter Studies," pp. 241–250. IAEA/FAO, Vienna.

Gadd, G. M., and Griffiths, A. J. (1978). Micro-organisms and heavy metal toxicity. *Microb. Ecol.* **4**, 303–317.

Gadet, R., and Soubies, L. (1965). Apparent and real balance (measured by ^{15}N) of inorganic nitrogen from fertilizers applied to the soil. *Meded. Landbouwogesch. Opzoekingstn. Staat Gent* **30**, 1241–1253.

Gasser, J. K. R. (1958). Use of deep freezing in the preservation and preparation of fresh soil samples. *Nature (London)* **181**, 1334–1335.

Gillard, P. (1967). Coprophagous beetles in pasture ecosystems. *J. Aust. Inst. Agric. Sci.* **33**, 30–34.

Goh, K. M., and Edmeades, D.C. (1979). Distribution and partial characterization of acid hydrolysable organic nitrogen in six New Zealand soils. *Soil Biol. Biochem.* **11**, 127–132.

Goodfellow, M., and Cross, T. (1974). Actinomycetes. *In* "Biology of Litter Decomposition" (C. H. Dickinson and G. J. F. Pugh, eds.), Vol. 2, pp. 269–302. Academic Press, New York.

Goring, C. A. I., and Laskowski, D. A. (1982). The effects of pesticides on nitrogen transformations in soils. *In* "Nitrogen in Agricultural Soils" (F. J. Stevenson, ed.), pp. 689–720. Am. Soc. Agron., Madison, Wisconsin.

Gosz, J. R. (1981). Nitrogen cycling in coniferous ecosystems. *In* "Terrestrial Nitrogen Cycles: Processes, Ecosystem Strategies and Management Impacts" (F. E. Clark and T. Rosswall, eds.), pp. 405–426. Ecological Bulletins, Stockholm.

Gosz, J. R., Likens, G. E., and Bormann, F. H. (1973). Nutrient release from decomposing leaf and branch litter in the Hubbard Brook Forest, New Hampshire. *Ecol. Monogr.* **43**, 173–191.

Greenfield, L. G. (1972). The nature of organic nitrogen of soils. *Plant Soil* **36**, 191–198.

Greenland, D. J. (1971). Interactions between humic and fulvic acids and clays. *Soil Sci.* **111,** 34–41.

Habte, M., and Alexander, M. (1977). Further evidence for the regulation of bacterial populations in soil by protozoa. *Arch. Microbiol.* **113,** 181–183.

Habte, M., and Alexander, M. (1978). Protozoa density and the coexistence of protozoan predators and bacterial prey. *Ecology* **59,** 140–146.

Haider, K., and Martin, J. P. (1981). Decomposition in soil of specifically ^{14}C-labelled model and cornstalk lignins and coniferyl alcohol over two years as influenced by drying, re-wetting, and additions of an available substrate. *Soil Biol. Biochem.* **13,** 447–450.

Haider, K., Frederick, L. R., and Flaig, W. (1965). Reactions between amino acid compounds and phenols during oxidation. *Plant Soil* **22,** 49–64.

Hamaker, J. W., and Thompson, J. M. (1972). Adsorption. *In* "Organic Chemicals in the Soil Environment" (C. A. I. Goring and J. W. Hamaker, eds.), Vol. 1, pp. 49–143. Dekker, New York.

Hanlon, R. D. G., and Anderson, J. M. (1979). The effects of collembola grazing on microbial activity in decomposing leaf litter. *Oecologia* **38,** 93–99.

Harding, D. J. L., and Stuttard, R. A. (1974). Microarthropods. *In* "Biology of Plant Litter Decomposition" (C. H. Dickinson and G. J. F. Pugh, eds.), Vol. 2, pp. 489–532. Academic Press, New York.

Harrison, A. F. (1971). The inhibitory effect of oak leaf litter tannins on the growth of fungi in relation to litter decomposition. *Soil Biol. Biochem.* **3,** 167–172.

Hartman, L. M. (1974). A preliminary report: fungal flora of the soil as conditioned by varying concentrations of heavy metals. *Am. J. Bot.* **61,** 23.

Hayes, A. J. (1965). Studies on the decomposition of coniferous litter. I. Physical and chemical changes. *J. Soil Sci.* **16,** 121–139.

Hayes, M. H. B., and Swift, R. S. (1978). The chemistry of soil organic colloids. *In* "The Chemistry of Soil Constituents" (D. J. Greenland and M. H. B. Hayes, eds.), pp. 179–320. Wiley, Chichester.

Haynes, R. J., and Goh, K. M. (1978). Ammonium and nitrate nutrition of plants. *Biol. Rev. Cambridge Philos. Soc.* **53,** 465–510.

Heal, O. W., Swift, M. J., and Anderson, J. M. (1982). Nitrogen cycling in United Kingdom forests: The relevance of basic ecological research. *Philos. Trans. R. Soc. London, Ser. B* **296,** 427–444.

Heilman, P. (1975). Effect of added salts on nitrogen release and nitrate levels in forests soils of the Washington coastal area. *Soil Sci. Soc. Am. Proc.* **39,** 778–782.

Henningsson, B. (1967). Microbial decomposition of unpeeled Birch and Aspen pulpwood during storage. *Stud. For. Suec.* **54,** 1–32.

Hodkinson, I. D. (1975). Dry weight loss and chemical changes in vascular plant litter of terrestrial origin, occurring in a beaver pond ecosystem. *J. Ecol.* **63,** 131–142.

Hole, F. D. (1981). Effects of animals on soil. *Geoderma* **25,** 75–112.

Howard, P. J. A., and Howard, D. M. (1974). Microbial decomposition of tree and shrub leaf litter. I. Weight loss and chemical composition of decomposing litter. *Oikos* **25,** 341–352.

Huntjens, J. L. M. (1971). The influences of living plants on mineralisation and immobilization of nitrogen. *Plant Soil* **35,** 77–94.

Huntjens, J. L. M. (1972). Amino acid composition of humic acid-like polymers produced by Streptomycetes and of humic acids from pastures and arable land. *Soil Biol. Biochem.* **4,** 339–345.

Huntjens, J. L. M., and Albers, R. A. J. M. (1978). A model experiment to study the influence of living plants on the accumulation of soil organic matter in pastures. *Plant Soil* **50**, 411–418.

Hurst, H. M., and Burges, N. A. (1967). Lignin and humic acids. *In* "Soil Biochemistry" (A. D. McLaren and G. H. Peterson, eds.), Vol. 1, pp. 260–286. Dekker, New York.

Ineson, P., Leonard, M. A., and Anderson, J. M. (1982). Effect of collembolan grazing upon nitrogen and cation leaching from decomposing leaf litter. *Soil Biol. Biochem.* **14**, 601–605.

Isirimah, N. O., and Keeney, D. R. (1973). Nitrogen transformations in aerobic and water-logged histosols. *Soil Sci.* **115**, 123–129.

Ivarson, K. C., and Schnitzer, M. (1979). The biodegradability of the "unknown" soil nitrogen. *Can. J. Soil Sci.* **59**, 59–67.

Jansson, S. L. (1958). Tracer studies on nitrogen transformations in soil with special attention to mineralisation–immobilisation relationships. *Lantbruksoegsk. Ann.* **24**, 101–361.

Jansson, S. L. (1971). Use of ^{15}N in studies of soil nitrogen. *In* "Soil Biochemistry" (A. D. McLaren and T. Skujins, eds.), Vol. 2, pp. 129–166. Dekker, New York.

Jansson, S. L., and Persson, J. (1982). Mineralization and immobilization of soil nitrogen. *In* "Nitrogen in Agricultural Soils" (F. J. Stevenson, ed.), pp. 229–252. Am. Soc. Agron., Madison, Wisconsin.

Jenkinson, D. S. (1965). Studies on the decomposition of plant material in soil. I. Losses of carbon from ^{14}C labelled ryegrass incubated with soil in the field. *J. Soil Sci.* **16**, 104–115.

Jenkinson, D. S. (1971). Studies on the decomposition of ^{14}C labelled organic matter in soil. *Soil Sci.* **111**, 64–70.

Jenkinson, D. S. (1977a). Studies on the decomposition of plant material in soil. IV. The effect of rate of addition. *J. Soil Sci.* **28**, 417–423.

Jenkinson, D. S. (1977b). Studies on the decomposition of plant material in soil. V. The effects of plant cover and soil type on the loss of carbon from ^{14}C labelled ryegrass decomposing under field conditions. *J. Soil Sci.* **28**, 424–434.

Jenkinson, D. S. (1981). The fate of plant and animal residues in soil. *In* "The Chemistry of Soil Processes" (D. J. Greenland and M. H. B. Hays, eds.), pp. 505–561. Wiley, New York.

Jenkinson, D. S., and Ayanaba, A. (1977). Decomposition of carbon-14 labelled plant material under tropical conditions. *Soil Sci. Soc. Am. J.* **41**, 912–915.

Jenkinson, D. S., and Ladd, J. N. (1981). Microbial biomass in soil: Measurement and turnover. *In* "Soil Biochemistry" (E. A. Paul and J. N. Ladd, eds.), Vol. 5, pp. 415–471. Dekker, New York.

Jenkinson, D. S., and Powlson, D. S. (1970) Residual effects of soil fumigation on soil respiration and mineralization. *Soil Biol. Biochem.* **2**, 99–108.

Jenny, H. (1941). "Factors of Soil Formation." McGraw-Hill, New York.

Jensen, V. (1974). Decomposition of angiosperm tree leaf litter. *In* "Biology of Plant Litter Decomposition" (C. H. Dickinson and G. J. F. Pugh, eds.), Vol. 1, pp. 69–104. Academic Press, New York.

Jones, L. H. P., and Jarvis, S. C. (1981). The fate of heavy metals. *In* "The Chemistry of Soil Processes" (D. J. Greenland and M. H. B. Hayes, eds.), pp. 593–620. Wiley, New York.

Kaarik, A. A. (1974). Decomposition of wood. *In* "Biology of Plant Litter Decomposition" (C. H. Dickinson and G. J. F. Pugh, eds.), Vol. 1, pp. 129–174. Academic Press, New York.

Kadirgamathaiyah, S., and MacKenzie, A. F. (1970). A study of soil nitrogen organic fractions and correlation with yield response of Sudan-Sorghum hybrid grass on Quebec soils. *Plant Soil* **33**, 120–128.

Kai, H., Ahmad, Z., and Harada, T. (1973). Factors affecting immobilisation and release of nitrogen in soil and chemical characteristics of the nitrogen newly immobilised. III. *Soil Sci. Plant Nutr.* **19**, 275–286.

Kanatchinova, M. K. (1969). Effect of molybdenum-containing industrial wastes on ammonification and nitrification. *Tr. Inst. Mikrobiol. Virusol., Akad. Nauk Kaz. SSR* **12**, 84–87.

Karenlampi, L. (1971). Weight loss of leaf litter on forest soil surface in relation to weather at Kevo Station, Finish Lapland. *Rep. Kevo Subarct. Res. Stn.* **8**, 101–103.

Keeney, D. R., and Bremner, J. M. (1964). Effect of cultivation on the nitrogen distribution in soils. *Soil Sci. Soc. Am. Proc.* **28**, 653–656.

Khandelwal, K. C., and Gauer, A. C. (1980). Degradation of humic acids, extracted from manure and soil, by some streptomycetes and fungi. *Zentralbl. Bakteriol., Parasitend., Infektionskr. Hyg., Abt. 2, Naturwiss: Mikrobiol. Landwirtsch. Technol. Umweltschutzes* **135**, 119–122.

King, H. G. C., and Heath, G. W. (1967). The chemical analysis of small samples of leaf material and the relationship between the disappearance and composition of leaves. *Pedobiologia* **7**, 192–197.

Kononova, M. M. (1966). "Soil Organic Matter." Pergamon, Oxford.

Kowalenko, C. G. (1978). Organic nitrogen, phosphorus and sulfur in soils. *In* "Soil Organic Matter" (M. Schnitzer and S. U. Khan, eds.), pp. 95–136. Am. Elsevier, New York.

Ladd, N. J. (1978). Origin and range of enzymes in soil. *In* "Soil Enzymes" (R. G. Burns, ed.), pp. 51–96. Academic Press, New York.

Ladd, J. N., and Butler, J. H. A. (1969). Inhibition of proteolytic enzyme activities by soil humic acids. *Aust. J. Soil Res.* **7**, 253–261.

Ladd, J. N., and Jackson, R. B. (1982). Biochemistry of ammonification. *In* "Nitrogen in Agricultural Soils" (F. J. Stevenson, ed.), pp. 173–228. Am. Soc. Agron., Madison, Wisconsin.

Ladd, J. N., and Paul, E. A. (1973). Changes in enzymatic activity and distribution of acid-soluble, amino acid-nitrogen in soil during nitrogen immobilisation and mineralisation. *Soil. Biol. Biochem.* **5**, 825–840.

Ladd, J. N., Brisbane, P. G., Butler, J. H. A., and Amato, M. (1976). Studies on soil fumigation. III. Effects on enzyme activities, bacterial numbers and extractable ninhydrin reactive compounds. *Soil Biol. Biochem.* **8**, 255–260.

Ladd, J. N., Oades, J. M., and Amato, M. (1981). Microbial biomass formed from ^{14}C, ^{15}N-labelled plant material decomposing in soils in the field. *Soil Biol. Biochem.* **13**, 119–126.

Lamb, D. (1975). Patterns of nitrogen mineralisation in the forest floor of stands of *Pinus radiata* on different soils. *J. Ecol.* **63**, 615–625.

Laura, R. D. (1974). Effects of neutral salts on carbon and nitrogen mineralisation of organic matter in soil. *Plant Soil* **41**, 113–127.

Laura, R. D. (1977). Salinity and nitrogen mineralisation in soil. *Soil Biol. Biochem.* **9**, 333–336.

Legg, J. O., and Stanford, G. (1967). Utilisation of soil and fertiliser–by oats in relation to the available N status of soils. *Soil Sci. Soc. Am. Proc.* **31**, 215–219.

Lewis, J. A., and Starkey, R. L. (1968). Vegetable tannins; their decomposition and effects on decomposition of some organic compounds. *Soil Sci.* **106**, 241–247.

Liang, C. N., and Tabatabai, M. A. (1977). Effects of trace elements on nitrogen mineralisation. *Environ. Pollut.* **12**, 141–147.

Liu, S. Y., Minard, R. D., and Bollag, J. M. (1981). Oligomerization of syringic acid, a lignin derivative, by phenoloxidase. *Soil Sci. Soc. Am. J.* **45**, 1100–1105.

Lofty, J. R. (1974). Oligochaetes. *In* "Biology of Plant Litter Decomposition" (C. H. Dickinson and G. J. F. Pugh, eds.), Vol. 2, pp. 467–488. Academic Press, New York.

McGill, W. B. (1971). Turnover of microbial metabolites during nitrogen mineralisation and immobilisation in soil. Ph.D. Thesis, University of Saskatchewan, Saskatoon.

Mack, A. R. (1962). Low temperature research on nitrate release from soil. *Nature (London)* **193**, 803–804.

Mack, A. R. (1963). Biological activity and mineralisation of nitrogen in three soils as induced by freezing and drying. *Can. J. Soil Sci.* **43**, 316–324.

McLaren, A. D., and Peterson, G. H. (1965). Physical chemistry and biological chemistry of clay mineral–organic nitrogen complexes. *In* "Soil Nitrogen" (W. V. Bartholomew and F. E. Clark, eds.), pp. 261–284. Am. Soc. Agron., Madison, Wisconsin.

Mahaptra, B., Patnaik, B., and Mishra, D. (1977). The exocellular urease in rice roots. *Curr. Sci.* **46**, 680–681.

Mahendrappa, M. K. (1978). Changes in the organic layers under a black spruce stand fertilised with urea and triple superphosphate. *Can. J. For. Res.* **8**, 237–242.

Marsh, J. A. P., and Greaves, M. P. (1979). The influence of temperature and moisture on the effects of the herbicide dalapon on nitrogen transformations in soil. *Soil Biol. Biochem.* **11**, 279–285.

Marshman, N. A., and Marshall, K. C. (1981). Bacterial growth on proteins in the presence of clay minerals. *Soil Biol. Biochem.* **13**, 127–134.

Martin, J. K. (1977). Factors influencing the loss of organic carbon from wheat roots. *Soil Biol. Biochem.* **9**, 1–7.

Martin, J. P., and Focht, D. D. (1977). Biological properties of soils. *In* "Soils for Management of Organic Wastes and Waste Waters," (L. F. Elliot and F. J. Stevenson, eds.), pp. 29–47. Am. Soc. Agron., Madison, Wisconsin.

Martin, J. P., and Haider, K. (1969). Phenolic polymers of *Stachybotrys atra*, *Stachybotrys chartarum* and *Epicoccum nigrum* in relation to humic acid formation. *Soil Sci.* **107**, 260–270.

Martin, J. P., and Haider, K. (1971). Microbial activity in relation to soil humus formation. *Soil Sci.* **111**, 54–63.

Martin, J. P., and Haider, K. (1977). Decomposition in soil of specifically ^{14}C-labelled DHP and cornstalk lignins, model humic acid-type polymers and coniferyl alcohols. *In* "Soil Organic Matter Studies," Vol. 2, pp. 23–32. IAEA/FAO, Vienna.

Martin, J. P., and Haider, K. (1979). Biodegradation of ^{14}C-labelled model and cornstalk lignins, phenols, model phenolase humic polymers, and fungal melanins as influenced by a readily available carbon source and soil. *Appl. Environ. Microbiol.* **38**, 283–289.

Martin, J. P., and Haider, K. (1980). A comparison of the use of phenolase and peroxidase for the synthesis of model humic acid-type polymers. *Soil Sci. Soc. Am. J.* **44**, 983–988.

Martin, J. P., Filip, Z., and Haider, K. (1976). Effect of montmorillonite and humate on growth and metabolic activity of some actinomycetes. *Soil Biol. Biochem.* **8**, 409–413.

Martin, J. P., Haider, K., and Linhares, L. (1979). Decomposition and stabilisation of ring-^{14}C-labelled catechol in soil. *Soil Sci. Soc. Am. J.* **43**, 100–104.

Martin, J. P., Haider, K., and Kassim, G. (1980). Biodegradation and stabilisation after 2 years of specific crop lignin and polysaccharide carbons in soils. *Soil Sci. Soc. Am. J.* **44**, 1250–1255.

Marumoto, T., Furukawa, K., Yoshida, T., Kai, H., Yamada, Y., and Harada, T. (1974). Contribution of microbial cells and their cell walls to an accumulation of the soil organic matter becoming decomposable due to drying a soil (Part 1). *J. Sci. Soil Manure, Jpn.* **45**, 23–28.

Marumoto, T., Kai, H., Yoshida, T., and Harada, T. (1977a). Relationship between an accumulation of soil organic matter becoming decomposable due to drying of soil and microbial cells. *Soil Sci. Plant Nutr.* **23**, 1–8.

Marumoto, T., Kai, H., Yoshida, T., and Harada, T. (1977b). Drying effect on mineralisation of microbial cells and their cell walls in soil and contribution of microbial cell walls as a source of decomposable soil organic matter. *Soil Sci. Plant Nutr.* **23**, 9–19.

Marumoto, T., Kai, H., Yoshida, T., and Harada, T. (1977c). Chemical fractions of organic nitrogen in acid hydrolysates given from microbial cells and their cell wall substances and characterisation of decomposable soil organic nitrogen due to drying. *Soil Sci. Plant Nutr.* **23**, 125–134.

Marumoto, T., Anderson, J. P. E., and Doms, K. H. (1982a). Decomposition of ^{14}C and ^{15}N-labelled microbial cells in soil. *Soil. Biol. Biochem.* **14**, 461–467.

Marumoto, T., Anderson, J. P. E., and Domsch, K. H. (1982b). Mineralisation of nutrients from soil microbial biomass. *Soil Biol. Biochem.* **14**, 469–475.

Mattson, S. (1932). The laws of colloidal behaviour. VII. Proteins and proteinated complexes. *Soil Sci.* **33**, 41–72.

Mayaudon, J. (1969). Stabilisation biologique des proteins ^{14}C dans le sol. *In* "Isotopes and Radiation in Soil Organic Matter Studies," pp. 177–188. IAEA/FAO, Vienna.

Mayaudon, J., Batistic, L., and Sarkar, J. M. (1975). Properties of proteolytically active extracts of fresh soil. *Soil Biol. Biochem.* **7**, 281–286.

Meints, V. W., and Peterson, G. A. (1977). The influence of cultivation on the distribution of nitrogen in soils of the Ustoll suborder. *Soil Sci.* **124**, 334–342.

Melillo, J. M., Aber, J. D., and Muratore, J. F. (1982). Nitrogen and lignin control of hardwood leaf litter decomposition dynamics. *Ecol. Monogr.* **63**, 621–626.

Miller, L. P., ed. (1973). "Phytochemistry" Vols. I, II, and III. Van Nostrand-Reinhold, Princeton, New Jersey.

Miller, R., and Johnson, D. D. (1964). The effect of soil moisture tension on carbon dioxide evolution, nitrification and nitrogen mineralisation. *Soil Sci. Soc. Am. Proc.* **28**, 644–647.

Monib, M., Hosny, I., Zohdy, L., and Khalafallah, M. (1981). Studies on humic acid decomposing Streptomycetes. III. Synthesis of amino acids and organic acids during humic acid decomposition. *Zentralbl. Bakteriol., Parasitend., Infektionskr. Hyg., Abt. 2, Naturwiss.: Mikrobiol. Landwirtsch., Technol. Umweltschutzes* **136**, 189–197.

Moore, A. W. (1974). Availability of Rhodesgrass (*Choris gayana*) of nitrogen in tops and roots added to soil. *Soil Biol. Biochem.* **6**, 249–255.

Moore, A. W., and Russell, J. S. (1970). Changes in chemical fractions of nitrogen during incubation of soils with histories of large organic matter increase under pasture. *Aust. J. Soil Res.* **8**, 21–30.

Morill, L. G., Mahilum, B. C., and Mohiuddin, H. C. (1982). "Organic Compounds in Soil: Sorption, Degradation and Persistence." Ann Arbor Sci. Publ., Ann Arbor, Michigan.

Mortland, M. M. (1970). Clay organic complexes and interactions. *Adv. Agron.* **22**, 75–117.

Mulvaney, R. L., and Bremner, J. M. (1981). Control of urea transformations in soils. *In*

"Soil Biochemistry" (E. A. Paul and J. N. Ladd, eds.), Vol. 5, pp. 153–196. Dekker, New York.

Munevar, F., and Wallum, A. G. (1977). Effects of the addition of phosphorus and inorganic nitrogen on carbon and nitrogen mineralisation in Andepts from Colombia. *Soil Sci. Soc. Am. J.* **41**, 540–545.

Myers, R. J. K. (1975). Temperature effects on ammonification and nitrification in a tropical soil. *Soil Biol. Biochem.* **7**, 83–86.

Myers, R. J. K., Campbell, C. A., and Weier, K. L. (1982). Quantitative relationship between net nitrogen mineralisation and moisture content of soils. *Can. J. Soil Sci.* **62**, 111–124.

Nelson, D. W., Martin, J. P., and Ervin, J. O. (1979). Decomposition of microbial cells and components in soil and their stabilisation through complexing with model humic acid-type phenolic polymers. *Soil Sci. Soc. Am. J.* **43**, 84–88.

Nommik, H. (1969). Nitrogen mineralisation and turnover in Norway spruce (*Picea abies* (L.) Karst.) raw humus as influenced by liming. *Trans. Int. Congr. Soil Sci., 9th, 1968*, Vol. 2, pp. 533–545.

Nommik, H. (1978). Mineralisation of carbon and nitrogen in forest humus as influenced by additions of phosphate and lime. *Acta Agric. Scand.* **28**, 221–230.

Novack, R. D., and Whittingham, W. F. (1968). Soil and litter microfungi of a maple–elm– ash flood plain community. *Mycologia* **60**, 776–787.

Nyborg, M., and Hoyt, P. B. (1978). Effects of soil acidity and liming on mineralisation of soil nitrogen. *Can. J. Soil Sci.* **58**, 331–338.

Nyhan, J. W. (1975). Decomposition of carbon-14-labelled plant materials in a grassland soil under field conditions. *Soil Sci. Soc. Am. Proc.* **39**, 643–648.

Nykvist, N. (1959). Leaching and decomposition of litter. I. Experiment of leaf litter of *Fraxinus excelsior*. *Oikos* **10**, 190–211.

Nykvist, N. (1963). Leaching and decomposition of water soluble organic substances from different types of leaf and needle litter. *Stud. For. Suec.* **3**, 1–31.

Oberlander, H. E. (1973). The fate of organic manures in soil as traced by means of radiocarbon. *Pontif. Acad. Sci. Scr. Varia* **38**, 1001–1071.

Olson, J. S. (1963). Energy storage and balance of producers and decomposers in ecological systems. *Ecology* **44**, 322–331.

Osborne, G. J. (1977). Chemical fractionation of soil nitrogen in six soils from southern New South Wales. *Aust. J. Soil Res.* **15**, 159–165.

Parkinson, D., Visser, S., and Whittaker, J. B. (1979). Effects of collembolan grazing on fungal colonisation of leaf litter. *Soil Biol. Biochem.* **11**, 529–535.

Parsons, J. W. (1981). Chemistry and distribution of amino sugars in soils and soil organisms. *In* "Soil Biochemistry" (E. A. Paul and J. N. Ladd, eds.), Vol. 5, pp. 197–227. Dekker, New York.

Parsons, J. W., and Tinsley, J. (1975). Nitrogenous substances. *In* "Soil Components" (J. E. Gieseking, ed.), pp. 263–304. Springer-Verlag, Berlin and New York.

Patrick, W. H. (1982). Nitrogen transformations in submerged soils. *In* "Nitrogen in Agricultural Soils" (F. J. Stevenson, ed.), pp. 449–466. Am. Soc. Agron., Madison, Wisconsin.

Paul, E. A. (1976). Nitrogen in terrestrial ecosystems. *In* "Environmental Biogeochemistry, Vol. 1: Carbon, nitrogen, phosphorus, sulphur and selenium cycles" (J. O. Nriagu, ed.), pp. 225–243. Ann Arbor Science, Ann Arbor, Michigan.

Paul, E. A. (1984). Dynamics of organic matter in soils. *Plant Soil* **76**, 275–285.

Paul, E. A., and Mathur, S. P. (1967). Cleavage of humic acids by *Penicillium frequentans*. *Plant Soil* **27**, 297–299.

Piper, T. J., and Posner, A. M. (1972). Humic acid nitrogen. *Plant Soil* **36**, 595–598.

Power, J. F. (1981). Nitrogen in the cultivated ecosystem. *In* "Terrestrial Nitrogen Cycles: Processes, Ecosystem Strategies and Management Impacts" (F. E. Clark and T. Rosswall, eds.), pp. 529–546. Ecological Bulletins, Stockholm.

Premi, P. R., and Cornfield, A. H. (1969). Effects of additives of copper, manganese, zinc and chromium compounds on ammonification and nitrification during incubation of soil. *Plant Soil* **31**, 345–352.

Pugh, G. J. F. (1974). Terrestrial fungi. *In* "Biology of Plant Litter Decomposition" (C. H. Dickinson and G. J. F. Pugh, eds.), Vol. 2, pp. 303–336. Academic Press, New York.

Quarishii, M. S. I., and Cornfield, A. H. (1973). Incubation study of nitrogen mineralisation and nitrification in relation to soil pH and level of copper (II) addition. *Environ. Pollut.* **4**, 159–163.

Rao, A. S., and Ghosh, A. B. (1981). Effect of continuous cropping and fertilizer use on the organic nitrogen fractions in a Typic Ustochrept soil. *Plant Soil* **62**, 377–383.

Reichle, D. E. (1977). The role of soil invertebrates in nutrient cycling. *In* "Soil Organisms as Components of Ecosystems" (U. Lohm and T. Persson, eds.), pp. 145–156. Ecological Bulletins, Stockholm.

Reichman, G. A., Grunes, D. L., and Viets, F. G. (1966). Effect of soil moisture on ammonification and nitrification in two northern plains soils. *Soil Sci. Soc. Am. Proc.* **30**, 363–366.

Reithel, F. J. (1971). Ureases. *In* "The Enzymes" (P. D. Boyers, ed.), 3rd ed., Vol. 4, pp. 1–21. Academic Press, New York.

Robinson, J. B. D. (1957). The critical relationship between soil moisture content in the region of the wilting point and mineralisation of native soil nitrogen. *J. Agric. Sci.* **49**, 100–105.

Rorison, I. H. (1980). The effects of soil acidity on nutrient availability and plant response. *In* "Effects of Acid Precipitation on Terrestrial Ecosystems" (T. C. Hutchinson and M. Havas, eds.), pp. 283–304. Plenum, New York.

Rother, J. A., Millbank, J. W., and Thornton, I. (1982). Effects of heavy-metal additions on ammonification and nitrification in soils contaminated with cadmium, lead and zinc. *Plant Soil* **69**, 239–258.

Rovira, A. D., and Greacen, E. L. (1957). The effect of aggregate disruption on the activity of micro-organisms in the soil. *Aust. J. Agric. Res.* **8**, 659–673.

Ruocco, J. J., and Barton, L. L. (1978). Energy-driven uptake of humic acids by *Aspergillus niger. Can. J. Microbiol.* **24**, 533–536.

Sabey, B. R. (1969). Influence of soil moisture tension on nitrate accumulation in soils. *Soil Sci. Soc. Am. Proc.* **33**, 263–266.

Sabey, B. R., Bartholomew, W. V., Shaw, R., and Pesek, J. (1956). Influence of temperature on nitrification in soils. *Soil Sci. Soc. Am. Proc.* **20**, 357–360.

Saiz-Jimenez, C., Haider, K., and Martin, J. P. (1975). Anthraquinones as intermediates in the formation of dark-coloured humic acid-like pigments by *Eurotium echinulatum. Soil Sci. Soc. Am. Proc.* **39**, 649–653.

Saphozhnikov, N. A., Nesterova, E. I., Rusinova, I. P., Sirota, L. B., and Vanova, T. K. L. (1969). The effect of fertiliser nitrogen on plant uptake of nitrogen from different podzolic soils. *Trans. Int. Congr. Soil Sci., 9th, 1968,* Vol. 2, pp. 467–474.

Sarkar, A. N., and Wyn Jones, R. G. (1980). Rhizosphere and its effect on the nutrient availability of plants—a review. *Agric. Rev.* **1**, 1–18.

Satchell, J. E., and Lowe, D. G. (1967). Selection of leaf litter by *Lumbricus terrestris. In* "Progress in Soil Biology" (O. Graff and J. E. Satchell, eds.), pp. 102–119. North-Holland Publ., Amsterdam.

Sauerbeck, D. (1966). A critical evaluation of incubation experiments on the priming effect of green manure. *In* "The Use of Isotopes in Soil Organic Matter Studies, Rep. FAO/IAEA Tech. Meet., pp. 209–221. Pergamon, Oxford.

Sauerbeck, D. R., and Gonzalez, M. A. (1977). Field decomposition of carbon-14 labelled plant residues in various soils of the Federal Republic of Germany and Costa Rica. *In* "Soil Organic Matter Studies," Vol. 1, pp. 159–170. IAEA/FAO, Vienna.

Sauerbeck, D. R., Johnen, B., and Massen, G. G. (1972). Der Abbau von ^{14}C-markiertem Pflanzenmaterial in verschiedenen Boden. *Agrochimica* **16**, 62–76.

Sequi, P., Guildi, G., and Petruzzelli, G. (1975). Distribution of amino acid and carbohydrate components in fulvic acid fractionated on polyamide. *Can. J. Soil Sci.* **55**, 439–445.

Shields, J. A., and Paul, E. A. (1973). Decomposition of ^{14}C-labelled plant material under field conditions. *Can. J. Soil Sci.* **53**, 297–306.

Shields, J. A., Paul, E. A., and Low, W. E. (1974). Factors influencing the stability of labelled microbial materials in soils. *Soil Biol. Biochem.* **6**, 31–37.

Sims, J. L., and Frederick, L. R. (1970). Nitrogen immobilisation and decomposition of corn residue in soil and sand as affected by residue particle size. *Soil Sci.* **109**, 355–361.

Singh, B. R., Agarwal, A. S., and Kanehiro, Y. (1969). Effect of chloride salts on ammonium nitrogen release in two Hawaiian soils. *Soil Sci. Soc. Am. Proc.* **33**, 557–560.

Singh, B. R., Uriyo, A. P., and Lontu, B. J. (1978). Distribution and stability of organic forms of nitrogen in forest soil profiles in Tanzania. *Soil Biol. Biochem.* **10**, 105–108.

Singh, B. R., Uriyo, A. P., and Tiisekwa, B. P. M. (1981). Forms of nitrogen in cultivated soil profiles in Tanzania. *Soil Biol. Biochem.* **13**, 441–446.

Sjoblad, R. D., and Bollag, J. M. (1981). Oxidative coupling of aromatic compounds by enzymes from soil micro-organisms. *In* "Soil Biochemistry" (E. A. Paul and J. N. Ladd, eds.), Vol. 5, pp. 113–152. Dekker, New York.

Smith, J. H., and Douglas, C. L. (1968). Influence of residual nitrogen on wheat straw decomposition in the field. *Soil Sci.* **106**, 456–459.

Smith, J. H., and Douglas, C. L. (1971). Wheat, straw decomposition in the field. *Soil Sci. Soc. Am. Proc.* **35**, 269–273.

Smith, S. J., Chichester, F. W., and Kissel, D. E. (1978). Residual forms of fertilizer nitrogen in field soils. *Soil Sci.* **125**, 165–169.

Sørensen, L. H. (1967). Duration of amino acid metabolites formed in soils during decomposition of carbohydrates. *Soil Sci.* **104**, 234–241.

Sørensen, L. H. (1972). Stabilisation of newly formed amino acid metabolites in soil by clay minerals. *Soil Sci.* **114**, 5–11.

Sørensen, L. H. (1974). Rate of decomposition of organic matter in soil as influenced by repeated air drying–rewetting and repeated additions of organic material. *Soil Biol. Biochem.* **6**, 287–292.

Sørensen, L. H. (1977). Factors affecting the biostability of metabolic materials in soil. *In* "Soil Organic Matter Studies," Vol. 2, pp. 3–14. IAEA/FAO, Vienna.

Sørensen, L. H. (1981). Carbon–nitrogen relationships during the humification of cellulose in soils containing different amounts of clay. *Soil Biol. Biochem.* **13**, 313–321.

Sørensen, L. H. (1982). Mineralisation of organically bound nitrogen in soil as influenced by plant growth and fertilisation. *Plant Soil* **65**, 51–61.

Soulides, D. A., and Allison, F. E. (1961). Effect of drying and freezing soils on carbon dioxide production, available mineral nutrients, aggregates and bacterial population. *Soil Sci.* **91**, 291–298.

Sowden, F. J., Chen, Y., and Schnitzer, M. (1977). The nitrogen distribution in soils formed under widely differing climatic conditions. *Geochim. Cosmochim. Acta* **41**, 1524–1526.

Staaf, H., and Berg, B. (1977). Mobilisation of plant nutrients in a Scots pine forest mor in Central Sweden. *Silva Fenn.* **11**, 210–217.

Staaf, H., and Berg, B. (1981). Plant litter input to soil. *In* "Terrestrial Nitrogen Cycles: Processes, Ecosystem Strategies and Management Impacts" (F. E. Clark and T. Rosswall, eds.), pp. 147–167. Ecological Bulletins, Stockholm.

Stanford, G., and Epstein, E. (1974). Nitrogen mineralization–water relations in soils. *Soil Sci. Soc. Am. Proc.* **38**, 103–107.

Stanford, G., Frere, M. H., and Schwaninger, D. H. (1973a). Temperature coefficient of soil nitrogen mineralization. *Soil Sci.* **115**, 321–323.

Stanford, G., Legg, J. O., and Smith, S. J. (1973b). Soil nitrogen availability evaluation potentials of soils and uptake of labelled and unlabelled nitrogen by plants. *Plant Soil* **39**, 113–124.

Stanford, G., Frere, M. H., and Vanderpol, R. A. (1975). Effect of fluctuating temperature on soil nitrogen mineralisation. *Soil Sci.* **119**, 222–226.

Steinbrenner, K., and Mundstock, I. (1975). Untersuchungen zum Bituminstaff Abbau durch Nokardien. *Arch. Acker-Pflanzenbau Bodenkd.* **19**, 243–255.

Sterritt, R. M., and Lester, J. N. (1980). Interactions of heavy metals with bacteria. *Sci. Total Environ.* **14**, 5–17.

Stevenson, F. J. (1965). Origin and distribution of nitrogen in soil. *In* "Soil Nitrogen" (W. V. Bartholomew and F. E. Clark, eds.), pp. 1–42. Am. Soc. Agron., Madison, Wisconsin.

Stevenson, F. J. (1982a). "Humus Chemistry. Genesis, Composition, Reactions." Wiley, New York.

Stevenson, F. J. (1982b). Organic forms of soil nitrogen. *In* "Nitrogen in Agricultural Soils" (F. J. Stevenson, ed.), pp. 67–122. Am. Soc. Agron., Madison, Wisconsin.

Stevenson, F. J., and Wagner, G. H. (1970). Chemistry of nitrogen in soils. *In* "Agricultural Practices and Water Quality," pp. 125–141. Iowa State Univ. Press, Ames.

Stewart, B. A., Johnson, D. D., and Porter, L. K. (1963). The availability of fertiliser nitrogen immobilised during decomposition of straw. *Soil Sci. Soc. Am. Proc.* **27**, 656–659.

Stotzky, G. (1967). Clay minerals and microbial ecology. *Trans. N. Y. Acad. Sci.* [2] **30**, 11–21.

Stout, J. D., Goh, K. M., and Rafter, T. A. (1981). Chemistry and turnover of naturally occurring resistant organic compounds in soil. *In* "Soil Biochemistry" (E. A. Paul and J. N. Ladd, eds.), Vol. 5. pp. 1–73. Dekker, New York.

Swift, M. J. (1973). The estimation of mycelial biomass by determination of hexosamine content of wood tissue decayed by fungi. *Soil Biol. Biochem.* **5**, 321–332.

Swift, M. J. (1976). Species diversity and the structure of microbial communities in terrestrial habits. *In* "The Role of Terrestrial and Aquatic Organisms in Decomposition Processes" (J. M. Anderson and A. Macfadyen, eds.), pp. 185–222. Blackwell, Oxford.

Swift, M. J. (1978). Growth of *Stereum hirsutum* during the long-term decomposition of oak branch-wood. *Soil Biol. Biochem.* **10**, 335–337.

Swift, M. J., Heal, O. W., and Anderson, J. M. (1979). "Decomposition in Terrestrial Ecosystems." Blackwell, Oxford.

Swift, R. S., and Posner, A. M. (1972). Autoxidation of humic acid under alkaline conditions. *J. Soil Sci.* **23**, 381–393.

Syers, J. K., Sharpley, A. N., and Keeney, D. R. (1979). Cycling of nitrogen by surface-casting earthworms in a pasture ecosystem. *Soil Biol. Biochem.* **11**, 181–185.

Taha, S. M., Zayed, M. N., and Zohdy, L. (1973). Studies in humic acids decomposing bacteria in soil. II. Isolation and identification of micro-organisms. *Z. Bakteriol., Parasitenkd., Infektionskr. Hyg., Abt. 2, Naturmiss.: Allg., Landwirtsch. Tech. Mikrobiol.* **128**, 168–172.

Tate, R. L. (1980). Microbial oxidation of organic matter of Histosols. *Adv. Microb. Ecol.* **4**, 169–201.

Theng, B. K. G. (1974). "The Chemistry of Clay–Organic Reactions." Hilger, London.

Theng, B. K. B. (1979). "Formation and Properties of Clay–Polymer Complexes." Elsevier, Amsterdam.

Toth, J. A., Papp, L. B., and Lenkey, B. (1974). Litter decomposition in an oak forest ecosystem (*Quercetum petraea cerris*) of Northern Hungary studied in the framework of "Sikfokut project." *In* "Biodegradation et Humification" (G. Kilbertus, O. Reisinger, A. Mouray, and J. A. Cansela de Fonesca, eds.), pp. 41–58. Pierron Editeur, Sarreguimenes.

Tsutsuki, K., and Kuwatsuka, S. (1978). Chemical studies on soil humic acids. III. Nitrogen distribution in humic acids. *Soil Sci. Plant Nutr.* **24**, 561–570.

Tyler, G. (1975). Heavy metal pollution and mineralisation of nitrogen in forest soils. *Nature (London)* **255**, 701–702.

Van Cleve, K., and Sprague, D. (1971). Respiration rates in the forest floor of birch and aspen stands in interior Alaska. *Arct. Alp. Res.* **3**, 17–26.

Van Schreven, D. A. (1968). Mineralisation of the carbon and nitrogen of plant material added to soil and of the soil humus during incubation following periodic drying and rewetting of the soil. *Plant Soil* **28**, 226–245.

Ventura, W., and Watanabe, I. (1978). Dry season soil conditions and soil nitrogen availability to wet season wetland rice. *Soil Sci. Plant Nutr.* **24**, 535–545.

Verma, L., Martin, J. P., and Haider, K. (1975). Decomposition of ^{14}C-labelled proteins, peptides, and amino acids; free and complexed with humic polymers. *Soil Sci. Soc. Am. Proc.* **39**, 279–283.

Volk, B. G., and Loeppert, R. H. (1982). Soil organic matter. *In* "Handbook of Soils and Climate in Agriculture" (V. J. Kilmer, ed.), pp. 211–268. CRC Press, Boca Raton, Florida.

von Pechmann, H., von Aufsess, H., Lecse, W., and Ammer, U. (1967). Investigations on red streak in Norway Spruce wood. *Forstwiss. Forsch.* **27.**

Wada, K., and Inoue, T. (1967). Retention of humic substances derived from rotted clover leaves in soils containing montmorillonite and allophane. *Soil Sci. Plant Nutr.* **13**, 9–16.

Wainwright, M. (1981). Mineral transformations by fungi in culture and soil. *Z. Pflanzenernaehr. Bodenkd.* **144**, 41–63.

Waksman, S. A., and Tenney, F. G. (1927). The composition of natural organic materials and their decomposition in soil. II. Influence of age of plant upon the rapidity and nature of its decomposition—rye plants. *Soil Sci.* **24**, 317–334.

Wallwork, J. A. (1976). "The Distribution and Diversity of Soil Fauna." Academic Press, New York.

Warembourg, F. R., and Billes, G. (1979). Estimating carbon transfers in the plant rhizosphere. *In* "The Soil–Root Interface" (J. L. Harley and R. Scott-Russell, eds.), pp. 183–196. Academic Press, New York.

Waring, S. A., and Bremner, J. M. (1964). Ammonium production in soil under waterlogged conditions as an index of nitrogen availability. *Nature (London)* **201**, 951–952.

Westerman, R. L., and Kurtz, L. T. (1973). Priming effect of ^{15}N-labelled fertilisers on soil nitrogen in field experiments. *Soil Sci. Soc. Am. Proc.* **37,** 725–727.

Westerman, R. L., and Tucker, T. C. (1974). Effects of salts and salts plus nitrogen-15-labelled ammonium chloride in mineralization of soil nitrogen, nitrification, and immobilization. *Soil Sci. Soc. Am. Proc.* **38,** 602–605.

Wetselaar, R. (1968). Soil organic nitrogen mineralization as affected by low soil water potentials. *Plant Soil* **29,** 9–17.

White, J. G. (1959). Mineralization of nitrogen and sulphur in sulphur-deficient soils. *N. Z. J. Agric. Res.* **2,** 255–258.

Will, G. M. (1967). Decomposition of *Pinus radiata* litter on the forest floor. Part I. Changes in dry matter and nutrient content. *N. Z. J. Sci.* **10,** 1030–1044.

Williams, S. T., and Gray, T. R. G. (1974). Decomposition of the litter on the soil surface. *In* "Biology of Plant Litter Decomposition" (C. H. Dickinson and G. J. F. Pugh, eds.), pp. 611–632. Academic Press, New York.

Williams, W. A., Mikkelsen, D. S., Mueller, K. E., and Ruckman, J. E. (1968). Nitrogen immobilization by rice straw incorporated in lowland rice production. *Plant Soil* **28,** 49–60.

Wilson, J. M., and Griffin, D. M. (1975). Water potential and the respiration of microorganisms in the soil. *Soil Biol. Biochem.* **7,** 199–204.

Witkamp, M. (1969). Environmental effects on microbial turnover of some mineral elements. I. Abiotic factors. *Soil Biol. Biochem.* **1,** 167–176.

Wood, T. G. (1974). Field investigations on the decomposition of leaves of *Eucalyptus delegatensis* in relation to environmental factors. *Pedobiologia* **14,** 343–371.

Woodmansee, R. G., Vallis, I., and Mott, J. J. (1981). Grassland nitrogen. *In* "Terrestrial Nitrogen Cycles: Processes, Ecosystem Strategies and Management Impacts" (F. E. Clark and T. Rosswall, eds.), pp. 443–462. Ecological Bulletins, Stockholm.

Yoshida, T. (1975). Microbial metabolism of flooded soils. *In* "Soil Biochemistry" E. A. Paul and A. D. McLaren, eds.), Vol. 3, pp. 83–112. Dekker, New York.

Zunino, H., Borie, F., Aguilera, S., Martin, J. P., and Haider, K. (1982). Decomposition of ^{14}C-labelled glucose, plant and microbial products and phenols in volcanic ash-derived soils of Chile. *Soil Biol. Biochem.* **14,** 37–43.

Chapter 3

Nitrification

R. J. HAYNES

I. INTRODUCTION

Nitrification is classically defined as the process whereby NH_4^+ is oxidized via NO_2^- to NO_3^-. The reactions are generally mediated in soil by the activities of two small groups of chemoautotrophic bacteria. One group, the NH_4^+ oxidizers, initiates the process with the formation of NO_2^-, while a second group, the NO_2^- oxidizers, completes the process by converting NO_2^- to NO_3^- as promptly as it is formed. The relatively narrow species diversity of the autotrophic nitrifier organisms means that the nitrification process can be greatly influenced by external factors.

Although the autotrophic nitrifiers are thought to be by far the most predominant agents of nitrification in the soil environment, several other minor pathways have been suggested. These include NO_2^- and NO_3^- production mediated by heterotrophs (Focht and Verstraete, 1977; Wainwright, 1981), oxidation of NH_4^+ to NO_2^- by methylotrophic bacteria (Dalton, 1977), and the chemical oxidation of NO_2^- to NO_3^- (Bartlett, 1981).

Agronomic studies have suggested that nitrification is quantitatively important in the N cycle of cultivated agricultural soils of moderate to high pH (Allison, 1973). Indeed, a high nitrification rate has often been considered an index of soil fertility by agriculturists. Nonetheless, ecological field studies have indicated that nitrification plays a minor role in the N cycle of many mature natural ecosystems (Bormann and Likens, 1979; Melillo, 1981). Inhibition of nitrification may well be an important strategy by which N is conserved within such ecosystems since nitrification can lead to gaseous losses of N as N_2 and N_2O and leaching losses as NO_3^-.

Thus, the desirability, or otherwise, of nitrification in terms of both environmental quality and plant growth is subject to much discussion.

In this chapter the process of nitrification is reviewed and the factors regulating it in both natural and agricultural ecosystems are discussed in detail.

II. PROCESSES OF NITRIFICATION

A. Chemoautotrophic Nitrification

1. Organisms and Diversity

Autotrophic nitrification is carried out by gram-negative bacteria of the family Nitrobacteraceae. All organisms of this family derive their energy from the oxidation of either NH_4^+ or NO_2^-. The currently recognized genera of the Nitrobacteraceae are shown in Table I. In soils, five genera are known to be able to oxidize NH_4^+ to NO_2^-: *Nitrosomonas, Nitrosococcus, Nitrosospira, Nitrosolobus,* and *Nitrosovibrio;* and one genus, *Nitrobacter* is known to oxidize NO_2^- to NO_3^-.

It is only recently that the diversity of nitrifier populations has been appreciated. Several serotypes of the common nitrifying species have now been identified from one soil sample (Belser, 1979; Schmidt, 1982). Serotypes refer to a number of antigenically distinguishable members of a single bacterial species and can be identified using the immunofluorescence (IF) technique as detailed by Bohlool and Schmidt (1980).

Table I

The Family Nitrobacteraceae[a]

Oxidation	Genus	Species	Habitat
Ammonium to nitrite	*Nitrosomonas*	*europa*	Soil, water, sewage
	Nitrosolobus	*multiformis*	Soil
	Nitrosovibrio	*tenuis*	Soil
	Nitrosospira	*briensis*	Soil
	Nitrosococcus	*nitrosus*	Soil
		oceanus	Marine
		mobilis	Marine
Nitrite to nitrate	*Nitrobacter*	*winogradskyi*	Soil, water
	Nitrospina	*gracilis*	Marine
	Nitrococcus	*mobilis*	Marine

[a] From Belser (1979).

a. Ammonium oxidizers. In agricultural soils *Nitrosospira* appears to be generally well represented, often accompanied by approximately equal numbers of *Nitrosomonas,* while *Nitrosolobus* is present in only low numbers (Belser and Schmidt, 1978a). *Nitrosomonas* is generally believed to be the dominant genus associated with sewage or manured agricultural land (Belser and Schmidt, 1978b; Walker, 1978).

A large serological diversity occurs among the genera *Nitrosomonas, Nitrosospira,* and *Nitrosolobus.* For example, Belser and Schmidt (1978c) observed that in one soil there were present at least four serotypes of *Nitrosomonas,* five of *Nitrosospira,* and one of *Nitrosolobus.* The existence of microsites in soils that differ in substrate concentration, pH, moisture content, and other environmental parameters presumably allows a diverse population of NH_4^+ oxidizers to coexist in multiple niches in the soil.

b. Nitrite oxidizers. *Nitrobacter* appears to be the only genus of NO_2^- oxidizer in soils even though NO_2^- oxidation generally occurs as promptly as the NO_2^- is formed (NO_2^- rarely accumulates in nature). However, serological diversity exists within the species *Nitrobacter winogradskyi* (Fliermans *et al.,* 1974; Rennie and Schmidt, 1977; Josserand and Cleyet-Marel, 1979; Josserand *et al.,* 1981; Stanley and Schmidt, 1981). At least two different serotypes of *N. winogradskyi* with different growth rates are known to coexist in the same soils (Rennie and Schmidt, 1977; Josserand *et al.,* 1981). Using the IF technique, Stanley and Schmidt (1981) identified 27 oxidizing isolates of the genus *Nitrobacter* originating from various soils and lakes.

2. Processes

The autotrophic nitrifiers are strict aerobes and depend on cytochrome systems for electron transport, and ultimately on oxygen. They can synthesize all of their cell constituents from CO_2 by way of the Calvin reductive pentose phosphate cycle, which also operates in plants and other autotrophic microorganisms. The driving force for the reduction of CO_2 is the production of ATP during the oxidation of NH_4^+ or NO_2^-.

a. Ammonium oxidation. The oxidation of NH_4^+ to NO_2^- by *Nitrosomonas* occurs as follows:

$$NH_4^+ + 1\tfrac{1}{2}O_2 \xrightarrow{\;6e^-\;} NO_2^- + 2H^+ + H_2O$$

There is a valency change from the $3-$ of NH_4^+ to the $3+$ of NO_2^- and the biochemical pathway involved is shown in Fig. 1. Hydroxylamine (NH_2OH) is a likely intermediate, while nitroxyl (NOH), or its dimer hyponitrite, is usually considered the most probable secondary intermediate (Nicholas, 1978). Two electrons from the dehydrogenation of NH_2OH are thought to pass through an electron transport chain involving cytochrome with the generation of ATP. The biochemistry of electron transport during NH_4^+ oxidation has been discussed in detail elsewhere (Hooper, 1978; Nicholas, 1978).

It is of considerable interest that N_2O gas is evolved during the process (Yoshida and Alexander, 1970; Nicholas, 1978) and the significance of this is discussed in Chapter 5. There are two possible ways in which N_2O could arise (Fig. 1). The presumed intermediate NOH, or its dimer hyponitrite, may dismutate chemically under reduced O_2 tensions to N_2O or the dissimilatory enzyme system, nitrite reductase, may yield N_2O when O_2 becomes limiting and NO_2^- replaces O_2 as an electron acceptor (Schmidt, 1982).

b. Nitrite oxidation. The oxidation of NO_2^- to NO_3^- by *Nitrobacter* occurs as follows:

$$NO_2^- + \tfrac{1}{2}O_2 \xrightarrow{2e^-} NO_3^-$$

There is a two-electron shift in oxidation state from $3+$ to $5+$. The reaction is mediated by a NO_2^- oxidase enzyme system with electrons carried to O_2 via cytochromes leading to the generation of ATP (Nicholas, 1978).

B. Heterotrophic Nitrification

Heterotrophic nitrification occurs when NO_2^- and/or NO_3^- are produced from inorganic or organic compounds by heterotrophic organisms through reactions that do not represent the sole sources of energy for the organisms. The demonstration of nitrification by these organisms in natural systems may be obscured by the fact that nitrification is not obligatory to their growth.

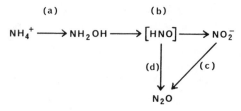

Fig. 1. Overall biochemical pathway for the oxidation of NH_4^+ to NO_2^-. (a) Oxygenase enzyme system; (b) hydroxylamine oxidoreductase enzyme system; (c) denitrifying nitrite reductase enzyme system; and (d) chemical dismutation of NOH.

1. Heterotrophic Organisms

A large number and variety of heterotrophic microorganisms (bacteria, fungi, and actinomycetes) can produce either NO_2^- or NO_3^- from NH_4^+ or other reduced forms of nitrogen when grown in culture (Schmidt, 1954; Eylar and Schmidt, 1959; Odu and Adeoye, 1970). For example, NO_2^- is formed by a wide range of heterotrophic bacteria and actinomycetes when they are grown in culture media containing NH_4^+ or amino N (Alexander, 1977; Focht and Verstraete, 1977). Several fungal species are able to oxidize NO_2^- to NO_3^- in culture, including *Aspergillus wentii* and *Penicillium* spp. (Focht and Verstraete, 1977). A limited number of heterotrophs, such as strains of the bacteria *Arthrobacter* and the fungal species *Aspergillus flavus* and related species, can produce NO_3^- from media containing NH_4^+ only. *Arthrobacter* and *Aspergillus flavus* can produce NO_3^- from media containing aliphatic organic nitrogenous substances while *Pseudomonas* spp. can produce NO_3^- from aromatic nitrogenous compounds (Focht and Verstraete, 1977). There is, however, no unequivocal evidence that any of the heterotrophs that nitrify in culture actually do so in their natural environment (Schmidt, 1982).

2. Processes of Heterotrophic Nitrification

The biochemical pathway of heterotrophic nitrification is subject to some controversy and might follow an organic, inorganic, or a combination of the two pathways. A summary of the possible pathways is presented in Fig. 2. The organic pathway has been suggested for *Aspergillus flavus* and other heterotrophs (Doxtader and Alexander, 1966) but there is also evidence implicating an inorganic pathway (Aleem *et al.,* 1964). Different genera may possess different pathways or different pathways may operate under different conditions; in any event, the pathways proposed by different workers are divergent (Focht and Verstraete, 1977).

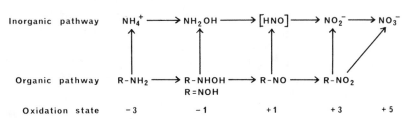

Fig. 2. Possible pathways of heterotrophic nitrification. [After Focht and Verstraete (1977).]

3. Significance of Heterotrophic Nitrification

The significance, or otherwise, of heterotrophic nitrification in relation to the formation of NO_3^- in soils is a subject of some controversy. The major lines of evidence supporting its importance are outlined below.

a. pH and temperature optima. Autotrophic nitrifiers cannot generally be isolated from acid soils (pH < 4.5) by the most probable number (MPN) technique (Ishaque and Cornfield, 1974; Cooper, 1975; Van de Dijk and Troelstra, 1980) yet NO_3^- is formed in such soils. This point is not unequivocal since the MPN method lacks accuracy and may recover only a small fraction of the *in situ* nitrifying populations (Belser, 1979).

The pH optimum for nitrification in some acid soils where heterotrophic nitrification is suspected is approximately 4.5 and in such soils nitrification can continue at temperatures of 50 to 60°C (Ishaque and Cornfield, 1972, 1974; Focht and Verstraete, 1977). In contrast, autotrophic nitrification is generally considered to have a pH optimum of pH 6 to 7 and an upper temperature limit of about 40°C (Sections III,B and D).

The possibility of the presence of indigenous autotrophs with pH and temperature optima in the range pH 4 to 5 and 50 to 60°C cannot, however, be ruled out.

b. Addition of ammonium or organic matter. As discussed later (Section III,A), the rate of ammonification often limits the rate of nitrification so that the addition of low concentrations of NH_4^+-N often stimulates autotrophic nitrification. However, in some soils, nitrate formation is related to the quantity of added organic N (substrate for heterotrophic nitrifiers), such as peptone, present and the addition of NH_4^+ inhibits or has no effect on nitrification (Weber and Gainey, 1962; Ishaque and Cornfield, 1974; Van de Dijk and Troelstra, 1980).

c. Nitrification inhibitors. Inhibitors of autotrophic NH_4^+ oxidation, such as nitrapyrin (Campbell and Aleem, 1965a), and autotrophic NO_2^- oxidation, such as chlorate (Lees and Simpson, 1957), or inhibitors of both processes, such as benomyl [methyl-1-(butylcarbamoyl)-2-benzimidazole carbamate] (Van Faassen, 1974), have been used to test for heterotrophic nitrification. Such studies have shown that in some soils, in which heterotrophic nitrification is suspected, the application of these inhibitors does not preclude the formation of NO_3^- (Ettinger-Tulczynska, 1969; Gowda *et al.*, 1976; Tate, 1977). While serotypes and species of autotrophs may differ in their response to such compounds (e.g., Belser and Schmidt, 1981), these results suggest that other organisms (e.g., heterotrophs) might mediate nitrification.

d. Perspective. The above and other studies provide circumstantial evidence that suggests that heterotrophic (mainly fungal) nitrification may occur. The environments in which heterotrophic nitrification has been postulated to occur are acid soils and/or soils with high temperatures (e.g., acid forest and woodland soils, histosols, and desert and tropical soils) (Focht and Verstraete, 1977; Wainwright, 1981). Nonetheless, Schmidt (1982) pointed out that unequivocal evidence relating the occurrence of a particular heterotroph in its natural environment to the progression of nitrification in that environment has yet to be provided.

C. Methylotrophic Nitrification

The methane (CH_4) -oxidizing bacteria (methylotrophs) occur in soils and waters in aerobic sites in contact with anaerobic, CH_4-generating sites. This morphologically diverse group of gram-negative, strictly aerobic bacteria use CH_4, CH_3OH, and CH_3OCH_3 as their major sources of C (Quayle, 1972; Smith and Hoare, 1977). They can incorporate an oxygen from O_2 into CH_4 by means of a CH_4 monooxygenase enzyme complex.

A special feature of these bacteria is that they have the ability to produce small quantities of NO_2^- when growing with NH_4^+ as an N source (Dalton, 1977; Romanovskaya *et al.*, 1977; Whittenbury and Kelly, 1977). The monooxygenase enzyme complex brings about the oxidation of NH_4^+ to NH_2OH which is subsequently oxidized by a dehydrogenase enzyme to NO_2^- (Dalton, 1977). The K_m (Michaelis-Menten constant) value of the monooxygenase enzyme for NH_4^+ is very high in comparison to that for CH_4 and, in fact, methylotrophs grow best with NO_3^- as an N source, while high concentrations of NH_4^+ (>200 mg N liter^{-1}) inhibit the process of CH_4 oxidation (Whittenbury *et al.*, 1970; Dalton, 1977).

In reality, nothing is known of what, if any, contribution the methylotrophs make to nitrification (Schmidt, 1982). It seems possible that in certain environments, where autotrophic nitrifiers are absent or inactive, methylotrophs could play a role. The high K_m value of methylotrophs for NH_4^+ means that it is unlikely that they could compete successfully with the ubiquitous autotropic nitrifiers for NH_4^+ (Verstraete, 1981a).

D. Chemical Nitrification

The possibility of nitrate formation in soils by chemical oxidation has been discussed in detail by Allison (1973) and more recently by Bartlett (1981).

Although NH_3 can be photochemically oxidized to NO_2^- in solution, the significance of photooxidation even in the surface of tropical soils is unknown. The chemical oxidation of NO_2^- to NO_3^- could, however, be of some significance in soils (Allison, 1973; Bartlett, 1981). Reuss and Smith (1965), for example, added $NaNO_2$ to two soils of pH 4.6 and 5.6 and they observed that after 24 hr about 25% of the NO_2^--N had been oxidized to NO_3^--N in the more acid soil and 14% in the soil of pH 5.6. Experiments with steam-sterilized soils gave identical results as unsterilized soils.

In very acid soils NO_2^- can be spontaneously oxidized to NO_3^- according to the reactions summarized below:

$$2NO_2^- + 2H^+ \rightleftharpoons \begin{matrix} NO + H_2O + NO_2 \\ \\ \downarrow \frac{1}{2}O_2 \\ \\ \longrightarrow NO_2 \end{matrix} \Bigg\} \rightleftharpoons NO_3^- + NO_2^- + 2H^+$$

The general validity of the above equation has been verified experimentally by Bartlett (1981). The NO_2^- serves as its own electron donor and acceptor, dismutating to form NO and NO_2 gases. The NO thus formed is oxidized to NO_2 in the presence of atmospheric O_2. The NO_2 gas formed by both sources dismutates to form NO_3^- and NO_2^-. Since the chemical oxidation of NO_2^- begins at about that pH at which autotrophic nitrification ceases (i.e., pH 4.5), Allison (1973) reasoned that chemical nitrification might be of considerable importance in acid soils.

Bartlett (1981) recently implicated Mn oxides with the nonmicrobial conversion of NO_2^- to NO_3^-. Nitrate formation in soils equilibrated with NO_2^- at 0.5°C was observed to be directly related to soil level of reactive Mn oxides. Furthermore, Bartlett (1981) showed that in the presence or absence of O_2 synthetic Mn oxides could stoichiometrically oxidize NO_2^- to NO_3^- and become reduced in the process (see Fig. 3). The equation below describes the NO_2^- to NO_3^- chemical transformation that might occur in soils of high Mn oxide content.

$$NO_2^- + 2MNO_2 \rightleftharpoons Mn_2O_3 + NO_3^-$$

Nonmicrobial conversion of NO_2^- to NO_3^- could possibly help explain why NO_2^- seldom accumulates in soils. It is, nevertheless, generally considered that chemical nitrification is of very minor importance in relation to the overall conversion of NH_4^+ to NO_3^- in most soils.

III. FACTORS REGULATING NITRIFICATION

In comparison with ammonification, which is mediated by the diverse heterotrophic biomass, nitrification is mediated predominantly by a small group of autotrophic bacteria. Thus, the latter process is generally influenced more strongly by external factors such as moisture, temperature,

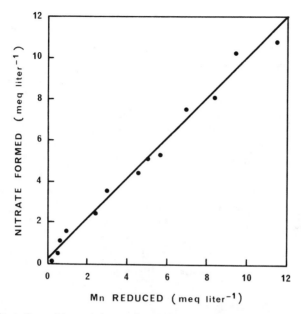

Fig. 3. Nitrate formed from nitrite as a function of Mn reduced in synthetic MnO_2 suspensions equilibrated with increasing levels of nitrite. [After Bartlett (1981). Reproduced from *Soil Sci. Soc. Am. J.* **45**, p. 1057 by permission of the Soil Science Society of America.]

and pH than the former. Nevertheless, as already outlined, considerable serological diversity exists within species of nitrifiers. Such diversity probably results in considerable variability in the response of nitrification to regulatory factors in different soils. Furthermore, the suggestion that heterotrophic nitrification may occur under environmental conditions that are apparently unsuitable for autotrophs has further complicated our understanding of the regulation of nitrification.

The more important factors are discussed below. These can be grouped into three broad categories: (1) ubiquitous factors (substrates and products, pH, aeration and moisture and temperature); (2) regulatory factors in natural ecosystems (allelopathy, limiting supply of NH_4^+ and other nutrient deficiencies); and (3) man-made factors (trace element toxicities, pesticide residues, and specific inhibitors).

A. Substrates and Products

1. Substrate Stimulation

The autotrophic nitrifiers are dependent on either NH_4^+ or NO_2^- as specific energy sources so that substrate concentration can be a very important factor influencing nitrifier activity. The K_m values (which indicate the substrate concentration that is required to give half the maximum velocity of the nitrification process) for NH_4^+ oxidation range from 1 to 10 mg N liter^{-1} between 20 and 30°C, while those for NO_2^- oxidation range from 5 to 8 mg N liter^{-1} (Table II). Except following heavy NH_4^+-N fertilizer applications, the K_m values quoted above are usually equal to, or considerably greater than, substrate concentrations in nature. Hence, populations and *in situ* activities of nitrifiers in soils are usually limited by the rate of production of NH_4^+ (i.e., the ammonification rate).

Indeed, several studies (McLaren, 1971; Ardakani *et al.*, 1974) have shown that the addition of NH_4^+ fertilizer to soils can increase populations of *Nitrosomonas* by several hundred million per gram of soil. Similar results for the population of *Nitrobacter*, following the addition of NO_2^- to soils, have also been observed (Ardakani *et al.*, 1973). The stimulation of nitrification in three Alberta soils following the addition of NH_4^+ at concentrations up to 200 μg N gm^{-1} is illustrated in Table III.

In contrast to the autotrophic nitrifiers, the heterotrophs can use a diversity of reduced organic and inorganic nitrogenous substances as substrates. Thus, as already discussed, if heterotrophic nitrifiers were active the addition of an organic nitrogenous compound such as peptone could result in an increase in heterotroph activity and a possible increase in the production of NO_3^- (e.g., Van de Dijk and Troelstra, 1980).

2. Substrate Repression

The maximum tolerable NH_4^+ concentrations in soils for nitrification to occur appear to vary between 400 (McIntosh and Frederick, 1958) and 800

Table II

Kinetic Constants of Nitrifying Organisms at pH 8[a]

Organism	K_m (substrate) (mg N liter^{-1})
Nitrosomonas	10 (30°C)
	3.5 (25°C)
	1.2 (20°C)
Nitrobacter	8 (32°C)
	5 (25°C)

[a] Data from Painter (1977). Reprinted with permission from Pergamon Press.

Table III

Nitrification Rates in Three Soils Incubated with
Different Concentrations of Ammonium N[a]

Concentration of ammonium (μg N gm^{-1})	Rate of nitrate formation (μg gm^{-1} day^{-1})[b]		
	Soil I[c]	Soil II	Soil III
50	2.5a	2.0a	1.9a
100	3.8b	3.3b	3.4b
200	5.9d	5.1d	4.9d
300	4.5c	4.1c	4.1c

[a] Data from Malhi and McGill (1982). Reprinted with permission from Pergamon Press.
[b] Soils incubated at 20°C and −33 kPa soil moisture.
[c] In each column, values are significantly different ($P \leq 0.05$) when followed by different letters.

μg N gm^{-1} (Broadbent *et al.*, 1957). A depression in nitrification at 300 μg N gm^{-1} of NH$_4^+$ is illustrated in Table III.

The depressing effect of high concentrations of NH$_4^+$ on nitrification has been attributed to toxic levels of NH$_3$ at high pH (Broadbent *et al.*, 1957; Stojanovic and Alexander, 1958), to an increase in salt content of the soil with increasing rates of NH$_4^+$ addition (Harada and Kai, 1968; Laura, 1977; Malhi and McGill, 1982), or to lowering of pH when (NH$_4$)$_2$SO$_4$ is added (Justice and Smith, 1962; Malhi and McGill, 1982). The mechanism of inhibition may therefore vary depending on soil conditions (particularly initial pH and the source of added NH$_4^+$).

The NH$_4^+$-oxidizing bacteria are characteristically less sensitive than *Nitrobacter* to high NH$_4^+$ concentrations. Jones and Hedlin (1970), for example, reported that the rate of NO$_3^-$ production increased with increasing NH$_4^+$ concentration from 50 to 800 μg N gm^{-1} but at higher NH$_4^+$-N concentrations NO$_2^-$ accumulated. Nakos and Wolcott (1972) reported similar results.

The greatest danger of encountering NH$_4^+$ toxicity is in limited volumes of soils, such as when anhydrous ammonia or ammonium fertilizers are applied as banded applications (Allison, 1973). Pang *et al.* (1975), for instance, demonstrated that NO$_2^-$ accumulated in a soil (pH 6.6) when urea or aqua ammonia were banded at rates of 200 and 800 kg N ha^{-1}.

3. End Product Repression

End product repression of nitrification can also occur. The concentration of NO_3^- exhibiting end product inhibition of *Nitrosomonas* in the logarithmic phase of growth is quoted variously as 2500 to 4200 mg N liter^{-1} (Painter, 1977). High concentrations of NO_3^- are also known to noncompetitively inhibit oxidation of NO_2^- by *Nitrobacter* (Boon and Laudelout, 1962). End product repression can be important when experiments are carried out in closed containers, especially when high rates of NH_4^+ are added and the reaction is allowed to proceed for a long period (i.e., weeks).

B. Soil pH

Soil pH is well known to be a limiting factor for nitrification. Generally, in culture, the optimal pH for the growth and metabolism of autotrophic nitrifiers is in the range pH 7 to 9. Nevertheless, in soils of pH above 7.5, toxic levels of NH_3 may result in the inhibition of the activity of *Nitrobacter* and in the accumulation of NO_2^- (Morrill and Dawson, 1967).

The lower limit for autotrophic nitrification is generally found to be around pH 4.5 (e.g., Sarathchandra, 1978; Sahrawat, 1982). The influence

Table IV

Total Nitrogen, Soil pH, and Mineralization[a] of Soil Nitrogen in 10 Tropical Soils[b]

	Total N (%)	pH (water)	Mineral N formed ($\mu g\ gm^{-1}$)	
			NH_4^+-N	NO_3^--N
Calahan sandy loam	0.11	3.4	89	0
Malinao loamy sand	0.09	3.7	93	0
Luisiana clay	0.18	4.4	102	0
Morong peat	0.56	5.6	242	5
Law Aw peat	1.2	6.1	404	116
Maahas clay	0.12	6.5	31	106
Quingua silty loam	0.12	6.5	18	115
Pila clay	0.19	7.5	21	123
Lipa loam	0.19	7.5	17	98
Maahas clay, alkalized	0.12	8.6	21	118

[a] Soils incubated aerobically at 30°C 4 weeks.
[b] Data from Sahrawat (1982).

of soil pH on nitrification in 10 tropical soils is illustrated in Table IV. Quantities of NO_3^- formed in soils were highly positively correlated with soil pH ($r = 0.86**$) but not with organic C or total N content of the soils (Sahrawat, 1982). Aluminum toxicity is suspected to be the major factor limiting nitrifier activity at low soil pH (Brar and Giddens, 1968). Generally, the liming of acid soils stimulates nitrification, often to a greater extent than ammonification (Chase et al., 1968; Nyborg and Hoyt, 1978).

It is commonly observed that nitrification can occur in soils of pH 4 to 5 (Weber and Gainey, 1962; Walker and Wickramasinghe, 1979; Matson and Vitousek, 1981; Vitousek et al., 1982; Federer, 1983; Olson and Reiners, 1983). It is possible that the sites at which bacteria perform their oxidation have a higher pH than that determined from bulk soil samples, while strains of nitrifiers may exist in acid soils that have adapted to acidic soil conditions. Walker and Wickramasinghe (1979), for instance, presented evidence that *Nitrosospira* mediated nitrification *in situ* in a soil of pH 4.1. Circumstantial evidence also suggests that in some acid soils (pH < 4.5) heterotrophic nitrification could be of some significance (Section II,B).

C. Aeration and Moisture

In general, the maximum rate of nitrification occurs at soil moisture potentials in the range of -10 (Miller and Johnson, 1964; Sabey, 1969) to -33 kPa (Justice and Smith, 1962; Malhi and McGill, 1982), presumably depending principally on soil physical properties. At 0 kPa, nitrification is either absent (Miller and Johnson, 1964; Sabey, 1969; Malhi and McGill, 1982) or occurs at a very slow rate (Dubey, 1968; Sabey, 1969) because of the shortage of O_2 in the soil system caused by excess water.

The influence of soil moisture tension on nitrification is illustrated in Fig. 4. With decreasing soil moisture potential, below -10 to -33 kPa, there is a general decrease in the relative rate of nitrification (Sabey, 1969; Malhi and McGill, 1982; Fig. 4). Nevertheless, appreciable nitrification usually occurs even at "permanent wilting point" (-1500 kPa) (Miller and Johnson, 1964; Dubey, 1968; Sabey, 1969). At soil moisture potentials below -1500 kPa the activity of nitrifier organisms appears to be inhibited to a greater extent than that of ammonifiers (Dommergues, 1966).

As noted in Chapter 2, when dry soils are rewetted, even by small amounts of precipitation, there is a characteristic flush of mineralization of native soil organic N. This is accompanied by a flush of nitrification and the temporary accumulation of NO_3^- in the soil (Campbell et al., 1975; Campbell and Biederbeck, 1982).

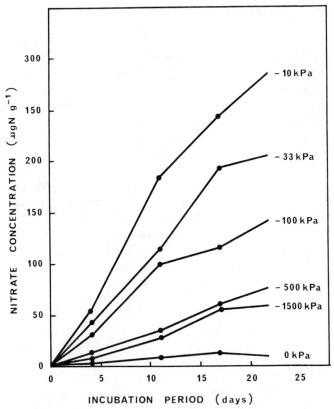

Fig. 4. Influence of soil moisture tension on nitrification of added ammonium in a silt loam incubated at 21°C. [Data from Sabey (1969). Reproduced from *Soil Sci. Soc. Am. Proc.* **33,** p. 264 by permission from the Soil Science Society of America.]

D. Temperature

The optimum temperature range for nitrification in soils is usually between 25 and 35°C (Justice and Smith, 1962; Thiagalingam and Kanehiro, 1973; Kowalenko and Cameron, 1976). The inhibitory effects of high and low temperatures are shown in Fig. 5.

It does, however, seem that indigenous nitrifiers have temperature optima adapted to their climatic regions (Mahendrappa *et al.,* 1966). For example, while Myers (1975) reported a temperature optimum of 35°C with nitrification proceeding up to 50°C in a tropical soil, Malhi and McGill (1982) found that soils from central Alberta (Canada) had temperature optima of 20°C and at 30°C nitrifier activity almost ceased. Malhi and

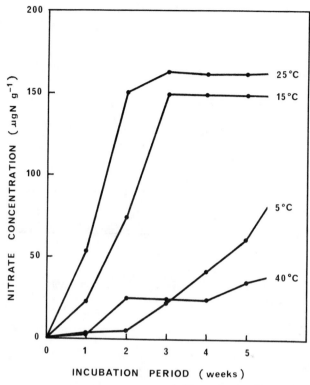

Fig. 5. Nitrification of added ammonium in a Hawaiian soil incubated at four different temperatures. [Data from Thiagalingam and Kanehiro (1973).]

Nyborg (1979) found that nitrification occurred in some Alberta soils even when they were frozen. The clearly different temperature optima for nitrification in three soils from locations with widely differing mean annual temperatures are shown in Fig. 6. As noted in Section II,B, at high temperatures (>40°C) heterotrophic nitrification may be of quantitative significance.

There are few studies of the effects of fluctuating temperatures on nitrification. However, several studies have shown that the nitrification rate under fluctuating low temperatures is less than that under a corresponding low mean temperature (Sabey *et al.,* 1956; Campbell *et al.,* 1971). Like drying and rewetting, freezing and thawing stimulates mineralization of native soil organic matter (see Chapter 2) and consequently there is a flush of nitrification (Campbell *et al.,* 1971; Biederbeck and Campbell, 1973).

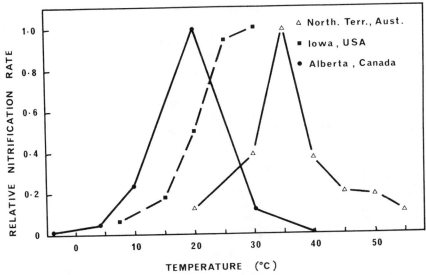

Fig. 6. Effect of soil temperature on relative nitrification rate in three soils from different climatic regions. Mean annual temperatures at the three locations are Northern Australia 25°C, Iowa 10°C, and Alberta 2.5°C. [Redrawn from Malhi and McGill (1982). Reprinted with permission from Pergamon Press.]

E. Allelopathic Substances

In general, allelopathy refers to the influence that one living organism has on another via secretion of chemical exudates. A variety of evidence (Rice, 1974, 1979) has led to the hypothesis that nitrification is inhibited in soils from climax ecosystems by allelopathic substances generated by plants that are characteristically present late in ecological succession.

1. Evidence for Allelopathy

The above hypothesis is strongly supported by the results of Rice and Pancholy (1972, 1973, 1974). These workers studied nitrification in two stages of old-field succession and climax in each of three types of vegetation in Oklahoma: oak–pine forest, past oak–blackjack oak forest, and tallgrass prairie. They found that soil NO_3^- concentrations and numbers of nitrifying bacteria (MPN) tended to decrease with increasing succession and that soil NH_4^+ concentrations increased. An example of their results for the tallgrass prairie succession is shown in Table V. They concluded that nitrification decreases as succession proceeds. Similar results have been observed along a South African grassland secondary succession

Table V

Concentrations of NH_4^+- and NO_3^--N and Numbers of Nitrifying Bacteria (MPN) in Soils (0–15 cm) of Two Successional Stages of Old Fields (S1 and S2) and of the Climax (C) of a Tallgrass Prairie Ecosystem[a]

	Ammonium N (μg gm^{-1})			Nitrate N (μg gm^{-1})		
	S1	S2[b]	C	S1	S2	C
April[c]	3.38	4.71a	6.06c	3.11	2.06a	1.50bc
June	2.77	5.03a	5.20c	2.85	2.37a	1.67bc
Aug.	3.97	4.09	5.00	2.95	1.06a	0.44bc

	Nitrosomonas (gm^{-1})			Nitrobacter (gm^{-1})		
	S1	S2	C	S1	S2	C
April	98	84	36	26	20	23
June	56	36	26	38	45	
Aug.	153	72	23	23	29	28

[a] Data from Rice and Pancholy (1973).
[b] a = Difference between S1 and S2 significant ($P \leq 0.05$). b = Difference between S2 and C significant ($P \leq 0.05$). c = Difference between S1 and C significant ($P \leq 0.05$).
[c] Soils sampled at three different times during 1972.

(Warren, 1965), a North Dakota mine spoil succession (Lodhi, 1979), an Arkansas upland forest succession (Wheeler and Donaldson, 1983), and in southern Appalachian forest successions (Todd *et al.*, 1975).

Rice and Pancholy (1973, 1974) observed that numerous organic compounds (e.g., tannins, phenolic acids, and phenolic glycosides) are produced by important plants in the intermediate and climax stages of old-field succession that are, at low concentrations (10^{-6} to 10^{-8} M), strongly inhibitory to autotrophic nitrifier organisms. Concentrations of these substances were found in generally increasing amounts in soils from progressively later stages of succession. Other research has indicated that organic substances of plant origin can inhibit nitrifier activity. Phenolic compounds extracted from the floor of a subalpine balsam fir forest greatly inhibited nitrification in the A_1 horizon below (Baldwin *et al.*, 1983; Olson and Reiners, 1983). Boquel *et al.* (1970) found that substances were present in fresh beech tree litter that exerted a strong inhibitory effect on *Nitrobacter*. Root extracts from several grass species (Munro, 1966; Moore and Waid, 1971) and forest trees (Melillo, 1977) have also been shown to have inhibitory effects on nitrification.

2. Significance of Allelopathy

Despite the above data there is no unequivocal evidence for allelopathic inhibition of nitrification. Most studies testing for inhibitors have involved addition of suspected sources of inhibitor (e.g., root washings) to incubated soil samples (e.g., Rice and Pancholy, 1973, 1974; Melillo, 1977) or to pure cultures of nitrifiers (e.g., Munro, 1966). Concentrations of plant extracts used are often considerably greater than would occur at any one time in the soil (Schmidt, 1982).

The addition of plant extracts to soils results in the unavoidable addition of a source of readily oxidizable C that could be decomposed by the heterotrophic biomass with the concomitant immobilization of NH_4^+. Thus nitrification could conceivably be suppressed by a lack of NH_4^+ rather than by allelopathic inhibition. Experiments involving pure cultures of nitrifiers avoid the immobilization effect but biotic and physical components of natural systems are excluded so meaningful ecological interpretations are difficult.

Several workers (Molina and Rovira, 1964; Odu and Akerele, 1973; Purchase, 1974a) have observed inhibitory effects of natural compounds in pure cultures of nitrifiers but stimulatory or no effect for the same compounds in soil incubations. Other researchers have found no indication of the inhibition of nitrifier activity in the rhizosphere of several grassland plants (Rennie et al., 1977; Smit and Woldendorp, 1981).

In conclusion, it can be said that the data regarding allelopathic inhibition of nitrification are rather conflicting and inconclusive.

F. Limiting Supply of Ammonium under Vegetation

As noted when discussing substrate stimulation of nitrification (Section III,A), it appears that the supply of NH_4^+ often limits the rate of nitrification. Thus where vegetation is present, strong competition for NH_4^+ between roots of vegetation (with associated mycorrhizal fungi) and the microbial biomass in the rhizosphere may leave little NH_4^+ available for autotrophic nitrifiers. Indeed, nitrifiers are generally poor competitors with the heterotrophic biomass for NH_4^+ (Jansson, 1958; Jones and Richards, 1977).

The production of carbonaceous materials in the rhizosphere of grasses (dead root hairs, dead root cells, and root excretions) encourages net immobilization of N (see Chapter 2) and this is thought to leave negligible NH_4^+ available for the nitrification process (Huntjens, 1971a,b; Huntjens and Albers, 1978). Any NH_4^+ not immobilized may be rapidly absorbed by the roots of the living plants. Thus, in general, levels of mineral N under grassland are low (NH_4^+ < 10 μg N gm^{-1} and NO_3^- < 1 μg N gm^{-1}) (Woodmansee et al., 1981). Huntjens (1971b), in fact, suggested that under grassland conditions the mineralization–immobilization cycle may proceed only as far as simple organic nitrogenous compounds such as amino acids since heterotrophic microorganisms assimilate amino acids more readily than inorganic N sources.

Similarly, in forest ecosystems competition between the root–mycorrhizal complex and the soil biomass for NH_4^+-N and simple organic nitrogenous compounds is thought to leave little NH_4^+ available for nitrifiers (Coats et al., 1976; Vitousek et al., 1979, 1982; Robertson and Vitousek, 1981).

Mycorrhizal fungi are known to prefer NH_4^+- to NO_3^--N and can also use a wide range of relatively simple organic compounds such as amides, amino acids, and nucleic acids (Smith, 1980; Bowen and Smith, 1981). Thus, as Verstraete (1981a) indicated, it seems possible that mycorrhizae may act as agents of biological control of nitrification under vegetated conditions.

G. Nutrient Deficiencies

Deficiencies of nutrients other than N, particularly P, can limit the activity of nitrifying bacteria. Purchase (1974a,b), for example, observed that in some soils from savanna grasslands the activity of NO_2^--oxidizing bacteria was limited by the concentration of available phosphate. Similarly, Melillo (1977) demonstrated much higher rates of nitrification in Hubbard Brook forest floor material that had been treated with NH_4^+-N plus phosphate than in those receiving NH_4^+ alone. Presumably, deficiencies of other nutrients can also limit nitrification.

The fact that P deficiency can limit nitrification has rather interesting ecological implications since with successional time, long-term weathering and leaching of nutrients results in a general decline in soil fertility (Jenny, 1980) and in particular the supply of available P decreases (Walker and Syers, 1976). Thus, in climax ecosystems nitrification could, to some extent, be limited by a deficiency of available phosphate.

H. Trace Element Toxicities

The contamination of the soil environment by trace elements originating from the application of waste products such as sewage sludge may inhibit nitrification processes (Wilson, 1977; Liang and Tabatabai, 1978; Chang and Broadbent, 1982).

The inhibitory effect that increasing concentrations of extractable trace metals in a soil have on nitrification is illustrated in Fig. 7. Small quantities of Cr, Cd, and Cu are obviously particularly detrimental. The addition of certain trace elements such as Ag, Ni, Co, Zn, Mn, Pb, As, B, Fe, As, Mo, and W to soils (5 μmol gm^{-1} soil) can inhibit *Nitrobacter* more than the NH$_4^+$ oxidizers, resulting in an accumulation of NO$_2^-$ (Liang and Tabatabai, 1978).

As noted when discussing the effect of trace element toxicity on N mineralization (Chapter 2), several variables, including adsorption processes at the surfaces of organic and clay colloids and soil pH, greatly influence the toxicity of applied trace elements. Thus, toxic rates of addition will depend greatly on soil properties.

The development of tolerance by nitrifiers in natural habitats in response to pollution and contamination is also a possibility. Rother *et al.*

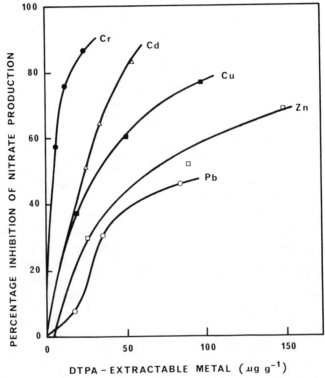

Fig. 7. Relationships between increasing levels of DTPA-extractable metals and percentage inhibition of nitrification in a silt loam. [Redrawn from Chang and Broadbent (1982). Reproduced from *J. Environ. Qual.* **11**, p. 4 by permission of the American Society of Agronomy.

(1982), in fact, found that the nitrifier organisms from heavy metal-contaminated soils showed clear signs of tolerance to considerable additions of Cd, Pb, and Zn although a small proportion of their population was killed or their activity inhibited for 1 or 2 days. Resistance could also be developed within a few days in nitrifier populations from uncontaminated soils (Rother et al., 1982).

I. Pesticides

The autotrophic nitrifying bacteria are considered to be among the most sensitive groups of soil organisms to soil-applied pesticides (fumigants, herbicides, insecticides, and fungicides) (Schmidt, 1982). The effect of pesticides on nitrification has been reviewed in detail elsewhere (Domsch and Paul, 1974; Goring and Laskowski, 1982). Since the activity of pesticides in soils is influenced by many factors, critical application rates will differ appreciably in different soils under differing environmental conditions.

The soil fumigants (e.g., carbon disulfide, chloropicrin, dazomet, DBCP, DD, metham-sodium, methyl bromide, and ethylene dibromide) are potent inhibitors of nitrification at rates at or below those recommended (e.g., Bremner and Bundy, 1974; Lebbink and Kolenbrander, 1974; Smith and Weeraratna, 1975; Goring and Scott, 1976; Ridge, 1976; Rovira, 1976; Elliot et al., 1977). The effect of chloropicrin in inhibiting nitrification and therefore promoting the accumulation of NH_4^+-N over a 29-day period is illustrated in Table VI.

At normal rates of soil application the vast majority of fungicides, insecticides, and herbicides are in general unlikely to affect nitrification

Table VI

Ammonium and Nitrate Concentrations in an Incubated Soil following Treatment with Chloropicrin[a,b]

Days after treatment	NH_4^+-N (μg gm^{-1})		NO_3^--N (μg gm^{-1})	
	Control	Chloropicrin	Control	Chloropicrin
7	2	14	11	9
14	1	18	12	8
29	1	27	14	8

[a] Chloropicrin applied at a rate equivalent to 220 kg ha^{-1}.
[b] Data from Rovira (1976). Reprinted with permission from Pergamon Press.

(Goring and Laskowski, 1982). At the upper end of recommended rates the dithiocarbamate fungicides (e.g., ferbam, maneb, nabam, zineb, and ziram) appear to inhibit nitrification (Jaques *et al.*, 1959; Chandra and Bollen, 1961; Mazur and Hughes, 1975). The herbicides most likely to inhibit nitrification appear to be the urea (e.g., diuron, fenuron, neburon, monuron, and monolinuron), carbamate (e.g., asulam, barban, chloropropham, and phenmidiphan), thiocarbamate (e.g., molinate, diallate, triallate, and thibencarb), and aminotriazole groups (Domsch and Paul, 1974; Goring and Laskowski, 1982).

At field rates, some herbicides, such as barban, methabenzthiazuron, metobromuron, monolinuron, and simazine, appear to inhibit the activities of NO_2^- oxidizers although much higher rates are required to inhibit NH_4^+ oxidation (Domsch and Paul, 1974; Goring and Laskowski, 1982). There is, therefore, the possibility of NO_2^- accumulation in soils if high rates of these herbicides are used.

J. Specific Inhibitors

In recent years there has been considerable interest in the use of chemicals that can directly regulate nitrification in the soil. The most commonly used nitrification inhibitor is 2-chloro-6-(trichloromethyl) pyridine, which has the common name nitrapyrin (Goring, 1962a,b). The compound acts principally by inhibiting the activity of the NH_4^+ oxidizers and *Nitrobacter* appears to be considerably less sensitive to its presence (Campbell and Aleem, 1965a,b). There is, however, considerable variation among genera and strains of nitrifier organisms in their sensitivity to nitrapyrin (Belser and Schmidt, 1981).

Other substances that are known to inhibit nitrification in soils include 4-amino-1,2,4-triazole (ATC), sodium or potassium azide, 2,4-diamino-6-trichloromethyl-S-triazine (CL-1580), dicyandiamide, 3-chloroacetanilide, 1-amidino-2-thiourea, 2-amino-4-chloro-6-methylpyridine (AM), and sulfathiazole (ST) (Bundy and Bremner, 1973).

The effectiveness of inhibition conferred by the most common inhibitor, nitrapyrin, is influenced greatly by a number of soil factors. Its effectiveness is greater in light-textured soils and at low soil temperatures (Bundy and Bremner, 1973) and higher concentrations are required with increasing pH and organic matter content of soils.

In view of the variability in sensitivity of NH_4^+ oxidizers to nitrapyrin, Belser and Schmidt (1981) suggested that soils that show a good response to low-level applications of nitrapyrin may have sensitive strains of nitrifiers present. Continued applications might, however, select for less sensitive strains, leading to a need for progressively higher application rates.

The agronomic applicability and significance of manufactured nitrification inhibitors are discussed in Chapter 7.

IV. ROLE OF NITRIFICATION IN ECOSYSTEMS

A. Significance of Nitrification

The overall significance and importance of nitrification in the plant–soil system are subject to great debate among scientists because it has both beneficial and detrimental effects on the soil and wider environment. In the following discussion, the major drawbacks and assets of nitrification are outlined.

1. Drawbacks of Nitrification

a. Expenditure of energy. Nitrification increases concentrations of soil NO_3^- and decreases those of NH_4^+ so that plants absorb relatively more NO_3^--N. Consequently, more energy may then be used within the ecosystem because while NH_4^+ can be directly channeled into protein synthesis, the assimilation of NO_3^- by plants requires a considerable amount of reducing equivalents and energy to bring about the reduction of NO_3^- to NH_4^+ (See Chapter 6).

In reality, the theoretical advantage of NH_4^+ over NO_3^- on plant growth is seldom observed primarily because NH_4^+ is toxic to plants at considerably lower concentrations than is NO_3^-. Furthermore, in the soil situation access of roots to NO_3^--N may be greater than that to equivalent amounts of NH_4^+-N because of the high mobility of NO_3^- in soil solution relative to that of NH_4^+.

b. Losses of nitrogen. Nitrification can certainly result in significant losses of N from ecosystems. For example, while the NH_4^+ cation is retained by the negatively charged soil colloids, the NO_3^- anion is very mobile in soils and is easily leached away (Chapter 4). Furthermore, during nitrification gaseous losses of N_2O occur (Chapter 5) and the NO_3^- anion itself can be denitrified with the release of N_2O and N_2 (Chapter 5).

These processes often give rise to losses in the vicinity of 10 to 50% of added fertilizer N on arable agricultural lands (Frissel, 1978). The economic significance of such losses is self-evident, especially for those areas where losses result in N becoming the limiting factor for plant growth.

c. Environmental and health hazards. Nitrate leaching results in the accumulation of NO_3^- in groundwater, which can contribute to the eutrophication process in lakes and streams (Kumm, 1976). High concentrations of NO_3^- in drinking waters can also be considered a health hazard since the conversion of NO_3^- to NO_2^- in the digestive tract can result in a blood disorder, methemoglobinemia, in infants and ruminants (see Chapter 4) and may also result in the formation of carcinogenic nitrosamines in the digestive tract.

Concern has also been expressed that the emission of N_2O into the atmosphere (through nitrification and denitrification) may contribute to the degradation of the stratospheric ozone layer (Crutzen, 1974).

High levels of NO_3^--N in soils can result in the accumulation of extremely high concentrations of NO_3^- in some plants (Maynard et al., 1976), particularly those belonging to the Chenopodiaceae family. Ingestion of such plants can have similar effects to that of drinking water containing high NO_3^-, i.e., the possibility of methemoglobinemia and/or the formation of nitrosamines.

d. Soil acidification. Nitrification results in acidification of the surrounding soil environment:

$$NH_4^+ + 2O_2 \rightleftharpoons NO_3^- + H_2O + 2H^+$$

The addition of NH_4^+-containing fertilizers that are rapidly nitrified to NO_3^--N adds two protons to the soil per NO_3^- anion accumulated. Nitrification of NH_4^+ ions originating from ammonification of organic N, however, results in the net accumulation of one proton per NO_3^- ion accumulated since during ammonification one proton is consumed (Helyar, 1976). Uptake of NO_3^--N by plants will also tend to counteract the acidifying effect of nitrification since for each NO_3^- ion absorbed by plant roots one OH^- or HCO_3^- ion is excreted (Nye, 1981).

Where significant downward movement of water occurs through soils in which NO_3^- has accumulated leaching of NO_3^- will occur. There will be an associated loss of cations (Ca^{2+}, Ma^{2+}, K^+, and Na^+) from the soil since these ions move downward as counterions with the NO_3^-. Thus nitrification and subsequent leaching of NO_3^- can be a major cause of a decrease in soil pH and a reduction in base saturation in agricultural soils (Pierre et al., 1971; Helyar, 1976; Haynes, 1981a,b).

e. Soil-borne diseases. A predominance of NH_4^+-N in the rooting medium may suppress the incidence of some soil-borne diseases in comparison with that when NO_3^--N predominates (Smiley, 1975; Henis, 1976). This effect appears to be at least partially a rhizosphere pH effect (Smiley, 1975) with NH_4^+ acting to reduce rhizosphere pH and NO_3^- acting to increase it, following their uptake by plant roots (Nye, 1981). Thus, suppression by NH_4^+-N and low pH has been reported for *Phymatotrichum omnivorum* in cotton, *Thieloviopsis basicola* in tobacco, and *Ophiobolus dahliae* and *Verticillium albo-atrum* in tomato, eggplant, and potato (Henis, 1976).

Conversely, suppression of diseases by NO_3^- nutrition and/or high pH has been reported for *Sclerotium rolfsii* on sugar beet and tomato and several diseases caused by *Fusarium* spp. (Henis, 1976). Thus, although nitrification is likely to favor root infections by some diseases it will, nevertheless, result in the suppression of infection by others.

2. Assets of Nitrification

a. Avoidance of ammonium toxicity. Nitrification decreases the levels of NH_4^+ in the soils and therefore reduces the chances of NH_4^+ toxicity to plants. The toxic effects of high ambient NH_4^+ on plant growth and metabolism are discussed in Chapter 6. Such a toxicity is potentially possible where high rates of NH_4^+ or urea N are applied to soils in which, for some reason, nitrification does not occur or occurs only slowly.

b. Decrease in ammonia volatilization. Obviously in the situation where NH_4^+ accumulation could potentially occur, nitrification reduces the likelihood of extensive volatilization loss of N as NH_3 (see Chapter 5). Nevertheless, it increases the likelihood of gaseous losses of N as N_2 and N_2O.

c. Availability of fixed ammonium. Fixation of NH_4^+ by clay minerals involves an equilibrium between the exchangeable and fixed fractions of soil NH_4^+ (see Chapter 4). Thus nitrification, which involves the removal of exchangeable NH_4^+ from the soil system, tends to bring about the release of fixed NH_4^+ (Nommik, 1981). In soils that fixed large amounts of fertilizer N (e.g., 40% or more), Kowalenko and Cameron (1978) found that nitrifiers were important intermediaries in making the fixed fraction available to plants.

3. Perspective

From an environmental viewpoint, nitrification certainly has considerable drawbacks, such as soil acidification and losses of N from the soil as leached NO_3^- and gaseous N_2O and N_2. Thus, Alexander (1965) concluded that "nitrification is a mixed blessing and possibly a frequent evil" while Verstraete (1981b) concluded that "nitrification qualifies as a process with few assets and many drawbacks."

Nevertheless, where large quantities of urea or NH_4^+-containing fertilizers are added to soils, nitrification results in the conversion of the potentially phytotoxic NH_4^+ ion to the less toxic NO_3^- ion. Nitrification can, therefore, generate soil conditions that are more conducive to crop-plant growth and hence it is often considered a desirable process in arable soils (Allison, 1973; Ahrens, 1977) or at least a necessary evil.

B. Nitrification in Natural Ecosystems

1. Occurrence of Nitrification

Most ecological field studies have indicated that nitrification normally plays a minor role in the N cycle of many natural ecosystems (Borman and Likens, 1979; Jordan et al., 1979; Gosz, 1981; Melillo, 1981; Woodmansee et al., 1981). Indeed, in most undisturbed ecosystems NO_3^- occurs in only very small concentrations (Clark, 1977; Woodmansee et al., 1978; Vitousek et al., 1979). Deserts can sporadically be exceptions to this general rule (Skujins, 1981). It has become generally recognized that the low level of nitrification encountered is a key process in N conservation in natural ecosystems (Likens et al., 1969; Vitousek et al., 1979) since leaching losses of NO_3^- and gaseous losses of N_2O and N_2 are minimized.

The results of "nitrification potential" measurements are often quoted in relation to nitrification in natural ecosystems (e.g., Ellenberg, 1971, 1977; Robertson and Vitousek, 1981; Robertson, 1982a). Such measurements involve field incubations of soil in buried bags (in which plant uptake, leaching, or replenishment of natural inhibitors cannot occur) or laboratory incubation of soils collected from the field. Results using such techniques do not necessarily reflect the magnitude of processes that would occur under undisturbed conditions.

Utilizing the field incubation technique Ellenberg (1971) classified European ecosystems into those possessing (1) NH_4^+-type, (2) mixed NH_4^+-plus NO_3^--type, and (3) NO_3^--type nitrogen economies (Table VII). Such data is an interesting application of the nitrification potential concept but since plant roots are not drawing upon the pool of NH_4^+ and the possibility that labile allelopathic inhibitors exist is excluded it is not known whether NO_3^- production would, in fact, occur under natural conditions in those ecosystems classified as NH_4^+/NO_3^- or NO_3^- type.

2. Regulation of Nitrification

The exact mechanisms by which nitrification is regulated in natural ecosystems are not well understood and they may well differ in importance in different ecosystems (Robertson, 1982a,b). In fact, in natural ecosystems several regulatory mechanisms of nitrification that were discussed in Section III have been suggested, including (1) an allochemical effect, (2) nutrient deficiencies, and (3) a limiting supply of NH_4^+.

Table VII

Ecosystem Types Classified by Field Nitrification Potential Estimates[a]

NH_4^+ type	NH_4^+/NO_3^- type	NO_3^- type
Taiga, dwarf-shrub tundra	Many temperate deciduous forests on loamy soil	Moist tropical lowland forest
Subalpine coniferous forest	Alluvial forest	Temperate deciduous forest on calcareous soil
Coniferous peat forest	Alder fen (*Aluus glutinosa*)	Fertilized meadows where soil is not wet
Oak–birch forest	Many grassland types	
Calluna heath	Dry grassland on calcareous soil	Most gardens
Many swamps		Ruderal formations
Raised sphagnum bogs	Tropical savanna	
	Some tropical forests	

[a] Classification from Ellenberg (1971).

It seems likely that in the long term the most important control over nitrification is the supply of NH_4^+ substrate. As noted previously, it appears that in most mature, vegetated ecosystems, the supply of NH_4^+ to nitrifiers is limited by competition between the heterotrophic biomass and the plant root–mycorrhizal complex. Any NO_3^- that was produced through nitrification would similarly be quickly used by the heterotrophic biomass or the root–mycorrhizal complex. Thus, when ecosystem disturbance (e.g., clear-cutting a forest) results in N mineralization in excess of immobilization and plant uptake, an increase in nitrification and a concomitant large increase in leaching of NO_3^- to groundwater often occur (Khanna, 1981; Vitousek, 1981).

Nevertheless, nitrification does not necessarily occur immediately. For example, in 8 out of 19 forests across the United States Vitousek *et al.* (1979) found significant delays in nitrification following clear-cutting despite increased concentrations of NH_4^+ in the soil. In such situations, an allochemical effect or a deficiency of a nutrient (e.g., P) could prevent or delay nitrification, as indeed could poor physicochemical conditions in the soil (e.g., cold, anaerobic, dry, or very acid soil). Another possible explanation is that populations of nitrifying organisms in the soil are initially very low or absent due to the very low levels of NH_4^+ that are normally present in the ecosystem (Coats *et al.*, 1976; Jordan *et al.*, 1979; Matson and Vitousek, 1981).

C. Nitrification in Agricultural Ecosystems

1. Occurrence

As already discussed, in natural ecosystems the supply of NH_4^+ for nitrification is severely limited because the soil microflora plus growing plants act as sinks for the ammonified N. Similarly, in most pastoral agricultural ecosystems nitrification will represent a minor component of the N cycle (Floate, 1981; Woodmansee et al., 1981).

In contrast, an abundant supply of NH_4^+ plus the lack of a complete plant cover during much of the year lead to extensive nitrification under arable cropping when environmental conditions are favorable (e.g., Campbell et al., 1975; Mahli and Nyborg, 1979; Hart and Goh, 1980; Campbell and Biederbeck, 1982). In the majority of agricultural studies, mineralization and nitrification are estimated by periodic sampling of soils for extractable NH_4^+ and NO_3^-. While such studies have clearly demonstrated that applied or mineralized N is rapidly nitrified, accurate measurements of nitrification are complicated by both leaching losses of NO_3^- and the upward movement of NO_3^- caused by upward water movement (e.g., Campbell and Biederbeck, 1982). Gaseous losses of N during nitrification and denitrification (see Chapter 5) further confuse interpretation of results.

The major sources of NH_4^+ in agricultural soils are fertilizer NH_4^+, urea, ammonification of organic residues and organic manures, and ammonification of soil organic N induced by tillage practices. Since nitrification is an important component of the N cycle of many arable agricultural ecosystems, losses of N from such systems through leaching of NO_3^- and gaseous losses of N_2O and N_2 are common. Such losses can, as already noted, account for in the vicinity of 10 to 50% of applied fertilizer N and detailed accounts of such losses are found in Chapters 4 and 5. Because these losses occur, efforts have been made to try to minimize nitrification in agricultural ecosystems.

2. Regulation

Complete inhibition of nitrification in agricultural soils where there is a large supply of NH_4^+-N would result in reduced growth of many crop plants due to NH_4^+ toxicity. Under some circumstances it might also increase gaseous losses of N through NH_3 volatilization. Nevertheless, if nitrification were to proceed at a slower, and ideally controllable, rate then crop productivity and fertilizer use efficiency, as well as environmental quality, would very likely be improved. The two major methods that have been employed in attempts to control the rate of nitrification are the use of slow-release fertilizers (control of the supply of added NH_4^+) and specific nitrification inhibitors (control of the rate of transformation of NH_4^+ to NO_3^-). The use of such compounds is discussed in more detail in Chapter 7. Nitrification can also be slowed by fertilizer placement (Nyborg and Mahli, 1979).

Slow-release fertilizers include organic formulations such as substituted ureas and inorganic preparations such as plastic-coated pellets. A sulfur-coated urea product is also available. In general, slow-release fertilizers are rather costly and any improvement in fertilizer efficiency does not often compensate for the extra cost (Hauck, 1972).

Both positive (Prasad, 1976; Huber *et al.*, 1977; Leyshon *et al.*, 1980) and negative (Goh and Young, 1975; Osborne, 1977; Hendrickson *et al.*, 1978) crop responses have been recorded following applications of specific nitrification inhibitors to soils. However, in general, nitrification inhibitors have had little effect on crop yields under field conditions (Hauck, 1972; Hendrickson *et al.*, 1978) and they are not normally used on a commercial scale.

Placing concentrated amounts of fertilizer N (e.g., urea) into limited volumes of soil can also slow nitrification and increase uptake of fertilizer N by subsequently planted crops (Nyborg and Malhi, 1979). This can be achieved by placing fertilizer in narrow bands in the soil, placing fertilizer at discrete points into the soil, or applying N as big pellets. The lower rates of nitrification are probably due to high pH and high concentrations of NH_3 produced after urea hydrolysis in the limited soil volume.

Thus, in contrast to natural ecosystems where nitrification is often naturally regulated, in man-made arable agricultural ecosystems nitrification can be an important component of the N cycle and practices aimed at its retardation have found only minor application.

V. CONCLUSIONS

Nitrification, the process whereby NH_4^+ is oxidized to NO_3^- via NO_2^-, occurs in virtually all soils where NH_4^+ is present and environmental conditions are favorable.

By far the most important pathway of nitrification is through the actions of the chemoautotrophic nitrifying bacteria. In soils five genera of autotrophs are known to be able to oxidize NH_4^+ to NO_2: *Nitrosomonas*, *Nitrosococcus*, *Nitrosospira*, *Nitrosolobus*, and *Nitrosovibrio*; and one genus, *Nitrobacter*, is known to oxidize NO_2^- to NO_3^-. Wide serological diversity exists among both the NH_4^+ and NO_2^- oxidizers.

Nitrification mediated by heterotrophic microorganisms (particularly fungi) might be of significance in acid soils (pH < 4.5) and/or soils with high temperatures ($>35°C$). Other possible pathways of nitrification include the oxidation of NH_4^+ to NO_2^- by methylotrophic bacteria and the chemical oxidation of NO_2^- to NO_3^-. Virtually nothing is known of the practical significance, if any, of these three possible pathways.

In many situations, the rate-limiting factor for nitrification appears to be the supply of NH_4^+-N. Applications of NH_4^+-containing fertilizers to soils can massively increase populations and activities of autotrophic nitrifiers. There is considerable evidence that a limiting supply of NH_4^+ regulates nitrification in many natural ecosystems. Strong competition between roots of vegetation (and associated mycorrhizal fungi) and the heterotrophic soil biomass for the pool of NH_4^+ in the soil may leave little available for use by the autotrophic nitrifiers. Some evidence also suggests that in mature natural ecosystems, allochemicals originating from the predominant plant species present might also inhibit nitrifier activity.

The narrow species diversity of organisms involved in nitrification results in nitrification being greatly influenced by environmental factors such as soil pH, moisture, aeration, and temperature. Nevertheless, serotypic diversity among nitrifiers and the adaptation of indigenous nitrifiers to their particular environmental conditions mean that no definite "cutoff" points can be identified.

The role and significance of nitrification are subject to much controversy. The major drawbacks of nitrification are (1) the process results in acidification of the soil, (2) during the process N is lost as N_2O, (3) NO_3^- can undergo denitrification in anoxic soil sites with the release of N_2 and N_2O gases, and (4) the NO_3^- ion, in contrast to that of NH_4^+, is very mobile in soils and thus easily lost through leaching processes.

However, particularly in agricultural soils that are often fertilized with NH_4^+-containing fertilizers or urea, nitrification also results in (1) low levels of phytotoxic NH_4^+-N existing in soils and thus a medium conducive to good plant growth and (2) a decrease in gaseous losses of N through NH_3 volatilization.

Hence, although on balance, particularly from an environmental viewpoint, nitrification is often considered to have few assets and many drawbacks, in agricultural ecosystems it can still be considered as an agronomically desirable process. Attempts to regulate nitrification with the use of slow-release NH_4^+ fertilizers or with specific nitrification inhibitors have met with little application.

To a small extent, man has unwittingly influenced nitrification. Contamination of soils by trace elements (e.g., Hg, Ag, Cd, Ni, As, Cr, Zn, and Cu) originating from application of waste products such as sewage sludge can inhibit nitrification while the application of pesticides to soils can also sometimes inhibit nitrifier activity.

REFERENCES

Ahrens, A. (1977). Beitrag zur Frage der Indikatorfunktion der Bodenmikroorganismen am Beispel von drei verschiedenen Nutzungsstufen eines Sandbodens. *Soil Biol. Biochem.* **9**, 185–191.

Aleem, M. I. H., Lees, H., and Lyric, R. (1964). Ammonium oxidation by cell-free extracts of *Aspergillus wentii. Can. J. Biochem.* **42**, 989–992.

Alexander, M. (1965). Nitrification. *In* "Soil Nitrogen" (W. V. Bartholomew and F. E. Clark, eds.), pp. 307–343. Am. Soc. Agron., Madison, Wisconsin.

Alexander, M. (1977). "Introduction to Soil Microbiology." Wiley, New York.

Allison, F. E. (1973). "Soil Organic Matter and Its Role in Crop Production." Am. Elsevier, New York.

Ardakani, M. S., Rehbock, J. T., and McLaren, A. D. (1973). Oxidation of nitrite to nitrate in a soil column. *Soil Sci. Soc. Am. Proc.* **37**, 53–56.

Ardakani, M. S., Schulz, R. K., and McLaren, A. D. (1974). A kinetic study of ammonium and nitrite oxidation in a soil field plot. *Soil Sci. Soc. Am. Proc.* **38**, 273–277.

Baldwin, I. T., Olsen, R. K., and Reiners, W. A. (1983). Protein binding phenolics and the inhibition of nitrification in subalpine balsam fir soils. *Soil Biol. Biochem.* **15**, 419–423.

Bartlett, R. J. (1981). Nonmicrobial nitrite-to-nitrate transformations in soils. *Soil Sci. Soc. Am. J.* **45**, 1054–1058.

Belser, L. W. (1979). Population ecology of nitrifying bacteria. *Annu. Rev. Microbiol.* **33**, 309–333.

Belser, L. W., and Schmidt, E. L. (1978a). Diversity on the ammonia-oxidising nitrifier population of a soil. *Appl. Environ. Microbiol.* **36**, 584–588.

Belser, L. W., and Schmidt, E. L. (1978b). Nitrification in soils. *In* "Microbiology—1978" (D. Schlessinger, ed.), pp. 348–351. Am. Soc. Microbiol., Washington, D.C.

Belser, L. W., and Schmidt, E. L. (1978c). Serological diversity within a terrestrial ammonia-oxidizing population. *Appl. Environ. Microbiol.* **36**, 589–593.

Belser, L. W. and Schmidt, E. L. (1981). Inhibitory effect of nitrapyrin on three genera of ammonia-oxidizing nitrifiers. *Appl. Environ. Microbiol.* **41**, 819–821.

Biederbeck, V. O., and Campbell, C. A. (1973). Soil microbial activity as influenced by temperature trends and fluctuations. *Can. J. Soil Sci.* **53**, 363–376.

Bohlool, B. B., and Schmidt, E. L. (1980). The immunofluorescence approach in microbiol ecology. *Adv. Microb. Ecol.* **4**, 203–214.

Boon, B., and Laudelout, H. (1962). Kinetics of nitrite oxidation by *Nitrobacter winogradskyi. Biochem. J.* **85**, 440–447.

Boquel, G., Bruckert, S., and Suavin, L. (1970). Inhibition de la nitrification par les extraits aqueux de la litière de hetre (*Fagus silvatica*). *Rev. Ecol. Biol. Sol.* **7**, 357–366.

Bormann, F. H., and Likens, G. E. (1979). "Pattern and Process in a Forested Ecosystem." Springer-Verlag, Berlin and New York.

Bowen, G. D., and Smith, S. E. (1981). The effects of mycorrhizas on nitrogen uptake by plants. *In* "Terrestrial Nitrogen Cycles: Processes, Ecosystem Strategies and Management Impacts" (F. E. Clark and T. Rosswall, eds.), pp. 237–247. Ecological Bulletins, Stockholm.

Brar, S. S., and Giddens, J. (1968). Inhibition of nitrification in Bladen grassland soil. *Soil Sci. Soc. Am. Proc.* **32**, 821–823.

Bremner, J. M., and Bundy, L. G. (1974). Inhibition of nitrification in soils by volatile sulphur compounds. *Soil Biol. Biochem.* **6**, 161–165.

Broadbent, F. E., Tyler, K. B., and Hill, G. N. (1957). Nitrification of ammonical fertilizers in some California soils. *Hilgardia* **27**, 247–267.

Bundy, L. G., and Bremner, J. M. (1973). Inhibition of nitrification in soils. *Soil Sci. Soc. Am. Proc.* **37**, 396–398.

Campbell, C. A., and Biederbeck, V. O. (1982). Changes in mineral N and numbers of bacteria and actinomycetes during two years under wheat-fallow in Southwestern Saskatchewan. *Can. J. Soil Sci.* **62**, 125–137.

Campbell, C. A., Biederbeck, V. O., and Warder, F. G. (1971). Influence of simulated fall and spring condition on the soil system. II. Effect on soil nitrogen. *Soil Sci. Soc. Am. Proc.* **35**, 480–483.

Campbell, C. A., Biederbeck, V. O., and Hinman, W. C. (1975). Relationships between nitrate in summer-fallowed surface soil and some environmental variables. *Can. J. Soil Sci.* **55**, 213–223.

Campbell, N. E. R., and Aleem, M. I. H. (1965a). The effect of 2-chloro-6-(trichloromethyl)-pyridine on the chemoautotrophic metabolism of nitrifying bacteria. I. Ammonia and hydroxylamine oxidation by *Nitrosomonas. Antonie van Leeuwenhoek* **31**, 124–129.

Campbell, N. E. R., and Aleem, M. I. H. (1965b). The effect of 2-chloro,6-(trichloromethyl) pyridine on the chemoautotrophic metabolism of nitrifying bacteria. II. Nitrite oxidation by *Nitrobacter. Antonie van Leeuwenhoek* **31**, 137–144.

Chandra, P., and Bollen, W. B. (1961). Effects of nabam and mylone on nitrification, soil respiration and microbial numbers in four Oregon soils. *Soil Sci.* **92**, 387–396.

Chang, F. H., and Broadbent, F. E. (1982). Influence of trace metals on some soil nitrogen transformations. *J. Environ. Qual.* **11**, 1–4.

Chase, F. E., Corke, C. T., and Robinson, J. B. (1968). Nitrifying bacteria in soil. *In* "The Ecology of Soil Bacteria" (T. R. G. Gray and D. Parkinson, eds.), pp. 593–611. Liverpool Univ. Press, Liverpool.

Clark, F. E. (1977). Internal cycling of [15]nitrogen in shortgrass prairie. *Ecology* **58**, 1322–1333.

Coats, R. B., Leonard, R. L., and Goldman, C. R. (1976). Nitrogen uptake and release in a forested watershed, Lake Tahoe basin, California. *Ecology* **57**, 995–1004.

Cooper, J. E. (1975). Nitrification in soils incubated with pig slurry. *Soil Biol. Biochem.* **7**, 119–124.

Crutzen, P. J. (1974). Estimation of possible variations in total ozone due to natural causes and human activities. *Ambio* **3**, 201–210.

Dalton, H. (1977). Ammonia oxidation by the methane oxidising bacterium *Methylococcus capsulatus* strain Bath. *Arch. Microbiol.* **114**, 273–279.

Dommergues, Y. R. (1966). "Biologie du Sol." Presses Universitaire de France, Paris.

Domsch, K. H., and Paul, W. (1974). Simulation and experimental analysis of the influence of herbicides on soil nitrification. *Arch. Microbiol.* **97**, 283–301.

Doxtader, K. G., and Alexander, M. (1966). Nitrification by heterotrophic soil microorganisms. *Soil Sci. Soc. Am. Proc.* **30**, 351–355.

Dubey, H. D. (1968). Effect of soil moisture levels on nitrification. *Can. J. Microbiol.* **14**, 1348–1350.

Ellenberg, H. (1971). Nitrogen content, mineralization and cycling. *In* "Productivity of Forest Ecosystems" (P. Duvigneaud, ed.), pp. 504–514. UNESCO, Paris.

Ellenberg, H. (1977). Stickstoffs als Standortsfaktor, inbesondere fur mitteleuropaische Pflanzengesellschaften. *Oecol. Plant.* **12**, 1–22.

Elliot, J. M., Marks, C. F., and Tu, C. M. (1977). Effects of certain nematicides on soil nitrogen, soil nitrifiers, and populations of *Pratylenchus penetrans* in flue-cured tobacco. *Can. J. Plant Sci.* **57**, 143–154.

Ettinger-Tulczynska, R. (1969). A comparative study of nitrification in soils from arid and semi-arid areas of Israel. *J. Soil Sci.* **20**, 307–317.

Eylar, O. R., and Schmidt, E. L. (1959). A survey of heterotrophic microorganisms from soil for ability to form nitrite and nitrate. *J. Gen. Microbiol.* **20**, 473–481.

Federer, C. A. (1983). Nitrogen mineralization and nitrification: Depth variation in four New England forest soils. *Soil Sci. Soc. Am. J.* **47**, 1008–1014.

Fliermans, C. B., Bohlool, B. B., and Schmidt, E. L. (1974). Autecological study of the chemoautotroph *Nitrobacter* by immunofluorescence. *Appl. Microbiol.* **30**, 676–684.

Floate, M. J. S. (1981). Effects of grazing by large herbivores on nitrogen cycling in agricultural ecosystems. *In* "Terrestrial Nitrogen Cycles: Processes, Ecosystem Strategies and Management Impacts" (F. E. Clark and T. Rosswall, eds.), pp. 585–601. Ecological Bulletins, Stockholm.

Focht, D. D., and Verstraete, W. (1977). Biochemical ecology of nitrification and denitrification. *Adv. Microb. Ecol.* **1**, 135–214.

Frissel, M. J., ed. (1978). "Cycling of Mineral Nutrients in Agricultural Ecosystems." Elsevier, Amsterdam.

Goh, K. M., and Young, A. W. (1975). Effects of fertilizer nitrogen and 2-chloro-6-(trichloromethyl) pyridine (N-Serve) on soil nitrification, yield and nitrogen uptake of "Arawa" and "Hilgendorf" wheats. *N. Z. J. Agric. Res.* **18**, 215–225.

Goring, C. A. I. (1962a). Control of nitrification by 2-chloro-6-(trichloromethyl) pyridine. *Soil Sci.* **93**, 211–218.

Goring, C. A. I. (1962b). Control of nitrification of ammonium fertilizers and urea by 2-chloro-6-(trichloromethyl) pyridine. *Soil Sci.* **93**, 431–439.

Goring, C. A. I., and Laskowski, D. A. (1982). The effects of pesticides on nitrogen transformations in soils. *In* "Nitrogen in Agricultural Soils" (F. J. Stevenson, ed.), pp. 689–720. Am. Soc. Agron., Madison, Wisconsin.

Goring, C. A. I., and Scott, H. H. (1976). Control of nitrification by soil fumigants and N-Serve nitrogen stabilizer. *Down Earth* **32**, 14–17.

Gosz, J. R. (1981). Nitrogen cycling in coniferous ecosystems. *In* "Terrestrial Nitrogen Cycles: Processes, Ecosystem Strategies and Management Impacts" (F. E. Clark and T. Rosswall, eds.), pp. 405–426. Ecological Bulletins, Stockholm.

Gowda, T. K. S., Siddaramappa, R., and Sethunathan, N. (1976). Heterotrophic nitrification and nitrite tolerance by *Aspergillus carneus* (van Tiegh) Blochwitz, a predominant fungus isolated from benomyl-amended soil. *Soil Biol. Biochem.* **8**, 435–437.

Harada, T., and Kai, H. (1968). Studies on the environmental conditions controlling nitrification in soils. I. Effects of ammonium and total salts in media on the rate of nitrification. *Soil Sci. Plant Nutr.* **14**, 20–26.

Hart, P. B. S., and Goh, K. M. (1980). Regression equations to monitor inorganic nitrogen changes in fallow and wheat soils. *Soil Biol. Biochem.* **12**, 147–151.

Hauck, R. D. (1972). Synthetic slow-release fertilizers and fertilizer amendments. *In* "Organic Chemicals in the Soil Environment" (C. A. I. Goring and J. W. Hamker, eds.), pp. 633–690. Dekker, New York.

Haynes, R. J. (1981a). Soil pH decrease in the herbicide strip of grassed-down orchards. *Soil Sci.* **132**, 274–278.

Haynes, R. J. (1981b). Laboratory study of nutrient leaching from the surface of cultivated, grassed and herbicided orchard soil. *Soil Tillage Res.* **1**, 281–288.

Helyar, K. R. (1976). Nitrogen cycling and soil acidification. *J. Aust. Inst. Agric. Sci.* **42**, 217–221.

Hendrickson, L. L., Keeney, D. R., Walsh, L. M., and Liegel, E. A. (1978). Evaluation of nitrapyrin as a means of improving nitrogen efficiency in irrigated sands. *Agron. J.* **70**, 699–703.

Henis, Y. (1976). Effects of mineral nutrients on soil-borne pathogens and host resistance. *Proc. Colloq. Int. Potash Inst.* **12**, 101–112.

Hooper, A. B. (1978). Nitrogen oxidation and electron transport in ammonia-oxidizing bacteria. *In* "Microbiology—1978" (D. Schlessinger, ed.), pp. 299–304. Am. Soc. Microbiol., Washington, D.C.

Huber, D. M., Warren, H. L., Nelson, D. W., and Tsai, C. Y. (1977). Nitrification inhibitors—new tools for food production. *BioScience* **27**, 523–529.

Huntjens, J. L. M. (1971a). Influences of living plants on immobilization of nitrogen in permanent pastures. *Plant Soil* **34**, 393–404.

Huntjens, J. L. M. (1971b). The influence of living plants on mineralization of nitrogen. *Plant Soil* **35**, 77–94.

Huntjens, J. L. M., and Albers, R. A. J. M. (1978). A model experiment to study the influence of living plants on the accumulation of soil organic matter in pastures. *Plant Soil* **50**, 411–418.

Ishaque, M., and Cornfield, A. N. (1972). Nitrogen mineralization and nitrification during incubation of East Pakistan "tea" soils in relation to pH. *Plant Soil* **37**, 91–95.

Ishaque, M., and Cornfield, A. N. (1974). Nitrogen mineralization and nitrification in relation to incubation temperature in an acid Bangladesh soil lacking autotrophic nitrifying organisms. *Trop. Agric. (Trinidad)* **51**, 37–41.

Jansson, S. L. (1958). Tracer studies on nitrogen transformations in soil with special attention to mineralization–immobilization relationships. *Ann. R. Agric. Coll. Swed.* **24**, 101–361.

Jaques, R. P., Robinson, J. B., and Chase, F. E. (1959). Effects of thiourea, ethyl urethane and some dithiocarbamate fungicides on nitrification in Fox sandy loam. *Can. J. Soil Sci.* **39**, 235–243.

Jenny, H. (1980). "The Soil Resource Origin and Behaviour." Springer-Verlag, Berlin and New York.

Jones, J. M., and Richards, B. N. (1977). Effect of reforestation on turnover of ^{15}N-labelled nitrate and ammonium in relation to changes in soil microflora. *Soil Biol. Biochem.* **9**, 383–392.

Jones, R. W., and Hedlin, R. A. (1970). Ammonium, nitrite and nitrate accumulation in three Manitoba soils as influenced by added ammonium sulfate and urea. *Can. J. Soil Sci.* **50**, 331–338.

Jordan, C. F., Todd, R., and Escalante, G. (1979). Nitrogen conservation in a tropical rain forest. *Oecologia* **39**, 123–128.

Josserand, A., and Cleyet-Marel, J. C. (1979). Isolation from soils of *Nitrobacter* and evidence for novel serotypes, using immunofluorescence. *Microb. Ecol.* **5**, 197–205.

Josserand, A., Gay, G., and Faurie, G. (1981). Ecological study of two *Nitrobacter* serotypes co-existing in the same soil. *Microb. Ecol.* **7**, 275–280.

Justice, J. K., and Smith, R. L. (1962). Nitrification of ammonium sulfate in a calcareous soil as influenced by a combination of moisture, temperature and levels of added N. *Soil Sci. Soc. Am. Proc.* **26**, 246–250.

Khanna, P. K. (1981). Leaching of nitrogen from terrestrial ecosystems—patterns, mechanisms and ecosystem responses. *In* "Terrestrial Nitrogen Cycles: Processes, Ecosystem Strategies and Management Impacts'' (F. E. Clark and T. Rosswall, eds.), pp. 343–352. Ecological Bulletins, Stockholm.

Kowalenko, C. G., and Cameron, D. R. (1976). Nitrogen transformations in an incubated soil as affected by combinations of moisture content and temperature and adsorption–fixation of ammonium. *Can. J. Soil Sci.* **56**, 63–77.

Kowalenko, C. G., and Cameron, D. R. (1978). Nitrogen transformations in soil–plant systems in three years of field experiments using tracer and non-tracer methods on an ammonium-fixing soil. *Can. J. Soil Sci.* **58**, 195–208.

Kumm, K. I. (1976). An economic analysis of nitrogen leaching caused by agricultural activities. *In* ''Nitrogen, Phosphorus and Sulfur—Global Cycles'' (B. H. Svensson and R. Söderlund, eds.), pp. 169–183. Ecological Bulletins, Stockholm.

Laura, R. D. (1977). Salinity and nitrogen mineralization in soil. *Soil Biol. Biochem.* **9**, 333–336.

Lebbink, G., and Kolenbrander, G. J. (1974). Quantitative effect of fumigation with 1,3-dichloropropene mixtures and with metham-sodium on the soil nitrogen status. *Agric. Environ.* **1**, 283–292.

Lees, H., and Simpson, J. R. (1957). The biochemistry of the nitrifying organisms. 5. Nitrite oxidation by *Nitrobacter*. *Biochem. J.* **65**, 297–305.

Leyshon, A. J., Campbell, C. A., and Warder, F. G. (1980). Comparison of the effect of NO_3^- and NH_4^+ on growth, yield and yield components of Manitou spring wheat and Conquest barley. *Can. J. Plant Sci.* **60**, 1063–1070.

Liang, C. N., and Tabatabai, M. A. (1978). Effects of trace elements on nitrification in soils. *J. Environ. Qual.* **7**, 291–293.

Likens, G. E., Bormann, F. N., and Johnson, N. M. (1969). Nitrification: Importance to nutrient losses from a cutover forested ecosystem. *Science* **163**, 1205–1206.

Lodhi, M. A. K. (1979). Inhibition of nitrifying bacteria, nitrification and mineralization in mine spoil soils as related to their successional stages. *Bull. Torrey Bot. Club* **106**, 284–291.

McIntosh, T. H., and Frederick, L. R. (1958). Distribution and nitrification of anhydrous ammonia in a Nicollet sandy clay loam. *Soil Sci. Soc. Am. Proc.* **22**, 402–405.

McLaren, A. D. (1971). Kinetics of nitrification in soil: Growth of the nitrifiers. *Soil Sci. Soc. Am. Proc.* **35**, 91–95.

Mahendrappa, M. K., Smith, R. L., and Christianson, A. T. (1966). Nitrifying organisms affected by climatic region in western United States. *Soil Sci. Soc. Am. Proc.* **30**, 60–62.

Malhi, S. S., and McGill, W. B. (1982). Nitrification in three Alberta soils: Effect of temperature, moisture and substrate concentration. *Soil Biol. Biochem.* **14**, 393–399.

Malhi, S. S., and Nyborg, M. (1979). Nitrate formation during winter from fall-applied urea. *Soil Biol. Biochem.* **11**, 439–441.

Matson, P. A., and Vitousek, P. M. (1981). Nitrification potentials following clearcutting in the Hoosier National Forest, Indiana. *For. Sci.* **27**, 781–791.

Maynard, D. N., Barker, A. F., Minotti, P. L., and Peck, N. H. (1976). Nitrate accumulation in vegetables. *Adv. Agron.* **28**, 71–118.

Mazur, A. R., and Hughes, T. D. (1975). Nitrogen transformations in the soil as affected by the fungicides benomyl, dyrene, and maneb. *Agron. J.* **67**, 775–778.

Melillo, J. M. (1977). Mineralization of nitrogen in northern forest ecosystems. Ph.D. Thesis, Yale University, New Haven, Connecticut.

Melillo, J. M. (1981). Nitrogen cycling in deciduous forests. ''Terrestrial Nitrogen Cycles: Processes, Ecosystem Strategies and Management Impacts'' (F. E. Clark and T. Rosswall, eds.), pp. 427–442. Ecological Bulletins, Stockholm.

Miller, R. D., and Johnson, D. D. (1964). The effect of soil moisture tension on CO_2 evolution, nitrification and nitrogen mineralization. *Soil Sci. Soc. Am. Proc.* **28**, 644–647.

Molina, J. A. E., and Rovira, A. D. (1964). The influence of plant roots on autotrophic nitrifying bacteria. *Can. J. Microbiol.* **10**, 249–256.

Moore, D. R. E., and Waid, J. S. (1971). The influence of washing of living roots on nitrification. *Soil Biol. Biochem.* **3**, 69–83.

Morrill, L. G., and Dawson, J. E. (1967). Patterns observed for the oxidation of ammonium to nitrate by soil organisms. *Soil Sci. Soc. Am. Proc.* **31**, 757–760.

Munro, P. E. (1966). Inhibition of nitrate-oxidizers by roots of grass. *J. Appl. Ecol.* **3**, 227–229.

Myers, R. J. K. (1975). Temperature effects on ammonification in a tropical soil. *Soil Biol. Biochem.* **7**, 83–86.

Nakos, G. G., and Wolcott, A. R. (1972). Bacteriostatic effect of ammonium on *Nitrobacter agilis* in mixed culture with *Nitrosomonas europaea*. *Plant Soil* **36**, 521–527.

Nicholas, D. J. D. (1978). Intermediary metabolism of nitrifying bacteria, with particular reference to nitrogen, carbon, and sulphur compounds. *In* "Microbiology—1978" (D. Schlessing, ed.), pp. 305–309. Am. Soc. Microbiol., Washington, D.C.

Nommik, H. (1981). Fixation and biological availability of ammonium in soil clay minerals. *In* "Terrestrial Nitrogen Cycles: Processes, Ecosystem Strategies and Management Impacts" (F. E. Clark and T. Rosswall, eds.), pp. 273–279. Ecological Bulletins, Stockholm.

Nyborg, M., and Hoyt, P. B. (1978). Effects of soil acidity and liming on mineralization of soil nitrogen. *Can. J. Soil Sci.* **58**, 331–338.

Nyborg, M., and Malhi, S. S. (1979). Increasing the efficiency of fall-applied urea fertilizer by placing in big pellets or in nests. *Plant Soil* **52**, 461–465.

Nye, P. H. (1981). Changes of pH across the rhizosphere induced by roots. *Plant Soil* **61**, 7–26.

Odu, C. T. I., and Adeoye, K. B. (1970). Heterotrophic nitrification in soils—a preliminary investigation. *Soil Biol. Biochem.* **2**, 41–45.

Odu, C. T. I., and Akerele, R. B. (1973). Effect of soil, grass and legume root extracts on heterotrophic bacteria, nitrogen mineralization and nitrification in soils. *Soil Biol. Biochem.* **5**, 861–867.

Olson, R. K., and Reiners, W. K. (1983). Nitrification in subalpine balsam fir soils: Tests for inhibitory factors. *Soil Biol. Biochem.* **15**, 413–418.

Osborne, G. F. (1977). Some effects of the nitrification inhibitor [2-chloro-(trichloromethyl) pyridine] on the use of fertilizer nitrogen and the growth of two wheat varieties. *Aust. J. Exp. Agric. Anim. Husb.* **17**, 645–651.

Painter, H. A. (1977). Microbial transformations of inorganic nitrogen. *Prog. Water Technol.* **8**, 3–29.

Pang, P. C., Cho, C. M., and Hedlin, R. A. (1975). Effects of pH and nitrifier population on nitrification of band-applied and homogenously mixed urea nitrogen in soils. *Can. J. Soil Sci.* **55**, 15–21.

Pierre, W. H., Webb, J. R., and Shrader, W. D. (1971). Quantitative effects of nitrogen fertilizer on the development and downward movement of soil acidity in relation to level of fertilization and crop removal in a continuous corn cropping system. *Agron. J.* **63**, 291–297.

Prasad, M. (1976). Nitrogen nutrition and yield of sugar cane as affected by N-Serve. *Agron. J.* **68**, 343–346.

Purchase, B. S. (1974a). Evaluation of the claim that grass root exudates inhibit nitrification. *Plant Soil* **41**, 527–539.

Purchase, B. S. (1974b). The influence of phosphate deficiency on nitrification. *Plant Soil* **41**, 541–547.

Quayle, J. R. (1972). The metabolism of one-carbon compounds by micro organisms. *Adv. Microb. Physiol.* **7**, 119–203.

Rennie, R. J., and Schmidt, E. L. (1977). Immunofluorescence studies of *Nitrobacter* populations in soils. *Can. J. Microbiol.* **23**, 1011–1017.

Rennie, R. J., Reyes, V. G., and Schmidt, E. L. (1977). Immunofluorescent detection of the effects of wheat and soybean roots on *Nitrobacter* in soil. *Soil Sci.* **124**, 10–15.

Reuss, J. O., and Smith, R. L. (1965). Chemical reactions of nitrites in acid soils. *Soil Sci. Soc. Am. Proc.* **29**, 267–270.

Rice, E. L. (1974). "Allelopathy." Academic Press, New York.

Rice, E. L. (1979). Allelopathy—an update. *Bot. Rev.* **45**, 15–109.

Rice, E. L., and Pancholy, S. K. (1972). Inhibition of nitrification by climax vegetation. *Am. J. Bot.* **59**, 1033–1040.

Rice, E. L., and Pancholy, S. K. (1973). Inhibition of nitrification by climax ecosystems. II. Additional evidence and possible role of tannins. *Am. J. Bot.* **60**, 691–702.

Rice, E. L., and Pancholy, S. K. (1974). Inhibition of nitrification by climax ecosystems. III. Inhibitors other than tannins. *Am. J. Bot.* **61**, 1095–1103.

Ridge, E. H. (1976). Studies on soil fumigation. II. *Soil Biol. Biochem.* **8**, 249–253.

Robertson, G. P. (1982a). Nitrification in forested ecosystems. *Philos. Trans. R. Soc. London, B Ser.* **296**, 445–457.

Robertson, G. P. (1982b). Factors regulating nitrification in primary and secondary succession. *Ecology* **63**, 1561–1573.

Robertson, G. P., and Vitousek, P. M. (1981). Nitrification potentials in primary and secondary succession. *Ecology* **62**, 376–387.

Romanovskaya, V. A., Malasheno, Y. R., Lyalko, V. I., Bogachenko, V. N., and Sokolev, I. G. (1977). Involvement of methane-utilizing micro organisms in accumulation and recirculation of readily mobile carbon and nitrogenous compounds. *In* "Soil Organisms as Components of Ecosystems" (U. Lohm and T. Persson, eds.), pp. 556–560. Ecological Bulletins, Stockholm.

Rother, J. A., Millbank, J. W., and Thornton, I. (1982). Effects of heavy-metal additions on ammonification and nitrification in soils contaminated with cadmium, lead and zinc. *Plant Soil* **69**, 239–258.

Rovira, A. D. (1976). Studies on soil fumigation. I. *Soil Biol. Biochem.* **8**, 241–247.

Sabey, B. R. (1969). Influence of soil moisture tension on nitrate accumulation in soils. *Soil Sci. Soc. Am. Proc.* **33**, 263–266.

Sabey, B. R., Batholomew, W. V., Shaw, R., and Pesek, J. (1956). Influence of temperature on nitrification in soils. *Soil Sci. Soc. Am. Proc.* **20**, 357–360.

Sahrawat, K. L. (1982). Nitrification in some tropical soils. *Plant Soil* **65**, 281–286.

Sarathchandra, S. U. (1978). Nitrification activities and the changes in the populations of nitrifying bacteria in soil perfused at two different H-ion concentrations. *Plant Soil* **50**, 99–111.

Schmidt, E. L. (1954). Nitrate formation by a soil fungus. *Science* **119**, 187–189.

Schmidt, E. L. (1982). Nitrification in soil. *In* "Nitrogen in Agricultural Soils" (F. J. Stevenson, ed.), pp. 253–288. Am. Soc. Agron., Madison, Wisconsin.

Skujins, J. (1981). Nitrogen cycling in arid ecosystems. *In* "Terrestrial Nitrogen Cycles:

Processes, Ecosystem Strategies and Management Impacts'' (F. E. Clark and T. Rosswall, eds.), pp. 477–491. Ecological Bulletins, Stockholm.

Smiley, R. W. (1975). Forms of nitrogen and pH in the root zone and their importance to root infections. *In* "Biology and Control of Soil-Borne Plant Pathogens" (G. W. Bruehl, ed.), pp. 55–62. Am. Phytopathol. Soc., St. Paul, Minnesota.

Smit, A. J., and Woldendorp, J. W. (1981). Nitrate production in the rhizosphere of *Plantago* species. *Plant Soil* **61,** 43–52.

Smith, A. J., and Hoare, D. S. (1977). Specialist phototrophs, lithotrophs and methylotrophs: A unity among a diversity of procaryotes? *Bacteriol. Rev.* **41,** 419–448.

Smith, M. S., and Weeraratna, C. S. (1975). Influence of some biologically active compounds on microbial activity on the availability of plant nutrients in soils. II. Nitrapyrin, dazomet, 2-chlorobenzamide and tributyl-3-chlorobenzyl ammonium bromide. *Pestic. Sci.* **6,** 605–615.

Smith, S. E. (1980). Mycorrhizas of autotrophic higher plants. *Biol. Rev. Cambridge Philos. Soc.* **55,** 475–510.

Stanley, P. M., and Schmidt, E. L. (1981). Serological diversity of *Nitrobacter* spp. from soil and aquatic habits. *Appl. Environ. Microbiol.* **41,** 1069–1071.

Stojanovic, B. J., and Alexander, M. (1958). Effect of inorganic nitrogen on nitrification. *Soil Sci.* **86,** 208–215.

Tate, R. L. (1977). Nitrification in histosols, a potential role for the heterotrophic nitrifier. *Appl. Environ. Microbiol.* **33,** 911–914.

Thiagalingam, K., and Kanehiro, Y. (1973). Effect of temperature on nitrogen transformation in four Hawaiian soils. *Plant Soil* **38,** 177–189.

Todd, R. L., Swank, W. T., Douglass, J. E., Kerr, P. C., Brockway, D. L., and Monk, C. D. (1975). The relationship between nitrate concentration in southern Appalachian mountain streams and terrestrial nitrifiers. *Agro-Ecosystems* **2,** 127–132.

Van de Dijk, S. J., and Troelstra, S. R. (1980). Heterotrophic nitrification in a heath soil demonstrated by an *in situ* method. *Plant Soil* **57,** 11–21.

Van Faassen, H. G. (1974). Effect of the fungicide benomyl on some metabolic processes, and on numbers of bacteria and actinomycetes in the soil. *Soil Biol. Biochem.* **6,** 131–133.

Verstraete, W. (1981a). Nitrification. *In* "Terrestrial Nitrogen Cycles: Processes, Ecosystem Strategies and Management Impacts" (F. E. Clark and T. Rosswall, eds.), pp. 303–314. Ecological Bulletins, Stockholm.

Verstraete, W. (1981b). Nitrification in agricultural systems; call for control. *In* "Terrestrial Nitrogen Cycles: Processes, Ecosystem Strategies and Management Impacts" (F. E. Clark and T. Rosswall, eds.), pp. 565–572. Ecological Bulletins, Stockholm.

Vitousek, P. M. (1981). Clear-cutting and the nitrogen cycle. *In* "Terrestrial Nitrogen Cycles: Processes, Ecosystem Strategies and Management Impacts" (F. E. Clark and T. Rosswall, eds.), pp. 631–642. Ecological Bulletins, Stockholm.

Vitousek, P. M., Gosz, J. R., Grier, C. C., Melillo, J. M., Reiners, J. M., and Todd, R. L. (1979). Nitrate losses from disturbed ecosystems. *Science* **204,** 469–474.

Vitousek, P. M., Gosz, J. R., Grier, C. C., Melillo, J. M., and Reiners, W. A. (1982). A comparative analysis of potential nitrification and nitrate mobility in forest ecosystems. *Ecol. Monogr.* **52,** 155–177.

Wainwright, W. (1981). Mineral transformations by fungi in culture and in soils. *Z. Pflanzenernaehr. Bodenkd.* **144,** 41–63.

Walker, N. (1978). Nitrification and nitrifying bacteria. *In* "Soil Microbiology" (N. Walker, ed.), pp. 133–146. Butterworth, London.

Walker, N., and Wickramasinghe, K. N. (1979). Nitrification and autotrophic nitrifying bacteria in acid tea soils. *Soil Biol. Biochem.* **11,** 231–236.

Walker, T. W., and Syers, J. K. (1976). The fate of phosphorus during pedogenesis. *Geoderma* **15,** 1–19.

Warren, M. (1965). A study of soil-nutritional and other factors operating in a secondary succession in highveld grassland in the neighbourhood of Johannesburg. Ph.D thesis, University of Johannesburg, South Africa.

Weber, D. F., and Gainey, P. L. (1962). Relative sensitivity of nitrifying organisms to hydrogen ions in soils and in solutions. *Soil Sci.* **94,** 138–145.

Wheeler, G. L., and Donaldson, J. M. (1983). Nitrification in an upland forest sere. *Soil Biol. Biochem.* **15,** 119–121.

Whittenbury, R., and Kelly, D. P. (1977). Autotrophy: A conceptual phoenix. *Symp. Soc. Gen. Microbiol.* **17,** 121–149.

Whittenbury, R., Phillips, K. C., and Wilkinson, J. F. (1970). Enrichment, isolation and some properties of methane-utilizing bacteria. *J. Gen. Microbiol.* **61,** 205–218.

Wilson, D. O. (1977). Nitrification in soil treated with domestic and industrial sewage sludge. *Environ. Pollut.* **12,** 73–82.

Woodmansee, R. G., Dodd, J. L., Bowman, R. A., Clark, F. E., and Dickinson, C. E. (1978). Nitrogen budget of a shortgrass prairie ecosystem. *Oecologia* **34,** 363–376.

Woodmansee, R. G., Vallis, I., and Mott, J. J. (1981). Grassland nitrogen. *In* "Terrestrial Nitrogen Cycles: Processes, Ecosystem Strategies and Management Impacts" (F. E. Clark and T. Rosswall, eds.), pp. 443–462. Ecological Bulletins, Stockholm.

Yoshida, T., and Alexander, M. (1970). Nitrous oxide formation by *Nitrosomonas europaea* and heterotrophic microorganisms. *Soil Sci. Soc. Am. Proc.* **34,** 880–882.

Chapter 4

Retention and Movement of Nitrogen in Soils

K. C. CAMERON AND R. J. HAYNES

I. INTRODUCTION

Organic N usually constitutes over 90% of total N in surface soil and its mobility in soils is generally low. Ammonium N is derived from the mineralization of soil organic N, added organic materials, or addition of urea or ammoniacal fertilizers. Ammonium is unlikely to be leached from soils because (1) NH_4^+ ions are held in the soil by the processes of cation exchange and fixation within clay lattices, (2) organic matter can fix considerable amounts of ammonia (NH_3), and (3) NH_4^+ can be readily immobilized by the microbial biomass or alternatively nitrified to NO_3^--N. In contrast to NH_4^+, there is little tendency for the NO_3^- anion to be absorbed by soil colloids, which commonly possess a net negative charge. Nitrate is thus susceptible to diffusion and mass transport with soil water.

Leaching losses of NO_3^- to groundwater occur principally when soil NO_3^- levels are high and downward water movement is large. The magnitude of such losses depends on factors such as rainfall, evaporation, soil type, and plant cover. On uneven terrain, surface runoff can become a problem when rain falls at a faster rate than water can infiltrate the soil surface. As the intensity and frequency of rainstorms increase, runoff is changed to water erosion. Losses of N can occur as soluble N in runoff or particulate N in sediments. Even under dry conditions there is a risk of erosion since wind can carry small soil particles considerable distances.

Most terrestrial ecosystems show a reasonably closed N cycle and only lose significant amounts of N through leaching, runoff, and erosion following severe disturbance such as burning, harvesting, irrigating, or fertilizing (Khanna, 1981; Vitousek, 1981). In contrast, losses can be large in agricultural ecosystems, which are often continually disturbed (Morisot, 1981; Power, 1981; Vitousek, 1981). Leaching is often the most important channel of N loss from cultivated field soils other than that accounted for in crop uptake (Allison, 1973; Legg and Meisinger, 1982).

The economic significance of such losses of N is self-evident especially in areas where the loss becomes a limiting factor for plant growth. These losses can also have environmental consequences since they can increase the productivity of surface waters, particularly lakes and estuaries. Over-enrichment (eutrophication) brings about many undesirable changes, including proliferation of algae, a decrease in water clarity, and a depletion of dissolved oxygen in bottom water [National Research Council (NRC), 1978]. A high level of nitrate in drinking water has also been considered as a potential health hazard (Shural and Gruener, 1977).

In this chapter, the processes of adsorption and fixation of mineral N in soils are outlined. The processes of runoff, erosion, and leaching are also reviewed and the factors affecting such losses of N are discussed.

II. ADSORPTION AND FIXATION PROCESSES

A. Adsorption of Ammonium

Soil clay and organic matter have a predominantly negative charge and are able to attract and hold positively charged cations such as Ca^{2+}, K^+, and NH_4^+ by the process of cation exchange (Thomas, 1977; Talibudeen, 1981). The net negative charge possessed by most soils is attributable to both inorganic and organic soil constituents. In the lattices of clay minerals, the source of negative charge arises from the substitution of Mg^{2+}, Al^{3+}, or Si^{4+} within the octahedral or tetrahedral sheets by isomorphic ions of smaller valency such as Li^+, Mg^{2+}, Fe^{2+}, and Al^{3+} (see Talibudeen, 1981). On soil organic matter the negative charges originate from dissociation of carboxyl (COOH) and phenolic OH groups (Hayes and Swift, 1978). The total negative charge on a soil represents its ability to hold positively charged ions (i.e., its cation exchange capacity, CEC).

Cation exchange is a reversible process in which cations in soil solution are in dynamic equilibrium with those held on exchange sites. Exchangeable cations are therefore readily available to plants. Thus, exchangeable NH_4^+ is a plant-available form of N that is, nevertheless, effectively protected against leaching by percolating waters. Leaching losses of NH_4^+ are only likely to be a problem in soils with an extremely low CEC.

B. Fixation of Ammonium by Clays

Ammonium can be held by 2:1 clay minerals (e.g., vermiculites and montmorillonites) in a nonexchangeable "fixed" form. The 2:1 clay minerals are made up of layers of structural units consisting of an octahedral Al—O—OH sheet sandwiched between two Si—O sheets (Borchardt, 1977; Brown *et al.*, 1978). The surfaces between the negatively charged layers consist of oxygen ions arranged hexagonally. The opening within the hexagon is approximately 2.8 Å and cations having a similar diameter (e.g., K^+, 2.66 Å, and NH_4^+, 2.86 Å) are able to fit into these openings. Such ions are tightly held within the interlayer space and the layers are able to approach and bind together, thus preventing reexpansion of the lattices. When the lattices are in contracted form, the K^+ and NH_4^+ held within the interlayers are considered to be in the fixed form (Fig. 1). Hydrated cations larger than 2.8 Å (e.g., Ca^{2+}, Mg^{2+}, Na^+, and H^+) cannot enter the hexagonal openings and are loosely held between the layers so that the lattice remains in an expanded state when these cations predominate.

Cations in an expanded crystal lattice are readily replaced by other cations in soil solution, which leaves the lattices in an expanded state, but not by cations that contract the lattices (Nommik and Vahtras, 1982). Similarly, cations present in a contracted crystal lattice may be slowly

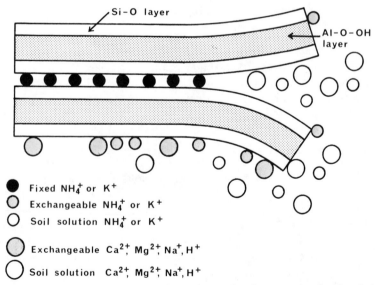

Fig. 1. Schematic diagram of the different forms of cations associated with a 2:1 clay mineral.

replaced by cations that expand the lattice but less readily replaced by cations that contract it. Nevertheless, added NH_4^+ has been shown to partially replace native fixed NH_4^+ (Kowalenko and Ross, 1980).

The equilibrium of soil NH_4^+ can be represented as shown below (Nommik and Vahtras, 1982):

Solution NH_4^+ $\xrightleftharpoons{\text{fast}}$ Exchangeable NH_4^+

$\xrightleftharpoons{\text{slow}}$ Intermediate NH_4^+ $\xrightleftharpoons{\text{very slow}}$ Fixed NH_4^+ (1)

The intermediate NH_4^+ ions are those that occupy the interlayer sites near the edge of the lattice. Ions held in this area of medium lattice closure are thus transitional between exchangeable and fixed. Intermediate NH_4^+ ions may be exchanged with H^+ or K^+ and with other cations if the lattice opens slightly (Fig. 1).

In view of the similarity of NH_4^+ and K^+ with respect to the fixation reaction, and the predominance of K^+ over NH_4^+ in soils, it is probably justifiable to insert the sum of NH_4^+ plus K^+, rather than NH_4^+ alone, in the above equilibrium equation (Nommik and Vahtras, 1982). When added in equivalent amounts, surface soils usually fix NH_4^+ and K^+ in the ratio of about 3 : 1 (Dissing-Nielsen, 1971; Sippola et al., 1973), although the ratio is generally higher for subsoils than for surface soils (Sippola, 1976).

1. Factors Affecting Fixation and Release

The absolute amount of ammonium fixed increases with increasing amounts of added ammonium (Nommik, 1965; Black and Waring, 1972; Sippola et al., 1973; Sowden et al., 1978), but the percentage fixation generally decreases with increasing NH_4^+ additions (Nommik and Vahtras, 1982). The actual rate of fixation is, of course, highest immediately after ammonium addition and slows as equilibrium is reached (Nommik, 1965; Dissing-Nielsen, 1971).

Since K^+ and NH_4^+ are fixed by the same mechanism, the amount of ammonium fixation can be greatly depressed by a prior addition of K^+ (Jansson, 1958; Osborne, 1976). Simultaneous application of K^+ and NH_4^+ can still result in considerable ammonium fixation but the amount fixed depends on the ratio of K^+ to NH_4^+ ions added. An addition of K^+ after NH_4^+ does not appreciably affect the amount of NH_4^+ fixed. However, addition of K^+ either in combination or after NH_4^+ addition tends to inhibit the release of fixed NH_4^+ (Axley and Legg, 1960; Walsh and Murdoch, 1963) since NH_4^+ is not likely to be released from interlayer positions as long as the concentration of soluble and exchangeable K^+ remains high (equation (1)). Ammonium fixation can also be reduced by organic substances entering the interlattice space and thereby preventing the entry of ammonium ions and/or the contraction of the lattice structure (Hinman, 1964; Porter and Stewart, 1970).

Ammonium fixation generally decreases with decreasing soil pH (Kaila, 1966; Nommik, 1965; Raju and Mukhopadhyay, 1976). It is thought that under acidic conditions hydroxyl Al groups partially fill the interlayer space of the 2:1 minerals restricting NH_4^+ ion entry and lattice collapse (Mortland and Wolcott, 1965). A second factor may be the relatively higher replacing power of the H^+ ion (Nommik and Vahtras, 1982).

Drying a moist soil increases its fixation capacity appreciably (Blasco and Cornfield, 1966). Black and Waring (1972) reported a 3- to 10-fold increase as a result of air drying a series of arable soils. Two mechanisms are responsible: (1) the concentration of NH_4^+ in soil solution is increased by the loss of water and (2) the dehydration of the interlayer space causes lattice contraction, which traps the NH_4^+ ions. Greatest ammonium fixation occurs when the soil undergoes alternate drying and wetting. Blasco and Cornfield (1966) reported average fixation ratios of 1:12:16 for air drying:oven drying (100°C):alternate drying plus wetting, respectively.

Similar effects can be expected with freezing of the soil as this also removes water and an increase in ammonium fixation has been reported by Walsh and Murdoch (1960, 1963) and Nommik (1965).

2. Amounts and Availability of Fixed Ammonium

Many soils contain native fixed ammonium in amounts that often exceed the total content of NO_3^- plus exchangeable NH_4^+. Soils with high amounts of fixed NH_4^+ are those with predominantly 2:1-type clay minerals (e.g., vermiculites and illites). The concentration of fixed NH_4^+ in soils does, however, vary greatly (see Table I) and can range from nil in sandy surface soils to over 1000 μg N gm^{-1} in clay subsoils (Young and Aldag, 1982). The amount of fixed NH_4^+ in topsoils of the United States and Europe may represent between 3 and 10% of the total N content of the soil (Bremner, 1959; Walsh and Murdock, 1960; Young, 1962; Kaila, 1966). In the subsoil the percentage of total N as fixed NH_4^+ can increase due to the lower organic matter and higher clay content. The total amount of fixed NH_4^+ within the root zone may range from 0 to over 1000 kg N ha^{-1} (Young, 1962; Hinman, 1964; Aldag, 1978).

While native clay-fixed NH_4^+ is generally considered to have very low availability to plants (Walsh and Murdock, 1963; Black and Waring, 1972; Mohammed, 1979), recently fixed NH_4^+, originating from fertilizer applications, appears to have a more dynamic nature (Black and Waring, 1972; Kowalenko, 1978, 1981; Kowalenko and Cameron, 1976; Kowalenko and Ross, 1980; Fischer et al., 1981; Kudeyarov et al., 1981; Mengel and Scherer, 1981; Preston, 1982). For example, in a field experiment Mengel

Table I

Quantities of Fixed Ammonium in Soils from Various World Locations[a]

Location	Concentration range ($\mu g\ gm^{-1}$)	Notes	References
Alberta, Canada	158–330	7–14% of N in topsoils	Moore (1965)
Pacific Northwest, U.S.A.	17–138	<1 to 10% of N in topsoils, 2 to 42% in subsoils	Young (1962); Young and McNeal (1964)
Finland	0–623		Kaila (1966)
England	52–252	4 to 8% of N in topsoils, 19 to 45% in subsoils	Bremner (1959); Bremner and Harada (1959)
France	25–130	Content positively correlated with amount of clay	Blanchet et al. (1963); Gouny et al. (1960)
Nigeria	32–220	2 to 6% of N in topsoils, 45 to 63% at depths of 1.5 to 2.2 m	Moore and Ayeke (1965); Opuwaribo and Odu (1974)
Israel	3–10	2 to 25% of N	Feigin and Yaalon (1974)
Punjab	10–68	5.5% of N in topsoils	Grewal and Kanwar (1967)
Australia	41–1076	5 to 90% of N	Martin et al. (1970); Osborne (1976); Black and Waring (1972)

[a] Adapted from Young and Aldag (1982).

and Scherer (1981) showed that there were significant changes in the content of fixed NH_4^+ during the growing season. Early in the growing season, the upper soil layers (0–60 cm) were depleted of fixed NH_4^+ while later fixed NH_4^+ at a depth of 60–90 cm was depleted. Several other workers have demonstrated considerable plant uptake of recently fixed NH_4^+ (Black and Waring, 1972; Kowalenko and Cameron, 1976; Osborne, 1976; Kowalenko, 1978; Preston, 1982).

It appears that nitrifying organisms are major intermediaries in making fixed NH_4^+ available to plants. Indeed, Kowalenko and Cameron (1976) observed that addition of nitrapyrin (a nitrification inhibitor) resulted in greatly reduced uptake of fixed $^{15}NH_4^+$ by plants. The consumption of NH_4^+ by nitrifiers will lower concentrations of soluble and exchangeable

NH_4^+ thus causing the release of some fixed NH_4^+ (Nommik and Vahtras, 1982; see equation (1)).

It is evident that from an agronomic viewpoint NH_4^+ fixation should not be considered as an entirely unfavorable phenomenon. Under certain soil and climatic conditions, the fixation of NH_4^+ may be a positive factor in preventing losses through leaching and ensuring a more even supply of N throughout the growing season (Nommik and Vahtras, 1982). Indeed, NH_4^+ that becomes fixed in 2:1 clay minerals is considerably more available to plants, at least in the short term, than is NH_4^+ that has recently been biologically immobilized into organic forms (Preston, 1982).

C. Adsorption of Ammonia

Initial adsorption reactions occur between NH_3 and soil colloids when aqueous or anhydrous NH_3 are injected into soils. Physical sorption occurs when the polar NH_3 molecule forms H bonds with oxide and hydroxide surfaces of clay minerals and humic colloids (James and Harward, 1964). This mechanism is only important in zones of high NH_3 concentration and in the absence of water since water molecules can readily replace NH_3 retained by H bonding (Nommik and Vahtras, 1982). Sorption can also occur through coordination of NH_3 with exchangeable metal cations (Russell, 1965; Mortland, 1969). Ammonia and water molecules compete for coordination positions around exchangeable cations and each can displace the other. Thus, the reaction of NH_3 is reversible and the positive reaction predominates as NH_3 is injected into the soil and the reverse reaction predominates as gas diffuses away from the placement (Nommik and Vahtras, 1982).

Injected NH_3 can also be chemisorbed by soils through the formation and retention of NH_4^+ ions following acquisition of a proton by the NH_3 molecule. Protons may be supplied by exchangeable H^+ ions on clay and organic colloid surfaces at low pH or from OH^- groups associated with silicon on the edges of clay minerals at high pH (Mortland, 1966; Ashworth and Pyman, 1979). The capacity of soils to retain NH_3 as exchangeable NH_4^+ is likely to be closely related to soil pH and buffering capacity (Nommick and Vahtras, 1982).

D. Fixation of Ammonia by Soil Organic Matter

Soil organic matter can fix considerable amounts of ammonia (NH_3) in nonexchangeable forms. Burge and Broadbent (1961) reported an average of 161 meq of gaseous ammonia fixed per 100 gm of carbon in some organic soils. Young (1964) reported the following regression equation describing NH$_3$ fixation in a range of soils, which illustrates the relative importance of the organic fraction:

$$\text{ppm NH}_3\text{-N} = 147 + 596 \, (\% \text{ organic C}) + 43.4 \, (\% \text{ clay}) \qquad (2)$$

The capacity of soil to fix NH$_3$ is variable and Young (1964) reported that between 2 and 28% of the total NH$_3$ retention of mineral soils from the Pacific Northwest was by the organic fraction.

The exact nature of NH$_3$ organic matter complexes is unknown although aromatic humic components (e.g., phenols and quinones) are thought to be primarily responsible for NH$_3$ fixation (Mortland and Wolcott, 1965; Broadbent and Stevenson, 1966; Nommik and Vahtras, 1982). The reaction of NH$_3$ with aromatic compounds is thought to depend on intense polymerization during which N is combined into bridging structures (Broadbent and Stevenson, 1966). An example of such a reaction mechanism is shown in Fig. 2. Other compounds, such as carbohydrates, can also fix NH$_3$ (Mortland and Wolcott, 1965).

The rate and amount of fixation are strongly pH dependent because reactive groups on organic matter become increasingly polarized and reactive with increasing alkalinity (see Mortland and Wolcott, 1965). Broadbent et al. (1961) reported an almost linear relationship between pH and the amount of NH$_3$ fixed at pH > 7, while at pH < 7 NH$_3$ fixation was reported to be low.

The extent of fixation is also controlled by the oxidation state of the system and Nommik (1970) reported 50% greater fixation under aerobic

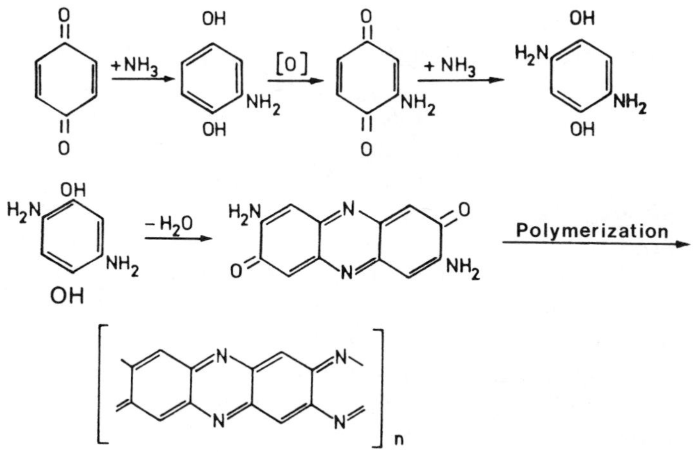

Fig. 2. Possible mechanisms of NH$_3$ fixation by aromatic humic substances. [After Nommik and Vahtras (1982).]

than anaerobic conditions. Similar results have been reported by others (Nyborg, 1969; Burge and Broadbent, 1961).

Although the stability of NH_3 fixed to organic matter can vary greatly, the majority of observations suggest that organically fixed NH_3 is extremely resistant to chemical hydrolysis and microbial attack (Burge and Broadbent, 1961; Mortland and Wolcott, 1965; Broadbent and Stevenson, 1966; Nommik, 1970).

The agronomic significance of NH_3 fixation is generally thought to be small. Nommik and Vahtras (1982) calculated that for a mineral soil with a carbon content of 2% a banded NH_3 application of 100 kg N ha^{-1} would result in less than 5% of the added N being fixed. It is unclear whether NH_3 fixation occurs when fertilizer materials such as urea are applied to soils, although if the pH and NH_3 concentration reach sufficiently high values then the potential for fixation exists.

E. Adsorption of Nitrate

A common observation in soils with predominantly 2 : 1-type clay minerals occurring in temperate regions is that NO_3^- moves freely through soils with rain or irrigation water (Wild and Cameron, 1980a,b).

However, NO_3^- can be nonspecifically adsorbed by electrostatic attraction to positively charged sites on soil minerals (Hingston et al., 1972; Mott, 1981). Soil minerals capable of developing positively charged sites include iron and aluminum oxides and hydroxides, 1 : 1 clay minerals (e.g., kaolinite), and allophane (Sumner and Reeves, 1966; Tweneboah et al., 1967; Hingston et al., 1972; Wada and Harward, 1974). Thus, soils that are rich in such minerals (mainly tropical and/or volcanic soils) can have a significant anion exchange capacity (AEC) so that NO_3^- is held by the soil and rapid NO_3^- leaching is prevented (Kinjo and Pratt, 1971a,b; van Raij and Camargo, 1974; Jones, 1975; Black and Waring, 1976a,b,c; Arora and Juo, 1982), Because of the many positive sites on allophane, volcanic ash soils generally retain NO_3^- more strongly than other soils (Kinjo and Pratt, 1971a; Kinjo et al., 1971).

The adsorption of NO_3^- by soils is concentration dependent and the amount adsorbed increases as the pH is lowered (Kinjo and Pratt, 1971a). The relative amount of NO_3^- adsorption is reduced at high solution concentrations (Singh and Kanehiro, 1969). The presence of organic matter can tend to decrease adsorption so that the extent of NO_3^- adsorption can be greatest in subsoils (Black and Waring, 1976b).

Nitrate adsorption is a mechanism that restricts free movement of NO_3^- with water under field conditions (Black and Waring, 1976a; Arora and

Juo, 1982). It tends to maintain NO_3^- in the root zone during wet periods and within the recall zone during the dry periods when water tends to move back toward the root zone (Kinjo et al., 1971). Arora and Juo (1982) calculated that a Nigerian kaolinitic ultisol had the capacity to adsorb 28–66 kg NO_3^--N ha^{-1} in the top 120 cm although the actual amount adsorbed was unknown due to the presence of competing anions such as Cl^- and SO_4^{2-}.

III. EROSION AND SURFACE RUNOFF

Accelerated erosion induced by man's activities is older than recorded history. Erosion has been severe from the earliest civilizations in the Middle East to the most recently cultivated lands in the Americas, Australia, and southern Africa (Troeh et al., 1980). Soil deterioration has sometimes been so great that land has been abandoned because it is no longer productive.

The loss of N by erosion and runoff is a serious problem in the United States and current estimates of erosion are on the order of 5 billion metric tons of soil annually, with 80% lost by water-borne sediments and the remainder by wind erosion (Legg and Meisinger, 1982). Approximately one-half to three-quarters of the eroded soil is from agricultural land (Pimentel et al., 1976; Wischmeier, 1976) and assuming a value of 3 billion metric tons of soil from cropland and a soil N content of 0.15% N (Willis and Evans, 1977) the erosion loss of N from crop land is about 4.5 million metric tons of N annually. Most of this N is in the organic (potentially available) form and is eventually deposited in streams, lakes, and oceans.

Erosion by water and wind has reduced the crop productivity of many soils (Buntly and Bell, 1976; Langdale et al., 1979a; Young, 1980; Frye et al., 1982). Short-term effects of erosion on soil productivity result from losses of the A horizon, namely, loss of organic matter and subsequent fertility, particularly N-supplying power (Englestad and Shrader, 1961), a reduction of available water-holding capacity (Frye et al., 1982), poor tilth, and reduced infiltration rate (Young, 1980). Long-term effects are caused principally by a reduction of crop rooting depth (Young, 1980).

Erosion can often represent a transfer of soil, and thus N, from one part of an ecosystem to another rather than a loss of N from the ecosystem. Nonetheless, the transfer of soluble and sediment N to surface waters does represent a net loss of N from the system. Wind erosion can cause anything from a transfer of soil from one side of a field to another to a reduction in crop root depth over large areas of cultivated lands.

A. Processes of Loss

1. Water Erosion

The potential for water erosion occurs whenever rainfall strikes bare soil or runoff flows over erodable and insufficiently protected soils. Runoff takes place whenever more rain falls than can infiltrate into the soil. Soluble nitrogenous substances may flow in runoff water but N is also transported as sediment.

Erosion by water is a process of particle detachment and transport that requires energy. Both rainfall and runoff have detachment potential although transport is mainly by runoff (Onstad and Moldenhauer, 1975). The erosiveness of rainfall is proportional to the energy of the falling drops and is influenced by the total amount of rain, the size of the drops, and their velocity of fall. The erosiveness of runoff is proportional to its energy and is influenced by the volume and velocity of flow.

Most of the kinetic energy of raindrops is dissipated at the surface of bare soil where the impacting drops detach soil particles. The detached particles are initially transported by splash action and shallow sheet flow. Raindrop impact on bare soil also tends to disperse soil aggregates and reduce surface roughness. Eventually a seal or crust forms over the soil surface. This greatly restricts infiltration and increases runoff and erosion (Wischmeier, 1973). As runoff increases, the water incises the soil, forming small channels (rills) and finally large channels (gullies). Although gully erosion is the most obvious form, it is sheet and rill erosion that are responsible for most of the erosion on cropland (Hayes and Kimberlin, 1978).

Soil properties such as texture and structure influence the ease or difficulty with which soil particles are detached and transported (erodability) and also influence infiltration and percolation rates. Vegetation and crop residues intercept rainfall before it reaches the soil and also slow the passage of runoff thus reducing the energy of rainfall and runoff.

The universal soil loss equation helps establish relationships between the amount of erosion and the factors influencing erosion (Wischmeier and Smith, 1961):

$$A = RKLSCP \tag{3}$$

where A = average annual soil loss (tons ha^{-1}); R = rainfall factor (erosivity) based on the number of erosion-index units in a normal year's rainfall for a specific location; K = soil erodability factor (tons ha^{-1}); L = slope length factor; S = land slope gradient factor; C = crop management factor; and P = erosion control factor.

Values for factors R, K, L, and S at a given location are based on prevailing conditions and are not readily subject to change. Erosion control practices normally involve manipulating values for C and/or P by crop management practices such as tillage and residue management and by erosion control factors such as terracing.

The universal soil loss equation often fails to predict losses of N from soils by water erosion (Gill *et al.*, 1976; Morisot, 1981) for two major reasons. First, sediments can have a higher N content than that of the soils they originate from (Baker, 1980) because light-weight organic material, rich in N, tends to be associated with the fine particles and is transported further than heavier particles. Second, not all N losses occur as sediment. Runoff water can contain significant quantities of soluble organic NH_4^+- and NO_3^--N (e.g., Alberts *et al.*, 1978; Langdale *et al.*, 1979b).

The concentration of organic N in the runoff water often exceeds that of inorganic N (NH_4^+ and NO_3^-) (Langdale *et al.*, 1979b; Schepers and Francis, 1982; Sharpley *et al.*, 1983) although this is not always the case (Burwell *et al.*, 1975; Timmons *et al.*, 1977). The ratio of $NH_4^+ : NO_3^-$ in the runoff water is extremely variable (Alberts *et al.*, 1978; Schepers and Francis, 1982; Sharpley *et al.*, 1983; McLeod and Hegg, 1984). Total N losses associated with eroded sediments are usually several-fold greater than soluble N losses in runoff (Burwell *et al.*, 1975; Neilsen and Mac-Kenzie, 1977; Alberts *et al.*, 1978).

2. Wind Erosion

Wind erosion is the process by which loose surface material is picked up and transported by the wind and surface material is abraded by windborne particles. It can be a problem when (1) soils are loose, dry, and reasonably finely divided, (2) there is a smooth soil surface on which vegetation cover is absent or sparse, (3) there is a large enough field width parallel to the wind, and (4) the wind is strong enough to move soil (Skidmore and Siddoway, 1978; Wilson and Cooke, 1980). It is a particular problem in agricultural areas that experience low, variable, and unpredictable rainfall, high temperatures, high wind velocity, or periodic droughts.

Sandy soils are extremely susceptible to wind erosion because of their low coherence between particles, small particle sizes, and rapid drying. Other soils are susceptible when they are dry and loose, such as when the soil has been finely divided by tillage. However, tillage operations that leave a rough, cloddy surface or maintain residue at the soil surface can help minimize wind erosion. Wind velocity, turbulence, gustiness, and direction all affect the severity of erosion. Soil erodability increases as the field width parallel to the wind direction increases. A mantle of growing plants or a mulch of crop residue is very effective in reducing losses of soil by wind erosion. Indeed the stubble mulch farming system was developed on the Great Plains of North America to prevent serious dust and sand storms that occurred during the prolonged droughts in the 1930s.

The wind erosion equation (Woodruff and Siddoway, 1965) relates the amount of wind erosion from a field to the factors affecting erosion:

$$A = f(ICKLV) \tag{4}$$

where A = potential annual wind erosion loss per unit area and is a function (f) of I = soil erodability; C = local wind erosion factor (varying directly with the cube of windspeed and inversely with the square of soil moisture content); K = soil surface roughness; L = unsheltered field length along the prevailing wind erosion direction; and V = equivalent quantity of vegetation cover. Tillage operations generally influence factors I and K while surface residues affect factor V of the equation.

There is little quantitative data on the amounts of N moved by wind erosion. However, some values for North American arid ecosystems have been presented (Fletcher et al., 1978) where losses in the range of 1 to 6 kg N ha^{-1} yr^{-1} were recorded. Legg and Meisinger (1982) estimated that about 20% of the erosion loss from soils in the United States occurs as wind erosion.

B. Factors Affecting Losses

Losses of N by wind erosion have not been studied to any great extent. Nonetheless, losses of N in runoff from agricultural lands have been studied in detail and the factors affecting losses of soluble and sediment N are well known. Below, the major factors influencing such losses are discussed.

1. Climate

As discussed previously, the detachment of sediment and the transportation of sediments and solutes depend on rainfall energy. Thus, both the distribution and total amount of rainfall are important and sediment and runoff losses show marked seasonal variations (Neilsen and MacKenzie, 1977; Alberts et al., 1978; Langdale et al., 1979a). Losses of N in runoff have been correlated with both the volume of rainfall (Taylor et al., 1971; Kilmer et al., 1974; Schreiber et al., 1976) and the intensity of rainfall (Jackson et al., 1973; Dunigan and Dick, 1980). Runoff losses can be influenced by the season in which fertilizers are applied; heavy fertilization in fall, before intense heavy rains, results in appreciable losses of NO_3^- in runoff (Klausner et al., 1974). Indeed, in general, the most important factor influencing the N content of runoff is the intervening time between fertilizer or manure applications and runoff events (Westerman and Overcash, 1980a; Sherwood and Fanning, 1981; Steenvoorden, 1981). Greatest losses occur when heavy rainfall occurs immediately after application (Dunigan and Dick, 1980; Westerman and Overcash, 1980a; Doran *et al.*, 1981; McLeod and Hegg, 1984). Westerman and Overcash (1980b), for example, observed that even one extra day between a manure application and a heavy rainfall caused a significant reduction in the N concentration of the resulting runoff. The number of rainfall events after fertilizer application can also be an important consideration. McLeod and Hegg (1984) observed that concentrations of N in runoff were reduced by 80% following two runoff events after fertilizer or manure applications.

Since rainfall can vary greatly from year to year, large differences in annual losses of N in runoff and sediment can occur (Jones *et al.*, 1977; Menzel *et al.*, 1978).

Snowmelt in spring can be a very important factor responsible for losses of N in surface runoff and sediment in some localities (Burwell *et al.*, 1975; Klausner *et al.*, 1976; Neilsen and MacKenzie, 1977; Nicholaichuk and Read, 1978). Neilsen and MacKenzie (1977) observed that 56 to 100% of the annual soluble N loss in surface runoff occurred during snowmelt in several Quebec and Ontario watersheds.

2. Soil Properties

Soil properties that affect infiltration rate and permeability to water affect the amount of erosion that occurs. In general, a moderate storm will produce more runoff and erosion from finer-textured soils than from sandy soils since the infiltration and permeability rates are lower in fine-textured soils. Hoyt *et al.* (1977) compared the amounts of soluble and sediment N in runoff as well as NO_3^- leaching from three contrasting soils. They reported that the trends in volumes of runoff and amounts of sediment transported were generally parallel and that therefore the amounts of soluble and sediment N were similarly related (Fig. 3). An inverse relationship existed between losses of N by runoff and leaching losses of N.

Soil water status at the time of the rainfall event is an important factor affecting infiltration and runoff. Kissel *et al.* (1976) found that runoff losses of NO_3^- were highest when the soil was near field capacity and lowest when large amounts of water infiltrated into dry soil immediately before runoff.

Organic matter content is another factor affecting N losses. Indeed, total N loss in sediments can be closely related to the organic N content of soils (Gambrell *et al.*, 1975; Neilsen and MacKenzie, 1977).

Other soil properties can influence erosion losses. Erodability depends

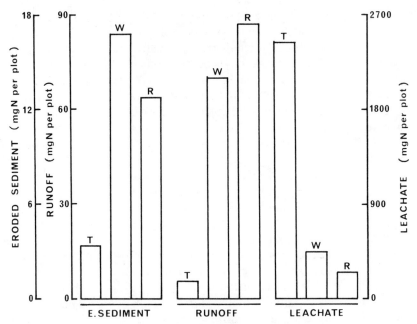

Fig. 3. Influence of soil type on the amounts of N moving with eroded sediments, runoff, and leachate from three cultivated soils. T = Toledo silty clay, W = Wausenon sandy loam, R = Rossmoyne silt loam. [Redrawn from Hoyt *et al.* (1977).]

on factors such as aggregate stability, grain size, organic matter content, and cohesiveness. Soil detachment, as well as sediment and solute transport, depends greatly on land characteristics (see equation (2)) such as slope angle and slope length (Onstad and Moldenhauer, 1975).

3. Land Management

a. Land use. Generally, any factor that reduces raindrop impact and obstructs runoff (e.g., plant cover) can prevent erosion (Lal, 1975). In the longer term, the presence of plants can also decrease runoff by increasing the infiltration capacity of soils (Kandiah, 1979). Plant cover results in a greater reduction in sediment losses of N than that of runoff (Lal, 1976; Hoyt *et al.*, 1977).

Thus under the fully developed plant cover found in most grasslands the adverse effects of steep topography, unstable soils, and high-intensity rain and wind are reduced and consequently the runoff and sediment losses of N are usually negligible (Kilmer *et al.*, 1974; Chichester, 1977; Woodmansee *et al.*, 1981). Where plant cover is reduced, by fire, drought, or intense grazing, nitrogen losses during intense rainfall events can be significant (Woodmansee et al., 1981). Chichester et al. (1979), for example, demonstrated that a winter-feeding practice on pastures caused a high degree of soil and plant cover disturbance and an increase in surface runoff N in comparison with a pasture grazed only in summer.

Losses of N in runoff are also negligible in most forest ecosystems (Gosz, 1981; Melillo, 1981) since the leaf canopies break the velocity of the raindrops and the organic layer at the forest floor protects the soil from erosion. Removal of forest vegetation by fire or clear-cutting can greatly increase losses of N by erosion (Vitousek, 1981; Woodmansee and Wallach, 1981; Chang et al., 1982). Removal of trees results in increased surface runoff and loss of associated sediments, increased stream-bank erosion, and greater particulate transport in streams (Vitousek, 1981). It can also cause increases in soil creep and various modes of slope failure (Swanston and Swanson, 1976). Following forest fires, the quantity of ash deposited at the soil surface is obviously an important factor determining potential losses of N through both runoff and wind erosion (Woodmansee and Wallach, 1981).

The potential for losses of N through both water and wind erosion is high in cultivated agricultural ecosystems because for at least part of the year the disturbed soil surface lacks vegetative cover. Indeed most of the loss of soluble and, more particularly, sediment N from cultivated soils occurs during the first few months after planting before the plant canopy is fully developed (Schuman et al., 1973; Gambrell et al., 1975; Hoyt et al., 1977).

A large proportion of the N lost during runoff from cultivated soils is associated with sediments (Schuman et al., 1973; Gambrell et al., 1975; Burwell et al., 1975; Hoyt et al., 1977) and losses were shown by Gambrell et al. (1975) to be largely related to the organic N content of soils rather than to fertilizer N application rates. The N content of sediments from cultivated watersheds can be greater than that from grassland watersheds (Ritchie et al., 1975), because under cultivation sediments are derived from sheet and rill erosion of fertile topsoil and under grassland a higher proportion of sediment comes from gullies, stream channels, and other similar sources of low N status.

Losses of N through erosion can also occur when vegetation is removed and the soil surface is disturbed by nonagricultural activities such as on residential construction sites or during highway construction (McLeese and Whiteside, 1977; Daniel et al., 1979).

   b. Conservation practices. Losses of N in runoff ranging from 20 to 150 kg N ha^{-1} yr^{-1} have been recorded for gently sloping cultivated watersheds (Schuman *et al.*, 1973; Burwell *et al.*, 1975, 1976; Alberts *et al.*, 1978). Sediment N accounted for 80 to 90% of such losses. Thus, the objective of conservation practices is generally to lower the losses of sediment from agricultural watersheds. As noted previously, soil cover can be a critical factor in erosion control and where erosion is a problem, early in the growing season, crops with a rapid development of ground cover are more suitable than crops with slow-growing canopies (Lal, 1976; Aina *et al.*, 1979). It is well known that crop residues left on the soil surface can greatly reduce wind and water erosion losses induced by cultivation (Lal, 1975; Power, 1981). Mulching with plant residues has been shown to reduce losses of soluble and sediment N (Schuman *et al.*, 1973; Olness *et al.*, 1975; Neilsen and MacKenzie, 1977; White *et al.*, 1977). Residues can, however, also act as a small source of soluble N in runoff (Timmons *et al.*, 1968; White, 1973).

One means of retaining a mulch on the soil surface is by practicing minimum or zero tillage (Unger and McCalla, 1980) and this can greatly reduce losses of sediment from agricultural lands (Mensah-Bonsu and Obeng, 1979; Gumbs and Lindsay, 1982; Lal, 1982). Romkins *et al.* (1973) compared the loss of N in runoff from five tillage methods. Coulter and chisel systems controlled soil loss but runoff water contained relatively high levels of soluble N from surface-applied fertilizer. Disk and till systems were less effective in controlling soil erosion, but resulted in lower concentrations of soluble N in runoff water. Conventional tillage, in which fertilizers were plowed under, had the highest losses of sediment N but losses of soluble N were small.

A variety of management practices can be employed to reduce the velocity of running water on sloping agricultural land and hence reduce sediment losses (Troeh *et al.*, 1980). These include the use of ridges produced by tillage and plant rows to form barriers to water movement (ridging), the rotation of crops in strips across the path of moving water (strip planting), tillage and planting across the slope along contour lines (contour planting), and the construction of terraces. In general, level terraces have been shown to be an extremely effective method of reducing losses of soluble and sediment N from cultivated watersheds (Alberts *et al.*, 1978; Langdale *et al.*, 1979a) and considerably more effective than contour planting (Schuman *et al.*, 1973; Burwell *et al.*, 1974; Alberts *et al.*, 1978). Table II shows clearly that losses of soluble, and particularly sediment, N were greatly reduced by terracing in the Missouri Valley. Indeed total losses of N in runoff from the terraced watershed were similar to those from a pasture.

As noted by Morisot (1981), management practices that reduce erosion losses of N tend to increase leaching losses of NO_3^--N by increasing

Table II

Effect of Management System on Losses (kg ha⁻¹ yr⁻¹) of Soluble and Sediment Nitrogen in Streamflow from Various Watersheds[a,b]

Watershed	Soluble N		Sediment N	Total N loss
	NH_4^+-N	NO_3^--N		
Contour corn	1.36	1.69	36.59	39.64
Contour corn	0.92	0.97	23.16	25.05
Terraced corn	0.24	0.18	2.62	3.04
Pasture	0.39	0.76	1.21	2.36

[a] Data from Schuman et al. (1973).
[b] Data are the means of a 3-year study.

infiltration of water. Burwell et al. (1976), for example, showed that terracing greatly reduced losses of sediment N in comparison with contour planting, but there was a complementary increase in subsurface discharge of N that resulted in a similar amount of total N being transported from the two watersheds.

4. Irrigation

Irrigation, like rainfall, can cause surface water runoff. The quantity of runoff will depend greatly on the type of irrigation being practiced (surface, sprinkler, or trickle irrigation) and the amount, rate, and frequency of water application. Surface (flood or furrow) irrigation can be a significant cause of erosion because large amounts of water flow across land. The problem is well known and methods of reducing runoff (irrigation return flow) have been outlined (Carter, 1976).

5. Fertilizer Applications

In forest, grassland, and cultivated ecosystems, increasing the rate of fertilizer N addition generally tends to result in increased losses of soluble N in runoff (Schuman et al., 1973; Kilmer et al., 1974; Olness et al., 1975; Dunigan and Dick, 1980; Sharpley et al., 1983; McLeod and Hegg, 1984) even though runoff losses are still frequently less than precipitation inputs (e.g., Schuman et al., 1973; Burwell et al., 1975; Schepers and Francis, 1982; Sharpley et al., 1983). Nevertheless, increases in the levels of soluble N in individual runoff events can occur if runoff occurs shortly after surface application (Timmons et al., 1973; Dunigan and Dick, 1980; Westerman and Overcash, 1980b, Owens et al., 1984). Upward movement of native or previously applied soil NO_3^- during dry spells, and its accumulation at the soil surface, can also result in significant losses of NO_3^- in runoff during the first subsequent rainfall (Kissel *et al.*, 1976; DeBoodt *et al.*, 1979). In general the total amount of N lost by surface runoff is usually less than 5% of the applied N (Baker, 1980). Incorporation of fertilizers into the soil can greatly reduce losses of N in runoff events soon after fertilizer application (Timmons *et al.*, 1973; Kissel *et al.*, 1976; Dunigan and Dick, 1980). Nonetheless, as already noted, the cultivation itself may cause an increase in the N loss as sediment.

Soil characteristics and tillage operations are the major factors influencing losses of N in sediments and fertilizer applications appear to have little effect on such losses (Olness *et al.*, 1975). Immediately following applications of NH_4^+-containing fertilizers sediments may be temporarily rich in NH_4^+-N (e.g., Langdale *et al.*, 1979a). In some situations fertilizer N applications could tend to reduce losses of N in sediments by stimulating rapid plant growth thus reducing surface runoff and sediment loss (Gambrell *et al.*, 1975).

6. *Organic Waste Applications*

Land disposal of animal wastes, originating principally from feedlots, has been a worldwide practice for many years. Although it recycles nutrients back to the plant–soil system, a major problem with animal waste disposal is contamination of groundwaters with nitrogen. While much of this contamination may occur through the process of NO_3^- leaching (see Section IV,D), surface runoff can also be an important contributor.

Since organic materials release mineral N relatively slowly during their decomposition, losses of soluble N in runoff can often be greater immediately following applications of readily soluble inorganic fertilizers than from organic manures applied at similar rates of N (Long, 1979; Dunigan and Dick, 1980; McLeod and Hegg, 1984). The application of organic manures to soils generally results in a marked increase in the proportion of soluble N in runoff present in organic forms (McLeod and Hegg, 1984) although concentrations of NH_4^+ and NO_3^- are also raised (Long, 1979; McLeod and Hegg, 1984). Surface applications of animal wastes can also cause an increase in the amount of sediment N lost in runoff (Hoyt *et al.*, 1977).

The overall effect that applications of organic wastes have on runoff losses is strongly related to their effects on soil physical properties. Repeated substantial applications generally increase the soil organic matter content, improve aggregation, decrease bulk density, increase water-holding capacity at both field capacity and wilting point (Khaleel *et al.*, 1981), and can increase infiltration rates (Cross and Fischbach, 1972; Mazurak *et al.*, 1975). Indeed, the improvement of physical conditions caused by incorporation of organic wastes can result in there being little loss of waste constituents in runoff (Khaleel *et al.*, 1981).

In contrast, however, some adverse effects of waste applications on infiltration rates have been reported (e.g., Manges *et al.*, 1974; Weil and Kroontje, 1979). High rates of application of wastes with high contents of Na and K can result in salt accumulation at the soil surface and the dispersion of soil aggregates (Hinrichs *et al.*, 1974; Manges *et al.*, 1974; Powers *et al.*, 1975). This reduces the movement of water into the soil surface and through the soil matrix and so runoff is increased.

Losses of N in runoff are a particular problem when animal wastes are spread on agricultural lands at high rates, especially where soils have a low permeability, and/or they are saturated with water (Sherwood and Fanning, 1981; Steenvoorden, 1981), or where wastes are spread onto snow-covered soil (Young and Mutchler, 1976; Uhlen, 1981).

Runoff is also a potential problem from feedlot areas. Swanson *et al.* (1971), for example, found that runoff from feedlots contained 2–30 times the NH_4^+-N concentration and up to four times the NO_3^--N concentration of runoff from fallow land. In fact, the Royal Commission on Environmental Pollution in the United Kingdom (1979) recommended that intensive livestock units be regarded as industrial rather than agricultural in nature and that as with other industries strict pollution control regulations should apply.

IV. LEACHING LOSSES OF NITRATE

Nitrate leaching is a particular problem on cultivated agricultural lands and it is often the most important channel of N loss from field soils. Losses normally range from 2 to 100 kg N ha^{-1} yr^{-1} (Wild and Cameron, 1980a; Hauck and Tanji, 1982). The NO_3^- originates from mineralization of soil organic matter and crop and animal residues, fertilizer N not used by crops, and, to a lesser extent, rainfall inputs. Additions of fertilizer N, which are essential to obtain high crop yields, commonly increase leaching losses. When high fertilizer rates are combined with heavy irrigation regimes on light-textured soils leaching losses of NO_3^- can be large.

The processes involved in NO_3^- leaching and the factors influencing losses have been studied extensively because of their economic and environmental significance.

A. Description of Solute Movement

Extensive literature exists on the theory and equations describing solute movement in porous media such as soils and the reader is referred to reviews by Gardner (1965), Biggar and Nielsen (1967, 1980), Frissel and Poelstra (1967), Fried and Combarnous (1971), Boast (1973), Nye (1974), Wild (1981), and Nielsen *et al.* (1982) and textbooks by Kirkham and Powers (1972), Marshall and Holmes (1979), and Hillel (1980) for comprehensive treatments. In this section the principles of solute movement are outlined in relation to the process of NO_3^- leaching.

1. Convection, Diffusion, and Dispersion

If it is assumed that steady-state water conditions exist in a homogeneous nonaggregated soil and that there is no interaction between the NO_3^- ion and the soil, then NO_3^- movement can be described by a combination of three processes: convection, diffusion, and dispersion.

a. Convection. Convection refers to solute transport due to mass flow of water alone. The water and solutes move in response to a hydraulic gradient and the rate of movement is dependent on the magnitude of the hydraulic gradient and the hydraulic conductivity of the soil. Such movement can be described as shown below:

$$\frac{\partial c}{\partial t} = -U \frac{\partial c}{\partial x} \tag{5}$$

where c = concentration of NO_3^- ($\mu g\ ml^{-1}$); t = time (days); U = average pore velocity (cm day^{-1}), which is obtained by dividing the rate of water flow by the volumetric water content of the soil; and x = linear distance in the direction of flow (cm).

The effect on the solute distribution in the soil profile is illustrated in Fig. 4a. In reality, however, the band of solute does not remain contiguous but tends to spread throughout the soil profile through the processes of diffusion and dispersion (Fig. 4b).

b. Diffusion. When there is an uneven distribution of solutes in soil solution there is a diffusive flux of solute from areas of high concentration to areas of low concentration. Movement by diffusion is described below:

$$\frac{\partial c}{\partial t} = D_s \frac{\partial^2 c}{\partial x^2} \tag{6}$$

where D_s = effective diffusion coefficient in soil (cm^2 day^{-1}).

In soil, solute diffusion can only occur in the fraction of soil volume occupied by water. The effective diffusion coefficient in a water-saturated

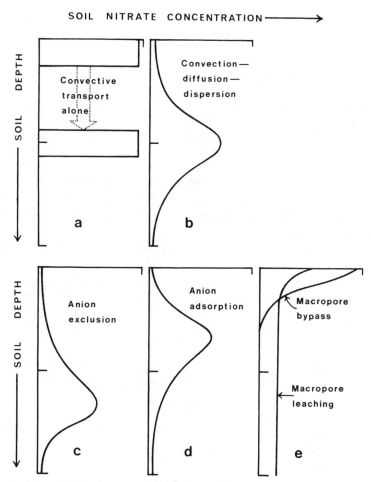

Fig. 4. Schematic diagram of various components of NO_3^- leaching: (a) convective transport alone, (b) convection–diffusion–dispersion, (c) anion exclusion, (d) anion adsorption, and (e) macropore bypass and macropore leaching.

soil is less than that in bulk water because of the smaller volume of soil solution available for diffusion and the increased path length due to the tortuosity of soil pore geometry. In an unsaturated soil, the effective diffusion coefficient is further reduced because of a decrease in volume-fraction of water and an increase in the tortuosity of the diffusion paths (Rowell *et al.*, 1967). The diffusion coefficient of NO_3^- in soil at -1.0 kPa is approximately 10^{-6} cm^2 sec^{-1} (Nye and Tinker, 1977) and an average NO_3^- ion would thus move about 0.5 cm per day.

Equations (5) and (6) can be combined to give a convection–diffusion equation:

$$\frac{\partial c}{\partial t} = D_s \frac{\partial^2 c}{\partial x^2} - U \frac{\partial c}{\partial x} \tag{7}$$

c. *Dispersion.* The mechanical action of a solution flowing through soil causes mixing and tends to equalize the solute distribution by a process commonly called "hydrodynamic dispersion." This process enhances the dispersive effect of diffusion and often completely masks it. Hydrodynamic dispersion occurs because (1) the flow velocity within a single pore is not uniform since it is fastest at the pore center, (2) the large variation in pore size within a soil results in an extremely wide range of pore water velocities, and (3) the path length of pores fluctuates greatly due to the tortuosity of pore geometry.

Mathematically, dispersion can be considered analogous to diffusion (equation (6)).

d. *Combined equation.* The combined effects of the convective–diffusive–dispersive mechanism can be described by

$$\frac{\partial c}{\partial t} = E \frac{\partial^2 c}{\partial x^2} - U \frac{\partial c}{\partial x} \tag{8}$$

where E = dispersion coefficient, often called the apparent diffusion coefficient, and is the sum of diffusion plus mechanical dispersion

$$E = D_s + mU \tag{9}$$

where m = dispersivity. The value of E thus depends on the flow velocity and tends to increase with increasing values of U (Nielsen and Biggar, 1963; Passioura and Rose, 1971).

2. Charge Characteristics of Soils

a. *Anion exclusion.* Soil surfaces generally carry a net negative charge and as a result cations in soil solution distribute themselves near the surfaces so that electroneutrality is maintained (i.e., they form a double layer) (Arnold, 1978). Because of electrostatic repulsion, nonspecifically adsorbed anions such as NO_3^- (see Section II,E) are excluded from this area close to the soil surfaces. Thus, a proportion of soil water does not participate in NO_3^- leaching and the effective pore volume is less (often up to 10–20%) than the water content of the soil (Wild, 1981). Nitrate ions therefore travel correspondingly faster than predicted by convection theory (Nielsen and Biggar, 1962; Thomas and Swoboda, 1970; Krupp *et al.*, 1972). The effect on the leaching pattern is shown in Fig. 4c.

b. Adsorption. Where soils have significant AEC, nonspecific adsorption of NO_3^- does reduce its rate of leaching (Kinjo and Pratt, 1971a,b; Black and Waring, 1976c). The effect of adsorption on leaching can be described by adding a retardation factor to equation (8) (Davidson and Chang, 1972):

$$\left[1 + \left(p\,\frac{N}{\theta}\right)\right] \tag{10}$$

where $N = NO_3^-$ distribution between soil and solution and the expression $p(N/\theta)$ effectively represents the apparent increase in pore volume as a result of adsorption. The effect on the leaching pattern is shown in Fig. 4d.

3. Macropore Movement

During infiltration and drainage significant volumes of water may flow through large cracks or channels in the soil (macropores). These channels conduct water under heavy rainfall or irrigation but under other conditions they are air filled. Although macropores that are not open at the soil surface will not usually flow, those that are open and are continuous may transmit water rapidly to depth in the soil (Ritchie *et al.*, 1972; Shuford *et al.*, 1977; Wild and Babiker, 1976b; Omoti and Wild, 1979; Smettem and Trudgill, 1983; Scotter and Kanchanasut, 1981; Kanachanasut and Scotter, 1982; Mohammed *et al.*, 1984).

Earthworm activity, root growth, freezing and thawing, and wetting and drying cycles can lead to the development of a network of surface connected pores. Even large spaces between aggregates may be considered macropores (Ritchie *et al.*, 1972). In a coarsely structured soil that receives a rapid application of water, over half the water may move through the macropore system (Quisenberry and Phillips, 1976; Thomas *et al.*, 1978).

Movement of water in soil macropores has two important implications to the process of NO_3^- leaching: (1) When the infiltrating water contains a high concentration of NO_3^- then macropore flow will lead to extensive leaching at a faster rate than predicted by equation (8), as illustrated in Fig. 4e. (2) When NO_3^- is present within the micropores of aggregates it may be bypassed by the bulk of flowing water and this leads to solute retention and a slower than predicted rate of leaching (Fig. 4e).

Thus, when heavy rainfall or irrigation occurs soon after an application of fertilizer N, some will be transported quickly downward in macropores (Wild and Babiker, 1976a). Nonetheless, between rainfall or irrigation events, NO_3^- will diffuse into soil aggregates and clods. Where soil NO_3^- is predominantly the result of mineralization and nitrification within aggregates or where NO_3^- has diffused into aggregates, then water moving in macropores will carry little NO_3^- with it (Wild, 1972).

4. Transformations and Plant Uptake of Nitrogen

As discussed in other chapters, biological transformations of NO_3^- such as mineralization–immobilization turnover and denitrification occur in soils. These processes are extremely difficult to describe mathematically because often they are spasmodic rather than steady state. Nevertheless, since they can act as major inputs and outputs of NO_3^- to the system some attempt to include them in a more complete description of solute behavior in soil has been considered necessary. An input–output term S can be included in equation (8) to account for the rate of NO_3^- production or disappearance:

$$\frac{\partial c}{\partial t} = E \frac{\partial^2 c}{\partial x^2} - U \frac{\partial c}{\partial x} + S \tag{11}$$

Mostly it is assumed that S may be approximated by a zero or first-order reaction but more complex descriptions have been used (Macura and Kunc, 1965; Cho, 1971; McLaren, 1970; Nye and Tinker, 1977).

Nitrate uptake by plant roots can be considered in a similar way to equation (11) but the equations required to represent the term S are even more complex (Tanji and Mehran, 1979), often making the mathematics of the combined equations intractable.

5. Soil Variability

Thus far the description of leaching has been deterministic in nature. That is, it has been based on the underlying principles of solute movement related through parameters obtained by experimentation (e.g., equation (8)). The use of such models assumes that values for the required parameters (e.g., concentration of soil NO_3^- and average pore velocity) can be determined that are applicable over the domains of time and space in which the processes are occurring.

The spatial and temporal variability of field soils, in fact, makes determination of such parameters extremely difficult. There is generally wide spatial variability in soil NO_3^- concentrations within a field (Biggar, 1978; Nielsen et al., 1979; Burns, 1980; Cameron and Wild, 1984; MacDuff and White, 1984). Similarly, the random spatial arrangement of soil pores combined with their extreme variation in size makes it difficult to determine an average value for either pore water velocity or dispersivity that is even approximately correct. Warrick and Nielsen (1980), for example, showed that in one particular field study 1500 samples would have been required to obtain an average value of hydraulic conductivity that was within 15% of the true value. Biggar and Nielsen (1976) reported that 100 observations were required to be within 50% of the true value of pore water velocity in a 150-ha field.

The recognition of the variability of soil properties has led to the use of stochastic models for prediction of solute movement in field soils or at least the inclusion of stochastic parameters in deterministic models (Nielsen *et al.*, 1982). A stochastic model is one in which the solutions are predictable in a statistical sense. Stochastic methods of analysis were developed to describe spatial variation in geological and geographic parameters and are often referred to as geostatistical techniques.

B. Prediction of Nitrate Leaching

Many workers have attempted to model and predict NO_3^- leaching with various degrees of success. The ability to predict NO_3^- leaching has obvious applications to the prediction of the fertilizer N requirement for cropping systems. Leaching models also represent an important component of larger models that simulate the complete N cycle (Tanji and Gupta, 1978; Tanji, 1982).

Leaching models have been reviewed by Gardner (1965), Biggar and Nielsen (1967), Frissel and Poelstra (1967), Boast (1973), Gupta *et al.* (1979), and Wild and Cameron (1980a). Approaches to modeling solute movement have been detailed by Nielsen *et al.* (1982) and Addiscott and Wagenet (1985). Principles of computer simulation modeling have been presented by Hillel (1977). The purpose of this section is to briefly introduce the three major approaches that have been used.

1. Deterministic Methods

Analytical and numerical solutions to the convective–dispersive–diffusive flow equation (equation (8)) are numerous (Kirkham and Powers, 1972; Bresler, 1973; Kirda *et al.*, 1973; De Smidt and Wierenga, 1978; Gupta *et al.*, 1979). Such models have proven accurate in the description of laboratory breakthrough curves, but have usually been less appropriate and difficult to use for field prediction. The problem is mainly associated with obtaining a suitable value for dispersivity (m), which is a component of the apparent diffusion coefficient according to equation (9).

The term mU, commonly called mechanical dispersion, remains virtually unknown for unsaturated conditions (Nielsen and Biggar, 1981) and as no direct method of calculating dispersivity has yet been devised it is necessary to conduct a leaching experiment to obtain the value. Nevertheless, once a suitable value has been obtained for an area of soil, the models can give accurate descriptions (e.g., Rose *et al.*, 1982a,b).

Good agreement between model predictions and $^{15}NO_3^-$ leaching patterns from monolith lysimeter studies was reported by Rose *et al.* (1982b) and good agreement was also obtained in a field tracer leaching experiment reported by Cameron and Wild (1982b). In both studies the dispersivity value appropriate for each soil was derived from the initial measured leaching data.

2. Stochastic Methods

Because the spatial and temporal variance of field soil properties is sufficient to render estimates of their means highly unreliable it is becoming clear that there is a need to analyze and simulate the behavior of field soils from a probabilistic viewpoint. A knowledge of the mean behavior of a field may be of less importance in some cases than that of its statistical or spatial variance (Nielsen *et al.*, 1983). For this reason, geostatistical methods of data analysis have been introduced into analysis of solute movement, and applications have allowed considerable progress in spatial interpolation, spatial averaging, and design of better sampling schemes (Nielsen *et al.*, 1982). The use of stochastic techniques has been reviewed elsewhere (Nielsen *et al.*, 1982, 1983; Webster and Burgess, 1983; Peck, 1984).

The classical flow equation (equation (8)) may be written as a mixed deterministic–stochastic model (e.g., Bresler and Dagan, 1979, 1981). Jury *et al.* (1982) introduced a transfer function model that can include the effects of spatially nonuniform rainfall or irrigation and nonuniform soil profiles. Our understanding and modeling of soil water/solute dynamics promise to benefit greatly from these emerging concepts.

3. Empirical Methods

Empirical methods have been developed that are not based directly on the classical flow equation but on the observed quantitative relationships among variables. Essentially the soil profile is considered as a series of layers that have parameters controlling the rate of water and solute movement through each layer to the one below. For example, in the computer simulation model developed by Burns (1974), a minimum water content, termed evaporation limit, and a storage maximum, field capacity, are used to characterize each layer. The initial values of the amounts of water and anion in each layer are changed by additions of rainfall or irrigation. When rainfall exceeds evaporation the net water excess is added to the top layer and is simulated to move down the profile if the input causes the water content to exceed field capacity. As water moves into each successive layer it is assumed to equilibrate with the nitrate present and thus invoke solute transfer from one layer to another. The model has been shown to account successfully for nitrate leaching in some sandy soils (Burns, 1975) but has been found to be less appropriate for more coarsely structured soils because of the assumption of instantaneous total equilibrium of water and solute in each layer (Cameron and Wild, 1982b). The model rules have been simplified to produce a series of simple leaching equations and a mechanical disk calculator for farmer use (Burns, 1976).

A further example of this type of approach is the model of Addiscott (1977), which accounts for the effects of soil structure, or more precisely the effects of variations in the soil pore size, by partitioning soil water into mobile and retained phases. The retained phase represents the water held in small pores and "dead-end" pores that are considered not to contribute to flow but that are accessible to solute only by molecular diffusion. The mobile phase represents the pore system that is involved in mass flow and therefore only this phase can be displaced. Solute equilibrium is established between mobile and retained phases only after flow has ceased. From the discussion of theory in Section IV,A it is obvious that this has a more sound physical basis. However, there are difficulties in defining the critical limit between mobile and retained phases. Such computer simulation models can be easily expanded to include N transformation processes.

C. Estimation of Leaching Losses

A number of approaches have been taken to study solute leaching. The methods employed deserve some description and appraisal to allow appreciation of the source and limitations of reported results. A detailed discussion of results obtained by the various methods has been presented elsewhere (Wild and Cameron, 1980a).

1. Field Soil Sampling

The profile distribution of nitrate can be measured by soil sampling and extraction of the soil with a salt solution (Bremner, 1965). Alternatively, porous ceramic cups can be used to extract the soil solution (Hansen and Harris, 1975). These data, when combined with measurements of the water flux, allow calculation of the loss in kg N ha^{-1}. The major constraints are that large numbers of measurements are required to be representative of an area and that deep samples may be difficult to obtain on stony soils. Spatial variability of NO_3^- in soils results in large variability in results of both soil solution NO_3^- levels (Nielsen et al., 1979) or extracted NH_4^+ and NO_3^- levels (Broadbent and Carlton, 1978; Cameron and Wild, 1984; MacDuff and White, 1984).

2. Borehole Sampling

Deep cores taken from the unsaturated zones above aquifers are difficult and expensive to obtain but provide information on the potential pollution of underground drinking water sources. Defining the rate of movement toward the aquifer of peak concentrations of nitrate and relating the occurrence of a peak to a previous land use activity is a major difficulty. One approach used has been to assume that the rate of nitrate leaching can be inferred from measured rates of movement of tritiated water (Young *et al.,* 1976; Young and Gray, 1978). The latter is calibrated against years of peak levels of atmospheric tritium caused by thermonuclear weapons testing. The assumption has been challenged by some (Mercer and Hill, 1975, 1976; Russell, 1978) but the data (Young *et al.,* 1976; Young and Gray, 1978) seem self-sufficient and have been supported by other field studies (Cameron and Wild, 1982a).

3. Catchment Studies

Collection of water draining from strictly confined catchment areas can provide an integrated measure of the leaching loss and thus avoid the variability problems associated with soil sampling. However, in studies of large catchments, where the intensity of management varies, it is difficult to relate land use to leaching losses of NO_3^-. Furthermore, they are restricted to soils with an impermeable substratum and it is usually not possible to separate losses of NO_3^- due to surface runoff into ditches from that actually collected by tiles draining into ditches. Large variations in relationships between losses and soil management are reported in the literature because of differences in experimental conditions (see Burwell *et al.,* 1976).

4. Tile Drain Studies

Water draining from field tiles can provide a relationship between a specific land use activity and the leaching loss. The major criticism of this approach is that in the majority of studies reported water samples have been collected on a time rather than volume flow basis. This intermittent sampling does not provide a reliable indicator of the amounts of nitrate loss (Cooke and Williams, 1970; Thomas and Barfield, 1974) and although useful comparative data have been reported, the tile effluent needs to be sampled in proportion to the water flow. A further problem is that tiles usually intercept only part of the water flow down the profile so that the water collected may not be representative of the total drainage (Wild and Cameron, 1980a).

5. Lysimeter Studies

Lysimeters allow quantitative measurements from a more or less defined soil volume and they generally avoid the large variations associated with field studies. The three principal types of lysimeters are (1) undisturbed soil blocks either walled in or monoliths fitted into boxes, (2) tanks filled with loose soil, and (3) field tension lysimeters based on placing a suction plate below the soil surface, i.e., they have no walls. The first two types may suffer from container edge effects leading to increased aeration and preferential drainage pathways (Wild and Cameron, 1980a) while the tension type avoids these problems but no longer confines the soil volume exactly.

6. Laboratory Column Studies

Complete cores of undisturbed soil taken in the field can be leached under controlled laboratory conditions and are essentially minilysimeters. However, most studies in the laboratory have been conducted on columns of repacked soil and other porous materials such as pumice, sand, and porcelain (e.g., Passioura and Rose, 1971). The experiments involve following the miscible displacement of one fluid by another through the column of material. The effluent is analyzed to produce a breakthrough curve (i.e., effluent solution concentrations versus time or cumulative drainage volume). The study of such curves has been the basis for the development of most current theories of leaching.

D. Factors Affecting Leaching Losses

1. Season and Climate

Variations in rainfall distributions and evapotranspiration patterns from year to year and from season to season affect the leaching pattern; nevertheless, some general statements can be made from the literature.

(1) The intensity as well as amount of rainfall is of great importance in determining the pattern and extent of leaching (Wild and Cameron, 1980a).

(2) Summer rainfall is generally used to satisfy the evapotranspiration deficit and leaching is therefore usually minimal. However, summer leaching can occur under intense heavy rainfalls due to macropore flow, but the amount of loss is dependent on time of fertilizer application (Williams, 1975).

(3) Autumn rainfall can leach any residual fertilizer nitrate left after

harvest or any nitrate released by mineralization at this time, provided that the soil is approaching or has reached field capacity.

(4) Winter rainfall readily leaches any nitrate present in the soil profile since there is a large excess of rainfall over evapotranspiration and a low rate of N uptake by crops (Shaw, 1962; Kilmer *et al.*, 1974; Dancer, 1975). An exception is when the soil is frozen (Baker *et al.*, 1975) or when conditions conducive for denitrification predominate.

(5) Spring rainfall determines whether freshly applied fertilizer or newly mineralized nitrate is quickly leached (Williams, 1975).

(6) A dry summer can result in the accumulation of soil nitrate due to poor crop uptake and significantly higher than average leaching losses then occur over the subsequent winter (Garwood and Tyson, 1977).

Results from a tile drain study in England (Williams, 1975) demonstrate the complex relationship between rainfall amount and intensity and losses of NO_3^- in drainage over autumn, winter, and spring (Fig. 5). Initial autumn rain satisfied the soil moisture deficit and therefore miminum drainage flow occurred. Nevertheless, there was a distinct peak in drainage NO_3^- concentration, which was attributed to macropore flow from intense rainfall on a dry, cracked soil. The amount of water transported was, however, small and consequently the total amount of NO_3^- leached was also small. In late autumn when the soil had reached field capacity, the drainage volume increased and fluctuated over winter and spring according to rainfall pattern. Total losses of NO_3^- during winter and spring were closely related to the volume of water flowing since NO_3^- concentrations in drainage tended to decline during the period with only slight fluctuations occurring. The results shown in Fig. 5 demonstrate that large protracted flows of low concentrations of NO_3^- can be considerably more significant to total loss than short periods with high NO_3^- concentrations.

In contrast to the results of Williams (1975) (Fig. 5), many workers have observed a close positive relationship between percolate volumes and N concentrations in percolates (e.g., Fig. 6); the yearly maximum for both often occurs during winter (Hood, 1976a; Chichester, 1977; Steele *et al.*, 1984). The relationship between leachate N concentrations and volumes will depend largely on the relative concentration of NO_3^- in the soil profile and the relative amount of water flowing through the soil.

In any event, total leaching losses of NO_3^- are characteristically highest during the winter period (Wild and Cameron, 1980a) and peak levels of NO_3^- in rivers from both agricultural and undisturbed forested watersheds have been observed during the winter months (Johnson *et al.*, 1969; Tomlinson, 1970; Troake *et al.*, 1976; Likens *et al.*, 1977; Greene, 1978).

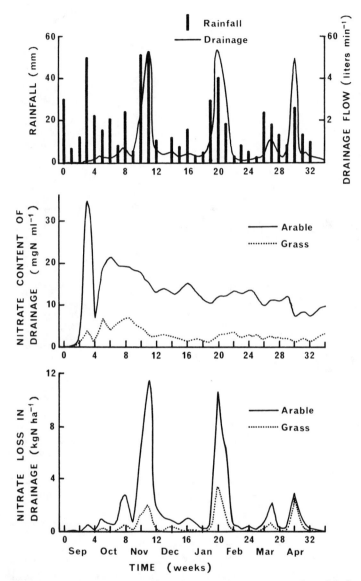

Fig. 5. Seasonal changes in rainfall, tile drainage, nitrate concentration of drainage, and the total quantity of nitrate lost in drainage from arable and grassed fields. [Redrawn from Williams (1975).]

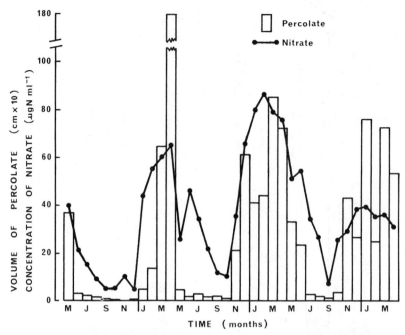

Fig. 6. Seasonal changes in percolate and nitrate concentration of percolate from a tilled corn-cropped lysimeter. [Redrawn from Chichester (1977).]

2. Soil Properties

The influence of anion exchange capacity of soil (i.e., NO_3^- adsorption) on leaching has already been discussed. Below, the effects that soil physical properties and soil organic N have on NO_3^- leaching are discussed.

a. Soil physical properties. Since the hydraulic conductivity and water storage capacity of a soil are directly related to its texture and structure, NO_3^- losses are normally greater from poorly structured sandy soils than from coarsely structured clay soils. The rate of denitrification is likely to be greater in wet clay soils (see Chapter 5) and this can further reduce the apparent leaching loss (Kolenbrander, 1972; Gambrell *et al.*, 1975).

Catchment (Kudeyarov *et al.*, 1981), borehole (Avnimelech and Raveh, 1976), lysimeter (Woldendorp *et al.*, 1965; Kolenbrander, 1969), and column studies (Sommerfeldt *et al.*, 1982) have all shown faster rates of nitrate leaching through coarse-textured than fine-textured soils. Avnimelech and Raveh (1976), for example, reported from a borehole study on irrigated orchards that approximately 50% of the applied fertilizer N (surplus over crop uptake) was leached in a sandy loam while only 12% was lost by leaching in a clay loam. Pratt *et al.* (1980) presented data that provide an approximate ratio of leaching loss of 5 : 1 for a silt loam compared to a clay loam.

Using a regression equation derived from leaching studies on soils in southeast England, Wild and Babiker (1976a) presented a table of calculated depths of leaching that clearly demonstrates the influence of soil texture (Table III).

b. Soil organic nitrogen. A considerable quantity of NO_3^- leached from agricultural soils can originate from mineralization and nitrification of soil organic N rather than directly from applied fertilizer N (Kolenbrander, 1981). Indeed, substantial leaching losses of NO_3^- can occur from unfertilized bare fallow soils (Low and Armitage, 1970; Low, 1973; Guiot, 1981). For example, at Rothamsted Experimental Station unfertilized bare fallow lysimeters lost 41.5 kg N ha^{-1} yr^{-1} between 1877 and 1887 and 28 kg N ha^{-1} yr^{-1} in the period 1905–1915 (Miller, 1906; Russell and Richards, 1920). Low and Armitage (1970) measured leaching losses of N from bare fallow lysimeters of 102 kg N ha^{-1} yr^{-1}, 92% of which was in the form of NO_3^--N.

Table III

Theoretical Leaching Depths of Nitrate Peaks according to Textural Classes[a]

Texture class	Volumetric water content, θ (cm^3 cm^{-3})[b]	Depth of nitrate peak (cm) for water inputs, Q (cm)[c]		
		15	20	30
Sand	0.09	99	131	194
Loamy sand	0.25	38	50	73
Sandy loam	0.27	36	46	67
Fine sandy loam	0.34	29	38	54
Loam	0.34	29	38	54
Clay loam	0.30	33	42	61
Silty clay loam	0.35	29	37	53
Clay	0.39	26	33	48

[a] Wild and Babiker (1976a).

[b] Values of θ from Salter and Williams (1965).

[c] Calculated from an experimentally based regression equation where the depth of leaching is given by $y = 0.57x + 4.1$, where $x = Q/\theta$, and Q is for winter rainfall in the Reading area, southeast England.

The proportion of NO_3^- originating from the mineralization of soil organic N and that originating from fertilizer N or mineralization of added organic wastes will differ greatly depending on environmental conditions, the rate of applied N, and the crop management system employed. Factors influencing mineralization and nitrification were discussed in Chapters 2 and 3. In particular, cultivation of soil is likely to stimulate mineralization of soil organic N and can result in subsequent NO_3^- leaching (Viets, 1971; Kolenbrander, 1973). The significance of soil organic matter as a source of NO_3^- in cultivated ecosystems is illustrated by the results of Dowdell *et al.* (1984), who found no significant difference in total amount of N leached from fertilized (80 and 120 kg N ha^{-1} yr^{-1}) and unfertilized soil cropped with barley (Table IV). The ^{15}N results in Table IV also show that the vast majority (over 90%) of the total amount of N leached originated from native soil N and that less than 10% was actually from the applied fertilizer. Even under ungrazed pasture conditions, Dowdell (1981) found that when a high rate of fertilizer was applied (400 kg N ha^{-1}) the loss of fertilizer N represented less than 59% of the total quantity of N leached (Table V).

It is, however, possible that the addition of fertilizer N may tend to stimulate mineralization of soil organic N and thus lead to leaching of native soil NO_3^- (i.e., have a "priming effect"; see Chapter 2). That is, due to mineralization–immobilization turnover, fertilizer N will be immo-

Table IV

Total Quantities of N Lost to Drainage and ^{15}N Recoveries in Drainage from Lysimeters Cropped with Barley[a]

	Fertilizer rate (kg N ha^{-1} yr^{-1})		
Fertilizer N applied (kg N ha^{-1} yr^{-1})[b]	0	80	120
Mean leaching loss (kg N ha^{-1} yr^{-1})	83	74	83
Total leaching loss (June, 1977–1981) (kg N ha^{-1})	332	297	335
Recovery of ^{15}N-labeled nitrogen (June, 1977–1981) (% of total N)	—	6.3	8.1
Recovery of ^{15}N-labeled nitrogen (June, 1977–1981) (% of applied)	—	6.6	6.3

[a] Data summarized from Dowdell *et al.* (1984).
[b] ^{15}N-fertilizer applied in first year only.

Table V

Loss of Nitrogen by Leaching (kg N ha^{-1} yr^{-1}) after ^{15}N Fertilizer Addition to Grassland Lysimeters[a]

	Type of soil	
	Clay soil	Silt loam
Unfertilized	4	5
Fertilized[b]		
Labeled N	25	22
Unlabeled N	19	16
Total	44	38
Labeled N as a percentage of Total N leached	57	58

[a] Data from Dowdell (1981).
[b] Nitrogen applied as ^{15}N-labeled urea at a rate of 400 kg N ha^{-1}.

bilized but at the same time soil organic N will be mineralized and possibly lost.

3. Land Management

Many terrestrial ecosystems fail to produce soil NO_3^- to any great extent despite the presence of apparently suitable environmental conditions for nitrification (see Chapter 3). There is considerable evidence that a limiting supply of NH_4^+ limits nitrification in many of these ecosystems. Strong competition between the roots of vegetation (and associated mycorrhizal fungi) and the heterotrophic soil microbial biomass for the pool of NH_4^+ in the soil probably leaves little available for use by the autotrophic nitrifying bacteria (Huntjens and Albers, 1978; Vitousek et al., 1979; Robertson, 1982). Any NO_3^- produced would be quickly used by the heterotrophic biomass or the growing vegetation. Thus, leaching losses of NO_3^- are low (Khanna, 1981). Indeed, as noted in Chapter 1, large amounts of N are cycled within mature natural ecosystems and both inputs and losses of N are normally small.

However, following ecosystem disturbance (e.g., fire, harvesting, fallowing, cultivating, or fertilizing) leaching losses of NO_3^- are often greatly increased (Khanna, 1981). Agricultural ecosystems tend to be continually disturbed and leaching losses can be large (Wild and Cameron, 1980a). Thus land management has a large effect on the quantities of NO_3^- leached from soils (Cameron, 1983).

  a. Forests. Nitrate leaching losses from undisturbed forest ecosystems are generally very low (Likens *et al.*, 1977; Henderson *et al.*, 1978; Gosz, 1981; Melillo, 1981). Typical values for coniferous forests are 0.5 to 1.5 kg N ha^{-1} yr^{-1} (Gosz, 1981) and for deciduous forests 3 to 4 kg N ha^{-1} yr^{-1} (Melillo, 1981). The amount of N lost by leaching is often less than the input from rainfall (Gosz, 1978; Melillo, 1981).

  Removal of vegetation by clear-cutting can result in large increases in net mineralization and nitrification rates (Vitousek, 1981) and a consequent increase in the amount of NO_3^- leached (Fredriksen *et al.*, 1975). Hornbeck *et al.* (1975), for example, reported that streamwater NO_3^- concentrations increased from 2 to 20 μg N ml^{-1} and the total N loss was increased by 342 kg N ha^{-1} over 3 years following clear-cutting of a North American hardwood forest. However, the extent of net mineralization and nitrification in clear-cut forest sites is very much dependent on the C:N ratio of the forest litter layer and of the remaining residues of dead plant material. In some cases where the C:N ratio is high the extent of NO_3^- leaching is minimal since net immobilization occurs (e.g., Tamm *et al.*, 1974; Sopper, 1975).

  The burning of forests has also been shown to increase leaching losses of NO_3^- (Lewis, 1974; Stark, 1977) but the magnitude of the loss depends on factors such as the intensity of the burn, amount of N left in the residues, and extent of damage to root systems, as well as factors already mentioned (Khanna, 1981; Woodmansee and Wallach, 1981).

  b. Permanent grasslands. In general, extensive pastoral systems lose very little NO_3^- by leaching (Kilmer, 1974) since plant uptake and rapid immobilization of NH_4^+ in the rhizosphere leaves little NO_3^- in the soil profile for leaching (Huntjens, 1971a,b; Huntjens and Albers, 1978). Indeed, in grassland soils levels of mineral N are normally low: NH_4^+-N < 10 μg gm^{-1}, NO_3^--N < 1 μg gm^{-1} (Woodmansee *et al.*, 1981). Thus, leaching losses of N from upland pasture catchments grazed with sheep can be in the range of 1 to 6 kg N ha^{-1} yr^{-1} (Crisp, 1966; Bargh, 1978) and losses are often less than rainfall inputs (Batey, 1982).

  Leaching losses of NO_3^- under legume pastures can be 8 to 10 times those under all-grass pastures and in general losses of N from grass–legume pastures also tend to be higher than those from all-grass pastures (Kilmer *et al.*, 1974). Legumes accelerate N leaching losses (Guiot, 1981) because they fix large amounts of N, some of which is released when top growth dies back and/or roots die and nodules slough off (Vallis, 1978). Low and Armitage (1970), for example, observed leaching losses from an unfertilized grass sward of 2.6 kg N ha^{-1} yr^{-1} while those from a growing clover sward were 30 kg N ha^{-1} yr^{-1} and 131 kg N ha^{-1} yr^{-1} after the clover had died.

Heavy fertilizer applications can also cause significant leaching of NO$_3^-$ below pasture sites. Hood (1976a,b), for instance, measured losses of 11 and 54 kg N ha^{-1} yr^{-1} for pasture sites fertilized with 250 and 750 kg N ha^{-1} yr^{-1}, respectively. Nonetheless, leaching losses of N are considerably lower from pasture than from arable sites receiving the same fertilizer N inputs (Kolenbrander, 1973; Kilmer, 1974). Shallow-rooted grasses with their less efficient extraction of soil NO$_3^-$ and water permit higher losses of fertilizer N than do comparatively deep-rooting species (Kilmer, 1974).

Leaching losses can be relatively large (i.e., 50–200 kg N ha^{-1} yr^{-1}) on intensively managed pastures where high fertilizer rates are combined with high stocking rates (Horne, 1980; Steele and Shannon, 1982; Ball and Ryden, 1984; Steele et al., 1984). Substantial leaching of NO$_3^-$ can occur from the unevenly distributed urine patches (O'Connor, 1974; Floate, 1981) since such patches can contain localized concentrations of N equivalent to approximately 500 kg N ha^{-1} for sheep and 950 kg N ha^{-1} for cattle (Steele, 1982). The effect of grazing animals on leaching losses was demonstrated by Ball and Ryden (1984), who observed leaching losses of 140–190 kg N ha^{-1} yr^{-1} below an intensively grazed, fertilized, ryegrass pasture in the United Kingdom, while losses from a similar, cut sward were only 35–40 kg N ha^{-1} yr^{-1}. Similarly, Kolenbrander (1981) calculated that grazing cattle (2.25 cattle units ha^{-1}) may increase N leaching losses on grassland by a factor of 3.5 to 5. In lysimeter studies on intensively cattle-grazed pastures in New Zealand receiving 0 or 172 kg N ha^{-1} yr^{-1}, Steele et al. (1984) recorded leaching losses of 88 and 193 kg N ha^{-1} yr^{-1}, respectively. Irrigated pastures have a higher potential for NO$_3^-$ leaching than dryland pastures due to the associated higher stocking rates and the greater volume of water throughput (Turner, 1976; Burden, 1982).

c. Cropping. Lack of vegetation for at least part of the year is a key factor stimulating NO$_3^-$ leaching from cropping systems. Where conditions are favorable for rapid mineralization and nitrification high levels of NO$_3^-$ will accumulate in the surface of soils lacking growing vegetation. If high rates of rainfall occur then NO$_3^-$ leaching is virtually inevitable. Thus leaching losses can be a particular problem where summer fallow is practiced and it is followed by heavy winter rainfall. In a 10-year comparison of NO$_3^-$ profiles under pasture, wheat–fallow, and continuous fallow Rennie et al. (1976) observed essentially no downward movement of NO$_3^-$ under pasture but about 500 kg NO$_3^-$-N ha^{-1} yr^{-1} was found in the upper

3.6 m of soil under wheat–fallow. Under the continuous fallow treatment 1082 kg of NO_3^--N ha^{-1} yr^{-1} was found in the top 3.6 m.

As already noted, pasture vegetation can greatly reduce NO_3^- leaching and it has been consistently reported that leaching losses of NO_3^- from cultivated cropped lands are considerably greater than those from grassland whether or not fertilizers have been applied (Kolenbrander, 1969; Power, 1970; Williams, 1975; Young et al., 1976; Chichester, 1977; Adams et al., 1979; Jaakkola, 1984). Williams (1975), for example (Fig. 5), found that NO_3^- losses in drainage were 51 kg N ha^{-1} yr^{-1} from cultivated crops and 18 kg N ha^{-1} yr^{-1} from grassland.

In crop rotation systems where grass is included for only a few years and then plowed under leaching losses can, however, be large. Deep borehole studies conducted in England have shown that the plowing of grassland has been responsible for the release of peak concentrations of nitrate that are now moving slowly through the unsaturated zones of aquifers and will eventually contaminate groundwater supplies (Young et al., 1976; Young and Gray, 1978). An example of a nitrate profile found in chalk is shown in Fig. 7; each major peak is associated with a date when

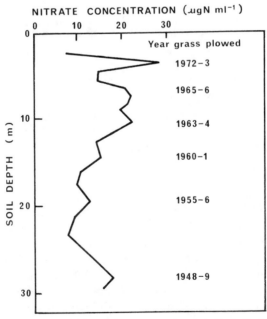

Fig. 7. A profile of nitrate concentration in a deep borehole in chalk. [From Young et al. (1976).]

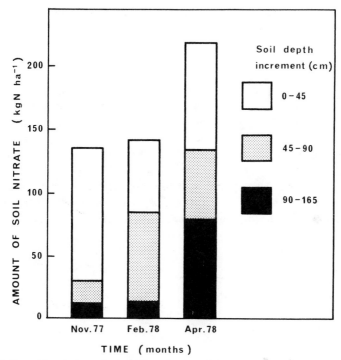

Fig. 8. Quantities of nitrate in the soil profile during the winter following fall-plowing of a 3-year-old grassland. [Data from Cameron and Wild (1984).]

grassland was plowed. Soil studies have confirmed that large amounts of nitrate are mineralized when grass is plowed in fall and that this nitrate is readily lost by leaching over the winter (Cameron and Wild, 1984). Results in Fig. 8 illustrate a large accumulation and downward movement of NO_3^- in the soil profile during winter following fall-plowing of a 3 year-old grassland. Similarly, plowing under of leguminous crops or pastures (e.g., alfalfa) results in high concentrations of NO_3^- being released into the soil and leached (Letey *et al.*, 1977; Robins and Carter, 1980).

Rooting habits of plants can exert a great influence on NO_3^- leaching through the root zone since plant roots remove both NO_3^- and water from the soil profile (Singh and Sekhon, 1979). Alfalfa, with its deep root system, has been shown to be an effective scavenger of NO_3^- that may have accumulated under prior crops (Mathers *et al.*, 1975; Muir *et al.*, 1976). Singh and Sekhon (1979) showed that maximum leaching of NO_3^- occurred from crop rotations with heavily fertilized shallow-rooted crops like potato. On the other hand, wheat and maize, when grown in rotation, absorbed a large fraction of the applied N due to their relatively deep and extensive root systems.

Tillage practice can also influence leaching loss. As already noted when discussing runoff (Section III,B), conservation tillage generally results in a greater infiltration capacity than under conventional tillage because of the presence of surface mulches and the larger number of continuous macropores that are open at the soil surface (Unger and McCalla, 1980). Initial leaching losses of surface-applied N can therefore be rapid under direct drilling if heavy rainfall occurs soon after fertilizer application and therefore results in significant movement of water and solutes through the macropores (McMahon and Thomas, 1976; Tyler and Thomas, 1977). Conversely, fertilizer that has had time to diffuse into aggregate micropores will be afforded greater protection from subsequent leaching under direct drilling due to the higher proportion of water flowing in the macropore system.

A considerable amount of work has been reported on leaching losses from different crops and selected information is presented in Table VI. It is difficult to rank crops in order of leaching potential because of the wide range of soils, climate, fertilizers, and experimental conditions reported. Nevertheless, in general, leaching losses are greater from horticultural crops than from arable crops. This is due primarily to the higher rates of fertilizer applied, although root depth and density, irrigation rate, and the fact that horticultural crops are usually grown on lighter soils will all have an effect.

d. Nonagricultural activities. River studies have indicated the relative importance of point sources of nitrate pollution, such as sewage or industrial effluent. Although high nitrate levels are reported at the point of intrusion (Olson *et al.,* 1974), their overall relative contribution is frequently less dominating. In a study of a number of catchments in England and Wales, Owens (1970) reported that the proportion of total N from sewage effluent was generally small (<17%) except in industrial areas and that the remainder of N was from other sources such as land drainage. The highest average N load came from arable land (13 kg ha^{-1} yr^{-1}), then permanent pasture (8 kg ha^{-1} yr^{-1}), and lastly urban areas (4 kg ha^{-1} yr^{-1}). Studies in the United States (Task Group Report, 1967) also indicate the relative importance of land drainage as the major source of nitrogen. Nevertheless, calculations do show that where urbanization has increased, the nitrogen load also increases (Owens, 1970).

4. Irrigation

Since irrigation increases crop growth it also increases N uptake. Therefore, application of the optimum amount of irrigation for crop

Table VI

Leaching Losses of Nitrate from Varied Cropping Systems and Fertilizer Rates

Land management	Soil type	N applied (kg ha^{-1} yr^{-1})	Average N leached (mg liter^{-1})	(kg ha^{-1} yr^{-1})	Methodology	Reference
Bare fallow	Heavy loam over chalk	0	9	34	Lysimeter	Miller (1906)
	Sandy loam (pH 5.2)	0	—	84	Lysimeter	Morgan et al. (1942)
		224	—	231	Lysimeter	
	Sandy loam (pH 6.3)	0	—	57	Lysimeter	
		224	—	218	Lysimeter	
Grass	Sandy loam over clay/chalk	250	max. = 7	6	Lysimeter	Garwood and Tyson (1973, 1977)
		500	max. = 120	128	Lysimeter	
	Sandy loam over clay	250	2	4	Catchment	Barraclough et al. (1983)
		500	13	27	Catchment	
		900	86	151	Catchment	
	Sandy soil	8 × 30	2	13	Lysimeter	Woldendorp et al. (1966)
	Heavy clay	3 × 50	1	5	Lysimeter	Kolenbrander (1969)
Corn, carrots	Sandy soil	396	29	155	Tile drain	Letey et al. (1977)
Lemons	Sandy soil	26	4	46	Tile drain	
Dates	Sandy soil	149	49	62	Tile drain	
Cotton	Clay soil	492	21	71	Tile drain	
Milo	Clay soil	224	92	119	Tile drain	
Cotton	Clay soil	169	16	35	Tile drain	
Potatoes	Sandy soil	0	16	43	Lysimeter	Pfaff (1963)
	Sandy soil	80	19	47	Lysimeter	
Winter rye	Sandy soil	0	23	61	Lysimeter	
	Sandy soil	80	32	74	Lysimeter	
Oats	Sandy soil	0	23	60	Lysimeter	
	Sandy soil	80	25	60	Lysimeter	

growth can reduce leaching losses (Pfaff, 1958; Bauder and Schneider, 1979). Hahne *et al.* (1977) reported a reduction from 48 to 5% in the amount of nitrate lost when optimum irrigation and fertilizer rates were applied.

However, in many cases, irrigation has been reported to increase leaching losses of NO_3^- because of excess water passing through the crop root zone (e.g., Bauder and Schneider, 1979; Timmons and Dylla, 1981). Timmons and Dylla (1981) applied supplemental irrigation to corn as either partial replenishment (2.5 cm) or full replenishment (5 cm) every time the available soil water decreased to about 5 cm (50% depletion). In comparison to nonirrigated fertilized corn, annual NO_3^- leaching losses increased by an average of 17 and 53%, respectively, for the partial and full replenishment irrigations. McNeil and Pratt (1978) reported average N leaching losses under irrigated croplands in southern California to be between 25 and 50% of applied N; higher losses occurred where there was excessive irrigation, particularly when this was combined with an excessive N application. In Nebraska, Olson *et al.* (1974) attributed a 24% increase in groundwater NO_3^- concentrations over a 10-year period to a 50% increase in area being irrigated during that period.

The method of irrigation can greatly affect the quantity of NO_3^- leached (see Viets *et al.*, 1967). Furrow irrigation generally removes nutrients from directly below the furrow, while with flood and sprinkler irrigation the downward movement of NO_3^- is more uniform due to the more extensive volume of soil that is leached.

5. Fertilizer Applications

Convincing evidence that NO_3^- originating from fertilizers accumulates in shallow groundwaters has been presented from studies in Rheingau, Germany (Sturm and Bibo, 1965), Arroya Grande Basin, California (Stout and Burau, 1967), San Joaquin Valley, California (Nightingale, 1970, 1972), Santa Anna River Basin, California (Ayers and Branson, 1973), and the central coastal region of Israel (Gruener and Shuval, 1970). As already noted, concentrations of NO_3^- in groundwater are generally higher from fertilized areas under irrigation than from unirrigated areas (Olson *et al.*, 1973; Muir *et al.*, 1976).

In general, if fertilizer N applications do not exceed crop requirements then there is little NO_3^- available for leaching (Pratt *et al.*, 1972; Fried *et al.*, 1976; Singh and Sekhon, 1979). Numerous studies have demonstrated that when fertilizer N is applied at rates higher than the optimum for crop production there is a considerable increase in the quantity of NO_3^- leached (e.g., Burwell *et al.*, 1976; Broadbent and Carlton, 1978; Gast *et al.*, 1978; Olson, 1979; Baker and Johnson, 1981). Quantities of NO_3^- leached generally increase as fertilizer application rates increase (Broadbent and Carlton, 1978; Gast *et al.*, 1978) although the actual amounts leached depend on the soil, climate, and other factors. Table VII demonstrates that when the N application rate is raised above the recommended rate the quantity of N removed in the crop may not be greatly affected but the amount of N lost by leaching may be increased considerably. The quantity of N remaining in the soil profile will also increase.

One important factor influencing NO_3^- leaching from agricultural lands is fertilizer use efficiency (percentage recovery of fertilizer N by a crop) (Fried *et al.*, 1976; Singh and Sekhon, 1979). One method of increasing fertilizer use efficiency is to apply fertilizer N in split applications. Gerwing *et al.* (1979) found that splitting a 179 kg N ha^{-1} addition into four applications increased the recovery of fertilizer N by 30 to 52% in a corn crop. While the split application had no significant effect on NO_3^--N concentrations in the aquifer, the one-time application increased concentrations by 7 μg ml^{-1}. Singh and Sekhon (1979) concluded that as the number of splits are increased, the susceptibility of applied NO_3^- to leaching decreases.

The form of N applied can also have some influence on leaching losses. Some workers have recorded greater leaching losses when NO_3^- salts (e.g., $Ca(NO_3)_2$) rather than NH_4^+ salts or urea are applied (Pratt *et al.*, 1967; Wiklander and Vahtras, 1975; Bauder and Montgomery, 1979; Koren'kov *et al.*, 1979). For example, in a lysimeter study, Wiklander and Vahtras (1975) showed that leaching losses of N in the first season followed the order Urea $<$ $(NH_4)_2SO_4$ $<$ Nitro Chalk $<$ $Ca(NO_3)_2$ (Fig. 9). Nonetheless, a subsequent experiment over the following growing season (Vahtras and Wiklander, 1977) showed increased leaching from urea treatments due to intensive nitrification of NH_4^+ remaining from the pre-

Table VII

Nitrogen Balance in a Corn Crop Fertilized with Four Rates of Nitrogen over a 3-Year Period[a]

Annual treatment (kg N ha^{-1} yr^{-1})	Total amount of N added 1973–1975 (kg N ha^{-1})	N removed in corn (kg N ha^{-1})	N loss from tile (kg N ha^{-1})	N remaining in 0–3 m soil profile fall 1975 (kg N ha^{-1})
20	60	105	41	54
112[b]	336	167	53	100
224	672	166	93	425
448	1344	196	180	770

[a] Data from Gast *et al.* (1978).
[b] Recommended rate.

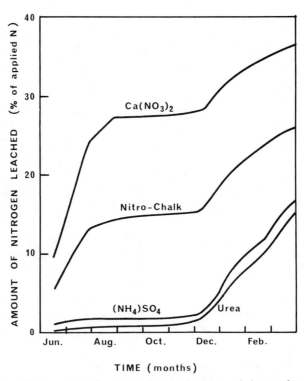

Fig. 9. Effect of form of fertilizer N on the leaching losses of nitrogen from soil blocks. Nitrogen was applied at a rate of 1.5 gm N per 5 liters of soil. [Redrawn from Wiklander and Vahtras (1975).]

vious season. Indeed, the extent of NO_3^- leaching from NH_4^+-based fertilizers will depend on the rate and time of nitrification relative to the rate of plant uptake and the period of leaching. Where nitrification is rapid, differences in leaching due to N form are expected to be minimal or nonexistent.

Some studies have shown that leaching of NO_3^- from applied NH_4^+ or urea fertilizers can be reduced by applying a nitrification inhibitor, such as nitrapyrin, along with the fertilizer (Soubiès *et al.,* 1962; Swoboda, 1977; Owens, 1981). Even so, the success of such a measure will depend greatly on the rate of N application and the periods of potential leaching and plant uptake. Slow-release N fertilizers can also reduce leaching losses in some situations (Terman and Allen, 1970; Jung and Dressel, 1974).

6. Organic Waste Applications

Spreading animal wastes, as solids or as slurries, on agricultural land at rates of N higher than the crop can utilize can result in leaching of NO_3^- (Mathers and Stewart, 1974; Haghiri *et al.*, 1978; Liebhardt *et al.*, 1979; Vetter and Steffens, 1981). The optimum rates, however, are not always easy to determine since the organically combined N in the manure must be mineralized before it is available to plants or subject to leaching. Mineralization rate is therefore the key factor required to estimate optimum application rates (Smith and Peterson, 1982). Yearly mineralization rates can be expressed as a decay series for different manures (Pratt *et al.*, 1973; Turner, 1976) and in general from 20 to 80% of manure N is mineralized in the first year, 10–25% in the second year, and 5–10% in the third year. Because of the residual N-supplying power of manures, decreasing amounts need be applied each year to meet annual crop N requirements.

The amounts of NO_3^- leached typically increase as the rate of waste application increases (Bielby *et al.*, 1973; Haghiri *et al.*, 1978; Liebhardt *et al.*, 1979; Weil *et al.*, 1979; Dunthion, 1981; Sherwood, 1981; Vetter and Steffens, 1981). Sherwood (1981), for instance, applied pig slurry to grassland over a 4-year period at rates of nil, 400, 700, and 1400 kg N ha^{-1} yr^{-1} and estimated leaching losses as 0.9, 18, 77, and 162 kg N ha^{-1} yr^{-1}, respectively. In some cases the percentage of applied N lost in leachate increases with increasing application rates (Spallacci, 1981) while in others it decreases (Dam Kofoed, 1979).

Leaching losses are particularly high when wastes are spread on fallow land in autumn and in such cases losses often amount to 20 to 30% of applied N on an annual basis (Dam Kofoed, 1979; Smilde, 1979; Vetter and Steffens, 1981). Losses are generally greatly reduced if wastes are spread in spring close to the onset of crop growth and after heavy rainfalls have passed (Vetter and Steffens, 1981).

Irrigating crops with sewage effluent can present particular problems since N is applied in a soluble, readily mineralizable form and irrigation is applied in accordance with the crop's requirement for water, not N. Leaching losses of NO_3^- can therefore be large following irrigation with effluent (Burton and Hook, 1979; Hook and Burton, 1979; Lund *et al.*, 1981). Hook and Burton (1979) found that irrigation of a pasture with 5 or 10 cm of sewage effluent per week resulted in losses of 19 and 44%, respectively, of the applied effluent N. In general, long-season, high-yielding sod crops are more suitable for irrigation with effluent than are annual crops since the former utilize more N and therefore reduce the leaching loss (Smith and Peterson, 1982).

V. SIGNIFICANCE OF RUNOFF AND LEACHING LOSSES

A. Consequences of Losses

Losses of N from agricultural lands through runoff of soluble and particulate N as well as NO_3^- leaching have several notable consequences. Where losses are large enough to cause decreases in crop yield the economic significance is obvious. However, it is the environmental effects of such losses that have recently received the most attention. The over-enrichment of lakes with nutrients (eutrophication) has many adverse effects on their aesthetic and recreational value while health problems have been associated with a high NO_3^- content of drinking water.

1. Economic Loss

Losses of soluble N in runoff from agricultural land do not generally exceed 5% of the fertilizer input and are often less than rainfall inputs (Baker, 1980). Such losses are therefore not normally of economic significance. In contrast, on sloping cultivated land, losses of N in sediment can be considerable, sometimes on the order of 20–70 kg N ha^{-1} yr^{-1} (e.g., Burwell et al., 1975, 1976; Alberts et al., 1978). Such losses obviously represent a significant loss of N (and soil) from the cultivated area. This N loss is principally in the form of organic N (native soil N plus immobilized fertilizer N) and continued losses will inevitably bring about a reduction in soil fertility and the requirement for greater fertilizer inputs. Methods of minimizing losses of sediment from agricultural land are well known and documented and have proven very effective in many cases (see Section III,B).

Leaching losses of NO_3^- are not normally large except when fertilizer inputs greatly exceed crop requirements or the ground is left fallow. Thus, in general, economic losses of N through leaching are likely to be significant only when high rates of fertilizer (above the rate of greatest economic return) are applied (see Section V,B). Leaching losses of NO_3^- from fallow land or following the plowing in of grasslands or other cover crops can be considered as economic losses since the NO_3^- could otherwise have been used by a subsequent crop. Such losses are, however, for the most part unavoidable unless farming practices are greatly altered. Leaching losses from intensively grazed pastures may be large, particularly if irrigation follows immediately after the grazing period.

2. Decreased Soil pH and Base Saturation

As noted in Chapter 3, nitrification and the subsequent leaching of NO_3^- has been shown to have an acidifying effect on the surface soil (Wolcott et *al.*, 1965; Pierre *et al.*, 1971). During the process of nitrification H_3O^+ ions are released:

$$NH_4^+ + 2O_2 + H_2O \longrightarrow 2H_3O^+ + NO_3^- \tag{12}$$

Exchangeable cations, displaced by the H_3O^+ ions, move downward as counterions with NO_3^- resulting in a decrease in pH and base saturation of the surface soil (Haynes, 1981a, 1983). In fertile soils Ca^{2+} is often the dominant balancing cation for leached NO_3^- (Terman, 1977; Haynes, 1981b; Steele *et al.*, 1984). Indeed, in nonsaline soils, even under irrigation leaching of NO_3^- is generally the dominant factor determining the quantity of exchangeable bases leached (Raney, 1960; Viets *et al.*, 1967; Haynes, 1984).

3. Eutrophication

Nitrate pollution can originate from point sources (e.g., sewage outfalls or industrial effluents) or nonpoint sources such as stormwater runoff and leaching and runoff from croplands. In terms of N inputs to surface and groundwaters, nonpoint sources make up by far the largest contribution (Loehr, 1974; McElroy *et al.*, 1976). Estimates indicate that more than 90% of the N entering surface waters originates from nonpoint sources and that more than 80% of that portion is from agricultural lands including livestock feedlots (NRC, 1978). Many studies have demonstrated that high levels of total N and/or NO_3^- in surface waters are generally related to agricultural activities (e.g., Omernick, 1976; Smith *et al.*, 1982). Although point sources of N are of minimal importance on a regional basis they can represent major sources on a local basis (NRC, 1978). Because the forms of N in aquatic systems are readily interconvertible (Keeney, 1973; Barica, 1977; Larsen, 1977) all N inputs to surface water (NH_4^+, NO_3^-, soluble organic N, and particulate organic N) rather than NO_3^- alone should be considered.

Since P and N are the nutrients limiting production in most lakes, these nutrients are the most important in stimulating eutrophication (Keeney, 1973). Most low-producing oligotrophic lakes (low in nutrients) are P- rather than N-limited (Keeney, 1973; Forsberg, 1977; Organization for Economic Cooperation and Development (OECD), 1982; Sonzogni *et al.*, 1982) due to the paucity of P in the biosphere compared to N. Nitrogen can, however, be a limiting element in some ultraoligotrophic lakes (Forsberg, 1977). The productivity of coastal and estuarine ecosystems is quite often limited by N (Goldman, 1976). In many already eutrophic lakes, biotic productivity is controlled by N because the N/P ratios of pollutants from many sources are far below the ratios required for plant growth.

Some functions of aquatic ecosystems can benefit from anthropogenic

NO_3^- inputs. In some oligotrophic lakes where N is the limiting nutrient, the input from groundwater, surface runoff, or precipitation may be essential to maintain biological productivity (Keeney, 1982). However, overenrichment of surface waters with nutrients results in a range of changes in water quality that are generally considered undesirable. The most common of these are a decrease in water clarity, the proliferation of "blooms" of algae and other aquatic plants, the depletion of dissolved oxygen in the bottom water with the concomitant loss of cold (bottom) water fisheries, and the general shortening of food chains (NRC, 1978; OECD, 1982).

Eutrophication decreases the recreational value of lakes through a general loss of aesthetic appeal, reduced boat access due to aquatic vegetation, and health problems such as ear, nose, and throat infections for swimmers (NRC, 1978). Eutrophication also results in a requirement for increased water treatment before domestic use due to the increased color, taste, and odor of the water and its increased chlorine demand. Eutrophication can partially block irrigation or drainage canals due to excessive growth of aquatic vegetation and in arid regions can result in increased water loss from irrigation canals because of evapotranspiration from the floating vegetation.

Levels of N that can enter lakes before eutrophication will occur will differ greatly depending on factors such as the size of the lake, its present N status, and whether N is limiting its productivity. Some workers have developed the concept of nutrient loading rates and presented graphs of critical available nutrient loading rates (gm N m^{-2} of lake surface per year) versus mean depth (e.g., Vollenweider, 1968; NRC, 1978). The level of 0.3 μg N ml^{-1} of inorganic N is widely quoted as the critical level of lake N above which nuisance algal growth may be stimulated (e.g., Vollenweider, 1968) but this will of course depend on whether N is limiting growth or not.

4. Health Problems

Nitrate *per se* is relatively nontoxic to domesticated animals and humans alike (NRC, 1978). Ingested nitrate can, however, be reduced to NO_2^- by gastrointestinal bacteria present in the tract of ruminant animals and in the human infant during the first few months of life (NRC, 1972; Shuval and Gruener, 1977; Taylor, 1975). Nitrite is rapidly absorbed from the stomach into the blood, where it readily oxidizes the iron of hemoglobin to the ferric state, forming methemoglobin. Methemoglobin cannot function in oxygen transport and cellular anoxia can result. If over 50% of the blood hemoglobin becomes oxidized, death is likely.

The large majority of cases of infant methemoglobinemia have been reported from households with a private well water supply containing more than 10 μg NO_3^--N ml^{-1} (Shuval and Gruener, 1977; NRC, 1978). The U.S. Public Health Service (1962) drinking water standard is 10 μg NO_3^--N ml^{-1} while the World Health Organization's (1970) recommended level is less than 11.3 μg NO_3^--N ml^{-1} and the accepted level is 11.3 to 22.6 μg NO_3^--N ml^{-1}. It is important to note, however, that in the majority of cases of methemoglobinemia that well water was also contaminated with bacteria (International Standing Committee, 1974) and that in the reduction of NO_3^- the gastrointestinal microflora are of paramount importance (Phillips, 1971). The actual "safe" level is debatable (Wild, 1977; Wild and Cameron, 1980b) but it does appear that the use of domestic water supplies containing greater than 20 μg NO_3^--N ml^{-1} is likely to significantly increase the number of young infants at risk (NRC, 1978; Burden, 1982).

The possibility of *in vivo* formation of carcinogenic nitrosamines by reaction of ingested amines with NO_2^- in the human stomach is thought to present a further health hazard after exposure to high levels of NO_3^-. A large number of *N*-nitroso compounds have been shown to induce tumors in test animals in many tissues (Shank, 1975; Crosby and Sawyer, 1976). No scientifically documented cause-and-effect data relating NO_3^- intake to cancer have yet been gathered (Doll, 1977) but some circumstantial data relating exposure to NO_3^- or NO_2^- to the incidence of cancer have been reported (e.g., Hill *et al.*, 1973; Armijo and Coulson, 1975).

For several decades, researchers have implied that NO_3^- may affect the cardiac function of man (Malberg *et al.*, 1978). A relation between high NO_3^- concentrations of drinking water and hypertension has been recorded (Morton, 1971) but other studies have failed to establish any relationship (Malberg *et al.*, 1978).

B. Methods to Control Losses

Factors affecting surface runoff losses of N and NO_3^- leaching have been discussed in previous sections. Some potential methods of controlling NO_3^- pollution of surface waters from croplands are summarized in Table VIII. A detailed discussion of such methods is presented by Keeney (1982). The basic principles are outlined below.

Where it is feasible, the containment of runoff and storage and treatment of nutrient-rich effluents from agricultural lands may be a viable control measure. Natural removal of N through uptake by aquatic vegetation and gaseous loss through nitrification–denitrification and NH_3 volatilization can account for over 70% of the N input to such detention reservoirs (Reddy, 1983).

Table VIII

Potential Methods of Controlling Nitrogen Pollution of Waters from Croplands

---

Containment and/or treatment or runoff
Management of cropping practices
    Soil conservation to minimize erosion
    Water conservation to optimize irrigation
    Use of cover crops to scavenge nitrogen
    Use of crop rotations
Management of fertilizer use
    Improved estimation of crop nutrient requirements
    Timing fertilizer applications to correspond with plant needs
    Fertilizer placement to improve fertilizer efficiency
    Use of slow-release fertilizers or nitrification inhibitors
    Foliar applications of fertilizers

Methods of soil conservation (see Section III,B) that decrease soil erosion and surface runoff reduce the amount of N lost from croplands through these processes. Soil conservation practices such as contouring, terracing, and conservation tillage are designed to hold the topsoil in place by increasing infiltration and reducing the velocity and quantity of runoff. However, they do not necessarily control other problems such as NO_3^- leaching. Indeed, control of runoff generally increases infiltration and percolation and tends to enhance leaching losses of NO_3^- (Thomas *et al.,* 1973; Burwell *et al.,* 1976).

Since the flux of N to groundwater via leaching and to surface water via runoff is a function of both the volume of water involved and its N content, water conservation can be important in reducing N losses (Keeney, 1982). By increasing the efficiency of crop water use, and therefore reducing the amount of water percolating below the root zone, it is possible to substantially reduce the amount of NO_3^- leached (Pratt, 1976; Saffigna *et al.,* 1977; Smika *et al.,* 1977; Smika and Watts, 1978). However, the concentrations of NO_3^- in the smaller volumes of percolate can be high since there is less of a dilution effect (Branson *et al.,* 1975; Devitt *et al.,* 1976).

Cover crops are sometimes planted for erosion control after the major crop has been removed. These crops can take up residual inorganic N and decrease winter leaching losses (NRC, 1978). The cover crop is incorporated during seedbed preparation and mineralization of residues results in some of the residual N being made available to the main crop. Crop rotations can also reduce N losses since some crop plants (e.g., soybeans, wheat, and barley) are more efficient in removing inorganic N from soils than others (e.g., corn and potatoes) (Singh and Sekhon, 1979). Soybeans can scavenge residual fertilizer N from a previous corn crop as well as use mineralized soil N and symbiotically fixed N (Johnson *et al.*, 1974) while deep-rooted alfalfa is a well-known scavenger of NO_3^- from the soil profile (Schertz and Miller, 1972).

Large leaching losses of NO_3^- generally occur when N inputs greatly exceed those that can be efficiently used by crops (Fried *et al.*, 1976; Singh *et al.*, 1978; Legg and Meisinger, 1982). Fertilizer use efficiency refers to the percentage recovery of applied N by the crop. The basic philosophy of improved fertilizer management is to apply N at such rates that use efficiency is high and the amount of unutilized N is reduced to environmentally acceptable levels.

The yield–response curve of crop plants to applied N (see Chapter 7) is generally such that each successive increment of fertilizer produces a lower increase in yield. The response curve flattens near the maximum yield and the optimum rate of fertilizer cannot usually be closely defined (Standford, 1966). In most cases, the point of greatest economic return to applied N is somewhere below the point of maximum yield (Singh *et al.*, 1978). Fertilizing for maximum yield can result in low fertilizer use efficiency and a sizable fraction of applied N can remain in the soil creating the potential for NO_3^- leaching (Singh and Sekhon, 1979). Indeed, in practice the efficiency of fertilizer recovery by crops rarely exceeds 70% even under favorable conditions and the average is probably nearer 50% (Allison, 1966; Legg and Meisinger, 1982). One possible way to minimize NO_3^- leaching is to restrict fertilizer applications to rates that do not exceed the economic optimum. However, the optimum rate of application is influenced by a multitude of site-specific factors that are difficult to predict and so accurate fertilizer N recommendations are extremely difficult to accomplish (see Chapter 7).

Methods of increasing fertilizer efficiency, and thus decreasing leaching losses, often involve supplying N as it is required by the crop. This is normally achieved by applications of fertilizer as split dressings although slow-release fertilizers and foliar applications have also been experimented with. The use of nitrification inhibitors to regulate the supply of NO_3^- from applied ammoniacal fertilizers has also been tried.

In view of the multitude of factors influencing NO_3^- leaching it is not surprising that amounts leached from croplands vary greatly (Baker, 1980; Keeney, 1982). However, on intensively managed croplands, concentrations of NO_3^--N in leachates seem to normally exceed 10 μg N ml^{-1} (Baker, 1980; Keeney, 1982). Keeney (1982) concluded that for many irrigated crops with good agronomic practices and profitable production about 20 μg N ml^{-1} in drainage effluent may be the lowest achievable.

On intensively grazed pastures, leaching losses of N can be significant due to the large concentrations of N deposited in urine patches (Section IV,D). Mean concentrations of NO_3^- in the range of 7 to 25 μg N ml^{-1} have been recorded under such pastures (Baber and Wilson, 1972; Steele *et al.*, 1984). There appears to be no short-term prospect of a change in pastoral farm management that will result in a marked decrease in NO_3^- leaching (Burden, 1982).

VI. CONCLUSIONS

Ammonium N is held in the soil by the processes of cation exchange and fixation within clay lattices. Cation exchange is a reversible process in which cations in soil solution are in dynamic equilibrium with those held on the negatively charged exchange sites on the soil colloids. Exchangeable NH_4^+ is therefore readily available to plants and nitrifying organisms but it is effectively protected against leaching by percolating waters. Fixation of NH_4^+ involves the NH_4^+ ion being held within the interlayer of 2:1 clay minerals in a nonexchangeable form. Soils vary greatly in their ability to fix NH_4^+ depending on their content of 2:1-type clay minerals. Potassium can also be fixed by such minerals and application of K^+ before NH_4^+ can greatly reduce the extent of NH_4^+ fixation. While native fixed NH_4^+ is generally considered to have a very low availability to plants, recently fixed fertilizer NH_4^+ appears to have a more dynamic nature.

When N is applied to soils as aqueous or anhydrous NH_3, adsorption and fixation reactions of NH_3 with soil components can be important. Adsorption reactions of NH_3 include the formation of hydrogen bonds with oxide and hydroxide surfaces and coordination of NH_3 with exchangeable metal cations. Soil organic matter can fix considerable amounts of NH_3 in nonexchangeable forms. The reaction mechanism is thought to involve reaction with phenolic humic components followed by intense polymerization in which the N is combined into bridging structures.

Unlike NH_4^+, nitrate is not normally adsorbed by most soils in the temperate region and it is therefore readily leached down the soil profile by percolating water. However, NO_3^- can be nonspecifically adsorbed by electrostatic attraction to positively charged sites on soil minerals. Thus, soils that possess a significant number of positively charged sites (particularly tropical soils) can retain NO_3^- in the root zone during wet periods and within the recall zone during dry periods when water tends to move back toward the root zone.

Processes of wind and water erosion have caused significant reductions in soil fertility in many areas where intensive cultivated agriculture has been practiced. Wind erosion is a problem where strong winds blow over large areas of loose, dry, finely divided soil. The potential for water erosion exists whenever rainfall strikes bare soil or surface runoff flows over erodable unprotected soils. The fully developed plant cover of natural ecosystems, such as grasslands or forests, protects the soil from erosional forces of wind and water. However, following ecosystem disturbance (e.g., clear-cutting a forest) losses of N through erosion can be significant. The most serious losses occur from agricultural ecosystems.

Losses of soluble N (NH_4^+, NO_3^-, and organic N) in runoff are generally small even from fertilized land and seldom exceed rainfall inputs. However, the potential for losses of sediment N from sloping cultivated agricultural lands is large because for at least part of the year the disturbed soil surface lacks vegetative cover. Thus, losses of sediment N can sometimes be an order of magnitude greater than losses of soluble N in runoff. Generally, any factor that reduces raindrop impact and obstructs runoff (e.g., plant cover or mulch) can prevent losses of sediment. Soil conservation practices that aim to prevent losses of sediment include minimum tillage, ridging, strip planting, contour planting, and construction of terraces. However, management practices that reduce losses of N through erosion generally increase infiltration of water and thus tend to increase leaching losses of NO_3^-.

For the process of NO_3^- leaching to occur, two prerequisites must be met: (1) there must be an accumulation of NO_3^- in the soil and (2) there must be appreciable downward movement of water in the soil profile. In most mature natural ecosystems NO_3^- does not normally accumulate in the soil and losses of N through leaching are consequently small. Following ecosystem disturbance (particularly where the vegetation is destroyed) mineralization may produce more NH_4^+ and consequently NO_3^- than the system requires and NO_3^- leaching can then be a problem.

Leaching of NO_3^- is a particular problem in cultivated agricultural ecosystems, where it is often the most important channel of N loss from field soils. The NO_3^- originates mainly from mineralization of soil organic N and crop residues, particularly following cultivation, and from fertilizer applications that are required to obtain high crop yields. Such losses have ecological as well as agronomic implications and NO_3^- pollution of ground and surface waters is causing increasing concern.

Because of the importance of NO_3^- leaching in agricultural soils many attempts have been made to model the process. Theories to describe the leaching process have been developed and provide the basis of our understanding of transport phenomena. However, when attempting to use the deterministic equations for predictions, specific difficulties are met in quantifying the components of the system and in accounting for the natural spatial heterogeneity of field soils. Stochastic approaches are now being developed to attempt to account for soil variability. Simpler empirical models have also appeared that use observed quantitative relationships among variables rather than being based on classic flow equations. There is room for such a varied approach because of the difficulty of predicting the behavior of soil NO_3^- under field conditions.

Leaching losses of N have been monitored by a variety of methods and the data regarding leaching losses of N from agricultural soils are voluminous. Differences in experimental design and method, crop and soil management, and fertilizer and soil type, as well as climate, make comparisons difficult. Nonetheless, the following generalizations can be made:

(1) Leaching losses are greater from sandy soils than from clay soils.

(2) Greatest leaching losses occur during winter but leaching can occur at any time if the soil is at, or approaching, field capacity and rainfall occurs.

(3) The intensity as well as the amount of rainfall or irrigation is important in determining the pattern and extent of leaching.

(4) Little NO_3^- is leached from grazed pastoral ecosystems except under heavy stocking rates where NO_3^- leaching occurs primarily from urine patches.

(5) Plowing in of pasture or other high-N crops, particularly in fall, can release large amounts of NO_3^- that can be readily leached.

(6) Lack of vegetation for at least part of the year is a key factor stimulating NO_3^- leaching from arable cropping systems. Where summer fallow is followed by heavy winter rainfall, leaching can be a particular problem.

(7) Irrigation often increases NO_3^- leaching because of excess water passing through the crop root zone. However, optimum water supplies can improve crop N uptake and reduce the leaching loss.

(8) When fertilizer N is applied in excess of that required by the crop, considerable amounts of NO_3^- can be leached. Factors that increase fertilizer use efficiency by the crop generally decrease leaching losses.

(9) Spreading of animal wastes on agricultural land at rates higher than the crop can utilize may also result in NO_3^- leaching.

REFERENCES

Adams, J. A., Campbell, A. S., McKeegan, W. A., McPherson, R. J., and Tonkin, P. J. (1979). Nitrate and chloride in groundwater, surface water and deep soil profiles in Central Canterbury, New Zealand. *Progress in Water Technology* **11**, 351–360.

Addiscott, T. M. (1977). A simple computer model for leaching in structured soils. *J. Soil Sci.* **28**, 554–563.

Addiscott, T. M., and Wagenet, R. J. (1985). Concepts of solute leaching in soils: a review of modelling approaches. *J. Soil Sci.* **36**, 411–424.

Aina, P. O., Lal, R., and Taylor, G. S. (1979). Effects of vegetative cover on soil erosion on an Alfisol. *In* "Soil Physical Properties and Crop Production in the Tropics" (R. Lal and D. J. Greenland, eds.), pp. 501–508. Wiley, New York.

Alberts, E. E., Schuman, G. E., and Burwell, R. E. (1978). Seasonal runoff losses of nitrogen and phosphorus from Missouri valley loess watershed. *J. Environ. Qual.* **7**, 203–208.

Aldag, R. W. (1978). Anteile des mineralisch fixierten Ammoniums am Amidstickstoff in Bodenhydrolysaten. *Mitt. Dtsch. Bodenkd. Ges.* **27**, 293–302.

Allison, F. E. (1966). The fate of nitrogen applied to soils. *Adv. Agron.* **18**, 219–258.

Allison, F. E. (1973). "Soil Organic Matter and Its Role in Crop Production." Elsevier, Amsterdam.

Armijo, R., and Coulson, A. N. (1975). Epidemiology of stomach cancer in Chile. The role of nitrogen fertilizer. *Int. J. Epidemiol.* **4**, 301–309.

Arnold, P. W. (1978). Surface–electrolyte interactions. *In* "The Chemistry of Soil Constituents" (D. J. Greenland and M. H. B. Hayes, eds.), pp. 355–404. Wiley, New York.

Arora, Y., and Juo, A. S. R. (1982). Leaching of fertilizer ions in a kaolinitic Ultisol in the high rainfall tropics: Leaching of nitrate in field plots under cropping and bare fallow. *Soil Sci. Soc. Am. J.* **46**, 1212–1218.

Ashworth, J., and Pyman, M. A. F. (1979). Reactions of ammonia with soil. III. Sorption of aqueous NH_3 by homoionic soil clays. *J. Soil Sci.* **30**, 17–27.

Avnimelech, Y., and Raveh, J. (1976). Nitrate leakage from soils differing in texture and nitrogen load. *J. Environ. Qual.* **5**, 79–82.

Ayers, R. S., and Branson, R. L. (1973). Nitrates in the upper Santa Ana river basin in relation to groundwater pollution. *Bull.—Calif. Agric. Exp. Stn.* **861**.

Axley, J. H., and Legg, J. O. (1960). Ammonium fixation in soils and the influence of potassium on nitrogen availability from nitrate and ammonium sources. *Soil Sci.* **90**, 151–156.

Baber, H. L., and Wilson, A. T. (1972). Nitrate pollution of groundwater in the Waikato region. *Chem. N. Z.* **36**, 179–183.

Baker, J. L. (1980). Agricultural areas as nonpoint sources of pollution. *In* "Environmental Impact of Nonpoint Source Pollution" (M. R. Overcash and J. M. Davidson, eds.), Ann Arbor Sci. Publ., pp. 275–310. Ann Arbor, Michigan.

Baker, J. L., and Johnson, H. P. (1981). Nitrate-nitrogen in tile drainage as affected by fertilization. *J. Environ. Qual.* **10**, 519–522.

Baker, J. L., Campbell, K. L., Johnston, H. P., and Hanway, J. J. (1975). Nitrate, phosphorus and sulphate in subsurface drainage water. *J. Environ. Qual.* **4**, 406–412.

Ball, R. P., and Ryden, J. C. (1984). Nitrogen relationships in intensively managed temperate grasslands. *Plant Soil* **76**, 23–33.

Bargh, B. J. (1978). Output of water, suspended sediment, and phosphorus and nitrogen forms from a small catchment. *N. Z. J. Agric. Res.* **21**, 29–38.

Barica, J. (1977). Nitrogen regime of shallow lakes on the Canadian prairies. *Prog. Water Technol.* **8**, 313–321.

Barraclough, D., Hyden, M. J., and Davis, G. P. (1983). Fate of fertilizer nitrogen applied to grassland. 1. Field leaching results. *J. Soil Sci.* **34**, 483–499.

Batey, T. (1982). Nitrogen cycling in upland pastures of the U.K. *Philos. Trans. R. Soc. London, Ser. B* **296**, 551–556.

Bauder, J. W., and Montgomery, B. R. (1979). Overwinter redistribution and leaching of fall applied nitrogen. *Soil Sci. Soc. Am. J.* **43,** 744–747.

Bauder, J. W., and Schneider, R. P. (1979). Nitrate-nitrogen leaching following urea fertilization and irrigation. *Soil Sci. Soc. Am. J.* **43,** 348–352.

Bielby, D. G., Miller, M. H., and Webber, L. R. (1973). Nitrate content of percolates from manured lysimeters. *J. Soil Water Conserv.* **28,** 124–126.

Biggar, J. W. (1978). Spatial variability of nitrogen in soils. *In* "Nitrogen in the Environment" (D. R. Nielsen and J. G. MacDonald, eds.), Vol. 1, pp. 201–211. Academic Press, New York.

Biggar, J. W., and Nielsen, D. R. (1967). Miscible displacement and leaching phenomenon. *In* "Irrigation of Agricultural Lands" (R. M. Hagan, H. R. Haise, and T. W. Edminster, pp. 254–274. Am. Soc. Agron., Madison, Wisconsin.

Biggar, J. W., and Nielsen, D. R. (1976). Spatial variability of the leaching characteristics of a field soil. *Water Resour. Res.* **12,** 78–84.

Biggar, J. W., and Nielsen, D. R. (1980). Mechanisms of chemical movement in soils. *In* "Agrochemicals in Soils" (A. Banin and V. Kafkafi, eds.), pp. 213–229. Int. Inf. Cent., Pergamon, Oxford.

Black, A. S., and Waring, S. A. (1972). Ammonium fixation and availability in some cereal producing soils in Queensland. *Aust. J. Soil Res.* **10,** 197–207.

Black, A. S., and Waring, S. A. (1976a). Nitrate leaching and adsorption in a Krasnozem from Redland Bay, Queensland. I. Leaching of banded ammonium nitrate in a horticultural rotation. *Aust. J. Soil Res.* **14,** 171–180.

Black, A. S., and Waring, S. A. (1976b). Nitrate leaching and adsorption in a Krasnozem from Redland Bay, Queensland. II. Soil factors influencing adsorption. *Aust. J. Soil Res.* **14,** 181–188.

Black, A. S., and Waring, S. A. (1976c). Nitrate leaching and adsorption in a Krasnozem from Redland Bay, Queensland. III. Effect of nitrate concentration on adsorption and movement in soil columns. *Aust. J. Soil Res.* **14,** 189–195.

Blanchet, R., Studer, R., Chaumont, C., and LeBlevenec, L. (1963). Principaux facteurs influençant la retrogradation de l'ammonium dans les conditions naturelles des sols. *C.R. Hebd. Seances Acad. Sci.* **256,** 2223–2225.

Blasco, M. L., and Cornfield, A. H. (1966). Fixation of added ammonium and nitrification of fixed ammonium in soil clays. *J. Sci. Food Agric.* **17,** 481–484.

Boast, C. W. (1973). Modelling the movement of chemicals in soil by water. *Soil Sci.* **115,** 224–30.

Borchardt, G. A. (1977). Montmorillonite and other smectite minerals. *In* "Minerals in Soil Environments" (J. B. Dixon, S. B. Weed, J. A. Kittrick, M. H. Milford, and J. E. White, eds.), pp. 293–330. Soil Sci. Soc. Am., Madison, Wisconsin.

Branson, R. L., Pratt, P. F., Rhoades, J. D., and Oster, J. D. (1975). Water quality in irrigated watersheds. *J. Environ. Qual.* **4,** 33–40.

Bremner, J. M. (1959). Determination of fixed ammonium in soil. *J. Agric. Sci.* **52,** 147–160.

Bremner, J. M. (1965). Inorganic forms of nitrogen. *In* "Methods of Soil Analysis" (C. A. Black, ed.), Part 2, pp. 1179–1273. Am. Soc. Agron., Madison, Wisconsin.

Bremner, J. M., and Harada, T. (1959). Release of ammonium and organic matter from soil by hydrofluoric acid and effect of hydrofluoric acid treatment on extraction of soil organic matter. *J. Agric. Sci.* **52,** 137–160.

Bresler, E. (1973). Simultaneous transport of solutes and water under transient unsaturated flow conditions. *Water Resour. Res.* **9,** 975–986.

Bresler, E., and Dagan, G. (1979). Solute dispersion in unsaturated heterogeneous soil at field scale. II. Applications. *Soil Sci. Soc. Am. J.* **43,** 467–472.

Bresler, E., and Dagan, G. (1981). Convective and pore scale dispersive solute transport in unsaturated heterogeneous fields. *Water Resour. Res.* **17**, 1683–1693.

Broadbent, F. E., and Carlton, A. B. (1978). Field trials with isotopically labelled nitrogen fertilizer. *In* "Nitrogen in the Environment" (D. R. Nielsen and J. G. Macdonald, eds.), pp. 1–63. Academic Press, New York.

Broadbent, F. E., and Stevenson, F. J. (1966). Organic matter interactions. *In* "Agricultural Anhydrous Ammonia: Technology and Use" (H. N. McVickar, W. P. Martin, I. E. Miles, and H. H. Tucker, eds.), pp. 169–187. Am. Soc. Agron., Madison, Wisconsin.

Broadbent, F. E., Burge, W. D., and Nakashima, T. (1961). Factors influencing the reaction between ammonia and soil organic matter. *Trans. Int. Congr. Soil Sci., 7th, 1960*, Vol. 2, pp. 509–516.

Brown, G., Newman, A. C. D., Rayner, J. H., and Weir, A. H. (1978). The structures and chemistry of soil clay minerals. *In* "The Chemistry of Soil Constituents" (D. J. Greenland and M. H. B. Hayes, eds.), pp. 29–178. Wiley, New York.

Buntly, G. J., and Bell, F. F. (1976). "Yield Estimates for the Major Crops Grown on the Soils of West Tennessee." Agric. Exp. Stn., University of Tennessee, Knoxville.

Burden, R. J. (1982). Nitrate contamination of New Zealand aquifers: A review. *N. Z. J. Sci.* **25**, 205–220.

Burge, W. D., and Broadbent, F. E. (1961). Fixation of ammonia by organic soils. *Soil Sci. Soc. Am. Proc.* **25**, 199–204.

Burns, I. G. (1974). A model for predicting the redistribution of salts applied to fallow soils after excess rainfall or evaporation. *J. Soil Sci.* **25**, 165–178.

Burns, I. G. (1975). An equation to predict the leaching of surface-applied nitrate. *J. Agric. Sci.* **85**, 443–454.

Burns, I. G. (1976). Equations to predict the leaching of nitrate uniformly incorporated to a known depth or uniformly distributed throughout a soil profile. *J. Agric. Sci.* **86**, 305–313.

Burns, I. G. (1980). Influence of the spatial distribution of nitrate on the uptake of N by plants: A review and a model for rooting depth. *J. Soil Sci.* **31**, 155–173.

Burton, T. M., and Hook, J. E. (1979). A mass balance study of application of municipal waste water to forests in Michigan. *J. Environ. Qual.* **8**, 589–596.

Burwell, R. E., Schuman, G. E., Piest, R. F., Spomer, R. G., and McCalla, T. M. (1974). Quality of water discharged from two agricultural watersheds in Southwestern Iowa. *Water Resour. Res.* **10**, 259–265.

Burwell, R. E., Timmons, D. R., and Holt, R. F. (1975). Nutrient transport in surface runoff as influenced by soil cover and seasonal periods. *Soil Sci. Soc. Am. Proc.* **39**, 523–528.

Burwell, R. E., Schuman, G. E., Saxton, K. E., and Heinemann, H. G. (1976). Nitrogen in subsurface discharge from agricultural watersheds. *J. Environ. Qual.* **5**, 325–329.

Cameron, K. C. (1983). Nitrate leaching: Some fundamentals. *Proc. Agron. Soc. N.Z.* **13**, 15–21.

Cameron, K. C., and Wild, A. (1982a). Comparative rates of leaching of chloride, nitrate and tritiated water under field conditions. *J. Soil Sci.* **33**, 649–657.

Cameron, K. C., and Wild, A. (1982b). Prediction of solute leaching under field conditions: An appraisal of three methods. *J. Soil Sci.* **33**, 659–669.

Cameron, K. C., and Wild, A. (1984). Potential aquifer pollution from nitrate leaching following the plowing of temporary grassland. *J. Environ. Qual.* **13**, 274–278.

Carter, D. L. (1976). Guidelines from sediment control in irrigation return flow. *J. Environ. Qual.* **5**, 119–124.

Chang, M., Roth, F. A., and Hunt, E. V. (1982). Sediment production under various forest-site conditions. *IAHS-AISH Publ.* **137**, 13–22.

Chichester, F. W. (1977). Effects of increased fertilizer rates on nitrogen content of runoff and percolate from monolith lysimeters. *J. Environ. Qual.* **6**, 211–217.

Chichester, F. W., Van Keuren, R. W., and McGuinness, J. L. (1979). Hydrology and chemical quality of flow from small pastured watersheds. II. Chemical quality. *J. Environ. Qual.* **8**, 167–171.

Cho, C. M. (1971). Convective transport of ammonium with nitrification in soil. *Can. J. Soil Sci.* **51**, 339–350.

Cooke, G. W., and Williams, R. J. B. (1970). Losses of nitrogen and phosphorus from agricultural land. *Water Treat. Exam.* **19**, 253–276.

Crisp, D. T. (1966). Input and output of minerals for an area of Pennine moorland. The importance of precipitation, drainage, peat erosion and animals. *J. Appl. Ecol.* **3**, 327–348.

Crosby, N. T., and Sawyer, R. (1976). *N*-Nitrosamines: A review of chemical and biological properties and their estimation in foodstuffs. *Adv. Food Res.* **22**, 1–71.

Cross, O. E., and Fischbach, P. E. (1972). Water intake rates on a silt loam soil with various manure applications. *Pap.—Am. Soc. Agric. Eng.* **72-218.**

Dam Kofoed, D. (1979). "Experiments on Heavy Application of Animal Manure to Land," Rep. EC-Meet. Oldenburg, Bad Zwischenahn.

Dancer, W. S. (1975). Leaching losses of ammonia and nitrate in the reclamation of sand spoils in Cornwall. *J. Environ. Qual.* **4**, 499–504.

Daniel, T. C., McGuire, P. E., Stoffel, D., and Miller, B. (1979). Sediment and nutrient yield from residential construction sites. *J. Environ. Qual.* **8**, 304–308.

Davidson, J. M., and Chang, R. K. (1972). Transport of picloram in relation to soil physical conditions and pore water velocity. *Soil Sci. Soc. Am. Proc.* **36**, 257–261.

DeBoodt, M., Van den Berghe, C., and Gabriels, D. (1979). Fertilizer losses associated with soil erosion. *In* "Soil Physical Properties and Crop Production in the Tropics" (R. Lal and D. J. Greenland, eds.), pp. 455–464. Wiley, New York.

De Smidt, F., and Wierenga, P. J. (1978). Approximate analytical solution for solute flow during infiltration and redistribution. *Soil Sci. Soc. Am. J.* **42**, 407–412.

Devitt, D., Letey, J., Lund, L. J., and Blair, J. W. (1976). Nitrate-nitrogen movement through soil as affected by soil profile characteristics. *J. Environ. Qual.* **5**, 283–288.

Dissing-Nielsen, J. (1971). Fixation and release of ammonium in Danish soils (in Danish). *Tidsskr. Planteavl* **75**, 239–255.

Doll, R. (1977). Strategy for detection of cancer hazard to man. *Nature (London)* **265**, 589–596.

Doran, J. W., Schepers, J. S., and Swanson, N. P. (1981). Chemical and bacteriological quality of pasture runoff. *J. Soil Water Conserv.* **36**, 166–171.

Dowdell, R. J. (1981). Introduction to general discussion. *In* "Nitrogen Losses and Surface Run-off from Landspreading of Manures" (J. C. Brogan, ed.), pp. 334–335. Martinus Nijhoff/Dr. W. Junk, The Hague.

Dowdell, R. J., Webster, C. P., Hill, D., and Mercer, E. R. (1984). A lysimeter study of the fate of fertilizer nitrogen in spring barley crops grown as shallow soil overlying Chalk: Crop uptake and leaching losses. *J. Soil Sci.* **35**, 169–183.

Dunigan, E. P., and Dick, R. P. (1980). Nutrient and coliform losses in runoff from fertilized and sewage sludge-treated soil. *J. Environ. Qual.* **9**, 243–250.

Dunthion, C. (1981). Nitrogen leaching after spreading pig manure. *In* "Nitrogen Losses and Surface Run-off from Landspreading of Manures" (J. C. Brogan, ed.), pp. 274–283. Martinus Nijhoff/Dr. W. Junk, The Hague.

Englestad, O. P., and Shrader, W. D. (1961). The effect of surface soil thickness on corn yields. II. As determined by an experiment using normal surface soil and artificially exposed subsoil. *Soil Sci. Soc. Am. Proc.* **25**, 497–499.

Feigin, A. and Yaalon, D. H. (1974). Non-exchangeable ammonium in soils of Israel and its relation to clay and parent materials. *Soil Sci.* **25**, 384–397.

Fischer, W. R., Pfanneberg, T., Niederbudde, E. A., and Medina, R. (1981). Transformations of ^{15}N-labelled ammonium in two soils differing in NH_4^+-fixing capacity. *J. Soil Sci.* **32**, 409–418.

Fletcher, J. E., Sorensen, D. L., and Porcella, D. B. (1978). Erosional transfer of nitrogen in desert ecosystems. *In* "Nitrogen in Desert Ecosystems" (N. E. West and J. Skujins, eds.), pp. 171–181. Dowden, Hutchinson & Ross, Stroudsburg, Pennsylvania.

Floate, M. J. S. (1981). Effects of grazing by large herbivores on nitrogen cycling in agricultural ecosystems. *In* "Terrestrial Nitrogen Cycles: Processes, Ecosystem Strategies and Management Impacts" (F. E. Clark and T. Rosswall, eds.), pp. 585–601. Ecological Bulletins, Stockholm.

Forsberg, C. (1977). Nitrogen as a growth factor in fresh water. *Prog. Water Technol.* **8**, 275–290.

Fredriksen, R. L., Moore, D. G., and Norris, L. A. (1975). The impact of timber harvest, fertilization, and herbicide treatment on stream water quality in western Oregon and Washington. *In* "Forest Soils and Forest Land Management" (B. Bernier and C. H. Winget, eds.), pp. 283–313. Les Presses de L'Université Laval, Quebec.

Fried, J. J., and Combarnous, M. A. (1971). Dispersion in porous media. *Adv. Hydrosci.* **7**, 169–282.

Fried, M., Tanji, K. K., and Van de Pol, R. M. (1976). Simplified long term concept for evaluating leaching of nitrogen from agricultural land. *J. Environ. Qual.* **5**, 197–200.

Frissel, M. J., and Poelstra, P. (1967). Chromatographic transport through soils. I. Theoretical evaluation. *Plant Soil* **26**, 285–302.

Frye, W. W., Ebelhar, S. A., Murdock, L. W., and Blevins, R. L. (1982). Soil erosion effects on properties and productivity of two Kentucky soils. *Soil Sci. Soc. Am. J.* **46**, 1051–1055.

Gambrell, R. P., Gilliam, J. W., and Weed, S. B. (1975). Nitrogen losses from soils of North Carolina coastal plain. *J. Environ. Qual.* **4**, 317–323.

Gardner, W. R. (1965). Movement of nitrogen in soils. *In* "Soil Nitrogen" (W. V. Bartholomew and F. E. Clark, eds.), pp. 550–572. Am. Soc. Agron., Madison, Wisconsin.

Garwood, E. A., and Tyson, K. C. (1973). Losses of nitrogen and other plant nutrients. *J. Agric. Sci.* **80**, 303–312.

Garwood, E. A., and Tyson, K. C. (1977). High loss of nitrogen in drainage from soil under grass following a prolonged period of low rainfall. *J. Agric. Sci.* **89**, 767–768.

Gast, R. G., Nelson, W. W., and Randall, G. W. (1978). Nitrate accumulation in soils and loss in tile drainage following nitrogen applications to continuous corn. *J. Environ. Qual.* **7**, 258–261.

Gerwing, J. R., Caldwell, A. C., and Goodroad, L. L. (1979). Fertilizer nitrogen distribution under irrigation between soil, plant and aquifer. *J. Environ. Qual.* **8**, 281–284.

Gill, A. C., McHenry, J. R., and Ritchie, J. C. (1976). Efficiency of nitrogen, carbon and phosphorus retention by small agricultural reservoirs. *J. Environ. Qual.* **5**, 310–315.

Goldman, J. C. (1976). Identification of nitrogen as a growth-limiting factor in wastewaters and coastal marine waters through continuous culture algal assays. *Water Res.* **10**, 97–104.

Gosz, J. R. (1978). Terrestrial contribution of nitrogen to stream water from forests along an elevation gradient in New Mexico. *Water Resour. Res.* **12,** 725–734.

Gosz, J. R. (1981). Nitrogen cycling in coniferous ecosystems. *In* "Terrestrial Nitrogen Cycles: Processes, Ecosystem Strategies and Management Impacts" (F. E. Clark and T. Rosswall, eds.), pp. 405–426. Ecological Bulletins, Stockholm.

Gouny, P., Mériaux, S., and Grosman, R. (1960). Importance de l'ion ammonium à l'état non èchangeable dans un profil de sol. *C.R. Hebd. Seances Acad, Sci.* **251,** 1418–1420.

Greene, L. A. (1978). Nitrates in water supply abstractions in the Anglican Region: Current trends and remedies under investigation. *Water Pollut. Control* **77,** 478–491.

Grewal, G. S., and Kanwar, J. S. (1967). Forms of nitrogen in Punjab soils. *J. Res. (Punjab Agric. Univ.)* **4,** 447–480.

Gruener, N., and Shuval, H. I. (1970). Health aspects of nitrates in drinking water. *In* "Developments in Water Quality Research" (H. I. Shuval, ed.), pp. 89–105. Ann Arbor Sci. Publ., Ann Arbor, Michigan.

Guiot, J. (1981). The nature and origin of leached nitrogen in cultivated land. *In* "Nitrogen Losses and Surface Run-off from Landspreading of Manures" (J. C. Brogan, ed.), pp. 289–306. Martinus Nijhoff/Dr. W. Junk, The Hague.

Gumbs, F. A., and Lindsay, J. I. (1982). Runoff and soil loss in Trinidad under difficult crops and soil management. *Soil Sci. Soc. Am. J.* **46,** 1264–1266.

Gupta, S. C., Shaffer, M. J., and Larson, W. E. (1979). Land treatment mathematical modelling. *Int. Symp. Proc. Land Treat. Waste Water, 1978,* Vol. 1, pp. 121–132.

Haghiri, F., Miller, R. H., and Logan, T. J. (1978). Crop response and quality of soil leachate as affected by land application of beef cattle waste. *J. Environ. Qual.* **7,** 406–412.

Hahne, H. C. H., Kroontje, W., and Lutz, J. A., Jr. (1977). Nitrogen fertilization. I. Nitrate accumulation and losses under continuous corn cropping. *Soil Sci. Soc. Am. J.* **41,** 562–567.

Hansen, E. A., and Harris, A. R. (1975). Validity of soil-water samples collected with porous ceramic cups. *Soil Sci. Soc. Am. Proc.* **39,** 528–536.

Hauck, R. D., and Tanji, K. K. (1982). Nitrogen transfers and mass balances. *In* "Nitrogen in Agricultural Soils" (F. J. Stevenson, ed.), pp. 891–925. Am. Soc. Agron., Madison, Wisconsin.

Hayes, M. H. B., and Swift, R. S. (1978). The chemistry of soil organic colloids. *In* "The Chemistry of Soil Constituents" (D. J. Greenland and M. H. B. Hayes, eds.), pp. 179–320. Wiley, New York.

Hayes, W. A., and Kimberlin, L. W. (1978). A guide for determining crop residue for water erosion control. *In* "Crop Residue Management Systems" (W. R. Oschwald, ed.), pp. 35–48. Am. Soc. Agron., Madison, Wisconsin.

Haynes, R. J. (1981a). Soil pH decrease in the herbicide strip of grassed-down orchards. *Soil Sci.* **132,** 274–278.

Haynes, R. J. (1981b). Laboratory study of nutrient leaching from the surface of cultivated, grassed and herbicided orchard soil. *Soil Tillage Res.* **1,** 281–288.

Haynes, R. J. (1983). Soil acidification induced by leguminous crops. *Grass Forage Sci.* **38,** 1–11.

Haynes, R. J. (1984). Lime and phosphate in the soil–plant system. *Adv. Agron.* **37,** 249–315.

Henderson, G. S., Swank, W. T., Waide, J. B., and Grier, C. C. (1978). Nutrient budgets of

Appalachian and Cascade region watersheds: A comparison. *For. Sci.* **24,** 385–397.

Hill, M. J., Hawksworth, G., and Tattersall, G. (1973). Bacteria, nitrosamines, and cancer of the stomach. *Br. J. Cancer* **28,** 562–567.

Hillel, D. (1977). "Computer Simulation of Soil Water Dynamics." Int. Dev. Res. Cent., Ottawa.

Hillel, D. (1980). "Fundamentals of Soil Physics." Academic Press, New York.

Hingston, F. J., Posner, A. M., and Quirk, J. P. (1972). Anion adsorption by geothite and gibbsite. I. The role of the proton in determining adsorption envelopes. *J. Soil Sci.* **23,** 177–192.

Hinman, W. C. (1964). Fixed ammonium in some Saskatchewan soils. *Can. J. Soil Sci.* **44,** 151–157.

Hinrichs, D. G., Mazurak, A. P., and Swanson, P. (1974). Effects of effluent from beef feedlot on the physical and chemical properties of the soil. *Soil Sci. Soc. Am. Proc.* **38,** 661–663.

Hood, A. E. M. (1976a). Leaching of nitrates from intensively managed grassland at Jealotts Hill. *Tech. Bull.—Minist. Agric., Fish Food (G.B.)* **32,** 201–222.

Hood, A. E. M. (1976b). Nitrogen, grassland and water quality in the United Kingdom. *Outlook Agric.* **8,** 320–327.

Hook, J. E., and Burton, T. M. (1979). Nitrate leaching from sewage-irrigated perennials as affected by cutting management. *J. Environ. Qual.* **8,** 496–502.

Hornbeck, J. W., Likens, G. E., Pierce, R. S., and Bormann, F. H. (1975). Strip cutting as a means of protecting site and streamflow quality when clearcutting northern hardwoods. *In* "Forest Soils and Forest Land Management" (B. Bernier and C. H. Winget, eds.), pp. 209–225. Les Presses de l'Université Laval, Quebec.

Horne, B. (1980). Soil, water and fertilizers. "Great House Experimental Husbandry Farm Annual Review," pp. 21–26. G. B. Minst. Agric. Fish. Food, London.

Hoyt, G. D., McLean, E. O., Reddy, G. Y., and Logan, T. J. (1977). Effects of soil cover crop, and nutrient source on movement of soil, water, and nitrogen under simulated rain-slope conditions. *J. Environ. Qual.* **6,** 285–290.

Huntjens, J. L. M. (1971a). Influences of living plants on immobilization of nitrogen in permanent pastures. *Plant Soil* **34,** 393–404.

Huntjens, J. L. M. (1971b). The influence of living plants on mineralization of nitrogen. *Plant Soil* **35,** 77–94.

Huntjens, J. L. M., and Albers, R. A. J. M. (1978). A model experiment to study the influence of living plants on the accumulation of soil organic matter in pastures. *Plant Soil* **50,** 411–418.

International Standing Committee on Water Quality and Treatment (ISC) (1974). Nitrates in water supplies. *Aqua* **1,** 5.

Jaakkola, A. (1984). Leaching losses of nitrogen from a clay soil under grass and cereal crops in Finland. *Plant Soil* **76,** 59–66.

Jackson, W. A., Asmussen, L. E., Hause, A. W., and White, A. W. (1973). Nitrate in surface and subsurface flow from a small agricultural watershed. *J. Environ. Qual.* **2,** 480–482.

James, D. W., and Harward, M. E. (1964). Competition of NH_3 and H_2O for adsorption sites on clay minerals. *Soil Sci. Soc. Am. Proc.* **28,** 636–640.

Jansson, S. L. (1958). Tracer studies on nitrogen transformations in soil with special attention to mineralisation–immobilization relationships. *Lantbrukshoegsk. Ann.* **24,** 101–361.

Johnson, J. W., Welch, L. F., and Kurtz, L. T. (1974). Soybean's role in nitrogen balance. *Ill. Res.* **16**, 6–7.

Johnson, N. M., Likens, G. E., Bormann, F. H., Fisher, D. W., and Pierce, R. S. (1969). A working model for the variation in the streamwater chemistry at the Hubbard Brook Experimental Forest, New Hampshire. *Water Resour. Res.* **5**, 1535.

Jones, L. A., Smeck, N. E., and Wilding, L. P. (1977). Quality of water discharged from three small agronomic watersheds in the Maumee river basin. *J. Environ. Qual.* **6**, 296–302.

Jones, M. J. (1975). Leaching of nitrate under maize at Samaru, Nigeria. *Trop. Agric. (Trinidad)* **52**, 1–10.

Jung, J., and Dressel, J. (1974). Uber des Auswaschungsverhalten verschledener N. Formen im Lysimeterversuch. *Z. Acker.- Pflanzenbau* **140**, 1–10.

Jury, W. A., Stolzy, L. H., and Shouse, P. (1982). A field test for the transfer function model for predicting solute transport. *Water Resour. Res.* **18**, 369–375.

Kaila, A. (1966). Fixation of ammonium in some Finnish soils. *J. Agric. Soc. Finl.* **34**, 107–114.

Kanchanasut, P., and Scotter, D. R. (1982). Leaching patterns in soil under pasture and crop. *Aust. J. Soil Res.* **20**, 193–202.

Kandiah, A. (1979). Influence of soil properties and crop cover on the erodability of soils. *In* "Soil Physical Properties and Crop Production in the Tropics" (R. Lal and D. J. Greenland, eds.), pp. 455–464. Wiley, New York.

Keeney, D. R. (1973). The nitrogen cycle in sediment–water systems. *J. Environ. Qual.* **2**, 15–29.

Keeney, D. R. (1982). Nitrogen management for maximum efficiency and minimum pollution. *In* "Nitrogen in Agricultural Soils" (F. J. Stevenson, ed.), pp. 605–649. Am. Soc. Agron., Madison, Wisconsin.

Khaleel, R., Reddy, K. R., and Overcash, M. R. (1981). Changes in soil physical properties due to organic waste applications: A review. *J. Environ. Qual.* **10**, 133–141.

Khanna, P. K. (1981). Leaching of nitrogen from terrestrial ecosystems: Patterns, mechanisms and ecosystem responses. *In* "Terrestrial Nitrogen Cycles: Processes, Ecosystem Strategies and Management Impacts" (F. E. Clark and T. Rosswall, eds.), pp. 343–352. Ecological Bulletins, Stockholm.

Kilmer, V. J. (1974). Nutrient losses from grasslands through leaching and runoff. *In* "Forage Fertilization" (D. A. Mays, ed.), pp. 341–362. Am. Soc. Agron., Madison, Wisconsin.

Kilmer, V. J., Gilliam, J. W., Lutz, J. F., Joyce, R. T., and Edlund, C. D. (1974). Nutrient losses from fertilized grassed watersheds in western North Carolina. *J. Environ. Qual.* **3**, 214–219.

Kinjo, T., and Pratt, P. F. (1971a). Nitrate adsorption. I. In some acid soils of Mexico and South America. *Soil Sci. Soc. Am. Proc.* **35**, 722–725.

Kinjo, T., and Pratt, P. F. (1971b). Nitrate adsorption. II. In competition with chloride, sulphate, and phosphate. *Soil Sci. Soc. Am. Proc.* **35**, 725–728.

Kinjo, T., Pratt, P. F., and Page, A. L. (1971). Nitrate adsorption. III. Desorption movement and distribution in Andepts. *Soil Sci. Soc. Am. Proc.* **35**, 728–732.

Kirda, C., Nielsen, D. R., and Biggar, J. W. (1973). Simultaneous transport of chloride and water during infiltration. *Soil Sci. Soc. Am. Proc.* **37**, 339–345.

Kirkham, D., and Powers, W. L. (1972). "Advanced Soil Physics." Wiley (Interscience), New York.

Kissel, D. E., Richardson, C. W., and Burnett, E. (1976). Losses of nitrogen in surface runoff in the blackland Prairie of Texas. *J. Environ. Qual.* **5**, 288–293.

Klausner, S. D., Zwerman, P. J., and Ellis, P. F. (1974). Surface runoff losses of soluble nitrogen and phosphorus under two systems of soil management. *J. Environ. Qual.* **3**, 42–46.

Klausner, S. D., Zwerman, P. J., and Ellis, D. F. (1976). Nitrogen and phosphorus losses from winter disposal of dairy manure. *J. Environ. Qual* **5**, 47–49.

Kolenbrander, G. J. (1969). Nitrate content and nitrogen loss in drain-water. *Neth. J. Agric. Sci.* **17**, 246–255.

Kolenbrander, G. J. (1972). Eutrophication from agriculture with special reference to fertilisers and animal waste. *Soils Bull.* **16**, 305–327.

Kolenbrander, G. J. (1973). Fertilizers, farming practice and water quality. *Proc.—Fert. Soc.* **135**, 36.

Kolenbrander, G. J. (1981). Leaching of nitrogen in agriculture. *In* "Nitrogen Losses and Surface Run-off from Landspreading of Manures" (J. C. Brogan, ed.), pp. 199–216. Martinus Nijhoff/Dr. W. Junk, The Hague.

Koren'kov, D. A., Rudelev, Ye. V., Fillimonov, D. A., and Sonina, K. I. (1979). Penetration of fertiliser and soil nitrogen into infiltration water. *Sov. soil Sci. (Engl. Transl.)* **11**, 295–301.

Kowalenko, C. G. (1978). Nitrogen transformations and transport over 17 months in field fallow microplots using ^{15}N. *Can. J. Soil Sci.* **58**, 69–76.

Kowalenko, C. G. (1981). Effect of immobilization on nitrogen transformations and transport in a field ^{15}N experiment. *Can. J. Soil Sci.* **61**, 387–395.

Kowalenko, C. G., and Cameron, D. R. (1976). Nitrogen transformation in an incubated soil as affected by combinations of moisture content and temperature and adsorption–fixation of ammonium. *Can. J. Soil Sci.* **56**, 63–70.

Kowalenko, C. G., and Ross, G. J. (1980). Studies on the dynamics of recently clay-fixed NH_4^+ using ^{15}N. *Can. J. Soil Sci.* **60**, 61–70.

Krupp, H. K., Biggar, J. W., and Nielsen, D. R. (1972). Relative flow rates of salt and water in soils. *Soil Sci. Soc. Am. Proc.* **36**, 412–417.

Kudeyarov, V. N., Bashkin, V. N., and Kudeyarova, A. YV. (1981). Losses of nitrogen, phosphorus and potassium from agricultural watersheds of minor rivers in the Oka Valley. *Water, Air, Soil Pollut.* **16**, 267–276.

Lal, R. (1975). "Role of Mulching Techniques in Tropical Soil and Water Management," Tech. Bull. No. 1. I.I.T.A. Ibadan, Nigeria.

Lal, R. (1976). Soil erosion on an Alfisol in Western Nigeria. I. Effects of crop rotation and residue management. *Geoderma* **16**, 363–375.

Lal, R. (1982). Effect of slope length and terracing on runoff and erosion on a tropical soil. *IAHS-AISH Publ.* **137**, 23–31.

Langdale, G. W., Box, J. E., Leonard, R. A., Barnett, A. P., and Fleming, W. G. (1979a). Corn yield reduction on eroded southern Piedmont soils. *J. Soil Water Conserv.* **34**, 226–228.

Langdale, G. W., Leonard, R. A., Fleming, W. G., and Jackson, W. A. (1979b). Nitrate and chloride movement in small upland Piedmont watersheds. II. Nitrogen and chloride transport in runoff. *J. Environ. Qual.* **8**, 57–63.

Larsen, V. (1977). Nitrogen transformations in lakes. *Prog. Water Technol.* **8**, 419–431.

Legg, J. O., and Meisinger, J. J. (1982). Soil nitrogen budgets. *In* "Nitrogen in Agricultural soils" (F. J. Stevenson, ed.), pp. 503–557. Am. Soc. Agron., Madison, Wisconsin.

Letey, J., Blair, J. W., Devitt, D., Lund, L. J., and Nash, P. (1977). Nitrate-nitrogen in effluent from agricultural tile drains in California. *Hilgardia* **45**, 289–319.

Lewis, W. M. (1974). Effects of fire on nutrient movement on a South Carolina pine forest. *Ecology* **55**, 1120–1127.

Liebhardt, W. C., Golt, C., and Tupin, J. (1979). Nitrate and ammonium concentrations of ground water resulting from poultry manure applications. *J. Environ. Qual.* **8**, 211–215.

Likens, G. E., Bormann, F. H., Pierce, R. S., Eaton, J. S., and Johnson, N. M. (1977). "Biogeochemistry of a Forested Ecosystem." Springer-Verlag, Berlin and New York.

Loehr, R. C. (1974). Characteristics and magnitude of nonpoint sources. *J. Water Pollut. Control Fed.* **46**, 1849–1872.

Long, F. L. (1979). Runoff water quality as affected by surface-applied dairy cattle manure. *J. Environ. Qual.* **8**, 215–218.

Low, A. J. (1973). Nitrate and ammonium nitrogen concentration in water draining through soil monoliths in lysimeters cropped with grass and clover or uncropped. *J. Sci. Food Agric.* **24**, 1489–1495.

Low, A. J., and Armitage, E. R. (1970). The composition of the leachate through cropped and uncropped soils in lysimeters compared with that of rain. *Plant Soil* **33**, 393–411.

Lund, L. J., Page, A. L., Nelson, C. O., and Elliott, R. A. (1981). Nitrogen balances for an effluent irrigation area. *J. Environ. Qual.* **10**, 349–352.

MacDuff, J. H., and White, R. E. (1984). Components of the nitrogen cycle measured for cropped and grassland soil–plant systems. *Plant Soil* **76**, 35–47.

McElroy, A. D., Chin, S. Y., Nebgen, J. W., Aliti, A., and Bennett, F. W. (1976). "Loading Functions for Assessment of Water Pollution from Nonpoint Sources," Rep. No. EPA-600/2-76-151. Office of Research and Development, U.S. Environmental Protection Agency, Washington, D.C.

McLaren, A. D. (1970). Temporal and vectoral reactions of nitrogen in soils: A review. *Can. J. Soil Sci.* **50**, 97–109.

McLeese, R. L., and Whiteside, E. P. (1977). Ecological effects of highway construction upon Michigan woodlots and wet lands: Soil relationships. *J. Environ. Qual.* **6**, 467–471.

McLeod, R. V., and Hegg, R. O. (1984). Pasture runoff water quality from application of inorganic and organic nitrogen sources. *J. Environ. Qual.* **13**, 122–126.

McMahon, M. A., and Thomas, G. W. (1976). Anion leaching in two Kentucky soils under conventional tillage and a killed sod mulch. *Agron. J.* **68**, 437–442.

McNeil, B. L., and Pratt, P. F. (1978). Leaching of nitrate from soils. *In* "Management in Irrigated Agriculture" (P. F. Pratt, ed.), pp. 195–230. University of California, Riverside.

Macura, J., and Kunc, F. (1965). Continuous flow method in microbiology. V. Nitrification. *Folia Microbiol. (Prague)* **10**, 125–134.

Malberg, J. W., Savage, E. P. and Osteryoung, J. (1978). Nitrates in drinking water and early onset of hypertension. *Environ. Pollut.* **15**, 155–160.

Manges, H. L., Eisenhauer, D. E., Stritzke, R. D., and Goering, E. H. (1974). Beef feedlot manure and soil water movement. *Pap. Am. Soc. Agric. Eng.* **74-2019.**

Marshall, T. J., and Holmes, J. W. (1979). "Soil Physics." Cambridge Univ. Press, London and New York.

Martin, A. E., Gilkes, R. J., and Skjemstud, J. O. (1970). Fixed ammonium in soils developed in some Queensland phyllites and its relation to weathering. *Aust. J. Soil Res.* **8**, 71–80.

Mathers, A. C., and Stewart, B. A. (1974). Corn silage yield and soil properties as affected by cattle feedlot manure. *J. Environ. Qual.* **3,** 143–147.

Mathers, A. C., Stewart, B. A., and Blair, B. (1975). Nitrate removal from soil profiles by alfalfa. *J. Environ. Qual.* **4,** 403–405.

Mazurak, A. P., Chesnin, L., and Tia'rks, A. E. (1975). Detachment of soil aggregates by simulated rainfall from heavily manured soils in Eastern Nebraska. *Soil Sci. Soc. Am. Proc.* **39,** 732–736.

Melillo, J. M. (1981). Nitrogen cycling in deciduous forests. *In* "Terrestrial Nitrogen Cycles: Processes, Ecosystem Strategies and Management Impacts" (F. E. Clark and T. Rosswall, eds.), pp. 427–442. Ecological Bulletins, Stockholm.

Mengel, K., and Scherer, H. W. (1981). Release of non exchangeable (fixed) soil ammonium under field conditions during the growing season. *Soil Sci.* **131,** 226–232.

Mensah-Bonsu and Obeng, H. B. (1979). Effect of cultural practices on soil erosion and maize production in the semi-deciduous rainforest and forest–savanna transitional zone of Ghana. *In* "Soil Physical Properties and Crop Production in the Tropics" (R. Lal and D. J. Greenland, eds.), pp. 507–519. Wiley, New York.

Menzel, R. G., Rhoades, E. D., Olness, A. E., and Smith, S. J. (1978). Variability of annual nutrient and sediment discharges in runoff from Oklahoma cropland and rangeland. *J. Environ. Qual.* **7,** 401–406.

Mercer, E. R., and Hill, D. (1975). The relative movement of water and nitrate through soil and chalk. *In* "Letcombe Annual Report 1975," pp. 50–51. Agricultural Research Council, England.

Mercer, E. R., and Hill, D. (1976). Movement of water, nitrate and chloride through soil and chalk. *In* "Letcombe Annual Report 1976," pp. 90–93. Agricultural Research Council, England.

Miller, N. H. J. (1906). The amount and composition of the drainage through unmanured and uncropped land, Barnfield, Rothamsted. *J. Agric. Sci.* **1,** 377–399.

Mohammed, I. H. (1979). Fixed ammonium in Libyan soils and its availability to barley seedlings. *Plant Soil* **53,** 1–9.

Mohammed, I. H., Scotter, D. R., and Gregg, P. E. H. (1984). The short-term fate of urea applied to barley in a humid climate. *Aust. J. Soil Res.* **22,** 173–180.

Moore, A. W. (1965). Fixed ammonium in some Alberta soils. *Can. J. Soil Sci.* **45,** 112–115.

Moore, A. W., and Ayeke, C. A. (1965). HF-extractable ammonium nitrogen in four Nigerian soils. *Soil Sci.* **99,** 335–338.

Morgan, M. F., Jacobson, J. G. M., and Street, D. E. (1942). The neutralization of acid-forming nitrogenous fertilizers in relation to nitrogen availability and soil bases. (A report of Windsor lysimeter series D). *Soil Sci.* **54,** 127–148.

Morisot, A. (1981). Erosion and nitrogen losses. *In* "Terrestrial Nitrogen Cycles: Processes, Ecosystem Strategies and Management Impacts" (F. E. Clark and T. Rosswall, eds.), pp. 353–361. Ecological Bulletins, Stockholm.

Mortland, M. M. (1966). Ammonia interactions with soil minerals. *In* "Agricultural Anhydrous Ammonia Technology and Use" (M. H. McVickar, W. P. Martin, I. E. Miles, and H. H. Tucker, eds.), pp. 188–197. Agric. Ammonia Inst., Memphis, Tennessee.

Mortland, M. M. (1969). Protonation of compounds at clay surfaces. *Trans. Int. Congr. Soil Sci., 9th, 1968,* Vol. 1, pp. 691–699.

Mortland, M. M., and Wolcott, A. R. (1965). Sorption of inorganic nitrogen compounds by soil materials. *In* "Soil Nitrogen" (W. V. Bartholomew and F. E. Clark, eds.), pp. 150–197. Am. Soc. Agron., Madison, Wisconsin.

Morton, W. E. (1971). Hypertension and drinking water, a pilot statewide ecological study in Colorado. *J. Chronic Dis.* **23,** 537–545.

Mott, C. J. B. (1981). Anion and ligand exchange. *In* "The Chemistry of Soil Processes" (D. J. Greenland and M. H. B. Hayes, eds.), pp. 179–219. Wiley, New York.

Muir, J., Boyce, J. S., Seim, E. C., Mosher, P. N., Deibert, E. J., and Olson, R. A. (1976). Influence of crop management practices on nutrient movement below the root zone in Nebraska soils. *J. Environ. Qual.* **5,** 255–259.

National Research Council (NRC) (1971). "Accumulation of nitrate." National Academy of Sciences, Washington, D.C.

National Research Council (NRC) (1978). "Nitrates: An Environmental Assessment." National Academy of Sciences, Washington, D.C.

Neilsen, G. H., and MacKenzie, A. F. (1977). Soluble and sediment nitrogen losses as related to land use and type of soil in eastern Canada. *J. Environ. Qual.* **6,** 318–321.

Nicholaichuck, W., and Read, D. W. L. (1978), Nutrient runoff from fertilized and unfertilized fields in western Canada. *J. Environ. Qual.* **7,** 542–544.

Nielsen, D. R., and Biggar, J. W. (1962). Miscible displacement. III. Theoretical considerations. *Soil Sci. Soc. Am. Proc.* **26,** 216–221.

Nielsen, D. R., and Biggar, J. W. (1963). Miscible displacement. 4. *Soil Sci. Soc. Am. Proc.* **27,** 10–13.

Nielsen, D. R., and Biggar, J. W. (1981). Implications of the vadose zone to water resources management. *In* "National Research Council, Geophysics Study Commission." National Academy Press, Washington, D.C.

Nielsen, D. R., Biggar, J. W., and Barrada, Y. (1979). Water and solute movement in field soils. *In* "Isotopes and Radiation Techniques in Research on Soil–Plant Relationships," pp. 165–183. IAEA, Vienna.

Nielsen, D. R., Biggar, J. W., and Wierenga, P. J. (1982). Nitrogen transport processes in soils. *In* "Nitrogen in Agricultural soils" (F. J. Stevenson, ed.), pp. 423–448. Am. Soc. Agron., Madison, Wisconsin.

Nielsen, D. R., Tillotson, P. M., and Vieira, S. R. (1983). Analyzing field-measured soil-water properties. *Agric. Water Manage.* **6,** 93–109.

Nightingale, H. I. (1970). Statistical evaluation of salinity and nitrate content and trends beneath urban and agricultural areas—Fresno, California. *Groundwater* **8,** 22–28.

Nightingale, H. I. (1972). Nitrates in soil and groundwater beneath irrigated and fertilized crops. *Soil Sci.* **114,** 300–311.

Nommik, H. (1965). Ammonium fixation and other reactions involving a nonenzymatic immobilization of mineral nitrogen in soil. *In* "Soil Nitrogen" (W. V. Bartholomew and F. E. Clark, eds.), pp. 200–251. Am. Soc. Agron., Madison, Wisconsin.

Nommik, H. (1970). Non-exchangeable binding of ammonium and amino nitrogen by Norway spruce raw humus. *Plant Soil* **33,** 581–595.

Nommik, H., and Vahtras, K. (1982). Retention and fixation of ammonium and ammonia in soils. *In* "Nitrogen in Agricultural Soils" (F. J. Stevenson, ed.), pp. 123–126. Am. Soc. Agron., Madison, Wisconsin.

Nyborg, M. (1969). Fixation of gaseous ammonia by soils. *Soil Sci.* **107,** 131–136.

Nye, P. H. (1974). A theoretical aperçu of the movement of nutrients in the soil profile. *J. Sci. Food Agric.* **25,** 709–716.

Nye, P. H., and Tinker, P. B. (1977). "Solute Movement in the Soil–Root System." Blackwell, Oxford.

O'Connor, K. F. (1974). Nitrogen in agrobiosystems and its environmental significance. *N. Z. Agric. Sci.* **8,** 137–148.

Olness, A., Smith, S. J., Rhoades, E. D., and Menzel, R. G. (1975). Nutrient and sediment discharge from agricultural watersheds in Oklahoma. *J. Environ. Qual.* **4,** 331–336.

Olson, R. A. (1979). Isotope studies on soil and fertilizer nitrogen. *In* "Isotopes and Radiation Techniques in Research on Soil Plant Relationships." IAEA, Vienna.

Olson, R. A., Seim, E. C., and Muir, J. (1973). Influence of agricultural practices on water quality in Nebraska: A survey of streams, groundwater and precipitation. *Water Resour. Bull.* **9,** 301–311.

Olson, R. A., Muir, J. H., Wesely, R. W., Peterson, G. A., and Boyce, J. S. (1974). Accumulation of inorganic nitrogen in soils and waters in relation to soil and crop management. *In* "Effects of Agricultural Production on Nitrates in Food and Water with Particular Reference to Isotope Studies," pp. 19–41. IAEA, Vienna.

Omernick, J. M. (1976). "The Influence of Land Use on Stream Nutrient Levels," Ecol. Res. Ser. Rep. No. EPA-600/3-76-014. Office of Research and Development, U.S. Environmental Protection Agency, Corvallis, Oregon.

Omoti, V., and Wild, A. (1979). Use of fluorescent dyes to mark the pathways of solute movement through soils under leaching conditions. *Soil Sci.* **128,** 28–33, 98–104.

Onstad, C. A., and Moldenhauer, W. C. (1975). Watershed soil detachment and transportation factors. *J. Environ. Qual.* **4,** 29–33.

Opuwaribo, E., and Odu, C. T. I. (1974). Fixed ammonium in Nigerian soils. I. Selection of a method and amounts of native fixed ammonium. *J. Soil Sci.* **25,** 256–264.

Organization for Economic Cooperation and Development (OECD) (1982). "Eutrophication of Waters: Monitoring, Cooperation and Control." OECD, Paris.

Osborne, G. J. (1976). The significance of intercalary ammonium in representative surface and subsoils from southern New South Wales. *Aust. J. Soil Res.* **14,** 381–388.

Owens, L. B. (1981). Effects of nitrapyrin on nitrate movement in soil columns. *J. Environ. Qual.* **10,** 308–310.

Owens, L. B., Edwards, W. M., and Van Keuren, R. W. (1984). Peak nitrate-nitrogen values in surface runoff from fertilized pastures. *J. Environ. Qual.* **13,** 310–313.

Owens, M. (1970). Nutrient balances in rivers. *Water Treat. Exam.* **19,** 239–252.

Passioura, J. B., and Rose, D. A. (1971). Hydrodynamic dispersion in aggregated media. 2. Effects of velocity and aggregate size. *Soil Sci.* **111,** 345–351.

Peck, A. J. (1984). Recent Advances in Soil Salinity Research. Aust. Soc. Soil Sci. Conference Proc. (1984), Brisbane.

Pfaff, C. (1958). Einfluss der Berenung aud die Nahrstoffauswaschung bei mahrjahrigen Gemusear bau. *Z. Pflanzenernaehr., Dueng., Bodenkd.* **80,** 93–108.

Pfaff, C. (1963). Das Verhalten des Stickstoffs im Boden nach langjuhrigen Lysimeterversuchen. I. *Z. Acker.- Pflanzenbau* **117,** 77–79.

Pierre, W. H., Webb, J. R., and Schrader, W. D. (1971). Quantitative effects of nitrogen fertilizer on the development and downward movement of soil acidity in relation to level of fertilization and crop removal in a continuous corn cropping. *Agron. J.* **63,** 291–297.

Pimentel, D., Terhune, E. C., Dyuson-Hudson, R., Rochereau, S., Samis, R., Smith, E. A., Denman, D., Reifschneider, D., and Shepherd, M. (1976). Land degradation: Effects on food and energy resources. *Science* **194,** 149–155.

Porter, L. K., and Stewart, B. A. (1970). Organic interferences in the fixation of ammonium by soils and clay minerals. *Soil Sci.* **100,** 229–233.

Power, J. F. (1970). Leaching of nitrate nitrogen under dryland agriculture in the Northern Great Plains. *In* "Relationship of Agriculture to Soil and Water Pollution," Proc.– Annu. Agric. Pollut. Conf., Rochester, New York, pp. 111–122. Cornell Univ. Press, Ithaca, New York.

Power, J. F. (1981). Nitrogen in the cultivated ecosystem. *In* "Terrestrial Nitrogen Cycles: Processes, Ecosystem Strategies and Management Impacts" (F. E. Clark and T. Rosswall, eds.), pp. 529–546. Ecological Bulletins, Stockholm.

Powers, W. L., Wallingford, G. W., and Murphy, L. S. (1975). "Research Status on Effects of Land Application of Animal Wastes," EPA-660/2-75-010. U.S. Environmental Protection Agency, Washington, D.C.

Pratt, P. F. (1976). Irrigation for minimal nitrate pollution. *In* "Research Applied to National Needs," Vol. VI, pp. 99–101. Natl. Sci. Found., Washington, D.C.

Pratt, P. F., Cannell, G. H., Garber, M. J., and Blair, F. L. (1967). The effect of three nitrogen fertilizers on gains, losses and distribution of various elements in irrigated lysimeters. *Hilgardia* **38**, 265–283.

Pratt, P. F., Jones, W. W., and Hunsaker, V. E. (1972). Nitrate in deep soil profiles in relation to fertilizer rates and leaching volume. *J. Environ. Qual.* **1**, 97–101.

Pratt, P. F., Broadbent, F. E., and Martin, J. P. (1973). Using organic wastes as nitrogen fertilizer. *Calif. Agric.* **27**, 10–13.

Pratt, P. F., Lund, L. J., and Warneke, J. E. (1980). Nitrogen losses in relation to soil profile characteristics. *In* "Agrochemicals in Soil" (A. Banin and V. Kafkafi, eds.), pp. 33–47. Int. Inf. Cent., Pergamon, Oxford.

Preston, C. M. (1982). The availability of residual fertilizer nitrogen immobilized as clay-fixed ammonium and organic N. *Can. J. Soil Sci.* **62**, 479–486.

Quisenberry, V. L., and Phillips, R. E. (1976). Percolation of surface applied water in the field. *Soil Sci. Soc. Am. J.* **40**, 484–489.

Raju, G. S. N., and Mukhopadhyay, A. K. (1976). Ammonium fixing capacities of West Bengal soils. *Indian Soc. Soil Sci. J.* **24**, 270–274.

Raney, W. A. (1960). The dominant role of nitrogen in leaching losses from soils in humid regions. *Agron. J.* **52**, 563–566.

Reddy, K. R. (1983). Fate of nitrogen and phosphorus in a waste-water retention reservoir containing aquatic macrophytes. *J. Environ. Qual.* **12**, 137–141.

Rennie, D. A., Racz, G. J., and McBeath, D. K. (1976). Nitrogen losses. *In* "Proceedings of Western Canada Nitrogen Symp. Calgary, Alberta, Canada," pp. 325–353. Alberta Agriculture, Edmonton.

Ritchie, J. T., Kissel, D. E., and Burnett, E. (1972). Water movement in undisturbed swelling clay soil. *Soil Sci. Soc. Am. Proc.* **36**, 874–879.

Ritchie, J. T., Gill, A. C., and McHenry, J. R. (1975). A comparison of nitrogen, phosphorus and carbon in sediments and soils of cultivated and noncultivated watersheds in the North Central States. *J. Environ. Qual.* **4**, 339–341.

Robertson, G. P. (1982). Factors regulating nitrification in primary and secondary succession. *Ecology* **63**, 1561–1573.

Robins, C. W., and Carter, D. L. (1980). Nitrate-nitrogen leached below the root zone during and following alfalfa. *J. Environ. Qual.* **9**, 447–450.

Romkins, M. J. M., Nelson, D. W., and Mannering, J. V. (1973). Nitrogen and phosphorus composition of surface runoff as affected by tillage method. *J. Environ. Qual.* **2**, 292–295.

Rose, C. W., Chichester, F. W., Williams, J. R., and Ritchie, J. T. (1982a). A contribution to simplified models of field solute transport. *J. Environ. Qual.* **11**, 146–150.

Rose, C. W., Chichester, F. W., Williams, J. R., and Ritchie, J. T. (1982b). Application of an approximate analytic method of computing solute profiles with dispersion in soils. *J. Environ. Qual.* **11**, 151–155.

Rowell, D. L., Martin, M. W., and Nye, P. H. (1967). The measurement and mechanism of ion diffusion in soils. III. The effect of moisture content and soil solution concentration on the self diffusion of ions in soils. *J. Soil Sci.* **18,** 204–222.

Royal Commission on Environmental Pollution (1979). "Agriculture and Pollution." H. M. Stationery Office, London.

Russell, E. J., and Richards, E. H. (1920). The washing out of nitrates by drainage water from uncropped and unmanured land. *J. Agric. Sci.* **10,** 22–43.

Russell, J. D. (1965). Infra-red study of the reactions of ammonia with montmorillonite and saponite. *Trans. Faraday Soc.* **61,** 2284–2294.

Russell, R. S. (1978). Nitrate "fears" and the farmer. *Big Farm Weekly,* Feb. 24, p. 10.

Saffigna, P. G., Keeney, D. R., and Tanner, C. B. (1977). Nitrogen, chloride and water balance with irrigated Russet Burbank potatoes in central Wisconsin. *Agron. J.* **69,** 251–257.

Salter, P. J., and Williams, J. B. (1965). The influence of texture on the moisture characteristics of soils. II. Available-water capacity and moisture release characteristics. *J. Soil Sci.* **16,** 310–317.

Schepers, J. S., and Francis, D. D. (1982). Chemical water quality of runoff from grazing land in Nebraska. I. Influence of grazing livestock. *J. Environ. Qual.* **11,** 351–354.

Schertz, D. L., and Miller, D. A. (1972). Nitrate-N accumulation in the soil profile under alfalfa. *Agron. J.* **64,** 660–664.

Schreiber, J. D., Duffy, P. D., and McClurkin, D. C. (1976). Dissolved nutrient losses in storm runoff from five southern pine watersheds. *J. Environ. Qual.* **5,** 201–205.

Schuman, G. E., Burwell, R. E., Piest, R. F., and Spomer, R. G. (1973). Nitrogen losses in surface runoff from agricultural watersheds on Missouri Valley loess. *J. Environ. Qual.* **2,** 299–302.

Scotter, D. R., and Kanchanasut, P. (1981). Anion movement under pasture. *Aust. J. Soil Res.* **19,** 299–307.

Shank, R. C. (1975). Toxicity of *N*-nitroso compounds. *Toxicol. Appl. Pharmacol.* **31,** 361–368.

Sharpley, A. N., Syers, J. K., and Tillman, R. W. (1983). Transport of ammonium- and nitrate-nitrogen in surface runoff from pasture as influenced by urea application. *Water, Air, Soil Pollut.* **20,** 425–430.

Shaw, K. (1962). Loss of mineral nitrogen from soil. *J. Agric. Sci.* **58,** 145–151.

Sherwood, M. (1981). Leaching of nitrogen from animal manures under Irish conditions. *In* "Nitrogen Losses and Surface Run-off from Landspreading of Manures" (J. C. Brogan, ed.), pp. 272–273. Martinus Nijhoff/Dr. W. Junk, The Hague.

Sherwood, M., and Fanning, A. (1981). Nutrient content of surface run-off water from land treated with animal wastes. *In* "Nitrogen Losses and Surface Run-off from Landspreading of Manures" (J. C. Brogan, ed.), pp. 5–17. Martinus Nijhoff/Dr. W. Junk, The Hague.

Shuford, J. W., Fritton, D. D., and Baker, D. E. (1977). Nitrate nitrogen and chloride movement through undisturbed field soil. *J. Environ. Qual.* **6,** 736–739.

Shuval, H. I., and Gruener, N. (1977). Infant methemoglobinemia and other health effects of nitrates in drinking water. *Prog. Water Technol.* **8,** 183–193.

Singh, B. R., and Kanehiro, Y. (1969). Adsorption of nitrate in amorphous and kaolinite Hawaiian soil. *Soil Sci. Soc. Am. Proc.* **33,** 681–683.

Singh, B. R., and Sekhon, G. S. (1979). Nitrate pollution from farm use of nitrogen fertilizers—a review. *Agric. Environ.* **4,** 207–225.

Singh, B. R., Biswas, C. R., and Sekhon, G. S. (1978). A rational approach for optimizing application rates of fertilizer nitrogen to reduce nitrate pollution of natural waters. *Agric. Environ.* **4,** 57–64.

Sippola, J. (1976). Fixation of ammonium and potassium applied simultaneously in Finnish soils. *Ann. Agric. Fenn.* **15,** 304–308.

Sippola, J., Erviö, R., and Eleveld, R. (1973). The effects of simultaneous addition of ammonium and potassium in their fixation in some Finnish soils. *Ann. Agric. Fenn.* **12,** 185–189.

Skidmore, E. L., and Siddoway, F. H. (1978). Crop residue requirements to control wind erosion. *In* "Crop Residue Management Systems" (W. R. Oschwald, ed.), pp. 17–33. Am. Soc. Agron., Madison, Wisconsin.

Smettem, K. B. J., and Trudgill, S. T. (1933). An evaluation of some fluorescent and non-fluorescent dyes in the identificaiton of water transmission routes in soils. *J. Soil Sci.* **34,** 45–57.

Smika, D. E., and Watts, D. G. (1978). Residual nitrate-N in fine sand as influenced by N fertilizer and water management practices. *Soil Sci. Soc. Am. J.* **42,** 923–926.

Smika, D. E., Heermann, D. F., Duke, H. R., and Batchelder, A. R. (1977). Nitrate-N percolation through irrigated sandy soils as affected by water management. *Agron. J.* **69,** 623–626.

Smilde, K. W. (1979). "Effects of Land Spreading of Large Amounts of Livestock Excreta on Crop Yield and Crop and Water Quality," Rep. EC-Meet. Oldenburg, Bad Zwischenahn.

Smith, J. H., and Peterson, J. R. (1982). Recycling of nitrogen through land application of agricultural, food processing, and municipal wastes. *In* "Nitrogen in Agricultural Soils" (F. J. Stevenson, ed.), pp. 791–831. Am. Soc. Agron., Madison, Wisconsin.

Smith, R. V., Stevens, R. J., Foy, R. H., and Gibson, C. E. (1982). Upward trend in nitrate concentration in rivers discharging into Lough Neagh for the period 1969–1979. *Water Res.* **16,** 183–188.

Sommerfeldt, T. G., Chang, C., and Carefoot, J. M. (1982). A laboratory study on the effect of soil moisture content, texture, and timing of leaching on N loss from the Southern Alberta soils. *Can. J. Soil Sci.* **62,** 407–413.

Sonzogni, W. C., Chapra, S. C., Armstrong, D. E., and Logan, T. J. (1982). Bioavailability of phosphorus inputs to lakes. *J. Environ. Qual.* **11,** 555–563.

Sopper, W. E. (1975). Effect of timber harvesting and related management practices on water quality in forested watersheds. *J. Environ. Qual.* **4,** 24–29.

Soubiès, L., Gadet, R., and Lanain, M. (1962). Possibilité de controler dans le sol la transformation de l'azote ammonical en azote nitrique. *C. R. Seances Acad. Agric. Fr.* **48,** 789–803.

Sowden, F. J., McLean, A. A., and Ross, J. G. (1978). Native clay-fixed ammonium content and the fixation of added ammonium of some soils of Eastern Canada. *Can. J. Soil Sci.* **58,** 27–38.

Spallacci, P. (1981). Nitrogen losses by leaching on different soils manured with pig slurry. *In* "Nitrogen Losses and Surface Run-off from Landspreading of Manures" (J. C. Brogan, ed.), pp. 284–288. Martinus Nijhoff/Dr. W. Junk, The Hague.

Stanford, G. (1966). Nitrogen requirements of crops for maximum yield. *In* "Agricultural Anhydrous Ammonia: Technology and Use" (M. H. McVickar, W. P. Martin, I. E. Miles, and H. H. Tucker, eds.), pp. 237–257. Agric. Ammonia Inst., Memphis, Tennessee.

Stark, N. M. (1977). Fire and nutrient cycling in a douglas-fir/larch forest. *Ecology* **58,** 16–30.

Steele, K. W. (1982). Nitrogen in grassland soils. *In* "Nitrogen Fertilizers in New Zealand Agriculture" (P. B. Lynch, ed.), pp. 29–45. N.Z. Inst. Agric. Sci., Wellington.

Steele, K. W., and Shannon, P. (1982). Concepts relating to the nitrogen economy of a Northland intensive beef farm. *In* "Nitrogen Balances in New Zealand Ecosystems" (P. W. Gander, ed.), pp. 85–90. New Zealand Department of Scientific and Industrial Research, Palmerston North.

Steele, K. W., Judd, M. J., and Shannon, P. W. (1984). Leaching of nitrate and other nutrients from a grazed pasture. *N. Z. J. Agric. Res.* **27**, 5–11.

Steenvoorden, J. H. A. M. (1981). Landspreading of animal manure and run-off: Comments on the draft guidelines. *In* "Nitrogen Losses and Surface Run-off from Landspreading of Manures" (J. C. Brogan, ed.), pp. 26–33. Martinus Nijhoff/Dr. W. Junk, The Hague.

Stout, P. R.,and Burau, R. G. (1967). The extent and significance of fertilizer buildup in soils as revealed by vertical distribution of nitrogenous matter between soils and underlying water reservoirs. *In* "Agriculture and the Quality of Our Environment" (N. C. Brady, ed.), pp. 283–310. Am. Soc. Adv. Sci., Washington, D.C.

Sturm, G., and Bibo, F. J. (1965). Nitratgehalte in Tinkwasser unter besonderer Buruchsichtigung der Verhältnisse in Rheingaukreis. *GWF, Gas- Wasserfach* **106**, 332–334.

Sumner, M. E., and Reeves, N. G. (1966). The effect of iron oxide impurities on the positive and negative adsorption of chloride by kaolinites. *J. Soil Sci.* **17**, 274–279.

Swanson, N. P., Mielke, L. N., Lorimor, J. C., McCalla, T. M., and Ellis, J. R. (1971). Transport of pollutants from sloping cattle feedlots as affected by rainfall intensity, duration, and recurrence. *In* "Livestock Waste Management and Pollution Abatement," pp. 51–55. Am. Soc. Agric. Eng., St. Joseph, Michigan.

Swanston, D. N., and Swanson, F. J. (1976). Timber harvesting, mass erosion, and steepland forest geomorphology. *In* "Geomorphology and Engineering" (D. R. Coats, ed.), pp. 199–221. Dowden, Hutchinson & Ross, Stroudsburg, Pennsylvania.

Swoboda, A. R. (1977). "The Control of Nitrate as a Water Pollutant," Rep. No. EPA-600/2-77-158. U.S. Environmental Protection Agency, Washington, D.C.

Talibudeen, O. (1981). Cation exchange in soils. *In* "The Chemistry of Soil Processes" (D. J. Greenland and M. H. B. Hayes, eds.), pp. 115–177. Wiley, New York.

Tamm, C. O., Holmen, H., Popovic, B., and Wiklander, G. (1974). Leaching of plant nutrients from soils as a consequence of forestry operations. *Ambio* **3**, 211–221.

Tanji, K. K. (1982). Modelling of the soil nitrogen cycle. *In* "Nitrogen in Agricultural Soils" (F. J. Stevenson, ed.), pp. 721–770. Am. Soc. Agron., Madison, Wisconsin.

Tanji, K. K., and Gupta, S. K. (1978). Computer simulation modelling for nitrogen in irrigated croplands. *In* "Nitrogen in the Environment" (D. R. Nielsen and J. G. MacDonald, eds.), Vol. 1, pp. 79–130. Academic Press, New York.

Tanji, K. K., and Mehran, M. (1979). Conceptual and dynamic models for nitrogen in irrigated croplands. *In* "Nitrate in Effluents from Irrigated Lands" (P. F. Pratt, ed.), Final Report to National Science Foundation, pp. 555–646. University California.

Task Group Report (1967). Sources of nitrogen and phosphorus in water supplies. *J. Am. Water Works Assoc.* **59**, 344–366.

Taylor, A. W., Edwards, W. M., and Simpson, E. C. (1971). Nutrients in streams draining woodland farmland near Coschocton, Ohio. *Water Resour. Res.* **7**, 81–89.

Taylor, N. (1975). Medical aspects of nitrate in drinking water. *Water Treat. Exam.* **24**, 194.

Terman, G. L. (1977). Quantitative relationships among nutrients leached from soils. *Soil Sci. Soc. Am. J.* **41**, 935–940.

Terman, G. L., and Allen, S. E. (1970). Leaching of soluble and slow-release N and K fertilizers from lakeland sand under grass and fallow. *Proc.—Soil Crop Sci. Soc. Fla.* **30**, 130–140.

Thomas, G. W. (1977). Historical developments in soil chemistry: Ion exchange. *Soil Sci. Soc. Am. J.* **41**, 230–238.

Thomas, G. W., and Barfield, B. J. (1974). The unreliability of tile effluent for monitoring subsurface nitrate-nitrogen losses from soils. *J. Environ. Qual.* **3**, 183–185.

Thomas, G. W., and Swoboda, A. R. (1970). Anion exclusion effects on chloride movement in soil. *Soil Sci.* **110**, 163–166.

Thomas, G. W., Blevins, R. L., Phillips, R. E., and McMahon, M. A. (1973). Effect of killed sod mulch on nitrate movement and corn yield. *Agron. J.* **65**, 736–739.

Thomas, G. W., Phillips, R. E., and Quisenberry, V. L. (1978). Characterization of water displacement in soils using simple chromatographic theory. *J. Soil Sci.* **29**, 32–37.

Timmons, D. R., and Dylla, A. S. (1981). Nitrogen leaching as influenced by nitrogen management and supplemental irrigation level. *J. Environ. Qual.* **10**, 421–426.

Timmons, D. R., Burwell, R. E., and Holt, R. F. (1968). Loss of crop nutrients through runoff. *Minn. Sci.* **24**, 16–18.

Timmons, D. R., Burwell, R. E., and Holt, R. F. (1973). Nitrogen and phosphorus losses in surface runoff from agricultural land as influenced by placement of broadcast fertilizer. *Water Resour. Res.* **9**, 658–667.

Timmons, D. R., Verry, E. S., Burwell, R. E., and Holt, R. F. (1977). Nutrient transport in surface runoff and interflow from an Aspen Birch forest. *J. Environ. Qual.* **6**, 188–192.

Tomlinson, T. E. (1970). Trends in nitrate concentrations in English rivers in relation to fertilizer use. *Water Treat. Exam.* **19**, 277–288.

Troake, R. P., Troake, L. E., and Walling, D. E. (1976). Nitrate loads of South Devon streams. *Tech. Bull.—Minist. Agric., Fish Food (G.B.)* **32**, 340–354.

Troeh, F. R., Hobbs, J. A., and Donahue, R. L. (1980). "Soil and Water Conservation for Productivity and Environmental Protection." Prentice-Hall, Englewood Cliffs, New Jersey.

Turner, D. O. (1976). Guidelines for manure application in the Pacific Northwest. *Ext. Bull.—Wash.* State Univ., *Coop. Ext. Serv.* **EM 4009**, 1–25.

Tweneboah, C. K., Greenland, D. J., and Oades, J. M. (1967). Changes in charge characteristics in soils after treatment with $0.5M$ calcium chloride at pH 1.5. *Aust. J. Soil Res.* **5**, 247–261.

Tyler, D. D., and Thomas, G. W. (1977). Lysimeter measurements of nitrate and chloride losses from soil under conventional cultivation and no-tillage corn. *J. Environ. Qual.* **6**, 63–66.

Uhlen, G. (1981). Surface run-off and the use of farm manure. *In* "Nitrogen Losses and Surface Run-off from Landspreading of Manures" (J. C. Brogan, ed.), pp. 34–43. Martinus Nijhoff/Dr. W. Junk, The Hague.

Unger, P. W., and McCalla, T. M. (1980). Conservation tillage systems. *Adv. Agron.* **33**, 1–58.

U.S. Public Health Service (1962). Drinking water standards. *U.S., Public Health Serv. Publ.* **956.**

Vahtras, K., and Wiklander, L. (1977). Leaching of plant nutrients in soils. III. Loss of nitrogen as influenced by the form of fertilizer and residual effects of N fertilizer. *Acta Agric. Scand.* **27**, 165–174.

Vallis, I. (1978). Nitrogen relationships in grass/legume mixtures. *In* "Plant Relations in Pastures" (J. R. Wilson, ed.), pp. 190–201. CSIRO, Canberra, Australia.

van Raij, B., and Camargo, de O. A. (1974). Nitrate elution from soil columns of three Oxisols and one Alfisol. *Trans. Int. Congr. Soil Sci., 10th, 1974,* Vol. 2, pp. 384–391.

Vetter, H., and Steffens, G. (1981). Leaching of nitrogen after the spreading of slurry. *In* "Nitrogen Losses and Surface Run-off from Landspreading of Manures" (J. C. Brogan, ed.), pp. 251–269. Marinus Nijhoff/Dr. W. Junk, The Hague.

Viets, F. G. (1971). Water quality in relation to farm use of fertiliser. *BioScience* **21,** 460–467.

Viets, F. G., Humbert, R. P., and Nelson, C. E. (1967). Fertilizers in relation to irrigation. *In* "Irrigation of Agricultural Lands" (R. M. Hagan, H. R. Haise, and T. W. Edminster, eds.), pp. 1009–1036. Am. Soc. Agron., Madison, Wisconsin.

Vitousek, P. M. (1981). Clear-cutting and the nitrogen cycle. *In* "Terrestrial Nitrogen Cycles: Processes, Ecosystem Strategies and Management Impacts" (F. E. Clark and T. Rosswall, eds.), pp. 631–642. Ecological Bulletins, Stockholm.

Vitousek, P. M., Gosz, J. R., Grier, C. C., Melillo, J. M., Reiners, W. A., and Todd, R. L. (1979). Nitrate losses from disturbed ecosystems. *Science* **204,** 469–474.

Vollenweider, R. A. (1968). "Scientific Fundamentals of the Eutrophication of Lakes and Flowing Waters, with Particular Reference to Nitrogen and Phosphorus as Factors in Eutrophication," Tech. Rep. DAS/CSI 6827. OECD, Paris.

Wada, K., and Harward, M. E. (1974). Amorphous clay constituents of soils. *Adv. Agron.* **26,** 211–260.

Walsh, L. M., and Murdock, J. T. (1960). Native fixed ammonium and fixation of applied ammonium in several Wisconsin soils. *Soil Sci.* **89,** 183–193.

Walsh, L. M., and Murdock, J. T. (1963). Recovery of fixed ammonium by corn in greenhouse studies. *Soil Sci. Soc. Am. Proc.* **27,** 200–204.

Warrick, A. W., and Nielsen, D. R. (1980). Spatial variability of soil physical properties in the field. *In* "Nitrogen in the Environment" (D. R. Nielsen and J. G. MacDonald, eds.), pp. 319–344. Academic Press, New York.

Webster, R., and Burgess, T. M. (1983). Spatial variation in soil and the role of Kriging. *Agric. Water. Manage.* **6,** 111–122.

Weil, R. R., and Kroontje, W. (1979). Physical condition of a Davidson clay loam after five years of heavy poultry manure applications. *J. Environ. Qual.* **8,** 387–392.

Weil, R. R., Kroontje, W., and Jones, G. D. (1979). Inorganic nitrogen and soluble salts in a Davidson Clay loam used for poultry manure disposal. *J. Environ. Qual.* **8,** 86–91.

Westerman, P. M., and Overcash, M. R. (1980a). Dairy open lot and lagoon-irrigated pasture runoff quantity and quality. *Trans. ASAE* **25,** 1157–1164.

Westerman, P. W., and Overcash, M. R. (1980b). Short-term attenuation of runoff pollution potential from land applied swine and poultry manure. *In* "Livestock Waste: A Renewable Resource," pp. 289–292. Am. Soc. Agric. Eng., St. Joseph, Michigan.

White, E. M. (1973). Water leachable nutrients from frozen or dried prairie vegetation. *J. Environ. Qual.* **2,** 104–107.

White, E. M., Williamson, E. J., and Kingsley, Q. (1977). Correlations between rain and runoff amounts and composition in eastern South Dakota. *J. Environ. Qual.* **6,** 251–254.

Wiklander, L., and Vahtras, K. (1975). Leaching of plant nutrients in soils. II. Loss of nitrogen as influenced by the form of fertilizer. *Acta Agric. Scand.* **25,** 33–41.

Wild, A. (1972). Nitrate leaching under bare fallow at a site in northern Nigeria. *J. Soil Sci.* **23,** 315–324.

Wild, A. (1977). Nitrate in drinking water: Health hazard unlikely. *Nature (London)* **268,** 197.

Wild, A. (1981). Mass flow and diffusion. *In* "The Chemistry of Soil Processes" (D. J. Greenland and M. H. B. Hayes, eds.), pp. 37–80. Wiley, New York.

Wild, A., and Babiker, I. A. (1976a). Winter leaching of nitrate at sites in southern England. *Tech. Bull.—Minist. Agric., Fish Food (G.B.)* **32**, 153–162.

Wild, A., and Babiker, I. A. (1976b). The asymmetric leaching pattern of nitrate and chloride in a loamy sand under field conditions. *J. Soil Sci.* **27**, 460–466.

Wild, A., and Cameron, K. C. (1980a). Soil nitrogen and nitrate leaching. *In* "Soils and Agriculture" (P. B. Tinker, ed.), Soc. Chem. Industry; Crit. Rep. Appl. Chem., Vol. 2, pp. 35–70. Blackwell, Oxford.

Wild, A., and Cameron, K. C. (1980b). Leaching of nitrate through soils and environmental considerations with special reference to recent work in the United Kingdom. *In* "Soil Nitrogen as Fertilizer or Pollutant," Proc. IAEA/FAO Symp., Piracicaba, Brazil, 1978, pp. 289–306. IAEA, Vienna.

Williams, R. J. B. (1975). The chemical composition of water from land drainage at Saxmundham and Woburn (1970–75). *Rothamsted Exp. Stn. Rep. 1975*, Part 2, pp. 37–135.

Willis, W. O., and Evans, C. E. (1977). Our soil is valuable. *J. Soil Water Conserv.* **32**, 258–259.

Wilson, S. J., and Cooke, R. U. (1980). Wind erosion. *In* "Soil Erosion" (M. J. Kirkby and R. P. C. Morgan, eds.), pp. 217–251. Wiley, New York.

Wischmeier, W. H. (1973). Conservation tillage to control water erosion. *In* "Conservation Tillage," Proc. Natl. Conf., pp. 133–141. Soil Conserv. Soc. Am., Ankeny, Iowa.

Wischmeier, W. H. (1976). Cropland erosion and sedimentation. *In* "Control of Water Pollution from Cropland," Vol. II, pp. 31–57. U. S. Govt. Printing Office, Washington, D.C.

Wischmeier, W. H., and Smith, D. D. (1961). A universal soil loss equation to guide conservation farm planning. *Trans Int. Congr. Soil Sci. 7th, 1960,* pp. 418–426.

Wolcott, A. R., Foth, H. D., David, J. F., and Schicluna, J. C. (1965). Nitrogen carriers. I. Soil effects. *Soil Sci. Soc. Am. Proc.* **19**, 164–167.

Woldendrop, J. W., Dilz, K., and Kilenbrander, G. J. (1966). The fate of fertiliser nitrogen on permanent grassland soils. *Nitrogen Grassl., Proc. Gen. Meet. Eur. Grassl. Fed. 1st, 1965,* pp. 53–68.

Woodmansee, R. G., and Wallach, L. S. (1981). Effects of fire regimes on biogeochemical cycles. *In* "Terrestrial Nitrogen Cycles: Processes, Ecosystem Strategies and Management Impacts" (F. E. Clark and T. Rosswall, eds.), pp. 649–669.

Woodmansee, R. G., Vallis, I., and Mott, J. J. (1981). Grassland nitrogen. *In* "Terrestrial Nitrogen Cycles. Processes, Ecosystem Strategies and Management Impacts" (F. E. Clark and T. Rosswall, eds.), pp. 443–462. Ecological Bulletins, Stockholm.

Woodruff, N. P., and Siddoway, F. H. (1965). A wind erosion equation. *Soil Sci. Soc. Am. Proc.* **29**, 602–608.

World Health Organization (1970). "European Drinking Standards." WHO, Geneva.

Young, C. P., and Gray, E. A. (1978). "Nitrate in Groundwater," Tech. Rep. 69. Water Res. Cent., Medmeham.

Young, C. P., Hall, E. S., and Oakes, D. B. (1976). "Nitrate in Groundwater Studies on the Chalk near Winchester, Hampshire," Tech. Rep. 31. Water Res. Cent., Medmeham.

Young, J. L. (1962). Inorganic soil nitrogen and carbon: Nitrogen ratios in some Pacific Northwest soils. *Soil Sci.* **93**, 397–404.

Young, J. L. (1964). Ammonia and ammonium reactions with some Pacific Northwest soils. *Soil Sci. Soc. Am. Proc.* **28**, 339–345.

Young, J. L., and Aldag, R. W. (1982). Inorganic forms of nitrogen in soil. *In* "Nitrogen in Agricultural Soils" (F. J. Stevenson, ed.), pp. 423–448. Am. Soc. Agron., Madison, Wisconsin.

Young, J. L., and McNeal, B. L. (1964). Ammonia and ammonium reactions with some layer-silicate minerals.

Young, K. K. (1980). The impact of erosion on the productivity of soils in the United States. *In* "Assessment of Erosion" (M. DeBoodt and D. Gabriels, eds.), pp. 295–303. Wiley, New York.

Young, R. A., and Mutchler, C. K. (1976). Pollution potential of manure spread on frozen ground. *J. Environ. Qual.* **5,** 174–179.

Chapter 5

Gaseous Losses of Nitrogen

R. J. HAYNES AND R. R. SHERLOCK

I. INTRODUCTION

An upsurge of interest in gaseous losses of nitrogen from the soil has occurred during the last decade. Much of this research was stimulated by evidence from agronomic nitrogen-balance studies that generally showed an unexplained 10 to 30% loss of applied fertilizer nitrogen (Allison, 1955; Legg and Meisinger, 1982). Experiments involving the use of ^{15}N-labeled fertilizers confirmed that a significant proportion of applied nitrogen is unaccountably lost from soils during cropping (Hauck, 1971; Hauck and Bremner, 1976). A number of processes contribute to gaseous losses of soil nitrogen; these include ammonia volatilization, bacterial denitrification, nitrification, and reactions of NO_2^- with soil components.

Ammonia volatilization may occur whenever free NH_3 is present near the soil surface. The quantities of NH_3 lost are highly variable depending on such factors as rate, type and method of fertilizer nitrogen application, soil pH, and environmental factors including temperature, moisture, and wind (Black et al., 1985a, 1985b). Plants can both absorb and evolve NH_3 from their leaf canopies (Freney et al., 1981) but the major factors that influence the relative magnitude of the two processes are, as yet, unclear.

Denitrification is a major biological process through which N from the soil is returned to the atmosphere (Payne, 1981; Firestone, 1982). The process, which is mediated principally by aerobic bacteria which are capable of anaerobic growth only in the presence of nitrogen oxides, yields nitrous oxide (N_2O) and dinitrogen (N_2) gases. The role of N_2O in stratospheric chemical reactions has generated great interest in the denitrification process. This is because the photochemical breakdown of N_2O in the stratosphere yields NO, which has a principal role in catalyzing the decomposition of stratospheric ozone (Crutzen and Ehhalt, 1977; McElroy *et al.,* 1977). The stratospheric ozone layer shields the biosphere from harmful exposures to UV radiation.

Nitrous oxide is also released from soil as a by-product of the nitrification pathway (see Chapter 3) although the exact mechanism of N_2O production is unclear (Schmidt, 1982). Under field conditions losses of N_2O through dentrification and nitrification are thought to often occur simultaneously.

Under conditions that favor the accumulation of NO_2^- in soils, chemodenitrification may contribute to gaseous losses of N (Nelson, 1982; Chalk and Smith, 1983). The presence of NO_2^- provides a mechanism for chemodenitrification since NO_2^- tends to react with soil components to form gases (e.g., N_2, N_2O, NO, and NO_2).

In this chapter, the processes involved in gaseous losses of N from the plant–soil system are reviewed and the major factors influencing such losses are discussed. The magnitude and significance of these losses are also considered.

II. AMMONIA VOLATILIZATION

Ammonia volatilization is the term commonly used to describe the process by which gaseous NH_3 is released from the soil surface to the atmosphere. The subject has been reviewed in depth by several workers (Terman, 1979; Freney *et al.,* 1981, 1983; Vlek and Craswell, 1981; Nelson, 1982).

A necessary prerequisite for NH_3 volatilization is a supply of free ammonia (i.e., $NH_{3(aq)}$ and $NH_{3(g)}$) near the soil surface. The source of NH_3 is usually soil NH_4^+. The supply can be from organic nitrogenous sources such as urine or feces of animals, plant residues, or native soil organic matter, all of which decompose to release NH_4^+-N. The factors affecting the decomposition of organic residues and the subsequent release of NH_4^+ are discussed in detail in Chapter 2.

A wide variety of NH_4^+- and NH_4^+-forming compounds are also applied to soils as fertilizers (e.g., $(NH_4)_2SO_4$, NH_4NO_3, $(NH_4)_2HPO_4$, NH_4Cl, aqua ammonia, and urea). Ammonium-containing fertilizers such as $(NH_4)_2SO_4$ dissolve in soil solution and NH_4^+ ions are produced. In the soil urea, from either animal urine or applied fertilizers, undergoes hydrolysis catalyzed by the enzyme urease to form $(NH_4)_2CO_3$:

$$(NH_2)_2CO + 2H_2O \longrightarrow (NH_4)_2CO_3 \tag{1}$$

This reaction causes localized areas of high pH close to the site of hydrolysis:

$$CO_3^{2-} + H_2O \rightleftharpoons HCO_3^- + OH^- \tag{2}$$

Ammonium ions interact with the cation exchange complex of the soil resulting in electrostatic binding of NH_4^+ ions to clay and organic colloids. Some of this NH_4^+ may become "fixed" in clay lattices (see Chapter 4). However, NH_4^+ ions in soil solution also enter into equilibrium reactions with NH_3.

Since all of the above-mentioned sources supply N as NH_4^+ rather than NH_3, it is the conversion of NH_4^+ to NH_3 that normally regulates the potential loss of NH_3 through volatilization. Nonetheless, N can also be added to soils as anhydrous NH_3, which is handled as a liquid but changes to a gaseous state during injection.

A. Processes

1. Well Aerated Soils

The basic equilibria that govern ammonia loss are shown in Fig. 1. These equilibria indicate that, in theory, the soil can act as both a source and sink for atmospheric $NH_{3(g)}$. The NH_3 flux (F) into or out of the soil surface may be represented as

$$F = k \, (NH_{3(g)soil} - NH_{3(g)atm}) \tag{3}$$

where $NH_{3(g)soil}$ is the $NH_{3(g)}$ concentration in equilibrium with the soil solution at the soil surface, $NH_{3(g)atm}$ is the $NH_{3(g)}$ concentration of the

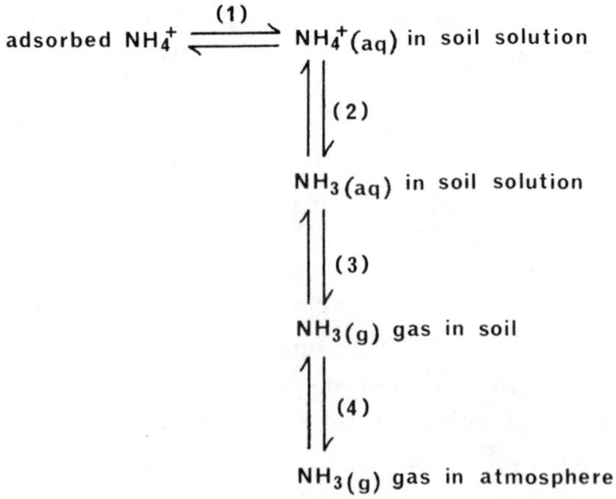

$$\text{adsorbed } NH_4^+ \underset{(1)}{\rightleftharpoons} NH_4^+(aq) \text{ in soil solution}$$

$$\updownarrow (2)$$

$$NH_3(aq) \text{ in soil solution}$$

$$\updownarrow (3)$$

$$NH_3(g) \text{ gas in soil}$$

$$\updownarrow (4)$$

$$NH_3(g) \text{ gas in atmosphere}$$

Fig. 1. The various equilibria that govern ammonia loss from soils.

bulk atmosphere, and k is an exchange coefficient whose value may vary with windspeed (Vlek and Craswell, 1981; Freney *et al.*, 1983). Whether $NH_{3(g)}$ is absorbed or volatilized is therefore largely determined by the difference in $NH_{3(g)}$ concentration between the soil surface and the atmosphere.

Atmospheric NH_3 concentrations, although variable, are usually very low, e.g., 2–6 μg NH_3-N m^{-3} (National Research Council, 1979), and there is no evidence that they seriously limit volatilization rates in the field (Vlek and Craswell, 1981; Freney *et al.*, 1983). No direct measurements of equilibrium $NH_{3(g)soil}$ concentrations have been reported but calculations by Vlek and Craswell (1981) show that for $NH_{3(g)atm}$ concentrations of 2–6 ppb, $NH_{3(aq)}$ concentrations of 0.5 ppm or greater are sufficient to promote volatilization. These workers maintained that where NH_3 volatilization is a problem, such levels of $NH_{3(aq)}$ are easily reached. Therefore, under these conditions $NH_{3(g)soil}$ is likely to greatly exceed $NH_{3(g)atm}$, whereupon equation (3) can be simplified to

$$F = k \, (NH_{3(g)soil}) \qquad (4)$$

Thus, ammoniacal N added to the soil from whatever source may be subject to loss as $NH_{3(g)}$. Also, the actual magnitude of any loss is likely to depend primarily on the concentration of $NH_{3(g)soil}$, which in turn depends on the total concentration of ammoniacal N species, the values of the individual equilibrium constants (Fig. 1), and the rate of attainment of equilibrium at each stage. Factors such as pH, temperature, etc., which can influence any or all of these separate equilibria, can influence the magnitude of NH_3 loss. Likewise, all strategies designed to limit volatilization losses attempt to manipulate these equilibria either directly or indirectly to reduce the $NH_{3(g)}$ concentration at the soil–air interface.

2. Flooded Soils

The fact that NH_3 is the most soluble gas known makes it tempting to suggest that a soil flooded with water would serve as an almost infinite sink for the gas and that any volatilization to the atmosphere would be negligible. This may be so for unfertilized flooded soil but can be demonstrably incorrect for fertilized systems. For example, Vlek and Craswell (1979) showed that up to 50% of urea surface applied to floodwater was lost as NH_3 within 2–3 weeks.

It is now clear that equation (3) and the equilibria described in Fig. 1 apply equally well to flooded and nonflooded soils (Freney *et al.*, 1983). Indeed, since it is relatively easy to measure the ammoniacal N concentration, pH, and temperature within the water overlying a flooded soil and also to obtain values for the exchange coefficient k for the transport of

NH_3 away from the floodwater surface, it has been suggested that the direct application of equation (2) may provide a simple way of predicting NH_3 losses from flooded rice paddies (Leuning *et al.*, 1984). However, in testing this hypothesis Leuning *et al.* (1984) found that because of temperature gradients within the water, the NH_3 concentration at the water–air interface was not generally characteristic of that in equilibrium with the bulk of the floodwater. This implies that for flooded soils transport processes in both the air and water are important in determining the rate of NH_3 volatilization.

Loss of NH_3 from flooded soils is also strongly influenced by wind through a mechanical mixing of the surface water. This and several other factors that influence NH_3 losses from flooded soils are discussed in following sections.

3. Calcareous Soils

The presence of $CaCO_3$ in soil is reported to stimulate volatilization of NH_3 from applied ammoniacal fertilizers (Freney *et al.*, 1981). For soils taken from various parts of the world, a strong correlation between NH_3 loss and $CaCO_3$ content has been reported (Lehr and Van Wesemael, 1961; Fenn and Kissel, 1975).

Apart from its effects on the alkalinity and buffering effect on soil pH, $CaCO_3$ also appears to have a more specific effect since when NH_4^+ fertilizers are applied to calcareous soils, the anion associated with the fertilizer can have a large effect on NH_3 loss. Fenn and Kissel (1973) showed that the NH_4^+ salts that produced the highest loss of NH_3 were those that formed insoluble precipitates with Ca (e.g., F^-, SO_4^{2-}, HPO_4^{2-}) whereas other salts having more soluble reaction products with Ca (e.g., NO_3^-, Cl^-, I^-) gave lower losses.

It has been suggested that applications of NH_4^+ compounds to calcareous soils result in the formation of $(NH_4)_2CO_3$ (Fenn and Kissel, 1973). However, NH_4HCO_3 seems a more likely intermediate since the high pH values required for $(NH_4)_2CO_3$ formation in soils are not generally observed (Feagley and Hossner 1978; Nelson, 1982). The sequence of reactions when $(NH_4)_2SO_4$ is added to a calcareous soil suggested by Nelson (1982) follows:

$$CaCO_3 + (NH_4)_2SO_4 \longrightarrow Ca^{2+} + 2OH^- + CaSO_4 + 2NH_4HCO_3 \qquad (5)$$

$$2NH_4HCO_3 \longrightarrow 2NH_3 + 2CO_2 + 2H_2O \qquad (6)$$

The Ca^{2+} and OH^- ions produced during hydrolysis of $CaCO_3$ may then react with $(NH_4)_2SO_4$ as follows:

$$Ca^{2+} + 2OH^- + (NH_4)_2SO_4 \longrightarrow CaSO_4 + 2NH_3 + 2H_2O \qquad (7)$$

The overall reaction (i.e., summation of equations (5), (6), and (7)) can be shown:

$$CaCO_3 + (NH_4)_2SO_4 \longrightarrow 2NH_3 + CO_2 + H_2O + CaSO_4 \qquad (8)$$

Thus, the formation of an insoluble Ca salt encourages the dissolution of $CaCO_3$ thereby generating the bases, HCO_3^- and OH^-, that act to deprotonate NH_4^+ to NH_3 and sustain volatilization.

B. Factors Affecting Volatilization

1. pH

The equilibrium between NH_4^+ and NH_3 can be represented as

$$NH_4^+ + OH^- \rightleftharpoons NH_3 + H_2O \qquad (9)$$

Thus, the concentrations of NH_4^+ and NH_3 are determined by the pH of the soil solution. An increase in pH (i.e., an increase in hydroxyl ion concentration) drives the equilibrium to the right thereby producing more NH_3. The proportion of aqueous ammoniacal N ($NH_{4(aq)}^+$ plus $NH_{3(aq)}$) present as $NH_{3(aq)}$ at pH 6, 7, 8, and 9 can be calculated as approximately 0.0004, 0.004, 0.04, and 0.3, respectively (Hales and Drewes, 1979). Many workers in both laboratory and field experiments have demonstrated that NH_3 losses increase as the soil pH increases (e.g., Wahhab *et al.*, 1956; Volk, 1959; Ernst and Massey, 1960; Watkins *et al.*, 1972; Lyster *et al.*, 1980). Nonetheless, the direct effects of soil pH are difficult to interpret since more often than not the original soil pH has been assumed to characterize the soil pH throughout the duration of NH_3 loss.

Such assumptions are not necessarily correct since equation (9) can also be represented as

$$NH_4^+ \longrightarrow NH_3 + H^+ \qquad (10)$$

$$H^+ + OH^- \longrightarrow H_2O \qquad (11)$$

Hence, volatilization is accompanied by net acidification of the system (Avnimelech and Laher, 1977), which tends to decrease the rate of volatilization. The original soil pH is, therefore, of prime importance in controlling the extent of volatilization only when the buffering capacity of the soil is high.

That factors other than soil pH can influence NH_3 loss is indicated by the fact that substantial NH_3 volatilization can occur from acid soils (e.g., Ernst and Massey, 1960; Blasco and Cornfield, 1966). Indeed, the pH in the solution immediately surrounding a urea or NH_4^+ salt granule may be considerably more important in determining NH_3 losses than the soil pH itself (Nelson, 1982; Sherlock and Goh, 1984; Black *et al.*, 1984, 1985a, 1985b).

For NH_3 volatilization to occur from flooded soils buffering substances need to be present to prevent acidification of the floodwater caused by the conversion of NH_4^+ to NH_3 (equation (10)). Bicarbonate (HCO_3^-) is the major proton acceptor normally present at typical pH values of floodwater (Vlek and Stumpe, 1978; Vlek and Craswell, 1981) and therefore volatilization in flooded systems can be represented as

$$NH_{4(aq)}^+ + HCO_{3(aq)}^- \longrightarrow NH_{3(g)} + CO_{2(g)} + H_2O \tag{12}$$

Volatilization can also be greatly influenced by the photosynthetic and respiratory balance of algal growth in the floodwater (Mikkelsen *et al.*, 1978; Vlek and Craswell, 1981). As a result of depletion of CO_2 in the water due to algal photosynthetic activity the pH of the floodwater may rise to 9 or above resulting in large losses of NH_3.

2. Temperature

Following applications of urea and NH_4^+ salts, both the instantaneous rate and ultimate extent of NH_3 volatilization increase with increasing temperature (Wahhab *et al.*, 1956; Volk, 1959; Ernst and Massey, 1960; Watkins *et al.*, 1972; Lyster *et al.*, 1980).

The effect of temperature on NH_3 volatilization can be explained at least in part by the temperature dependence of the equilibrium constants (K) for equilibria (2) and (3) (Fig. 1) (Vlek and Stumpe, 1978; Hales and Drewes, 1979; Vlek and Craswell, 1981; Sherlock and Goh, 1984, 1985a). The higher the temperature the greater the proportion of $NH_{3(aq)}$ (equilibrium (2)) present and the greater the proportion of $NH_{3(ag)}$ (equilibrium (3)) present and hence the greater is the potential for NH_3 loss.

Under field conditions the rate of emission of NH_3 follows a marked diurnal cycle that roughly follows that of solar radiation (McGarity and Rajaratnam, 1973; Denmead *et al.*, 1974, 1978; Beauchamp *et al.*, 1978; Freney *et al.*, 1981; Hoff *et al.*, 1981; Vallis *et al.*, 1982; Black *et al.*, 1985a). The diurnal pattern appears to be predominantly related to temperature fluctuations as illustrated clearly in Fig. 2 although the effects of evaporation of water and windspeed cannot be ignored (Denmead *et al.*, 1978).

The extent of losses can also follow a seasonal pattern. Ball and Keeney (1983), for example, found losses of NH_3 from urine patches that averaged 5, 16, and 66% of added urine N under cool moist (winter), warm moist (spring), and warm dry (summer) conditions, respectively.

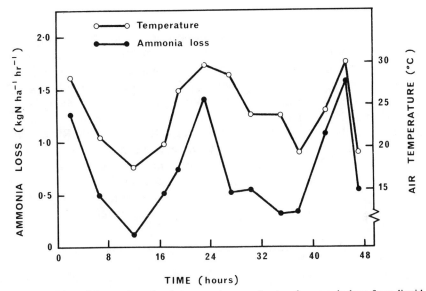

Fig. 2. Diurnal fluctuations in air temperatures and rate of ammonia loss from liquid swine manure. [Redrawn from Hoff *et al.* (1981). Reproduced from *J. Environ. Qual.* **10,** p. 93 by permission of American Society of Agronomy.]

3. Ammonium Concentration

The amount of NH_4^+ added to the soil must have a direct effect on the NH_3 evolved as predicted by equation (10) if all other factors are held constant. A linear relationship between the rate of fertilizer application and total NH_3 loss has been shown in a number of studies (Chao and Kroontje, 1964; Hargrove *et al.*, 1977; Hoff *et al.*, 1981). In other studies percentage losses increased as rates of application increased (Wahhab *et al.*, 1956; Volk, 1959; Kresge and Satchell, 1960; Lyster *et al.*, 1980; Black *et al.*, 1985b). Such nonlinear relationships occur mainly from applied urea and are the result of the increasing soil-surface pH induced by urea hydrolysis.

Factors that influence the NH_4^+ concentration in soil solution will also influence the potential for losses of NH_3 through volatilization. Many mechanisms can induce changes in the NH_4^+ concentration and thereby affect the chain of equilibria that determine the extent of NH_3 loss. The more obvious include plant uptake, nitrification, denitrification, leaching, immobilization, and the fixation of NH_4^+ by clay minerals in exchangeable and nonexchangeable forms. All these mechanisms would tend to decrease the NH_4^+ concentration in soil solution and so reduce NH_3 losses.

4. Soil Characteristics

Of major importance in determining the soil solution NH_4^+ concentration is the cation exchange capacity (CEC) of the soil. The adsorption of the positively charged NH_4^+ ion onto the exchange complex of soils reduces the amount of NH_4^+ and therefore NH_3 in soil solution at a given pH. Hence, many workers have observed a negative relationship between soil CEC and NH_3 volatilization (Wahhab *et al.*, 1956; Gasser, 1964; Ryan and Keeney, 1975; Fenn and Kissel, 1976; Lyster *et al.*, 1980; Ryan *et al.*, 1981). The effect of increasing CEC in reducing NH_3 volatilization is illustrated in Fig. 3. When other soluble cations are applied along with ammoniacal fertilizer, competition for the exchange sites can result. For example, soluble Ca^{2+} may depress normal adsorption of NH_4^+ on exchange sites leading to enhanced NH_3 losses (Fenn *et al.*, 1982).

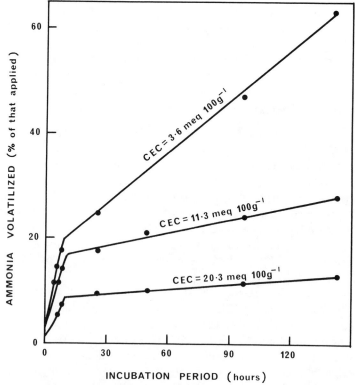

Fig. 3. Rate of ammonia volatilization as influenced by the cation exchange capacity of soil–sand mixtures. [Data from Daftardar and Shinde (1980), by courtesy of Marcel Dekker, Inc.]

As noted earlier when discussing the effect of soil pH, the buffering capacity of the soil can be an important factor influencing NH_3 volatilization because the dissociation of NH_4^+ ions releases H^+ ions as well as NH_3. Thus, volatilization is normally more prolonged in soils of high base status where the acidity produced can be neutralized by carbonate or other forms of alkalinity. Avnimelech and Laher (1977) showed that at a given pH, NH_3 losses increase with increasing buffer capacity. To some extent such an effect may confuse the negative effect of CEC described above since, in general, the higher the CEC of a soil the greater its buffering capacity.

The presence of organic residues has been reported to accelerate (Moe, 1967; Rashid, 1977), decrease (Tripathi, 1958), or not affect (Verma and Sarkar, 1974) NH_3 volatilization from added urea. Such results are not altogether surprising, since while organic matter with a low C : N ratio will be mineralized with the release of NH_4^+, that with a high C : N ratio will immobilize NH_4^+-N from the surrounding soil (Chapter 2). Partially decomposed organic matter will also have a CEC that will influence soil solution NH_4^+ levels.

5. Soil Moisture Content and Moisture Loss

Soil moisture content has an important influence on the rate of NH_3 volatilization since it affects the concentration of NH_4^+ and therefore NH_3 in soil solution. Ammoniacal N concentrations in solution at high moisture contents are likely to be lower than those at low moisture contents leading to lower net losses of NH_3 from wetter soils. This has been shown in a number of studies (Martin and Chapman, 1951; Wahhab et al., 1956; Fenn and Escarzaga, 1976).

However, interpretation of results can be confused by simultaneous losses of water. Indeed, the largest amounts of volatilized NH_3 are normally obtained from soils of high moisture content (below saturation) that are allowed to dry (Martin and Chapman, 1951; Wahhab et al., 1956; Fenn and Escarzaga, 1977). Loss of water promotes NH_3 evolution by increasing, or at least maintaining, ammoniacal concentrations in soil solution over time, which leads to greater losses than if no soil drying occurred. Drying also results in the upward movement of water, which helps transport dissolved NH_4^+ and NH_3 to the soil surface (Freney et al., 1981). Some workers have therefore reported that NH_3 losses increase with increasing initial moisture content up to field capacity (Volk, 1959; Ernst and Massey, 1960; Kresge and Satchell, 1960). Although moisture loss promotes NH_3 evolution, the volatilization of NH_3 can occur without concurrent loss of water (Ernst and Massey, 1960; Terry et al., 1978).

In the case of dry fertilizer materials (e.g., urea prills) a low initial soil moisture content, or rapid drying immediately after application, can slow the rate of fertilizer dissolution and urea hydrolysis and result in small losses of NH_3 (Ernst and Massey, 1960; Volk, 1966). Similarly, when fertilizer solutions or urine are added to very dry soils, losses of NH_3 are often small (Fenn and Escarzaga, 1977; Ball and Keeney, 1983; Vallis *et al.*, 1982). In dry soils, dissolved NH_4^+ may be adsorbed onto soil colloids wherever the solution moves, while in initially wet soils $NH_{4(aq)}^+$ may tend to remain in macropores. Convection to the soil surface would tend to proceed through macropores thus transporting more $NH_{4(aq)}^+$ to the surface of initially wet soils.

Time of rainfall or water application can also be an important factor influencing losses of NH_3. For example, before its hydrolysis urea moves rapidly into the soil when water is applied and losses from surface-applied urea decrease with increasing amounts of applied water (Fenn and Miyamoto, 1981; see Table I). Rainfall immediately after surface applications of urea can significantly reduce losses of NH_3 (Carrier and Bernier, 1971; Morrison and Foster, 1977; Black and Sherlock, 1985). However, as illustrated in Table I, once hydrolysis has proceeded (by 24 hr), the effectiveness of water applications is markedly reduced.

6. Windspeed

Estimates of NH_3 volatilization resulting from fertilizer applications have often been carried out in laboratory studies using unrealistically low air exchange rates that fail to simulate field conditions (Nelson, 1982).

Table I

Percentage of Applied Urea N Lost as Ammonia After 7 Days as Affected by the Amount of Water Applied and Its Time of Application[a]

Water applied (mm)	Time of water application (hours after urea application)[b]			
	0–3	8–10	24–26	48–50
0	28	32	31	32
4	8	24	27	n.d.
16	2	10	22	29
48	0.5	n.d.	21	n.d.

[a] Data from Black and Sherlock (1985).

[b] Urea granules were applied to a pasture surface at a rate of 100 kg N ha^{-1}. n.d., not determined.

Under such conditions NH_3 volatilization can be directly proportional to airflow (Watkins *et al.*, 1972; Kissel *et al.*, 1977) and a number of workers have reported that the rate of NH_3 volatilization increases with an increase in the rate of airflow over the samples (e.g., Overrein and Moe, 1967; Terry *et al.*, 1978; Vlek and Stumpe, 1978).

Increasing windspeed should tend to increase the volatilization rate by promoting more rapid transport of NH_3 away from the air–soil interface. However, when Beauchamp *et al.* (1978, 1982) used an aerodynamic procedure to measure NH_3 losses from surface-applied sewage sludge and liquid dairy manure under field conditions, they found no discernable relationship between windspeed and $NH_{3(g)}$ flux. These workers suggested that volatilization from the soil was diffusion controlled and was limited by depletion of ammoniacal N at sites from which volatilization was possible. Windspeed presumably had little effect on this diffusion process.

It has, however, been clearly demonstrated that increasing windspeed over a flooded soil surface increases the NH_3 volatilization rate (Bouwmeester and Vlek, 1981; Denmead *et al.*, 1982; Moeller and Vlek, 1982). Denmead *et al.* (1982) suggested that there may be considerable resistance to transport of NH_3 in the liquid phase and that the enhanced volatilization in high winds is due to better mechanical mixing of the surface water. Such mixing would also avoid the development at the floodwater surface of a region depleted of NH_3 that might limit the volatilization rate.

7. Form and Placement of Applied Fertilizer

Since anhydrous NH_3 is applied to soils as a gas one might expect a tendency for it to escape to the atmosphere. It has generally been shown, however, that regardless of soil type, losses of N during and immediately following injection of anhydrous NH_3 are small if it is applied at an adequate depth (5 to 13 cm) and provided soil moisture conditions and soil physical properties are such that the injection channel is rapidly sealed (Ernst and Massey, 1960; Khan and Haque, 1965; Parr and Papendick, 1966; Nelson, 1982). Thus, anhydrous NH_3 is normally retained in the soil due to its reactions with soil components (see Chapter 4).

As already noted, NH_4^+ salts that produce the largest losses of NH_3 from calcareous soils are those that form insoluble precipitates with Ca (e.g., $(NH_4)_2SO_4$ or $(NH_4)_2HPO_4$). In acidic and moderately acidic soils surface applications of alkaline fertilizers such as NH_4OH or urea generally result in larger losses of NH_3 through volatilization than do applications of neutral or acidic fertilizers such as $NH_4H_2PO_4$ or $(NH_4)_2SO_4$ (Terman *et al.*, 1968; Matocha, 1976). The effect of N form when applied to a pasture on pH around the fertilizer granules and on NH_3 volatilization from a pasture surface is shown in Fig. 4. The localized rise in pH, due to urea hydrolysis, and the consequently larger loss of NH_3 from urea are obvious.

Several methods have been examined to minimize NH_3 losses from applied urea. Mixing neutral ammonium salts or acidifying agents (e.g., NH_4Cl, $NH_4H_2PO_4$, HNO_3, or H_3PO_4) with urea prior to application can

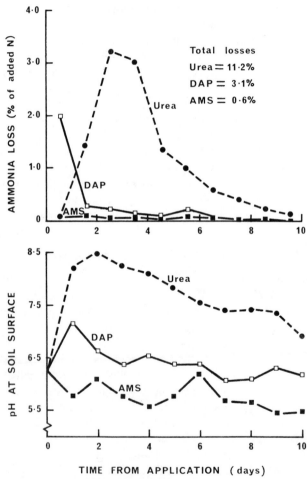

Fig. 4. Effect of form of applied nitrogen (urea, diammonium phosphate (DAP), or ammonium sulfate (AMS)) on daily ammonia volatilization rate from a pasture and pH of the soil surface at the granule site. Rate of applied $N = 30$ kg N ha^{-1}. [Data from Black, A. S., Sherlock, R. R., and Smith, N. P. (unpublished).]

markedly reduce losses from surface applications (Terman, 1979). Other techniques include the use of slow-release forms of urea (e.g., sulfur-coated urea) (Matocha, 1976; Presad, 1976; Vlek and Craswell, 1979) or urease inhibitors (Moe, 1967; Bremner and Douglas, 1973). Such techniques work by attempting to retard the rate of urea hydrolysis and so prevent the rapid buildup of ammoniacal N.

Another method of reducing losses of NH_3 from applications of NH_4^+ salts or urea is by placement of fertilizers below the soil surface or by thoroughly incorporating them into the topsoil (Ernst and Massey, 1960; Overrein and Moe, 1967; Fenn and Kissel, 1976; Vlek and Craswell, 1979; Hoff et al., 1981). This technique effectively reduces the ammoniacal N concentration of the soil solution at the soil surface thereby reducing losses of NH_3. However, in some circumstances it is common practice to broadcast all fertilizers on the soil surface (e.g., on pastures or under minimum tillage) in which case the opportunity for NH_3 volatilization is enhanced (Black et al., 1984).

As indicated previously, applications of nitrogenous fertilizers to flooded soils can result in considerable losses of NH_3 (Vlek and Craswell, 1979). Placement of fertilizers into soils before flooding can markedly reduce volatilization in comparison with broadcasting the fertilizer over the floodwater (Macrae and Ancajas, 1970; Craswell et al., 1981; Mikkelsen et al., 1978).

8. Presence of Animals

a. Grazing animals. In grazed grassland ecosystems the presence of animals can greatly influence the cycling of N within the system. Indeed it has been estimated that 85 to 95% of N ingested by grazing herbivores is excreted (Henzell and Ross, 1973) and most of this is voided as urine in localized patches on the soil surface (Doak, 1952). Urine is a concentrated N solution (approx. 10 gm N liter^{-1} of which 80–90% is urea) and the effective rate of application within urine patches is often greater than the equivalent of 500 kg N ha^{-1}. Such application rates are considered to be much too high for efficient plant utilization (Ball and Keeney, 1983; Carran et al., 1982). The urea is rapidly hydrolyzed to NH_4^+-N, which results in localized areas with both high pH and high ammoniacal concentrations. It is evident that urine patches provide concentrated focal points within a pasture from which significant NH_3 volatilization will occur. Such losses have been reported to be in the region of 20 to 60% of the urine N (Denmead et al., 1974; Ball et al., 1979; Ball and Keeney, 1983; Carran et al., 1982; Sherlock and Goh, 1984).

b. Feedlots. Modern animal-feeding practices in which large numbers of animals are concentrated in small areas have led to problems in the disposal of animal wastes (Loehr, 1974; Tunney, 1980). Appreciable NH_3 volatilization may occur during the storage of both heaps of manure and liquid slurries (Tunney, 1980). Apart from the offensive odors produced, the chief concern with such emissions is the reduction in fertilizer value of the materials since losses on the order of 30 to 80% of the N originally present are not uncommon (Vanderholm, 1975). Ammonia volatilization from feedlot areas can be 10 to 30 times that from surrounding areas (Hutchinson and Viets, 1969; Elliott *et al.*, 1971; Luebs *et al.*, 1973).

Volatile aliphatic amines of different molecular weights are also emitted from animal manures. Mosier *et al.* (1973) identified seven basic aliphatic organic nitrogenous compounds, namely, methyl, dimethyl, ethyl, *n*-propyl, isopropyl, *n*-butyl, and *n*-amyl amines, emanating from a high-density cattle feedlot. It was estimated that these amines constituted only 2 to 6% of the volatilized ammoniacal N with the balance released as NH_3.

Significant losses of NH_3 occur when animal manures are surface-applied to agricultural lands, particularly in the summer months (Lauer *et al.*, 1976; Hoff *et al.*, 1981). Approximately 50% of the added N can be lost as NH_3 in a 3- to 5-day period (Vanderholm, 1975) although incorporation or injection of the manure into the soil immediately following application can markedly reduce losses (Hoff *et al.*, 1981).

9. Presence of Plants

a. Passive role. Crop height and density can be important factors in determining the fraction of applied fertilizer in solution or, in the case of a grazed pasture, the amount of voided urine that reaches the soil surface. Intercepted solution may undergo a number of transformations including direct absorption by foliage (see Chapter 6). Leaf surfaces also possess considerable urease activity and direct volatilization of the hydrolysis products of intercepted aqueous urea and urine has been demonstrated by several workers (Doak, 1952; Volk, 1959; Simpson and Melsted, 1962; McGarity and Hoult, 1971; Watkins *et al.*, 1972). Since leaf surfaces have only a limited CEC and low buffering capacity it seems possible that interception of nitrogenous materials by plant cover may result in increased losses of applied N through NH_3 volatilization (Sherlock and Goh, 1985b).

b. Active role. The absorption and evolution of NH_3 from plants have been reviewed in detail elsewhere (Farquhar *et al.*, 1980, 1983). A finite partial pressure of NH_3 is maintained in the substomatal cavities of plant leaves. When this partial pressure exceeds that of the atmosphere, net evolution of NH_3 occurs. Factors that favor net evolution include high temperatures, low atmospheric partial pressures of NH_3, and high stomatal conductances. Stomatal conductance is greatest under conditions favoring CO_2 assimilation: high light intensity, ample moisture, and high levels of nutrition.

Factors that influence the partial pressure of NH_3 in the substomatal cavity are unclear. Ammonia assimilation in plants is discussed in Chapter 6. Except at high tissue ammonia concentrations, where glutamate dehydrogenase may play a role, the majority of ammonia assimilation occurs through the combined action of the glutamine synthetase and glutamate synthase enzymes. The combined action of these two enzymes is particularly important in the refixation of the massive amounts of ammonia produced during photorespiration. Thus levels of NH_4^+ and NH_3 in plant tissue are normally extremely low. The partial pressure of NH_3 in the substomatal cavities is thought to be maintained by small amounts of NH_4^+ supplied in the transpiration stream plus small amounts present in surrounding leaf cells (Farquhar et al., 1983). During leaf senescence, photorespiration declines but proteolysis (with the release of NH_4^+) increases. Losses of NH_3 from plants may, therefore, be greater during senescence (Farquhar et al., 1983).

A number of workers have measured losses of NH_3 from healthy plant canopies (Martin and Ross, 1968; Stutte and Weiland, 1978; Stutte et al., 1979; Weiland and Stutte, 1979, 1980; Weiland et al., 1979; Farquhar et al., 1980; Lemon and van Houtte, 1980; Hooker et al., 1980) and senescing plants (Farquhar et al., 1979; Hooker et al., 1980). Some circumstantial evidence from N-balance studies also indicates that annual cereal crop plants can lose N when approaching maturity (Wetselaar and Farquhar, 1980). Some of this could be lost as NH_3.

The magnitude of gaseous losses of NH_3 from plants is uncertain and estimates vary greatly. Stutte and co-workers, using pyrochemiluminescent N detection, suggested losses averaging 9 nmol $(m^2$ leaf surface$)^{-1}$ sec^{-1} but losses estimated by other methods are an order of magnitude less (Farquhar et al., 1979; Hooker et al., 1980).

Volatile amines also appear to be liberated from growing plants particularly during flowering (Richardson, 1966; Farquhar et al., 1983).

When the partial pressure of NH_3 in the ambient atmosphere exceeds that in the substomatal cavity then net absorption of NH_3 will occur (Farquhar et al., 1980, 1983). Indeed, NH_3 can be absorbed in the vapor phase through open leaf stomata of leaf canopies and it may also dissolve in water films on the plant leaf surfaces and be subsequently absorbed and metabolized (Denmead et al., 1976). Many workers have demonstrated the uptake of NH_3 by leaves placed in NH_3-enriched atmospheres (Hutchinson et al., 1972; Porter et al., 1972; Rogers and Aneja, 1980; Cowling and Lockyer, 1981). Faller (1972) showed in a long-term experiment that plants can absorb and utilize NH_3 as the only source of N without affecting their normal growth.

It is evident that plants can both absorb and release NH_3 from their canopies. Indeed, field measurements of NH_3 flux within and above the canopy of several crops have clearly indicated that NH_3 released at the soil surface can be absorbed within the leaf canopy while some NH_3 may also be simultaneously released from the top of the canopy (Denmead *et al.*, 1976, 1978; Lemon and van Houtte, 1980). The major factors that determine whether net absorption or evolution of NH_3 occurs are, as yet, unclear. Nonetheless, an increase in the height and density of the crop canopy appear to be important factors that tend to reduce NH_3 losses (Denmead *et al.*, 1982).

III. BIOLOGICAL GENERATION OF GASEOUS NITROGENOUS PRODUCTS

Gaseous nitrogenous products can be produced by three groups of organisms: dissimilatory denitrifying bacteria, nondenitrifying fermentative bacteria and fungi, and autotrophic nitrifying bacteria. Denitrifying bacteria are thought to be the most important organisms contributing to losses of nitrogenous gases from soils under anaerobic conditions (Letey *et al.*, 1981; Payne, 1981; Firestone, 1982) but under oxidized conditions nitrifying bacteria could well be important agents (Freney *et al.*, 1979; Bremner *et al.*, 1981).

The organisms and processes involved in the biological production of gaseous N in soils are discussed below and the major factors that are thought to affect such production are reviewed.

A. Processes

1. Dissimilatory Denitrification

Dissimilatory denitrification is a respiratory process that is present in a limited number of aerobic bacteria whereby they can grow in the absence of O_2 while reducing NO_3^- or NO_2^- to N_2 and/or N_2O. The majority of such bacteria are heterotrophs and obtain their energy and cellular C from organic substrates. In the absence of O_2, NO_3^- or oxides derived from it serve as terminal electron acceptors for respiratory electron transport during the oxidation of the organic substrate and a more reduced N oxide or N_2 is produced. The process is described as a dissimilatory reduction since the products of NO_3^- reduction, N_2 and N_2O, are not assimilated but are released to the atmosphere.

a. Denitrifying bacteria. The capacity to use N oxides as electron acceptors in place of O_2 with the evolution of N_2O and/or N_2 has been reported in approximately 20 genera of bacteria (Payne, 1973, 1981; Focht and Verstraete, 1977; Firestone, 1982; Knowles, 1982). These are listed in Table II. The presence of denitrifiers in surface soils may be regarded as ubiquitous (Payne, 1981) since their density frequently exceeds one million per gram of soil (e.g., Jacobsen and Alexander, 1980) and higher concentrations are present in the rhizosphere (Alexander, 1977).

The most common bacteria used for physiological and biochemical studies of denitrification are *Paracoccus denitrificans, Pseudomonas denitrificans,* and *Pseudomonas perfectomarinus.* Nevertheless, it is not known which denitrifying organisms are either numerically or functionally most important in soils since laboratory methods of isolating and enumerating these bacteria are likely to favor some organisms to the detriment of others (Firestone, 1982).

Denitrifying bacteria are biochemically as well as taxonomically diverse. Most are chemoheterotrophs: they use carbonaceous compounds as electron donors (reductants) and as sources of cellular C and chemical energy sources. Some grow as chemolithotrophs, oxidizing H_2 (e.g., *Paracoccus denitrificans* and *Alcaligenes* spp.) (John and Whatley, 1975; Thauer *et al.*, 1977) or reduced sulfur compounds (e.g., *Thiobacillus denitrificans*) (Ishaque and Alaem, 1973; Baldensperger and Garcia, 1975). One group is photosynthetic (e.g., *Rhodopseudomonas sphaeroides*) (Satoh, 1977; Sawada *et al.*, 1978).

A few N_2-fixing organisms are known to have the ability to denitrify under anaerobic conditions. These include a considerable number of strains of *Azospirillum brasilense,* which are commonly associated with the roots of many tropical grain and forage grasses (Eskew *et al.*, 1977; Neyra and van Berkum, 1977; Neyra *et al.*, 1977; Scott and Scott, 1978). Several strains of *Rhizobium,* including *R. japonicum, R. meliloti,* and

Table II

The Reported Genera of Denitrifying Bacteria

Acinetobacter	*Halobacterium*	*Rhizobium*
Alcaligenes	*Hyphomicrobium*	*Rhodopseudomonas*
Azospirillum	*Micrococcus*	*Spirillum*
Bacillus	*Moraxella*	*Thiobacillus*
Cytophaga	*Paracoccus*	*Vibrio*
Flavobacterium	*Propionobacterium*	*Xanthomonas*
Gluconobacter	*Pseudomonas*	

most of the slow-growing rhizobia, also possess denitrifying capabilities in their free-living state under anaerobic conditions and can produce N_2 or N_2O (Zablotowicz et al., 1978; Zablotowicz and Focht, 1979; Daniel et al., 1980, 1982). Possession of a denitrifying pathway may help free-living bacteria survive anaerobic conditions in the soil (Daniel et al., 1980) and might also be important to rhizobium–legume symbiosis by maintaining nodule integrity under anoxic conditions (Rigaud et al., 1973).

b. *Pathways.* Detailed discussions of the physiology and biochemistry of the denitrification process have been presented elsewhere (Payne, 1973, 1981; Firestone, 1982; Knowles, 1982; Fillery, 1983) and only the major features are outlined below. The pathway of N oxide reduction is generally represented as

$$\underset{NO_3^-}{\overset{(+5)}{}} \longrightarrow \underset{NO_2^-}{\overset{(+3)}{}} \longrightarrow \underset{NO}{\overset{(+2)}{}} \longrightarrow \underset{N_2O}{\overset{(+1)}{}} \longrightarrow \underset{N_2}{\overset{(0)}{}}$$

The enzymes responsible for the reductions are NO_3^- reductase, NO_2^- reductase, NO reductase, and N_2O reductase. The obligatory participation of NO and NO reductase in the sequence of reductions is still debatable (see Bryan, 1980; Knowles, 1982) and nitroxyl (NOH) has been suggested as an alternative intermediate (Garber and Hollocher, 1982).

While most denitrifying bacteria possess all the reductase enzyme complexes necessary to reduce NO_3^- to N_2, some lack NO_3^- reductase and are thus "NO_2^- dependent," some lack N_2O reductase and yield N_2O as the terminal product, and others possess N_2O reductase but lack the ability to reduce NO_2^- to N_2O (Knowles, 1981, 1982). Still other groups are sometimes referred to as partial denitrifiers (Ingraham, 1981). These include those lacking NO_2^- reductase and N_2O reductase and can therefore reduce NO_3^- to NO_2^- and NO to N_2O and organisms that lack NO_2^-, NO, and N_2O reductases and are capable of only limited reduction of NO_3^- to NO_2^-.

Characteristics of enzymes involved in the reductions vary depending on the bacterial species involved and the methods of purification used. Some common characteristics can be noted. Dissimilatory nitrate reductase is a membrane-bound enzyme that generally consists of multiple subunits and contains Mo, Fe, and labile sulfide groups (Firestone, 1982; Knowles, 1982). Nitrite reductase catalyzes the reduction of NO_2^- to gaseous products although there is still considerable controversy as to whether NO is the *in vivo* product. There appear to be two main types of nitrite reductase; copper-containing metalloflavoproteins and the apparently more common hemoproteins of cytochrome type *cd* (Knowles, 1982). Nitrite reductase appears to be membrane-associated but readily solubilized (Knowles, 1982). The reactive nature of NO makes isolation and characterization of NO reductase difficult and unequivocal evidence for its role in denitrification has yet to be provided. Little is known about N_2O reductase although it is thought to be membrane-associated, possibly contains Cu, and is linked to electron transport through cytochromes of types b and c (Knowles, 1982).

2. Fermentative Nitrite Dissimilation

Nitrous oxide can also be produced in soils by the actions of a miscellany of "nondenitrifying" fungi and bacteria (Yoshida and Alexander, 1970; Bollag and Tung, 1972; Smith and Zimmerman, 1981). These nondenitrifying organisms are only able to respire NO_3^- anaerobically as far as NO_2^-, but growing fermentatively they can further dissimilate NO_2^- to NH_4^+ (Sorenson, 1978; Caskey and Tiedje, 1979; Cole and Brown, 1980). Nitrous oxide is produced as a minor product.

The physiological function, if any, of N_2O production by these organisms is not clear. Nitrous oxide production does not appear to be directly related to growth or energy generation (Smith and Zimmerman, 1981) as is the case for the fermentative reduction of NO_2^- to NH_4^+ (Cole and Brown, 1980).

The significance of nondenitrifying NO_3^- reducers as a source of N_2O evolution from soils is unknown. Some research (Smith and Zimmerman, 1981) suggests that these organisms are more numerous in soils than denitrifiers. However, since only a small proportion of the NO_3^- reduced by nondenitrifiers is released as N_2O they are generally thought to be of minor agronomic importance.

3. Autotrophic Nitrification

The NH_4^+-oxidizing bacteria *Nitrosomonas, Nitrosospira,* and *Nitrosolobus* have the capacity to produce N_2O from NH_4^+ or hydroxylamine (an intermediate in the oxidation of NH_4^+ to NO_2^- by these microorganisms) under most conditions (Yoshida and Alexander, 1970, 1971; Ritchie and Nicholas, 1972, 1974; Bremner and Blackmer, 1980; Blackmer *et al.,* 1980; Goreau *et al.,* 1980). The mechanisms that are thought to be responsible for the production of N_2O by these organisms were outlined in Chapter 3.

Laboratory (Bremner and Blackmer, 1978; Freney *et al.,* 1978; Goodroad and Keeney, 1984) and field evidence (Denmead *et al.,* 1979; Breitenbeck *et al.,* 1980; Mosier *et al.,* 1981, 1982; Smith *et al.,* 1982) has shown that losses of N_2O can occur from soils during nitrification. However, less certain is the significance of these N_2O emissions in comparison with those from denitrification.

It has also been suggested that NO_x emissions may result during nitrification (Verstraete, 1981). Lipschultz *et al.* (1981) showed that cultures of

Nitrosomonas europaea liberated both NO and N_2O during the oxidation of NH_4^+. The ratio of NO produced relative to N_2O rose as the O_2 content within the medium decreased and a mean value of 7.5 mol NO per mole of N_2O produced was observed. Indeed, emissions of NO_x are often observed during nitrification of NH_4^+ fertilizers when applied at high rates (e.g., band applications) although they are thought to occur principally as a result of NO_2^- accumulation and subsequent chemodenitrification (see Section IV).

B. Factors Affecting Biological Gaseous Losses

Most research dealing with factors influencing gaseous losses of N_2O and N_2 from soils has been centered on the process of dissimilatory denitrification. An unknown proportion of N_2O emitted during such experiments is likely to have originated from nitrification of native soil NH_4^+. A detailed discussion of the factors influencing nitrification was presented in Chapter 3. Conditions that favor nitrification will obviously tend to promote N_2O losses from native or applied fertilizer NH_4^+ since the ratio of N_2O evolved to NO_3^- produced during nitrification appears to be reasonably constant (Goodroad and Keeney, 1984). The factors that are generally known to influence denitrification are discussed below and, where appropriate, factors known to influence losses of N_2O through nitrification are also noted.

1. Aeration and Moisture

The activity and synthesis of all the N oxide reductase enzyme systems involved in denitrification are repressed by O_2 (Firestone, 1982; Knowles, 1982). The presence of oxygen also inhibits the activity of preformed reductases (Payne, 1973; Stouthamer, 1976). Indeed, it appears that under aerobic soil conditions the N oxide reductase enzymes are present in repressed form and when O_2 is removed from the soil there are rapid increases in the absolute and relative activities of these enzymes and denitrification commences almost immediately (Smith and Tiedje, 1979a).

Denitrification can probably occur even in well structured aerobic soils due to the occurrence of anaerobic microsites (Firestone, 1982). Anaerobic pockets in soils may often be localized areas of intense respiratory activity where O_2 demand exceeds the supply (Craswell and Martin, 1975; Smith, 1980) rather than areas of passive anaerobiosis. Clearly, factors such as the rate of O_2 consumption and O_2 diffusion rate and structural considerations such as pore geometry and degree of soil compaction are important (Smith, 1977; Ryden and Lund, 1980). In well aerated soils emissions of N_2O are likely to originate, at least partially, through nitrification (Freney *et al.,* 1978).

Soil moisture is obviously an important factor that influences aeration since with increasing moisture content, air in soil pores is displaced with water. Hence, as illustrated in Fig. 5, with increasing soil moisture contents the rate of denitrification generally increases (Bremner and Shaw, 1958; Pilot and Patrick, 1972; Bailey and Beauchamp, 1973; Craswell and Martin, 1974; Ryden and Lund, 1980). Following periods of intense irrigation or rainfall, soils may become saturated with water at the soil surface for brief periods. During such periods short bursts of intense denitrification occur (Ryden *et al.,* 1979; Ryden and Lund, 1980). Increases in soil moisture content up to about -33 to -10 kPa also increase the rate of nitrification (Chapter 3) and thus the release of N_2O from applied NH_4^+ (Goodroad and Keeney, 1984).

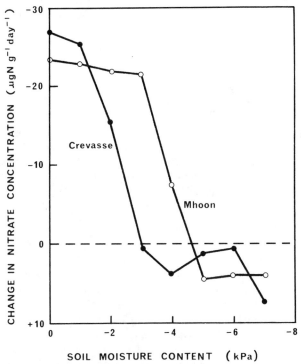

Fig. 5. Rate of nitrate reduction in two soils (Crevasse and Mhoon) as influenced by soil moisture tension. [Redrawn from Pilot and Patrick (1972). Reprinted with permission from The Williams and Wilkins Co., Baltimore.]

Although drainage should reduce denitrification losses by improving soil aeration, it also transfers more dissolved inorganic N from the soil to water courses, ditches, and rivers (Dowdell, 1982). Drainage water can, in fact, contain measurable amounts of dissolved N_2O (Dowdell *et al.*, 1979b, Dowdell, 1984) while NO_3^- transported in water is subject to denitrification during its journey in field drains and within stream courses (e.g., Swank and Caskey, 1982).

The later reductase enzymes in the denitrification sequence appear to be more sensitive to oxygen than are the earlier reductases (Krul and Veeningen, 1977; Betlach and Tiedje, 1981) so that with increasing oxygen concentrations there is an increase in the mole fraction of N_2O emitted (Focht, 1974).

Fluctuating moisture contents can influence the ratio of $N_2O : N_2$ evolved. The mole fraction of N_2O produced is generally high immediately following the onset of anaerobiosis and the proportion of N_2 produced generally increases with time (Rolston *et al.*, 1976, 1978; Letey *et al.*, 1979; Ryden *et al.*, 1979). This effect has been attributed to high concentrations of NO_3^- being initially present in the soil (Rolston, 1981). As noted later, high NO_3^- concentrations tend to inhibit the reduction of N_2O to N_2.

2. Organic Carbon

The most abundant denitrifiers are heterotrophs, which require organic compounds as electron donors and as a source of cellular material. Thus the availability of organic matter is an important factor moderating both the rate and total extent of denitrification. High levels of readily decomposable organic matter can also indirectly enhance the potential for denitrification through a general stimulation of microbial respiration, causing rapid O_2 consumption and an acceleration of the onset of anaerobiosis.

A general relationship between total soil organic C or N and denitrification has been observed by several workers (Bremner and Shaw, 1958; McGarity, 1961; Reddy *et al.*, 1982). However, it is the quantity of readily available soil organic carbon that is of particular importance. Thus, rates of denitrification are highly correlated with "available" soil C as evaluated by extractable reducing sugars (Stanford *et al.*, 1975), by water-soluble organic C (Bremner and Shaw, 1958; Burford and Bremner, 1975; Reddy *et al.*, 1982), or by readily mineralizable C (Burford and Bremner, 1975; Reddy *et al.*, 1982). The linear relationship between water-soluble C and the denitrification capacity of 17 soils is shown in Fig. 6. Factors that increase the levels of available C in soils (e.g., drying and rewetting or freezing and thawing) have been shown to increase the capacity of soils to denitrify added NO_3^- (McGarity, 1962; Patten *et al.*, 1980). Denitrification

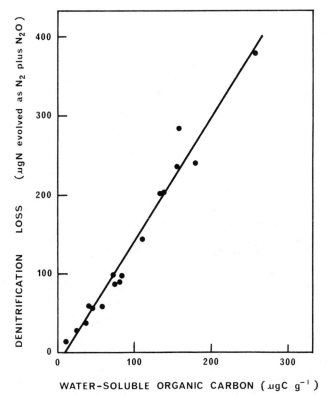

Fig. 6. Relationship between denitrification capacity and water-soluble organic carbon in 17 soils. [Redrawn from Burford and Bremner (1975). Reprinted with permission from Pergamon Press.]

is generally stimulated by additions of organic compounds and residues to soils (Nommik, 1956; Bowman and Focht, 1974; Guenzi *et al.*, 1978; Reddy *et al.*, 1978; Rolston *et al.*, 1982). Indeed, in many situations, the relationship between total soil C and denitrification is only of academic interest since readily available C sources are likely to provide the major C source for denitrification.

The dependence on C availability results in higher denitrification potentials being generally found in surface soils rather than in subsoils (Khan and Moore, 1968). Denitrification activity and populations of denitrifiers generally decrease with soil depth (Bailey and Beauchamp, 1973; Brar *et al.*, 1978; Cho *et al.*, 1979) although significant denitrification can occur down to 60 to 70 cm (Myers and McGarity, 1972; Rolston *et al.*, 1976;

Gilliam *et al.*, 1978). Organic soils generally have high potentials for denitrification (Bartlett *et al.*, 1979; Reddy *et al.*, 1980; Terry *et al.*, 1981).

Carbon availability can also influence the proportion of N_2O and N_2 produced. With increasing available C there is generally a more complete reduction of NO_3^- and therefore less N_2O production in relation to that of N_2 (Delwiche, 1959; Focht and Verstraete, 1977; Rolston *et al.*, 1978; Smith and Tiedje, 1979b).

3. Nitrate Supply

The apparent K_m values (which indicate the NO_3^- concentration required to give half the maximum velocity of denitrification) for the dissimilatory reduction of nitrogen oxides, determined *in vivo* or with purified enzymes, are normally in the range of 5 to 290 μM (see Knowles, 1982). Firestone (1982) noted that a K_m value of 15 μM NO_3^- for denitrifying bacteria would be equivalent to a concentration of 0.04 μg NO_3^--N per gram of soil that had a moisture content of 20%.

Thus it is not surprising that at relatively high concentrations of NO_3^- (greater than 40 to 100 μg N gm^{-1}) the rate of denitrification in soils has been shown to be independent of NO_3^- concentration, i.e., denitrification follows zero-order kinetics (Kohl *et al.*, 1976; Focht and Verstraete, 1977; Blackmer and Bremner, 1978). However, in soils the diffusion of NO_3^- to the sites of denitrification can become an important limiting factor (Phillips *et al.*, 1978; Reddy *et al.*, 1978) so that denitrification reactions are frequently reported to be first order up to 40 to 100 μg N gm^{-1} in soils (Stanford *et al.*, 1975; Starr and Parlange, 1975; Ryzhova, 1979). Indeed, presumably because of the limiting effect of the diffusion of NO_3^-, K_m values for NO_3^- reduction in soil are much higher than those obtained in cultures and range from 130 to 12,000 μM NO_3^- (Bowman and Focht, 1974; Yoshinari *et al.*, 1977).

High levels of NO_3^- can inhibit the reduction of N_2O causing an increase in the ratio of N_2O to N_2 in the product gases (Nommik, 1956; Blackmer and Bremner, 1978: Firestone *et al.*, 1980; Letey *et al.*, 1980; Terry and Tate, 1980; Gaskell *et al.*, 1981). The effect of NO_3^- concentration interacts with soil pH such that the inhibitory effect of NO_3^- on N_2O reduction increases markedly with a decrease in soil pH (Blackmer and Bremner, 1978; Firestone *et al.*, 1980).

4. Nitrifiable N

As already noted, in some studies greater losses of N_2O have accrued from soils after the application of nitrifiable N. In laboratory incubation experiments, soils amended with nitrifiable N (e.g., $(NH_4)_2SO_4$, urea, or the amino acid alanine) yielded more N_2O than similar nonamended or

Table III

Quantities of Nitrous Oxide Released from Well Aerated Soils Treated with Different Forms of Nitrogen[a]

Form	Rate	N_2O-N released (ng gm^{-1} soil)	
		8 days	30 days
None	0	2	5
$(NH_4)_2SO_4$	50	52	58
Urea	50	58	64
KNO_3	50	4	5
$(NH_4)_2SO_4$	100	146	153
Urea	100	118	124
KNO_3	100	4	7

[a] Data from Bremner and Blackmer (1978).

NO_3^--treated soils (Table III) (Bremner and Blackmer, 1978, 1980, 1981). In these experiments, N_2O production often increased linearly with nitrifiable N and, furthermore, losses were markedly reduced by addition of nitrapyrin (a compound that selectively inhibits autotrophic NH_4^+ oxidation) (Bremner and Blackmer, 1978). Under field conditions, however, the expected relationship between soil NH_4^+ concentrations and N_2O fluxes is frequently complicated by simultaneous denitrification (Mosier et al., 1982; Smith et al., 1982).

5. pH

In pure cultures and in soils, the overall rate of denitrification is often positively related to pH and has an optimum in the range of pH 7.0 to 8.0 (Nommik, 1956; Van Cleemput and Patrick, 1974; Muller et al., 1980). Generally, in the neutral pH range of soils (pH 6 to 8) there is little effect of pH (Burford and Bremner, 1975; Stanford et al., 1975) but at soil pH values below 6.0 denitrification can be strongly inhibited (Klemedtsson et al., 1978; Muller et al., 1980). At pH levels below 5.5, toxic levels of soil Al and Mn could well limit microbial activity. Nevertheless, several workers have reported significant denitrifier activity at soil pH values below 5.0 (e.g., Gilliam and Gambrell, 1978; Muller et al., 1980; Koskinen and Keeney, 1982). In short-term laboratory studies, increasing pH may, to some extent, increase denitrifier activity by temporarily increasing the solubility of soil organic matter (Fillery, 1983).

The lower rates of denitrification often found in acidic soils (Gilliam and Gambrell, 1978; Muller *et al.*, 1980) may be due to a small population of denitrifiers in microsites having a higher pH than the surrounding soil and/or because of denitrifier populations with a low pH optimum or wide pH tolerance (Focht and Joseph, 1974). The optimum pH for nitrification normally appears to be pH 6 to 7 (Chapter 3), and Goodroad and Keeney (1984) observed a greater rate of nitrification of added NH_4^+, and concomitant N_2O emission, from a limed soil (pH 6.7) than from an unlimed control (pH 4.7).

It appears that the N_2O reductase enzyme system is more sensitive than the other reductases to low pH such that, as noted in the previous section, the mole fraction of N_2O produced increases as the pH falls (Blackmer and Bremner, 1978; Tiedje *et al.*, 1981) and at pH 4.0 N_2O may be the major product (Nommik, 1956). Such an effect appears to occur only in the presence of added NO_3^- (Firestone *et al.*, 1980).

6. Temperature

Denitrification is markedly influenced by temperature. Below 10°C rates are low and in the range of 10 to 35°C the rate of denitrification is very temperature dependent with a Q_{10} value (the ratio by which denitrification increases for a 10°C rise in temperature) of about 2.0 (Dawson and Murphy, 1972; Bailey and Beauchamp, 1973; Stanford *et al.*, 1975). Rates increase up to a maximum at 60 to 75°C and then rapidly decline (Table IV) (Bremner and Shaw, 1958; Keeney *et al.*, 1979).

The unusually high optimum temperature reported for denitrification may, in part, be the result of the presence of thermophilic *Bacillus* spp. (Focht and Verstraete, 1977). However, above 50°C chemical decomposition reactions of NO_3^- increase in importance (Keeney *et al.*, 1979) so that the high optimum may not be wholly of biological origin. To some extent, indigenous denitrifiers may have temperature optima adapted to their climatic regions (Gamble *et al.*, 1977). The optimum temperature for nitrification is often 25 to 30°C (Chapter 3), and Goodroad and Keeney (1984) showed that N_2O production via nitrification increased as the temperature was raised from 10 to 30°C.

Several workers have observed marked diurnal variability in the rate of N_2O emission from soils that appears to be related to soil temperature (Ryden *et al.*, 1978; Denmead *et al.*, 1979; Blackmer *et al.*, 1982; Lensi and Chalamet, 1982). Maximum rates generally occur in the afternoon and minimum rates during the night. Such results presumably reflect the temperature dependence of both denitrification and nitrification.

Denitrification appears to follow a seasonal trend with losses of N_2O plus N_2 being markedly higher in summer than in winter (Rolston *et al.*,

Table IV

Effect of Temperature on Gaseous N Loss and the $N_2O/(N_2O + N_2)$ Ratio Evolved[a]

Temperature (°C)	Total incubation time (days)	Gaseous N (% of initial NO_3^--N in system)[b]	$N_2O/(N_2O + N_2)$ (%)
7	16	11	44
15	16	12	49
25	16	44	19
40	4	63	69
50	4	125	0
60	4	134	0
65	4	127	0
67	4	143	0
70	4	109	87
75	4	0	—

[a] Data from Keeney et al. (1979). Reproduced from Soil Sci. Soc. Am. J. **43**, p. 1126 by permission of the Soil Science Society of America.

[b] Initial NO_3^--N in the soil system = 122 μg N gm^{-1}.

1978). Bremner et al. (1980b) observed similar seasonal fluctuations in N_2O fluxes from unfertilized agricultural soils.

In general, increasing temperature tends to increase the proportion of N_2 to N_2O in the products of denitrification (Nommik, 1956; Bailey, 1976; Keeney et al., 1979) although in some cases temperature appears to have little effect (Bailey and Beauchamp, 1973).

7. Plants

Plant roots have several effects on the rhizosphere soil that may influence the potential for denitrification. First, roots release carbonaceous materials (organic substrate for denitrifiers) into the rhizosphere by excretion of soluble compounds, sloughing off root surface and root cap cells, and production of mucigel polysaccharide (Warembourg and Billies, 1979). Thus, large populations of denitrifiers frequently exist in the rhizosphere (Woldendorp, 1963), where they may be 10 to 100 times more numerous than in the root-free soil (Netti, 1955). The metabolism of the carbonaceous material by the rhizosphere microflora will tend to deplete the soil of O_2, as, indeed, will root respiration.

It is evident that if denitrification is limited by O_2 or C supply then the presence of plant roots will tend to stimulate denitrification. Many studies have confirmed that the presence of plant roots enhances the denitrification of added NO_3^- (Fig. 7) (Woldendrop, 1963; Stefanson, 1972a,b,c; Brar, 1972; Garcia, 1975; Bailey, 1976; Volz *et al.,* 1976) and sometimes causes a decrease in the mole fraction of N_2O produced (Stefanson, 1972a,b). However, Stefanson (1976) observed that while wheat roots stimulated denitrification in soils of low organic matter content they had no effect in soils high in organic matter.

The second major effect of roots is that they absorb NO_3^- and so deplete the soil of substrate for denitrification. Thus, if a supply of NO_3^- is limiting denitrification then plants tend to decrease the rate of denitrification (Guenzi *et al.,* 1978; Buresh *et al.,* 1981). Smith and Tiedje (1979b) confirmed that when soil NO_3^- concentrations are high denitrification rates are increased in the rhizosphere, whereas when NO_3^- concentrations are low denitrification rates are decreased in the presence of roots.

8. Animals

In grassland ecosystems the urine patches formed by grazing animals are recognized as focal points for the loss of N via NH_3 volatilization

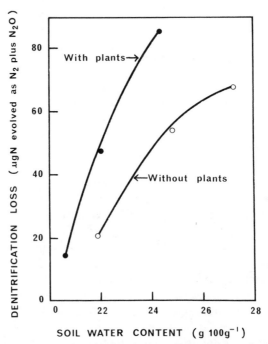

Fig. 7. Effect of plant growth and soil moisture content on the total amount of N (N_2 + N_2O) evolved from a soil. [Redrawn from Stefanson (1972b).]

(Catchpoole *et al.*, 1983; Sherlock and Goh, 1984) and leaching (Ball *et al.*, 1979). However, deposition of urine also results in an immediate release of N_2O, which does not occur from additions of urea of equivalent N content (Sherlock and Goh, 1983). The reason for this rapid production and release of N_2O is unknown although several mechanisms have been suggested (Sherlock and Goh, 1983), including stimulation of denitrification due to rapid onset of anaerobiosis caused by concomitant inputs of readily available C and rapid urea hydrolysis.

9. Tillage Method

In comparison with conventional tillage, the lack of soil disturbance and the presence of a surface mulch under zero tillage result in increased bulk density, a reduction in large pores, reduced aeration, larger but less aerobic aggregates, and a generally higher moisture content in the surface soil (Dowdell *et al.*, 1979a; Lin and Doran, 1984). Levels of water-soluble C can also be higher in surface soils under zero tillage (Lin and Doran, 1984). Such soil conditions obviously tend to favor the activity of denitrifiers and denitrifier populations are generally greater in the surface soil under zero rather than conventional tillage (Aulakh *et al.*, 1984a; Lin and Doran, 1984; Broder *et al.*, 1984). Studies have also shown that losses of N_2O (Burford *et al.*, 1981) or N_2O plus N_2 (Rice and Smith, 1982; Aulakh *et al.*, 1984a,b) are greater from zero tilled rather than conventionally tilled fields although the ratio of $N_2O : N_2$ emitted is not changed measurably. Aulakh *et al.* (1984a) estimated N_2O plus N_2 losses from cropped, conventionally tilled and zero tilled fields as 3–7 and 12–16 kg N ha^{-1} yr^{-1} respectively.

The generally lower rate of mineralization, and therefore nitrification, and the smaller populations of nitrifiers under zero tillage (Rice and Smith, 1983; Broder *et al.*, 1984) suggest that greater losses of N_2O under zero as compared to conventional tillage are the result of greater denitrification rather than nitrification.

IV. CHEMODENITRIFICATION

Chemodenitrification is the term commonly used to describe various chemical reactions of NO_2^- ions within soils that result in the emission of a variety of nitrogenous gases (e.g., N_2, NO, NO_2, and sometimes N_2O). Such gases are of nonbiological origin since they are also evolved from sterilized soil to which NO_2^- has been added. Normally, a higher proportion of added NO_2^--N is converted to (NO plus NO_2)-N than to N_2 (Nelson and Bremner, 1970b; Bollag *et al.*, 1973). Also the [(NO + NO_2)-N : N_2]

ratio usually varies from 2 : 1 at soil pH values of 5.0 to 5.8 to about 1 : 1 at pH values above 5.8 (Nelson, 1982). Small amounts of N_2O are also sometimes evolved from soils following treatment with NO_2^- (Reuss and Smith, 1965; Smith and Clark, 1980a,b).

A. Nitrite Accumulation

Chemodenitrification is likely to be significant only where NO_2^- accumulates in soils. The factors that favor the accumulation of NO_2^- in soils and thus chemodenitrification have been reviewed in detail elsewhere (Nelson, 1982; Chalk and Smith, 1983) and are outlined below.

In well aerated, unfertilized soils autotrophic oxidation of NO_2^- to NO_3^- proceeds at a faster rate than the conversion of NH_4^+ to NO_2^- (see Chapter 3). Consequently, NO_2^- is not normally present in amounts greater than 1 μg gm^{-1}. High concentrations may, however, accumulate when N fertilizers that form alkaline solutions upon hydrolysis are band-applied to soils. Urea, ammonium carbonate, diammonium phosphate, urea ammonium phosphate, and anhydrous ammonia all hydrolyze to produce an alkaline environment. In the fertilizer band of such materials soil pH may reach 10 and the N concentration may be several thousand μg N gm^{-1} (Parr and Papendick, 1966; Chalk et al., 1975). The activity of the NO_2^- oxidizer *Nitrobacter* is more greatly inhibited by high pH and high NH_4^+ levels than is that of the NH_4^+ oxidizers. Thus, in fertilizer bands, NO_2^- can accumulate up to several hundred μg N gm^{-1} (Chalk et al., 1975).

Nitrite has been shown to accumulate during nitrification of urea, NH_4^+ salts, and anhydrous and aqua ammonia (Hauck and Stephenson, 1965; Wetselaar et al., 1972; Pang et al., 1973, 1975; Chalk et al., 1975). Nitrite is particularly reactive under acidic conditions, which may occur around the periphery of fertilizer granules or bands where nitrification of the fertilizer is complete. Nitrite that accumulates in the alkaline fertilizer granule or band may, therefore, be unstable in the peripheral acid portion of the soil environment, leading to gaseous loss of N (Hauck and Stephenson, 1965). Nitrite can also accumulate in alkaline soils treated with acid-hydrolyzing NH_4^+ fertilizers such as $(NH_4)_2SO_4$ (Bezdicek et al., 1971) and in urine patches on grazed pastures (Vallis et al., 1982) where large amounts of urea N are deposited over a small surface area. Accumulation of NO_2^- has also been noted in NO_3^--treated soils during biological denitrification (Cady and Bartholomew, 1960; Doner et al., 1975; Cooper and Smith, 1963; Bailey, 1976; Volz and Starr, 1977). During the freezing of soils, NO_2^- levels can be concentrated in the unfrozen water resulting in a greater potential for chemodenitrification (Christianson and Cho, 1983).

In laboratory studies, a number of workers have observed large deficits during nitrification of urea or NH_4^+ fertilizers in soils and have attributed such losses to chemical reactions of NO_2^- (Hauck and Stephenson, 1965; Steen and Stojanovic, 1971). It is, however, noted that gaseous losses of N as N_2O, and apparently NO, can occur during autotrophic nitrification of NH_4^+-N (see Section III).

B. Mechanisms

A variety of reactions have been proposed to account for gaseous losses of N by chemodenitrification. The major reactions are discussed below.

1. Decomposition of Nitrous Acid

Nitrous acid is produced when NO_2^- is added to, or formed in, acid soils:

$$NO_2^- + H^+ \longrightarrow HNO_2 \tag{13}$$

At pH 5, 4, and 3, the proportions of the nitrite N present as undissociated nitrous acid are 1.9, 16, and 74%, respectively (Chalk and Smith, 1983). Nitrous acid can undergo spontaneous decomposition as shown below:

$$2HNO_2 \longrightarrow NO + NO_2 + H_2O \tag{14}$$

In closed incubation vessels, which are often used to study these reactions in the laboratory, the products actually obtained depend on a number of additional factors. In an aerobic system, NO is usually oxidized to NO_2 and both gases may then be absorbed by the moist soil as HNO_3 (Nelson, 1982). The overall reaction then becomes

$$2HNO_2 + O_2 \longrightarrow 2HNO_3 \tag{15}$$

In an anaerobic system, NO_2 is normally adsorbed as before but NO appears in the atmosphere. Under these conditions, the overall equation is

$$3HNO_2 \longrightarrow 2NO + HNO_3 + H_2O \tag{16}$$

The proportion of added NO_2^- evolved as (NO plus NO_2)-N is not related to soil organic matter content but increases with decreasing pH because of the large proportion of HNO_2 present at low pH (Bremner and Nelson, 1969; Nelson and Bremner, 1970b; Bollag et al., 1973). Significant amounts of NO and NO_2 are produced when NO_2^- is added to soils of pH greater than 5.5 (Porter, 1969; Nelson and Bremner, 1970a). This may indicate that self-decomposition of HNO_2 occurs at colloid surfaces, where the pH is considerably lower than that of the measured bulk soil pH (Nelson and Bremner, 1970a).

Under field conditions, the extent to which any of these decomposition reactions takes place is not well documented. Several workers have questioned the importance of HNO_2 decomposition in relation to gaseous losses of N from soils (e.g., Broadbent and Clark, 1965; Allison, 1966; Broadbent and Stevenson, 1966). Despite this, a number of workers have recorded emissions of (NO plus NO_2)-N from untreated soils and from soils treated with urea and NH_4^+ fertilizers (Steen and Stojanovic, 1971; Kim, 1973; Galbally and Roy, 1978; Smith and Chalk, 1980a; Johansson and Granat, 1984).

2. Reactions of Nitrous Acid with Organic Matter

Several researchers have shown a positive relationship between soil organic matter content and the rate of NO_2^- decomposition, with the concomitant emission of N_2 and N_2O, when NO_2^- is added to soils (e.g., Reuss and Smith, 1965; Nelson and Bremner, 1970b). It is the phenolic constituents of soil organic matter that are largely, if not entirely, responsible for such formation of N_2 and N_2O (Bremner and Nelson, 1969; Stevenson et al., 1970).

The mechanisms involved in the formation of gaseous products by reaction of HNO_2 with phenolic constituents are only partially understood. The reactions are known as nitrosation reactions and involve the addition of the nitroso group (—N=O) to an organic molecule by reaction with nitrous acid. Two possible mechanisms thought to be responsible for the reactions of HNO_2 with phenols are shown in Fig. 8.

Nitrosation reactions also result in fixation of NO_2^- by soil organic matter through the formation of nitroso groups on phenolic rings (Bremner, 1957; Bremner and Fuhr, 1966; Smith and Chalk, 1980b). Bremner and Fuhr (1966) found that when NO_2^- was added to soils with pH values ranging from 3 to 7, part of the N was fixed by organic matter (10–28%) and part was converted to gaseous forms (33–79%). The NO_2^- that is fixed to organic matter is resistant to biological decomposition (mineralization) (Bremner and Fuhr, 1966; Smith and Chalk, 1979).

As well as N_2 and N_2O, NO and nitromethane (CH_3ONO) have been detected as reaction products of the reactions of NO_2^- with organic matter (Stevenson and Swaby, 1964; Edwards and Bremner, 1966; Stevenson et al., 1970; Steen and Stojanovic, 1971). Several reaction mechanisms have been suggested to explain such emissions (see Chalk and Smith, 1983).

3. Reactions of Nitrous Acid with Compounds Containing Free Amino Groups

The reaction between HNO_2 and compounds containing free amino groups (e.g., amino acids, urea, and amines) has long been suggested as a

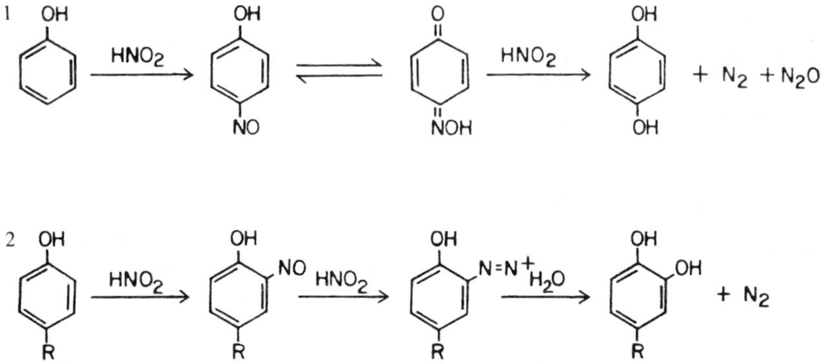

Fig. 8. Two possible reactions of phenols with nitrous acid. Mechanism (1) involves formation of p-nitrosophenol, tautomerization of this product to a quinone monoxime, and formation of N_2 and N_2O by reaction of the oxime with HNO_2. Mechanism (2) involves formation of an o-nitrosophenol and production of N_2 through decomposition of the diazo group in the diazonium compound formed by reaction of this o-nitrosophenol with HNO_2. [After Nelson (1982). Reproduced from "Nitrogen in Agricultural Soils" (F. J. Stevenson, ed.) Agronomy Mono. **22**, p. 353 by permission of the American Society of Agronomy. Madison, Wisconsin.]

possible mechanism for gaseous N loss from soil. This "Van Slyke" reaction only takes place at low pH and the N_2 gas evolved is derived in equal quantities from the two reactants:

$$R\text{-}NH_2 + HNO_2 \longrightarrow R\text{-}OH + H_2O + N_2 \tag{17}$$

Although this reaction is generally considered to be of limited importance as a mechanism for gaseous losses of N from soils (Nelson, 1982), several workers have suggested it as responsible for at least part of the N_2 produced when NO_2^- is added to acid soils (Reuss and Smith, 1965; Stevenson et al., 1970; Smith and Chalk, 1980b). Smith and Chalk (1980b), for instance, found that the ^{15}N enrichment of evolved N_2 from sterilized soils was approximately one-half that of the ^{15}N enrichment of the added NO_2^--N. Christianson et al. (1979) observed similar results. Such results suggest that Van Slyke-type reactions were involved since nitrosation reactions alone would result in no isotopic dilution of the evolved N_2.

4. Reaction of Nitrite with Ammonium

Solid ammonium nitrite (NH_4NO_2) explodes on heating to 60 to 70°C to produce N_2 gas (Weast, 1977). The same reaction proceeds much more slowly from concentrated solutions of NH_4NO_2 at low pH (pH < 5.2) (Smith and Clark, 1960). Since applications of NH_4^+ or NH_4^+-forming fertilizers can result in the accumulation of both NH_4^+ and NO_2^- in soils, some workers have suggested that significant gaseous loss of N might occur by chemical decomposition of NH_4NO_2 (e.g., Allison, 1963; Ewing and Bauet, 1966):

$$NH_4^+ + NO_2^- \longrightarrow NH_4NO_2 \longrightarrow N_2 + H_2O \qquad (18)$$

In general, NH_4NO_2 decomposition does not occur during incubation or air drying of acidic soils containing NH_4^+ and NO_2^- but some decomposition of NH_4NO_2 can occur when light-textured, neutral, and alkaline soils treated with NH_4^+ and NO_2^- are air dried (Wahhab and Uddin, 1954; Bremner and Nelson, 1969; Jones and Hedlin, 1970).

Thus, it is thought that the reaction is not of general significance in regard to gaseous losses of N from soils (Nelson, 1982) except perhaps when neutral or alkaline soils containing high concentrations of NH_4^+ and NO_2^- are subjected to drying conditions.

5. Reaction of Nitrous Acid with Hydroxylamine

A number of workers (e.g., Arnold, 1954; Wijler and Delwiche, 1954; Vine, 1962) have speculated that the chemical reaction of hydroxylamine (NH_2OH) with HNO_2 might generate N_2O:

$$NH_2OH + HNO_2 \longrightarrow N_2O + 2H_2O \qquad (19)$$

Although it has been shown that NH_2OH can be quantitatively decomposed by HNO_2 (Nelson, 1978), Bremner et al. (1980a) found that when NH_2OH was added to soils, large amounts of N_2O were formed in the absence of HNO_2. It was suggested by Bremner et al. (1980a) that N_2O is formed in soils through other nonbiological transformations of NH_2OH and very little is generated by the reaction of NH_2OH with HNO_2. Such nonbiological transformations were postulated to involve oxidized forms of Mn and Fe (e.g., MnO_2 and Fe_2O_3).

The fact that NH_2OH has not been detected in soils makes the significance of the above reactions questionable (Nelson, 1982). Hydroxylamine is, however, a postulated intermediate of both the biological reduction of NO_3^- to NH_4^+ (Alexander, 1977; Yordy and Ruoff, 1981) and the oxidation of NH_4^+ to NO_3^- (see Chapter 3). The fact that it is not present in soils may be a consequence of its rapid decomposition.

6. Other Reactions

Several other reactions have been suggested as possible pathways of chemodenitrification including reactions of HNO_2 with clay minerals and transition metal cations. Such reactions seem unlikely to be significant sources of gaseous N loss from soils although Nelson (1982) suggested that Fe^{2+} may promote decomposition of NO_2^- formed by microbial reduction of NO_3^- in waterlogged soils. This suggestion deserves further research since significant amounts of Fe^{2+} are often present in anaerobic soils. Indeed, Moraghan and Buresh (1977) showed that chemical decomposition of NO_2^-, with the evolution of N_2 and N_2O, occurred anaerobically in a high Fe^{2+} ion environment while Van Cleemput and Baert (1984) found that soil conditions promoting the formation of Fe^{2+} also promoted NO_2^- decomposition and NO emissions.

V. EXTENT, SIGNIFICANCE, AND FATE OF LOSSES

Global estimates of gaseous losses of N from the plant–soil system were presented in Chapter 1. In this section some recent field measurements of gaseous losses of N are reported and the fate and significance of such emissions are outlined.

A. Ammonia Volatilization

1. Extent of Losses

The amounts of NH_3 volatilized from applications of NH_4^+- or NH_4^+-yielding fertilizers are extremely variable (Terman, 1979) and depend on such factors as type, rate, and method of fertilizer application, soil pH, and environmental factors such as temperature and moisture. Losses of fertilizer N applied to the surface of grassland or bare soil often appear to be in the range of 0 to 25% (Hargrove and Kissel, 1979; Hoff *et al.*, 1981; Craig and Wollum, 1982; Catchpoole *et al.*, 1983; Black *et al.*, 1985b). Significant, but variable, losses of NH_3 can occur following fertilizer applications to flooded soils (Vlek and Craswell, 1981) or when fertilizer is applied in the irrigation water (Denmead *et al.*, 1982).

Volatilization of NH_3 from organic amendments (e.g., animal manures or sewage sludge) applied to soils can be large but also variable (Hoff *et al.*, 1981; Beauchamp *et al.*, 1982; Beauchamp, 1983). Losses are often in the range of 10–60% of applied N. Losses from urine patches on grazed pastures can similarly be relatively high, ranging from 10 to 60% of applied urea N (Harper *et al.*, 1983; Simpson and Steele, 1983; Sherlock and Goh, 1984).

2. Fate of Ammonia Emissions

Ammonia is present in the atmosphere as a gas and in the form of ammonium in water droplets and solid particles. Concentrations of NH_3

and NH_4^+ vary widely in the atmosphere due to inhomogeneity of the sources and sinks (e.g., biological sources and precipitation scavenging). The major removal mechanisms for atmospheric ammonia are by wet and dry deposition. These processes are very rapid and the mean atmospheric lifetime of NH_3 is in the region of 7 to 14 days (Hahn and Crutzen, 1982; Galbally and Roy, 1983).

Gaseous deposition of NH_3 is an important mechanism for the return of volatilized NH_3 to the biosphere since plants, land surfaces, lakes, and oceans can all act as sinks as well as sources for atmospheric NH_3 (Calder, 1972; Dawson, 1977; Georgii and Gravenhorst, 1977; Farquhar *et al.*, 1983). Ammonia volatilized at a particular site can therefore be reabsorbed by direct gaseous uptake nearby. However, the fraction of NH_3 that is converted to an atmospheric aerosol can travel to more distant locations.

Ammonia is extremely soluble in water and once dissolved it ionizes to NH_4^+:

$$NH_3 + H_2O \rightleftharpoons NH_4^+ + OH^- \tag{20}$$

Thus, in the troposphere, NH_3 quickly dissolves in water droplets in clouds with the formation of NH_4^+-containing aerosols. Atmospheric aerosols of H_2SO_4 (often from industrial sources) can be quickly ammoniated to NH_4HSO_4 and $(NH_4)_2SO_4$ under most tropospheric conditions (Huntzicker *et al.*, 1980). Much of the emitted NH_3 is therefore present in the atmosphere as aerosols in the form of NH_4^+ salts (Taylor *et al.*, 1983) such as NH_4NO_3 or $(NH_4)_2SO_4$. It is removed from the atmosphere predominantly by wet deposition. Ammonium found in rainwater can originate from sources hundreds or thousands of kilometers away (Lenhard and Gravenhorst, 1980). Upon evaporation of water, aerosol particles may also be returned by dry deposition.

B. Denitrification and Nitrification

1. Extent of Losses

Few studies have measured directly total losses of applied ^{15}N ($^{15}N_2$ plus $^{15}N_2O$) through denitrification (and nitrification) occurring under field conditions (Rolston, 1978; Rolston and Broadbent, 1977; Rolston *et al.*, 1978, 1982). Rolston *et al.* (1978) measured N_2 plus N_2O losses from plots at two temperatures, two water contents, cropped and uncropped, and manured or unmanured. Losses ranged from zero (for the uncropped driest treatments) to about 75% (for the wettest treatments with manure added). Rolston and Broadbent (1977) measured N_2 plus N_2O losses over an entire growing season and calculated a loss of about 13 kg N ha^{-1}, which represented approximately 9% of applied fertilizer N.

Several workers (e.g., Ryden *et al.,* 1979; Ryden, 1981; Colbourn *et al.,* 1984) have indirectly measured total denitrification losses by measuring N_2O fluxes in the presence of acetylene (which inhibits further reduction of N_2O to N_2 in the soil). The injection of acetylene into the soil also inhibits nitrification (Hynes and Knowles, 1978; Walter *et al.,* 1979) and thus N_2O emissions from that source. Ryden (1981) estimated denitrification losses of 11 and 29 kg N ha^{-1} yr^{-1} from grassed plots receiving 250 and 500 kg of fertilizer N ha^{-1}, respectively. Other workers (Colbourn and Dowdell, 1984; Colbourn *et al.,* 1984) estimated losses of 18–38 kg N ha^{-1} yr^{-1} from a grassland receiving 210 kg N ha^{-1} and 7–13 kg N ha^{-1} yr^{-1} from winter wheat receiving a fertilizer addition of 70 kg N ha^{-1}.

From the small amount of data available it is evident that gaseous losses can vary considerably. Colbourn and Dowdell (1984), however, generalized that direct and indirect estimates of losses of N_2 plus N_2O from soils range from 0 to 20% of fertilizer N applied to arable soils and 0 to 7% on grassland soils.

While the mole ratio of N_2O produced during denitrification is exceedingly variable (Rolston *et al.,* 1978, 1982; Ryden *et al.,* 1979; Rolston, 1981; Aulakh *et al.,* 1984a), in general, the quantity of N_2 produced during denitrification is much greater than that of N_2O. For example, field experiments have yielded time-averaged N_2O mole fractions of 0.12–0.18 (Ryden *et al.,* 1979) and 0.20–0.30 (Rolston *et al.,* 1982).

The extent of losses of N_2O through the nitrification pathway is not known although field studies have confirmed that significant amounts of N_2O are emitted during nitrification of applied NH_4^+ fertilizers (Hutchinson and Mosier, 1979; Breitenbeck *et al.,* 1980; Bremner *et al.,* 1981; Cochran *et al.,* 1981; Mosier and Hutchinson, 1981; Mosier *et al.,* 1981). Losses of N_2O following applications of urea or NH_4^+ fertilizers have ranged from 0.2 to 0.6% of applied N (Breitenbeck *et al.,* 1980; Mosier *et al.,* 1981) while losses following injection of anhydrous NH_3 have ranged from less than 0.1% (Cochran *et al.,* 1981) to 4.0 to 6.8% of applied N (Bremner *et al.,* 1981). Although losses of N_2O in the above experiments may have originated predominantly from nitrification, emissions caused by denitrification in anoxic microsites cannot be ruled out.

Although N_2O is usually assumed to be lost from soil only to the atmosphere, during winter it can also leave the soil dissolved in drainage water. Measured losses range from 0.25 to 4.4 kg N ha^{-1} from agricultural soils (Dowdell *et al.,* 1979b; Harris *et al.,* 1984), which were comparable with gaseous losses of N_2O over the same period.

2. Nitrous Oxide and Stratospheric Ozone

Much research on denitrification, and more recently nitrification, has been prompted by a concern that N_2O released into the atmosphere by these processes may increase the rate of reactions in the stratosphere that lead to the destruction of the ozone (O_3) layer (Crutzen and Ehhalt, 1977; McElroy *et al.*, 1977; National Research Council, 1978). The stratospheric ozone layer shields the biosphere from harmful UV radiation and also influences the vertical temperature profile and thus earth surface temperatures (Ramanathan *et al.*, 1976; Wang *et al.*, 1976; Hahn, 1979).

Atmospheric photochemistry in relation to the role of nitrogen oxides has been reviewed in detail elsewhere (Crutzen, 1981, 1983; Hahn and Crutzen, 1982). The low solubility of N_2O in water means that there is no significant removal of atmospheric N_2O from the troposphere by precipitation and it penetrates, almost unimpeded, into the stratosphere. Atmospheric destruction of N_2O occurs through photochemical reactions in the stratosphere:

$$N_2O + h\nu \longrightarrow N_2 + O \tag{21}$$

$$N_2O + O('D) \longrightarrow N_2 + O_2 \tag{22}$$

$$N_2O + O('D) \longrightarrow 2NO \tag{23}$$

The electronically excited $O('D)$ atom is produced by photolysis of ozone in the stratosphere. Approximately 10% of stratospheric N_2O is thought to be converted to NO by reaction (23). Direct transport of NO_x into the stratosphere from the earth's surface is unlikely because of the short atmospheric residence time of NO_x, which is quickly converted to HNO_3 aerosols and thermally unstable organic nitrates (e.g., peroxyacetyl nitrates—PAN) and removed by wet and dry deposition (Crutzen, 1981).

One of the major sinks of O_3 is reaction with NO_x, which catalyzes the destruction of O_3 above 25 km in the stratosphere (Crutzen, 1981, 1983). However, below 25 km, NO_x protects O_3 from destruction (Logan *et al.*, 1978; Zahniser and Howard, 1979). Thus, the major effect of increased production of NO_x in the stratosphere is likely to be a lowering of the center of gravity of the stratospheric ozone by a transfer of mass to altitudes below about 25 km (Crutzen, 1981). Nonetheless, Crutzen (1983) calculated that increasing N_2O emissions will tend to enhance O_3 loss through net catalysis of its destruction in the entire stratosphere; a doubling of atmospheric N_2O abundance might yield a 12% decrease in total O_3.

It is interesting to note that, overall, the global source of N_2O from N fertilizer applications is probably smaller than a few Tg N yr^{-1} (Crutzen,

1983). Thus, the impact of the increasing use of N fertilizers on stratospheric ozone is unlikely to be great. Weiss (1981), for example, observed that the global upward trend in atmospheric N_2O concentrations is about 0.2% per year, which can be explained in terms of the 3.5% per year increase in global N_2O emissions caused by fossil fuel combustion.

3. Fate of N_2 and N_2O

The quantities of N_2 evolved from the earth's surface during denitrification are extremely small in relation to the atmospheric content. The dinitrogen molecule is very stable and its atmospheric lifetime is of the order of millions of years. N_2 constitutes 79% of the atmospheric mass.

Upon release of N_2O from soils, an unknown portion is believed to be removed by gaseous deposition to vegetation, soil, and water (Rasmussen *et al.*, 1975). However, under conditions of high soil NO_3^- concentrations, when potentially high rates of denitrification may occur, the reduction of N_2O to N_2 is usually inhibited (Firestone *et al.*, 1980) (Table V) and it seems unlikely that the soil will act as a major sink for N_2O (Freney *et al.*, 1978).

The approximate stratospheric lifetime of N_2O is 100 to 150 yr (Galbally and Roy, 1983). The only known photochemical reactions that lead to removal of N_2O from the stratosphere were presented earlier (equations (21), (22), and (23)). Thus, N_2O in the stratosphere is converted to N_2 and NO. More than 90% of the N_2O is thought to be transformed to N_2 in the stratosphere and the remainder is converted to NO (Nicolet and Peetermans, 1972). As discussed in the next section, the NO produced can be

Table V

Effect of Nitrate N Concentration on the Denitrification Rate and the Proportion of Gas Evolved as N_2 and N_2O[a]

Concentration of added NO_3^--N ($\mu g\ gm^{-1}$)	Denitrification rate ($\mu g\ N\ gm^{-1}\ hr^{-1}$)	Percentage of total ^{15}N gas evolved	
		N_2	N_2O
0	—	95.2	4.8
0.5	0.54	93.9	6.1
2.0	0.73	89.8	10.2
20.0	1.15	85.4	14.6

[a] Data from Firestone *et al.* (1979). Reproduced from *Soil Sci. Soc. Am. J.* **43,** p. 1143 by permission of the Soil Science Society of America.

rapidly transformed to NO_2 and thence to HNO_3. These substances are eventually returned to the troposphere and then to the earth's surface by wet and dry deposition.

C. Chemodenitrification

1. Extent of Losses

High rates of chemodenitrification are likely to occur principally as a result of NO_2^- accumulation during nitrification of banded NH_4^+ or NH_4^+-yielding fertilizers, which form alkaline solutions following hydrolysis. Field studies are, however, required to quantify such losses of N_2, N_2O, and NO_x and identify the factors affecting such losses. Emission of N_2O and possibly NO during the nitrification process would tend to confound such results.

Few attempts have been made to directly measure losses of NO_x from soils. Galbally and Roy (1978) measured losses of NO from soils in the range 0.06 to 5×10^{-11} kg N m^{-2} sec^{-1}, which convert to a global source of NO over land of about 1 kg N ha^{-1} yr^{-1} (Galbally and Roy, 1983). Johansson and Granat (1984) measured annual emissions of NO from unfertilized and fertilized arable land of about 0.2 and 0.6 kg N ha^{-1}, respectively.

2. Fate of NO_x Emissions

As discussed in Chapter 1, the two major sources of atmospheric NO_x are emissions from soils through chemodenitrification and combustion sources (automobiles, furnaces, forest fires, etc.). The quantities produced by the two sources are thought to be comparable on a global scale.

Nitric oxide released from the soil surface is quickly converted to NO_2 by O_3 in the lower atmosphere:

$$NO + O_3 \longrightarrow NO_2 + O_2 \tag{24}$$

This NO_2 can be absorbed by the local plant and soil surface (Galbally, 1974; Rogers et al., 1979; Elkiey and Ormrod, 1981; Galbally and Roy, 1978) or transported long distances in the atmosphere (Galbally and Roy, 1983). Nitric oxide can also be absorbed by plants but much more slowly than NO_2 (Galbally and Roy, 1983).

The major sink identified for NO_x in the troposphere is the dissolution of soluble species in cloud and rain droplets with subsequent removal by precipitation. In the stratosphere NO_2 reacts with OH radicals to yield HNO_3.

Under atmospheric conditions, the conversion of NO_x to HNO_3 occurs within a few days and the mean atmospheric lifetime of NO_x is thought to be about 1.5 days (Hahn and Crutzen, 1982). Aerosol NO_3^- can readily form since gaseous HNO_3 is attached to, or dissolved in, preexisting aerosol particles (e.g., H_2SO_4, $(NH_4)_2SO_4$, or NH_4HSO_4) through heterogeneous condensation (Taylor *et al.*, 1983). Indeed, SO_4^{2-} and NO_3^- combined with NH_4^+ are usually dominant inorganic species found in atmospheric aerosols (Stevens *et al.*, 1978; Scott and Laulainen, 1979; Huebert and Lazrus, 1980).

The NO_3^--containing aerosols are then removed principally by washout and rainout processes (Fowler, 1978). If vaporization of droplets occurs then the NO_3^- salts may be removed by dry deposition.

VI. CONCLUSIONS

There are several mechanisms that lead to gaseous losses of N from the plant–soil system. These include ammonia volatilization, biological denitrification, nitrification, and chemodenitrification.

Volatilization of NH_3 to the atmosphere is a complex process affected by a combination of physical, chemical, and biological factors. A necessary prerequisite for NH_3 volatilization is a supply of free ammonia (i.e., $NH_{3(aq)}$ and $NH_{3(g)}$) near the soil surface. The conversion of NH_4^+ ions to NH_3 is thus a major process regulating the potential loss of NH_3 from soils. Sources of NH_4^+ in the soil include native soil organic matter, plant residues, animal excretions, added organic materials, or added NH_4^+-containing or -yielding fertilizers. The equilibrium between NH_4^+ and NH_3 is affected by pH, temperature, water loss from the soil, buffering capacity and CEC of the soil, and fixation of NH_3 and NH_4^+ by clay minerals or organic matter.

Most of the NH_3 emitted to the atmosphere from the soil surface is quickly returned to the earth's surface via wet and dry deposition. The lifetime of NH_3 in the atmosphere is only 1 to 2 weeks. Ammonia can be returned to the biosphere via gaseous deposition since plants, land surfaces, and water bodies can all act as sinks for atmospheric NH_3. In the troposphere, the emitted NH_3 quickly dissolves in water droplets in clouds with the formation of NH_4^+ ions. The NH_4^+ is returned via wet deposition as NH_4^+ salts dissolved in rainwater, or the water in the aerosol may evaporate and particles are returned via dry deposition.

Growing plants can act as either sources or sinks for atmospheric NH_3. Absorption of NH_3, emitted from the soil surface, by the leaf canopy above can greatly reduce losses of NH_3 from the plant–soil system.

There are several pathways for the biological generation of gaseous nitrogenous products. Dissimilatory denitrification is carried out by a limited number of aerobic bacteria that can grow in the absence of molecular oxygen while reducing NO_3^- or NO_2^- to gaseous products (N_2O and N_2). These bacteria are biochemically and taxonomically diverse although most are chemoheterotrophs and use carbonaceous compounds as electron donors and sources of cellular C and chemical compounds as energy sources. The denitrification process is promoted by anaerobic conditions, high levels of soil NO_3^-, and a readily available source of carbon and, in general, is positively related to soil pH and temperature. The quantity of N_2O emitted during denitrification is normally considerably less than that of N_2 although the mole ratio of N_2O produced is influenced by many factors and can vary from 0 to 1.0. The ratio is generally raised under conditions of high NO_3^- levels and low pH and lowered by high temperatures and increasing anoxia.

There is a group of "nondenitrifying" bacteria and fungi that is able to respire NO_3^- anaerobically as far as NO_2^- and when growing fermentatively they can further reduce NO_2^- to NH_4^+ with N_2O being produced as a minor product. The magnitude of the contribution that these organisms make to N_2O evolution from soils is unclear although it is generally thought to be small.

The autotrophic NH_4^+-oxidizing bacteria *Nitrosomonas, Nitrosospira,* and *Nitrosolobus* have the capacity to produce N_2O, and apparently NO, during the oxidation of NH_4^+ to NO_2^-. The exact mechanisms through which these gases are produced are unknown. The potential for losses of N_2O through nitrification is greatest when NH_4^+-containing or -yielding fertilizers are applied to aerobic soils.

The atmospheric lifetime of N_2 is millions of years and that for N_2O is about 150 yr. A major sink of N_2O from the troposphere is diffusion into the stratosphere, where the major part forms N_2 and a small fraction forms NO_x. The nitrogen oxides (NO_x) catalyze destruction of ozone (O_3) in the upper stratosphere but protect it from destruction in the lower stratosphere. Hence, the net result of increased emissions of N_2O from the earth's surface is likely to be a transfer of O_3 mass to lower altitudes rather than destruction of the ozone layer. The NO_x formed in the stratosphere is returned to the lower atmosphere at a low rate and then to the earth's surface through wet and dry deposition.

Chemodenitrification is a term that encompasses the processes responsible for gaseous loss of N from soils through chemical reactions of NO_2^-. Accumulation of NO_2^- does not normally occur except when nitrogenous fertilizers that form alkaline solutions upon hydrolysis (urea, NH_4^+ salts, and anhydrous and aqua NH_3) are band-applied to soils or in NO_3^--treated soils as an intermediate of denitrification. A variety of gases can be evolved from soils treated with NO_2^- including N_2, N_2O, NO, and NO_2.

Several mechanisms are thought to be involved in chemodenitrification. These include decomposition of nitrous acid with the emission of NO and NO_2, reactions of HNO_2 with phenolic constituents of soil organic matter with the formation of N_2, N_2O, and NO, reactions of HNO_2 with compounds containing free amino groups to liberate N_2, reactions of NO_2^- with NH_4^+ and hydroxylamine with the release of N_2 and N_2O, respectively, and reactions of HNO_2 with metallic cations to form NO and N_2. Although research has established that such reactions can occur, their significance and magnitude under field conditions have yet to be established.

Nitric oxide released from the soil surface can be quickly converted to NO_2 by O_3 in the lower atmosphere. Atmospheric NO_2 can be taken up by the local plant and soil surface or transported into the atmosphere. Nitrogen dioxide is highly soluble in water and reacts with OH radicals to yield HNO_3. The combined atmospheric lifetime of NO, NO_2, and HNO_3 is in the range of 1 to 2 weeks. Heterogeneous aerosol particles are formed in the atmosphere by interaction by gaseous HNO_3 with preformed aerosols and the nitrates are returned to the earth's surface by wet and dry deposition.

REFERENCES

Alexander, M. (1977). "Incubation to Soil Microbiology." Wiley, New York.

Allison, F. E. (1955). The enigma of soil nitrogen balance sheets. *Adv. Agron.* **7**, 213–250.

Allison, F. E. (1963). Losses of gaseous nitrogen from soils by chemical mechanisms involving nitrous acid and nitrites. *Soil Sci.* **96**, 404–409.

Allison, F. E. (1966). The fate of nitrogen applied to soils. *Adv. Agron.* **18**, 219–258.

Arnold, P. W. (1954). Losses of nitrous oxide from soil. *J. Soil Sci.* **5**, 116–128.

Aulakh, M. S., Rennie, D. A., and Paul, E. A. (1984a). Gaseous nitrogen losses from soils under zero-till as compared with conventional-till management systems. *J. Environ. Qual.* **13**, 130–136.

Aulakh, M. S., Rennie, D. A., and Paul, E. A. (1984b). The influence of plant residues on denitrification rates in conventional and zero tilled soils. *Soil Sci. Soc. Am. J.* **48**, 790–794.

Avnimelech, Y., and Laher, M. (1977). Ammonia volatilization from soils: Equilibrium considerations. *Soil Sci. Soc. Am. J.* **41**, 1080–1084.

Bailey, L. D. (1976). Effects of temperature and root on denitrification in a soil. *Can. J. Soil Sci.* **56**, 79–87.

Bailey, L. D., and Beauchamp, E. G. (1973). Effects of moisture, added NO_3^-, and macerated roots on NO_3^- transformation and redox potential in surface and subsurface soils. *Can. J. Soil Sci.* **53**, 219–230.

Baldensperger, J., and Garcia, J. L. (1975). Reduction of oxidised inorganic nitrogen compounds by a new strain of *Thiobacillus denitrificans*. *Arch. Microbiol.* **103,** 31–36.

Ball, P. R., and Keeney, D. R. (1983). Nitrogen losses from urine-affected areas of New Zealand pasture, under contrasting seasonal conditions. *Proc. Int. Grassl. Congr., 14th, 1981,* pp. 342–344.

Ball, P. R., Keeney, D. R., Theobald, P. W., and Nes, P. (1979). Nitrogen balance in urine-affected areas of a New Zealand pasture. *Agron. J.* **71,** 309–314.

Bartlett, M. S., Brown, L. C., Hanes, N. B., and Nickerson, N. H. (1979). Denitrification in freshwater wetland soil. *J. Environ. Qual.* **8,** 460–464.

Beauchamp, E. G. (1983). Nitrogen loss from sewage sludges and manures applied to agricultural lands. *In* "Gaseous Loss of Nitrogen from Plant–Soil Systems" (J. R. Freney and J. R. Simpson, eds.), pp. 181–194. Martinus Nijhoff/Dr. W. Junk, The Hague.

Beauchamp, E. G., Kidd, G. E., and Thurtell, G. (1978). Ammonia volatilization from sewage sludge applied in the field. *J. Environ. Qual.* **7,** 141–146.

Beauchamp, E. G., Kidd, G. E., and Thurtell, G. (1982). Ammonia volatilization from liquid dairy cattle manure in the field. *Can. J. Soil Sci.* **62,** 11–19.

Betlach, M. R., and Tiedje, J. M. (1981). Kinetic explanation for accumulation of nitrite, nitric oxide, and nitrous oxide during bacterial denitrification. *Appl. Environ. Microbiol.* **42,** 1074–1084.

Bezdicek, D. F., MacGregor, J. M., and Martin, W. P. (1971). The influence of soil–fertilizer geometry on nitrification and nitrite accumulation. *Soil Sci. Soc. Am. Proc.* **35,** 997–1000.

Black, A. S., and Sherlock, R. R. (1985). Ammonia loss from nitrogen fertiliser. *N.Z. Fert. J.* **68,** 12.

Black, A. S., Sherlock, R. R., Smith, N. P., Cameron, K. C., and Goh, K. M. (1984). Effect of previous urine application on ammonia volatilisation from 3 nitrogen fertilisers. *N.Z. J. Agric. Res.* **27,** 413–416.

Black, A. S., Sherlock, R. R., Cameron, K. C., Smith, N. P., and Goh, K. M. (1985a). Comparison of three field methods for measuring ammonia volatilization from urea granules broadcast on to pasture. *J. Soil Sci.* **36,** 271–280.

Black, A. S., Sherlock, R. R., and Smith, N. P. (1985b). Ammonia volatilisation from nitrogenous fertilisers broadcast onto pastures: Effects of application time and rate. *N.Z. J. Agric. Res.* **28,** 469–474.

Blackmer, A. M., and Bremner, J. M. (1978). Inhibitory effect of nitrate on reduction of N_2O to N_2 by soil microorganisms. *Soil Biol. Biochem.* **10,** 187–191.

Blackmer, A. M., Bremner, J. M., and Schmidt, E. L. (1980). Production of nitrous oxide by ammonia-oxidizing chemoautotrophic microorganisms in soil. *Appl. Environ. Microbiol.* **40,** 1060–1066.

Blackmer, A. M., Robbins, S. G., and Bremner, J. M. (1982). Diurnal variability in rate of emission of nitrous oxide from soils. *Soil Sci. Soc. Am. J.* **46,** 937–942.

Blasco, M. L., and Cornfield, A. H. (1966). Volatilization of nitrogen as ammonia from acid soils. *Nature (London)* **212,** 1279–1280.

Bollag, J. M., and Tung, G. (1972). Nitrous oxide release by soil fungi. *Soil Biol. Biochem.* **4,** 271–276.

Bollag, J. M., Drzymala, S., and Kardos, L. T. (1973). Biological versus chemical nitrite decomposition in soil. *Soil Sci.* **116,** 44–50.

Bouwmeester, R. J. B., and Vlek, P. L. G. (1981). Rate control of ammonia volatilization from rice paddies. *Atmos. Environ.* **15,** 130–140.

Bowman, R. A., and Focht, D. D. (1974). The influence of glucose and nitrate concentrations upon denitrification rates in sandy soil. *Soil Biol. Biochem.* **6**, 297–301.

Brar, S. S. (1972). Influence of roots on denitrification. *Plant Soil* **36**, 713–715.

Brar, S. S., Miller, R. H., and Logan, T. J. (1978). Some factors affecting denitrification in soils irrigated with wastewater. *J. Water Pollut. Control. Fed.* **50**, 709–717.

Breitenbeck, G. A., Blackmer, A. M., and Bremner, J. M. (1980). Effects of different nitrogen fertilizers on emission of nitrous oxide from soil. *Geophys. Res. Lett.* **7**, 85–88.

Bremner, J. M. (1957). Studies on soil humic acids. II. Observations on the estimation of free amino groups. Reactions of humic acid and lignin preparations with nitrous acid. *J. Agric. Sci.* **48**, 352–360.

Bremner, J. M., and Blackmer, A. M. (1978). Nitrous oxide: Emission from soils during nitrification of fertilizer nitrogen. *Science* **199**, 295–296.

Bremner, J. M., and Blackmer, A. M. (1980). Mechanisms of nitrous oxide production in soils. *In* "Biochemistry of Ancient and Modern Environments" (P. A. Trudinger, M. R. Walter, and R. J. Ralph, eds.), pp. 279–291. Aust. Acad. Sci., Canberra.

Bremner, J. M., and Blackmer, A. M. (1981). Terrestrial nitrification as a source of atmospheric nitrous oxide. *In* "Denitrification, Nitrification and Atmospheric Nitrous Oxide" (C. C. Delwiche, ed.), pp. 151–170. Wiley, New York.

Bremner, J. M., and Douglas, L. A. (1973). Effects of some urease inhibitors on urea hydrolysis in soils. *Soil Sci. Soc. Am. Proc.* **37**, 225–226.

Bremner, J. M., and Fuhr, F. (1966). Tracer studies of the reaction of soil organic matter with nitrite. *In* "The Use of Isotopes in Soil Organic Matter Studies," pp. 337–346. Pergamon, Oxford.

Bremner, J. M., and Nelson, D. W. (1969). Chemical decomposition of nitrite in soils. *Trans. Int. Congr. Soil Sci., 9th, 1968,* Vol. 2, pp. 495–503.

Bremner, J. M., and Shaw, K. (1958). Denitrification in soil. II. Factors affecting denitrification. *J. Agric. Sci.* **51**, 39–52.

Bremner, J. M., Blackmer, A. M., and Waring, S. A. (1980a). Formation of nitrous oxide and dinitrogen by chemical decomposition of hydroxlamine in soils. *Soil Biol. Biochem.* **12**, 263–269.

Bremner, J. M., Robbins, S. G., and Blackmer, A. M. (1980b). Seasonal variability in emission of nitrous oxide from soil. *Geophys. Res. Lett.* **7**, 641–644.

Bremner, J. M., Breitenbeck, G. A., and Blackmer, A. M. (1981). Effect of anhydrous ammonia fertilization on emission of nitrous oxide from soil. *J. Environ. Qual.* **10**, 77–80.

Broadbent, F. E., and Clark, F. E. (1965). Denitrification. *In* "Soil Nitrogen" (W. V. Bartholomew and F. E. Clark, eds.), pp, 344–359. Am. Soc. Agron., Madison, Wisconsin.

Broadbent, F. E., and Stevenson, F. J. (1966). Organic matter interactions. *In* "Agricultural Anhydrous Ammonia Technology and Use" (M. H. McVickar, W. P. Martin, I. E. Miles, and H. H. Tucker, eds.), pp. 169–187. Agric. Ammonia Inst., Memphis, Tennessee.

Broder, M. W., Doran, J. W., Peterson, G. A., and Fenster, C. R. (1984). Fallow tillage influence on spring populations of soil nitrifiers, denitrifiers, and available nitrogen. *Soil Sci. Soc. Am. J.* **48**, 1060–1067.

Bryan, B. A. (1980). Cell yield and energy characteristics of denitrification with *Pseudomonas stutzeri* and *Pseudomonas aeruginosa*. Ph.D. Thesis, University of California, Davis.

Buresh, R. J., De Laune, R. D., and Patrick, W. H. (1981). Influence of *Spartina alterniflora* on nitrogen loss from marsh soil. *Soil Sci. Soc. Am. J.* **45**, 660–661.

Burford, J. R., and Bremner, J. M. (1975). Relationships between the denitrification capacities of soils and total, water soluble and readily decomposable soil organic matter. *Soil Biol. Biochem.* **7**, 389–394.

Burford, J. R., Dowdell, R. J., and Crees, R. (1981). Emission of nitrous oxide to the atmosphere from direct-drilled and ploughed clay soils. *J. Sci. Food Agric.* **32**, 219–223.

Cady, F. B., and Bartholomew, W. V. (1960). Sequential products of anaerobic denitrification in Norfolk soil material. *Soil Sci. Soc. Am. Proc.* **24**, 477–482.

Calder, K. L. (1972). Absorption of ammonia from atmospheric plumes by natural water surfaces. *Water Air Soil Pollut.* **1**, 375–380.

Carran, R. A., Ball, P. R., Theobald, P. W., and Collins, M. E. G. (1982). Soil nitrogen balances in urine-affected areas under two moisture regimes in Southland. *N. Z. J. Exp. Agric.* **10**, 377–381.

Carrier, D., and Bernier, B. (1971). Loss of nitrogen by volatilization of ammonia after fertilizing in jack pine forest. *Can. J. For. Res.* **1**, 69–79.

Caskey, W. H., and Tiedje, J. M. (1979). Evidence for clostridia as agents of dissimilatory reduction of nitrate to ammonium in soils. *Soil Sci. Soc. Am. J.* **43**, 931–935.

Catchpoole, V. R., Harper, L. A., and Myers, R. J. K. (1983). Annual losses of ammonia from a grazed pasture fertilized with urea. *Proc. Int. Grassl. Congr., 14th, 1981,* pp. 344–347.

Chalk, P. M., and Smith, C. J. (1983). Chemodenitrification. *In* "Gaseous Loss of Nitrogen from Plant–Soil Systems" (J. R. Freney and J. R. Simpson, eds.), pp. 65–89. Martinus Nijhoff/Dr. W. Junk, The Hague.

Chalk, P. M., Keeney, D. R., and Walsh, L. M. (1975). Crop recovery and nitrification of fall and spring applied anhydrous ammonia. *Agron. J.* **67**, 33–37.

Chao, T. T., and Kroontje, W. (1964). Relationship between ammonia volatilization, ammonia concentration, and water evaporation. *Soil Sci. Soc. Am. Proc.* **28**, 393–395.

Cho, C. M., Sakdinan, L., and Chang, C. (1979). Denitrification intensity and capacity of three irrigated Alberta soils. *Soil Sci. Soc. Am. J.* **43**, 949–950.

Christianson, C. B., and Cho, C. M. (1983). Chemical denitrification of nitrite in frozen soils. *Soil Sci. Soc. Am. J.* **47**, 38–42.

Christianson, C. B., Hedlin, R. A., and Cho, C. M. (1979). Loss of nitrogen from soil during nitrification of urea. *Can. J. Soil Sci.* **59**, 147–154.

Cochran, V. L., Elliott, L. F., and Papendick, R. I. (1981). Nitrous oxide emissions from a fallow field fertilized with anhydrous ammonia. *Soil Sci. Soc. Am. J.* **45**, 307–310.

Colbourn, P., and Dowdell, R. J. (1984). Denitrification in field soils. *Plant Soil* **76**, 213–226.

Colbourn, P., Iqbal, M. M., and Harper, I. W. (1984). Estimation of the total gaseous nitrogen losses from clay soils under laboratory and field conditions. *J. Soil Sci.* **35**, 11–22.

Cole, J. A., and Brown, C. M. (1980). Nitrite reduction to ammonia by fermentative bacteria: A short circuit in biological nitrogen cycle. *FEMS Microbiol. Lett.* **7**, 65–72.

Cooper, G. S., and Smith, R. L. (1963). Sequence of products formed during nitrification in some diverse western soils. *Soil Sci. Soc. Am. Proc.* **27**, 659–662.

Cowling, D. W., and Lockyer, D. R. (1981). Increased growth of ryegrass exposed to ammonia. *Nature (London)* **292**, 337–338.

Craig, J. R., and Wollum, A. G. (1982). Ammonia volatilization and soil nitrogen changes after urea and ammonium nitrate fertilization of *Pinus taeda* L. *Soil Sci. Soc. Am. J.* **46**, 409–414.

Craswell, E. T., and Martin, A. E. (1974). Effect of moisture content on denitrification in a clay soil. *Soil Biol. Biochem.* **6**, 127–129.

Craswell, E. T., and Martin, A. E. (1975). Isotopic studies of the nitrogen balance in a cracking clay. I. Recovery of added nitrogen from soil and wheat in the glasshouse and gas lysimeter. *Aust. J. Soil Res.* **13**, 43–52.

Craswell, E. T., DeDatta, S. K., Obcemea, W. N., and Hartantyo, M. (1981). Time and mode of nitrogen fertilizer application to tropical wetland rice. *Fert. Res.* **2**, 247–259.

Crutzen, P. J. (1981). Atmospheric chemical processes of the oxides of nitrogen, including nitrous oxide. *In* "Denitrification, Nitrification and Atmospheric Nitrous Oxide" (C. C. Delwiche, ed.), pp. 17–44. Wiley, New York.

Crutzen, P. J. (1983). Atmospheric interactions—homogenous gas reactions of C, N, and S containing compounds. *In* "The Major Biogeochemical Cycles and Their Interactions" (B. Bolin and R. B. Cook, eds.), pp. 67–114. Wiley, New York.

Crutzen, P. J., and Ehhalt, D. H. (1977). Effects of nitrogen fertilizers and combustion on the stratospheric ozone layer. *Ambio* **6**, 112–117.

Daftardar, S. Y., and Shinde, S. A. (1980). Kinetics of ammonia volatilization of anhydrous ammonia applied to a Vertisol as influenced by farm yard manure, sorbed cations and cation exchange capacity. *Commun. Soil Sci. Plant Anal.* **11**, 135–145.

Daniel, R. M., Steele, K. W., and Limmer, A. W. (1980). Denitrification by rhizobia, a possible factor contributing to nitrogen losses from soils. *N. Z. J. Agric. Sci.* **14**, 109–112.

Daniel, R. M., Limmer, A. W., Steele, K. W., and Smith, I. M. (1982). Anaerobic growth, nitrate reduction and denitrification in 46 rhizobial strains. *J. Gen. Microbiol.* **128**, 1811–1815.

Dawson, G. A. (1977). Atmospheric ammonia from undisturbed land. *JGR, J. Geophys. Res.* **82**, 3125–3133.

Dawson, R. N., and Murphy, K. L. (1972). The temperature dependency of biological denitrification. *Water Res.* **6**, 71–83.

Delwiche, C. C. (1959). Production and utilization of nitrous oxide by *Pseudomonas denitrificans*. *J. Bacteriol.* **77**, 55–59.

Denmead, O. T., Simpson, J. R., and Freney, J. R. (1974). Ammonia flux into the atmosphere from a grazed pasture. *Science* **185**, 609–610.

Denmead, O. T., Freney, J. R., and Simpson, J. R. (1976). A closed ammonia cycle within a plant canopy. *Soil Biol. Biochem.* **8**, 161–164.

Denmead, O. T., Nulsen, R., and Thurtell, G. W. (1978). Ammonia exchange over a corn crop. *Soil Sci. Soc. Am. J.* **42**, 840–842.

Denmead, O. T., Freney, J. R., and Simpson, J. R. (1979). Studies of nitrous oxide emission from a grass sward. *Soil Sci. Soc. Am. J.* **43**, 726–728.

Denmead, O. T., Freney, J. R., and Simpson, J. R. (1982). Dynamics of ammonia volatilization during furrow irrigation of maize. *Soil Sci. Soc. Am. J.* **46**, 149–155.

Doak, B. W. (1952). Some chemical changes in the nitrogenous constituents of urine when voided on pasture. *J. Agric. Sci.* **42**, 162–171.

Doner, H. E., Volz, M. G., Belser, L. W., and Loken, J. P. (1975). Short term nitrate losses and associated microbial populations in soil columns. *Soil Biol. Biochem.* **7**, 261–263.

Dowdell, R. J. (1982). Fate of nitrogen applied to agricultural crops with particular reference to denitrification. *Philos. Trans. R. Soc. London, Ser. B* **296**, 363–373.

Dowdell, R. J., Crees, R., Burford, J. R., and Cannell, R. G. (1979a). Oxygen concentrations in a clay soil after ploughing or direct drilling. *J. Soil Sci.* **30**, 239–245.

Dowdell, R. J., Burford, J. R., and Crees, R. (1979b). Losses of nitrous oxide dissolved in drainage water from agricultural soil. *Nature (London)* **278**, 342–343.

Edwards, A. P., and Bremner, J. M. (1966). Formation of methyl nitrite in reaction of lignin with nitrous acid. *In* "The Use of Isotopes in Soil Organic Matter Studies," pp. 347–348. Pergamon, Oxford.

Elkiey, T., and Ormrod, D. P. (1981). Sorption of O_3, SO_2, NO_2 or their mixture by nine *Poa pratensis* cultivars of differing pollutant sensitivity. *Atmos. Environ.* **15**, 1739–1743.

Elliott, L. F., Schuman, G. E., and Viets, F. G. (1971). Volatilization of nitrogen-containing compounds from beef cattle areas. *Soil Sci. Soc. Am. Proc.* **35**, 752–755.

Ernst, J. W., and Massey, H. F. (1960). The effects of several factors on volatilization of ammonia formed from urea in soil. *Soil Sci. Soc. Am. Proc.* **24**, 87–90.

Eskew, D. L., Focht, D. D., and Ting, I. P. (1977). Nitrogen fixation, denitrification, and pleomorphic growth in a highly pigmented *Spirillum lipoferum*. *Appl. Environ. Microbiol.* **34**, 582–585.

Ewing, G. J., and Bauer, N. (1966). An evaluation of nitrogen losses from the soil due to the reaction of ammonium ions with nitrous acid. *Soil Sci.* **102**, 64–69.

Faller, V. N. (1972). Sulphur dioxide, hydrogen sulphide, nitrous gases and ammonia as sole source of sulphur and nitrogen for higher plants. *Z. Pflanzenernaehr. Bodenkd.* **131**, 120–130.

Farquhar, G. D., Wetselaar, R., and Firth, P. M. (1979). Ammonia volatilization from senescing leaves of maize. *Science* **203**, 1257–1258.

Farquhar, G. D., Firth, P. M., Wetselaar, R., and Weir, B. (1980). On the gaseous exchange of ammonia between leaves and the environment: Determination of the ammonia compensation point. *Plant Physiol.* **66**, 710–714.

Farquhar, G. D., Wetselaar, R., and Weir, B. (1983). Gaseous nitrogen losses from plants. *In* "Gaseous Loss of Nitrogen from Plant–Soil Systems" (J. R. Freney and J. R. Simpson, eds.), pp. 159–180. Martinus Nijhoff/Dr. W. Junk, The Hague.

Feagley, S. E., and Hossner, L. R. (1978). Ammonia volatilization reaction mechanisms between ammonium sulfate and carbonate systems. *Soil Sci. Soc. Am. J.* **42**, 364–367.

Fenn, L. B., and Escarzaga, R. (1976). Ammonia volatilization from surface applications of ammonium compounds on calcareous soils. V. Soil water content and method of nitrogen application. *Soil Sci. Soc. Am. J.* **40**, 537–541.

Fenn, L. B., and Escarzaga, R. (1977). Ammonia volatilization from surface applications of ammonium compounds to calcareous soils. VI. Effects of initial soil water content and quantity of applied water. *Soil Sci. Soc. Am. J.* **41**, 358–362.

Fenn, L. B., and Kissel, D. E. (1973). Ammonia volatilization from surface applications of ammonium compounds on calcareous soils. I. General theory. *Soil Sci. Soc. Am. Proc.* **37**, 855–859.

Fenn, L. B., and Kissel, D. E. (1975). Ammonia volatilization from surface applications of ammonium compounds on calcareous soils. IV. Effect of calcium carbonate content. *Soil Sci. Soc. Am. Proc.* **39**, 631–633.

Fenn, L. B., and Kissel, D. E. (1976). The influence of cation exchange capacity and depth of incorporation on ammonia volatilization from ammonium compounds applied to calcareous soils. *Soil Sci. Soc. Am. J.* **40**, 394–398.

Fenn, L. B., and Miyamoto, K. (1981). Ammonia loss and associated reactions of urea in calcareous soils. *Soil Sci. Soc. Am. J.* **45**, 537–540.

Fenn, L. B., Matocha, J. E., and Wu, E. (1982). Soil cation exchange capacity effects on ammonia loss from surface-applied urea in the presence of soluble calcium. *Soil Sci. Soc. Am. J.* **46**, 78–81.

Fillery, I. R. P. (1983). Biological denitrification. *In* "Gaseous Loss of Nitrogen from Plant–Soil Systems" (J. R. Freney and J. R. Simpson, eds.), pp. 33–64. Martinus Nijhoff/Dr. W. Junk, The Hague.

Firestone, M. K. (1982). Biological denitrification. *In* "Nitrogen in Agricultural Soils" (F. J. Stevenson, ed.), pp. 289–318. Am. Soc. Agron., Madison, Wisconsin.

Firestone, M. K., Smith, M. S., Firestone, R. B., and Tiedje, J. M. (1979). The influence of nitrate, nitrite, and oxygen on the composition of the gaseous products of denitrification in soil. *Soil Sci. Soc. Am. J.* **43**, 1140–1144.

Firestone, M. K., Firestone, R. B., and Tiedje, J. M. (1980). Nitrous oxide from soil denitrification: Factors controlling its biological production. *Science* **208**, 749–751.

Focht, D. D. (1974). The effect of temperature, pH, and aeration on the production of nitrous oxide and gaseous nitrogen—a zero-order kinetic model. *Soil Sci.* **118**, 173–179.

Focht, D. D., and Joseph, H. (1974). Degradation of 1,1-diphenylethylene by mixed cultures. *Can. J. Microbiol.* **20**, 631–635.

Focht, D. D., and Verstraete, W. (1977). Biochemical ecology of nitrification and denitrification. *Adv. Microb. Ecol.* **1**, 135–214.

Fowler, D. (1978). Wet and dry deposition of sulfur and nitrogen compounds from the atmosphere. *In* "Effects of Acid Precipitation on Terrestrial Ecosystems" (T. C. Hutchinson and M. Havas, eds.), pp. 9–27. Plenum, New York.

Freney, J. R., Denmead, O. T., and Simpson, J. R. (1978). Soil as a source or sink for atmospheric nitrous oxide. *Nature (London)* **273**, 530–532.

Freney, J. R., Denmead, O. T., and Simpson, J. R. (1979). Nitrous oxide emission from soils at low moisture contents. *Soil Biol. Biochem.* **11**, 167–173.

Freney, J. R., Simpson, J. R., Denmead, O. T. (1981). Ammonia volatilization. *In* "Terrestrial Nitrogen Cycles: Processes, Ecosystem Strategies and Management Impacts" (F. E. Clark and T. Rosswall, eds.), pp. 291–302. Ecological Bulletins, Stockholm.

Freney, J. R., Simpson, J. R., and Denmead, O. T. (1983). Volatilization of ammonia. *In* "Gaseous Loss of Nitrogen from Plant–Soil Systems" (J. R. Freney and J. R. Simpson, eds.), p. 1–32. Martinus Nijhoff/Dr. W. Junk, The Hague.

Galbally, I. E. (1974). Gas transfer near the earth's surface. *Adv. Geophys.* **18B**, 329–339.

Galbally, I. E., and Roy, C. R. (1978). Loss of fixed nitrogen from soils by nitric oxide exhalation. *Nature (London)* **275**, 734–735.

Galbally, I. E., and Roy, C. R. (1983). The fate of nitrogen compounds in the atmosphere. *In* "Gaseous Loss of Nitrogen from Plant–Soil Systems" (J. R. Freney and J. R. Simpson, eds.), pp. 265–284. Martinus Nijhoff/Dr. W. Junk, The Hague.

Gamble, T. N., Betlach, M. R., and Tiedje, J. M. (1977). Numerically dominant denitrifying bacteria from world soils. *Appl. Environ. Microbiol.* **33**, 926–939.

Garber, E. A. E., and Hollocher, T. C. (1982). Positional isotopic equivalence of nitrogen in N_2O produced by the denitrifying bacterium. *Pseudomonas stutzeri. J. Biol. Chem.* **257**, 4705–4708.

Garcia, J. L. (1975). La dénitrification dans les sols. *Bull. Inst. Pasteur (Paris)* **73**, 167–193.

Gaskell, J. F., Blackmer, A. M., and Bremner, J. M. (1981). Comparison of effects of nitrate, nitrite, and nitric oxide on reduction of nitrous oxide to dinitrogen by soil microorganisms. *Soil Sci. Soc. Am. J.* **45**, 1124–1127.

Gasser, J. K. R. (1964). Some factors affecting the losses of ammonia from urea and ammonium sulfate applied to soils. *J. Soil Sci.* **15**, 258–272.

Georgii, H. W., and Gravenhorst, G. (1977). The ocean as source or sink of reactive trace-gases. *Pure Appl. Geophys.* **115**, 503–511.

Gilliam, J. W., and Gambrell, R. P. (1978). Temperature and pH as limiting factors in loss of nitrate from saturated Atlantic Coastal Plain soils. *J. Environ. Qual.* **7**, 526–532.

Gilliam, J. W., Dasberg, S., Lund, L. J., and Focht, D. D. (1978). Denitrification in four California soils: Effect of soil profile characteristics. *Soil Sci. Soc. Am. J.* **42**, 61–66.

Goodroad, L. L., and Keeney, D. R. (1984). Nitrous oxide production in aerobic soils under varying pH, temperature and water content. *Soil Biol. Biochem.* **16**, 39–43.

Goreau, T. J., Kaplan, W. A., Wofsy, S. C., McElroy, M. B., Valois, F. W., and Watson, S. W. (1980). Production of NO_2^- and N_2O by nitrifying bacteria at reduced concentrations of oxygen. *Appl. Environ. Microbiol.* **40**, 526–532.

Guenzi, W. D., Beard, W. E., Watanabe, F. S., Olsen, S. R., and Porter, L. K. (1978). Nitrification and denitrification in cattle manure-amended soil. *J. Environ. Qual.* **7**, 196–202.

Hahn, J. (1979). Man-made perturbation of the nitrogen cycle and its possible impact on climate. *In* "Man's Impact on Climate" (W. Bach, J. Pankrath, and W. W. Kellog, eds.), pp. 193–213. Elsevier, Amsterdam.

Hahn, J., and Crutzen, P. J. (1982). The role of fixed nitrogen in atmospheric photochemistry. *Philos. Trans. R. Soc. London, Ser. B* **296**, 521–541.

Hales, J. M., and Drewes, D. R. (1979). Solubility of ammonia at low concentrations. *Atmos. Environ.* **13**, 1133–1147.

Hargrove, W. L., and Kissel, D. E. (1979). Ammonia volatilization from surface applications of urea in the field and laboratory. *Soil Sci. Soc. Am. J.* **43**, 359–363.

Hargrove, W. L., Kissel, D. E., and Fenn, L. B. (1977). Field measurements of ammonia volatilization from surface applications of ammonium salts to a calcareous soil. *Agron. J.* **69**, 473–476.

Harper, L. A., Catchpoole, V. R., and Vallis, I. (1983). Ammonia loss from fertilizer applied to tropical pastures. *In* "Gaseous Loss of Nitrogen from Plant–Soil Systems" (J. R. Freney and J. R. Simpson, eds.), pp. 195–214. Martinus Nijhoff/Dr. W. Junk, The Hague.

Harris, G. L., Goss, M. J., Dowdell, R. J., Howse, K. R., and Morgan, P. (1984). A study of mole drainage with simplified cultivation for autumn-sown crops on a clay soil. *J. Agric.* **102**, 561–581.

Hauck, R. D. (1971). Quantitative estimates of nitrogen cycle processes: Concept and review. *In* "Nitrogen–15 Soil–Plant Studies," IAEA-PL-341/6. IAEA, Vienna, Austria.

Hauck, R. D., and Bremner, J. M. (1976). Use of tracers for soil and fertilizer nitrogen research. *Adv. Agron.* **28**, 219–266.

Hauck, R. D., and Stephenson, H. F. (1965). Nitrification of nitrogen fertilizers. Effect of nitrogen source, size and pH of the granule, and concentration. *J. Agric. Food Chem.* **13**, 486–492.

Henzell, E. F., and Ross, P. J. (1973). The nitrogen cycles of pasture ecosystems. *In* "Chemistry and Biochemistry of Herbage" (G. W. Butler and R. W. Bailey, eds.), Vol. 2, pp. 227–246. Academic Press, New York.

Hoff, J. D., Nelson, D. W., and Sutton, A. L. (1981). Ammonia volatilization from liquid swine manure applied to cropland. *J. Environ. Qual.* **10**, 90–95.

Hooker, M. L., Sander, D. H., Peterson, G. A., and Daigger, L. A. (1980). Gaseous N losses from winter wheat. *Agron. J.* **72,** 789–792.

Huebert, B. J., and Lazrus, A. L. (1980). Bulk composition of aerosols in the remote troposphere. *JGR, J. Geophys. Res.* **85,** 7337–7344.

Huntzicker, J. J., Cary, R. A., and Ling, C. S. (1980). Neutralization of sulfuric acid aerosol by ammonia. *Environ. Sci. Technol.* **14,** 819–824.

Hutchinson, G. L., and Mosier, A. R. (1979). Nitrous oxide emissions from an irrigated corn field. *Science* **205,** 1125–1127.

Hutchinson, G. L., and Viets, F. G. (1969). Nitrogen enrichment of surface water by absorption of ammonia volatilized from cattle feedlots. *Science* **166,** 514–515.

Hutchinson, G. L., Millington, R. J., and Peters, D. B. (1972). Atmospheric ammonia: Absorption by plant leaves. *Science* **175,** 771–772.

Hynes, R. K., and Knowles, R. (1978). Inhibition by acetylene of ammonia oxidation in *Nitrosomonas europaea*. *FEMS Microbiol. Lett.* **4,** 319–321.

Ingraham, J.L. (1981). Microbiology and genetics of denitrifiers. *In* "Denitrification, Nitrification and Atmospheric Nitrous Oxide" (C. C. Delwiche, ed.), pp. 45–65. Wiley, New York.

Ishaque, M., and Aleem, M. I. H. (1973). Intermediates of denitrification in the chemoautotroph *Thiobacillus denitrificans*. *Arch. Microbiol.* **94,** 269–282.

Jacobsen, S. N., and Alexander, M. (1980). Nitrate loss from soil in relation to temperature, carbon source and denitrifier populations. *Soil Biol. Biochem.* **12,** 501–505.

Johansson, C., and Granat, L. (1984). Emission of nitric oxide from arable land. *Tellus* **36B,** 25–37.

John, P., and Whatley, F. R. (1975). *Paracoccus denitrificans* and the evolutionary origin of the mitochondrion. *Nature (London)* **254,** 495–498.

Jones, R. W., and Hedlin, R. A. C. (1970). Nitrite instability in three Manitoba soils. *Can. J. Soil Sci.* **50,** 339–345.

Keeney, D. R., Fillery, I. R., and Marx, G. P. (1979). Effect of temperature on the gaseous nitrogen products of denitrification in a silt loam soil. *Soil Sci. Soc. Am. J.* **43,** 1124–1128.

Khan, D. H., and Haque, M. Z. (1965). Volatilization loss of nitrogen from urea added to some soils of East Pakistan. *J. Sci. Food Agric.* **16,** 725–729.

Khan, M. F. A., and Moore, A. W. (1968). Denitrification capacity of some Alberta soils. *Can. J. Soil Sci.* **48,** 89–91.

Kim, C. M. (1973). Influence of vegetation types on the intensity of ammonia and nitrogen dioxide liberation from soil. *Soil Biol. Biochem.* **5,** 163–166.

Kissel, D. E., Brewer, H. L., and Arkin, G. F. (1977). Design and test of a field sampler for ammonia volatilization. *Soil Sci. Soc. Am. J.* **41,** 1133–1138.

Klemedtsson, L., Svensson, B. H., Lindberg, T., and Rosswall, T. (1978). The use of acetylene inhibition of nitrous oxide reductase in quantifying denitrification in soils. *Swed. J. Agric. Res.* **7,** 179–185.

Knowles, R. (1981). Denitrification. *In* "Terrestrial Nitrogen Cycles: Processes, Ecosystem Strategies and Management Impacts" (F. E. Clark and T. Rosswall, eds.), pp. 315–329. Ecological Bulletins, Stockholm.

Knowles, R. (1982). Denitrification. *Microbiol. Rev.* **46,** 43–70.

Kohl, D. H., Vithayathil, F., Whitlow, P., Shearer, G., and Chien, S. H. (1976). Denitrification kinetics in soil systems: The significance of good fits of data to mathematical forms. *Soil Sci. Soc. Am. J.* **40,** 249–253.

Koskinen, W. C., and Keeney, D. R. (1982). Effect of pH on rate of gaseous products of denitrification in silt loam soil. *Soil Sci. Soc. Am. J.* **46**, 1165–1167.

Kresge, C. B., and Satchell, D. P. (1960). Gaseous loss of ammonia from nitrogen fertilizers applied to soil. *Agron. J.* **52**, 104–107.

Krul, J. M., and Veeningen, R. (1977). The synthesis of the dissimilatory nitrate reductase under aerobic conditions in a number of denitrifying bacteria, isolated from activated-sludge and drinking water. *Water Res.* **11**, 39–43.

Lauer, D. A., Bouldin, D. R., and Klausner, S. D. (1976). Ammonia volatilization from dairy manure spread on the soil surface. *J. Environ. Qual.* **5**, 134–141.

Legg, J. O., and Meisinger, J. J. (1982). Soil nitrogen budgets. *In* "Nitrogen in Agricultural Soils" (F. J. Stevenson, ed.), pp. 503–566. Am. Soc. Agron., Madison, Wisconsin.

Lehr, I. J., and Van Wesemael, J. C. (1961). The volatilization of ammonia from lime rich soils. *Landbouwk. Tijdschr.* **73**, 1156–1168.

Lemon, E., and van Houtte, R. (1980). Ammonia exchange at the land surface. *Agron. J.* **72**, 876–883.

Lenhard, U., and Gravenhorst, G. (1980). Evaluation of ammonia fluxes into the free atmosphere over Western Germany. *Tellus* **32**, 48–55.

Lensi, R., and Chalamet, A. (1982). Denitrification in water-logged soils: *In situ* temperature-dependent variations. *Soil Biol. Biochem.* **14**, 51–55.

Letey, J., Veloras, N., Hadas, A., and Focht, D. D. (1979). Ratio of $N_2O : N_2$ evolution as affected by soil conditions during denitrification. *In* "Nitrates in Effluents from Irrigated Lands," pp. 807–811. Final Report to the National Science Foundation, University of California, Riverside.

Letey, J., Valoras, N., Hadas, A., and Focht, D. D. (1980). Effect of air-filled porosity, nitrate concentration and time on the ratio of N_2O/N_2 evolution during denitrification. *J. Environ. Qual.* **9**, 227–231.

Letey, J., Valoras, N., Focht, D. D., and Ryden, J. C. (1981). Nitrous oxide production and reduction during denitrification as affected by redox potential. *Soil Sci. Soc. Am. J.* **45**, 727–730.

Leuning, R., Denmead, O. T., Simpson, J. R., and Freney, J. R. (1984). Processes of ammonia loss from shallow floodwater. *Atmos. Environ.* **18**, 1583–1592.

Lin, D. M., and Doran, J. W. (1984). Aerobic and anaerobic microbial populations in no-till and plowed soil. *Soil Sci. Soc. Am. J.* **48**, 794–799.

Lipschultz, F., Zafiriou, O. C., Wofsy, S. C., McElroy, M. B., Valois, F. W., and Watson, S. W. (1981). Production of NO and N_2O by soil nitrifying bacteria. *Nature (London)* **294**, 641–643.

Loehr, R. C. (1974). "Agricultural Waste Management." Academic Press, New York.

Logan, J. A., Prather, M. J., Wofsy, S. C., and McElroy, M. B. (1978). Atmospheric chemistry: Response to human influence. *Philos. Trans. R. Soc. London, Ser. A* **290**, 187–234.

Luebs, R. E., Davis, K. R., and Laag, A. E. (1973). Enrichment of the atmosphere with nitrogen compounds volatilized from large dairy area. *J. Environ. Qual.* **2**, 137–141.

Lyster, S., Morgan, M. A., and O'Toole, P. (1980). Ammonia volatilization from soils fertilized with urea and ammonium nitrate. *J. Life Sci.* **1**, 167–176.

McElroy, M. B., Wofsy, S. C., and Yung, Y. L. (1977). The nitrogen cycle: Perturbations due to man and their impact on atmospheric N_2O and O_3. *Philos. Trans. R. Soc. London, Ser. B* **277**, 159–181.

McGarity, J. W. (1961). Denitrification studies on some South Australian soils. *Plant Soil* **14**, 1–21.

McGarity, J. W. (1962). Effect of freezing of soil on denitrification. *Nature (London)* **196**, 1342–1343.

McGarity, J. W., and Hoult, E. H. (1971). The plant component as a factor in ammonia volatilization from pasture swards. *J. Br. Grassl. Soc.* **26**, 31–34.

McGarity, J. W., and Rajaratnam, J. A. (1973). Apparatus for the measurement of losses of nitrogen as gas from the field and simulated field environments. *Soil Biol. Biochem.* **5**, 121–131.

Macrae, I. C., and Ancajas, R. (1970). Volatilization of ammonia from submerged tropical soils. *Plant Soil* **33**, 97–103.

Martin, A. E., and Ross, P. J. (1968). A nitrogen-balance study using labelled fertilizer in a gas lysimeter. *Plant Soil* **28**, 182–186.

Martin, J. P., and Chapman, H. D. (1951). Volatilization of ammonia from surface fertilized soils. *Soil Sci.* **71**, 25–34.

Matocha, J. E. (1976). Ammonia volatilization and nitrogen utilization from sulfur-coated ureas and conventional nitrogen fertilizers. *Soil Sci. Soc. Am. J.* **40**, 597–601.

Mikkelsen, D. S., De Datta, S. K., and Obcemea, W. N. (1978). Ammonia volatilization losses from flooded rice soils. *Soil Sci. Soc. Am. J.* **42**, 725–730.

Moe, P. G. (1967). Nitrogen losses from urea as affected by altering soil urease activity. *Soil Sci. Soc. Am. Proc.* **31**, 380–382.

Moeller, M. B., and Vlek, P. L. G. (1982). The chemical dynamics of ammonia volatilization from aqueous solution. *Atmos. Environ.* **16**, 709–717.

Moraghan, J. L., and Buresh, R. J. (1977). Chemical reduction of nitrite and nitrous oxide by ferrous ion. *Soil Sci. Soc. Am. J.* **41**, 47–50.

Morrison, I. K., and Foster, N. W. (1977). Fate of urea fertilizer added to a boreal forest *Pinus bankesiana* Lamb stand. *Soil Sci. Soc. Am. J.* **41**, 441–448.

Mosier, A. R., and Hutchinson, G. L. (1981). Nitrous oxide emissions from cropped fields. *J. Environ. Qual.* **10**, 169–173.

Mosier, A. R., Andre, C. E., and Viets, F. G. (1973). Identification of aliphatic amines volatilized from a cattle feedyard. *Environ. Sci. Technol.* **7**, 642–644.

Mosier, A. R., Stillwell, M., Parton, W. J., and Woodmansee, R. G. (1981). Nitrous oxide emissions from a native shortgrass prairie. *Soil Sci. Soc. Am. J.* **45**, 617–619.

Mosier, A. R., Hutchinson, G. L., Sabey, B. R., and Baxter, J. (1982). Nitrous oxide emissions from barley plots treated with ammonium nitrate or sewage sludge. *J. Environ. Qual.* **11**, 78–81.

Muller, M. M., Sundman, V., and Skujins, J. (1980). Denitrification in low pH Spodosols and peats determined with the acetylene inhibition method. *Appl. Environ. Microbiol.* **40**, 235–239.

Myers, R. J. K., and McGarity, J. W. (1972). Denitrification in undisturbed cores from a solodized solonetz B horizon. *Plant Soil* **37**, 81–89.

National Research Council (NRC) (1978). "Nitrates: An Environmental Assessment." Natl. Acad. Sci., Washington, D.C.

National Research Council (NRC), Subcommittee on Ammonia (1979). "Ammonia." Baltimore Univ. Press, Baltimore, Maryland.

Nelson, D. W. (1978). Transformations of hydroxylamine in soils. *Proc. Indiana Acad. Sci.* **87**, 409–413.

Nelson, D. W. (1982). Gaseous losses of nitrogen other than through denitrification. *In*

"Nitrogen in Agricultural Soils" (F. J. Stevenson, ed.), pp. 327–363. Am. Soc. Agron., Madison, Wisconsin.

Nelson, D. W., and Bremner, J. M. (1970a). Role of soil minerals and metallic cations in nitrite decomposition and chemodenitrification in soils. *Soil Biol. Biochem.* **2,** 1–8.

Nelson, D. W., and Bremner, J. M. (1970b). Gaseous products of nitrite decomposition in soils. *Soil Biol. Biochem.* **2,** 203–215.

Netti, I. T. (1955). Denitrifying bacteria of the oak rhizosphere. *Mikrobiologiya* **24,** 429–434.

Neyra, C. H., and van Berkum, P. (1977). Nitrate reduction and nitrogenase activity in *Spirillum lipoferum. Can. J. Microbiol.* **23,** 306–310.

Neyra, C. A., Dobereiner, J., Lalande, R., and Knowles, R. (1977). Denitrification by N_2-fixing *Spirillum lipoferum. Can. J. Microbiol.* **23,** 300–305.

Nicolet, M., and Peetermans, W. (1972). The production of nitric oxide in the stratosphere by oxidation of nitrous oxide. *Ann. Geophys.* **28,** 751–762.

Nommik, H. (1956). Investigations on denitrification in soil. *Acta Agric. Scand.* **6,** 195–228.

Overrein, L. N., and Moe, P. G. (1967). Factors affecting urea hydrolysis and ammonia volatilization in soil. *Soil Sci. Soc. Am. Proc.* **31,** 57–61.

Pang, P. C., Hedlin, R. A., and Cho, C. M. (1973). Transformation and movement of band-applied urea, ammonium sulphate and ammonium hydroxide during incubation in several Manitoba soils. *Can. J. Soil Sci.* **53,** 331–341.

Pang, P. C., Cho, C. M., and Hedlin, R. A. (1975). Effect of nitrogen concentration on the transformation of band-applied nitrogen fertilizers. *Can. J. Soil Sci.* **55,** 23–27.

Parr, J. F., and Papendick, R. I. (1966). Retention of ammonia in soils. *In* "Agricultural Anhydrous Ammonia Technology and Use" (M. H. McVickar, W. P. Martin, I. E. Miles, and H. H. Tucker, eds.), pp. 213–236. Agric. Ammonia Inst., Memphis, Tennessee.

Patten, D. K., Bremner, J. M., and Blackmer, A. M. (1980). Effects of drying and air-dry storage of soils on their capacity for denitrification of nitrate. *Soil Sci. Soc. Am. J.* **44,** 67–70.

Payne, W. J. (1973). Reduction of nitrogenous oxides by microorganisms. *Bacteriol. Rev.* **37,** 409–452.

Payne, W. J. (1981). "Denitrification." Wiley, New York.

Phillips, R. E., Reddy, K. R., and Patrick, W. H. (1978). The role of nitrate diffusion in determining the order and rate of denitrification in flooded soil. II. Theoretical analysis and interpretation. *Soil Sci. Soc. Am. J.* **42,** 272–278.

Pilot, L., and Patrick, W. H. (1972). Nitrate reduction in soils: Effect of soil moisture tension. *Soil Sci.* **114,** 312–316.

Porter, L. K. (1969). Gaseous products produced by anaerobic reaction of sodium nitrite with oxime compounds and oximes synthesized from organic matter. *Soil Sci. Soc. Am. Proc.* **33,** 696–702.

Porter, L. K., Viets, F. G., and Hutchinson, G. L. (1972). Air containing ^{15}N ammonia: Foliar absorption by corn seedlings. *Science* **175,** 759–761.

Presad, M. (1976). Gaseous loss of ammonia from sulfur-coated urea, ammonium sulfate, and urea applied to calcareous soil (pH 7.3). *Soil Sci. Soc. Am. J.* **40,** 130–134.

Ramanathan, V., Callis, L. B., and Boughner, R. E. (1976). Sensitivity of surface temperature and atmospheric temperature to perturbations in the stratospheric concentration of ozone and nitrogen dioxide. *J. Atmos. Sci.* **33,** 1092–1112.

Rashid, G. H. (1977). The volatilization losses of nitrogen from added urea in some soils of Bangladesh. *Plant Soil* **48,** 549–556.

Rasmussen, K. H., Taheri, M., and Kabel, R. L. (1975). Global emissions and natural processes for removal of gaseous pollutants. *Water, Air, Soil Pollut.* **4,** 33–64.

Reddy, K. R., Patrick, W. H., and Phillips, R. E. (1978). The role of nitrate diffusion in determining the order of denitrification in flooded soil. I. Experimental results. *Soil Sci. Soc. Am. J.* **42,** 268–272.

Reddy, K. R., Sacco, P. D., and Graets, D. A. (1980). Nitrate reduction in an organic soil–water system. *J. Environ. Qual.* **9,** 283–288.

Reddy, K. R., Rao, P. S. C., and Jessup, R. E. (1982). The effect of carbon mineralization on denitrification kinetics in mineral and organic soils. *Soil Sci. Soc. Am. J.* **46,** 62–67.

Reuss, J. O., and Smith, R. L. (1965). Chemical reactions of nitrites in acid soils. *Soil Sci. Soc. Am. Proc.* **29,** 267–270.

Rice, C. W., and Smith, M. S. (1982). Denitrification in no-till and plowed soils. *Soil Sci. Soc. Am. J.* **46,** 1168–1173.

Rice, C. W., and Smith, M. S. (1983). Nitrification of fertilizer and mineralized ammonium in no-till and plowed soil. *Soil Sci. Soc. Am. J.* **47,** 1125–1129.

Richardson, M. (1966). Studies on the biogenesis of some simple amines and quaternary ammonium compounds in higher plants. *Phytochemistry* **5,** 23–30.

Rigaud, J. F., Bergersen, F. J., Turner, G. L., and Daniel, R. M. (1973). Nitrate dependent anaerobic acetylene-reduction and nitrogen-fixation by soybean bacteroids. *J. Gen. Microbiol.* **77,** 137–144.

Ritchie, G. A. F., and Nicholas, D. J. D. (1972). Identification of the sources of nitrous oxide produced by oxidative and reductive processes in *Nitrosomonas europea.* *Biochem. J.* **126,** 1181–1191.

Ritchie, G. A. F., and Nicholas, D. J. D. (1974). The partial characterization of purified nitrite reductase and hydroxylamine oxidase from *Nitrosomonas europaea.* *Biochem. J.* **138,** 471–480.

Rogers, H. H., and Aneja, V. P. (1980). Uptake of atmospheric ammonia by selected plant species. *Environ. Exp. Bot.* **20,** 251–257.

Rogers, H. H., Jeffries, H. E., and Witherspoon, A. M. (1979). Measuring air pollutant uptake by plants: Nitrogen dioxide. *J. Environ. Qual.* **8,** 551–557.

Rolston, D. E. (1978). Application of gaseous-diffusion theory to measurement of denitrification. *In* "Nitrogen in the Environment" (D. R. Nielsen and J. G. MacDonald, eds.), Vol. 1, pp. 309–335. Academic Press, New York.

Rolston, D. E. (1981). Nitrous oxide and nitrogen gas production in fertilizer loss. *In* "Denitrification, Nitrification and Atmospheric Nitrous Oxide" (C. C. Delwiche, ed.), pp. 127–149. Wiley, New York.

Rolston, D. E., and Broadbent, F. E. (1977). "Field Measurement of Denitrification," EPA-600/2-77-233. U.S. Environmental Protection Agency, Ada, Oklahoma.

Rolston, D. E., Fried, M., and Goldhamer, D. A. (1976). Denitrification measured directly from nitrogen and nitrous oxide gas fluxes. *Soil Sci. Soc. Am. J.* **40,** 259–266.

Rolston, D. E., Hoffman, D. L., and Toy, D. W. (1978). Field measurement of denitrification. I. Flux of N_2 and N_2O. *Soil Sci. Soc. Am. J.* **42,** 863–869.

Rolston, D. E., Sharpley, A. N., Toy, D. W., and Broadbent, F. E. (1982). Field measurements of denitrification. III. Rates during irrigation cycles. *Soil Sci. Soc. Am. J.* **46,** 289–296.

Ryan, J. A., and Keeney, D. R. (1975). Ammonia volatilization from surface-applied wastewater sludge. *J. Water Pollut. Control Fed.* **47,** 386–393.

Ryan, J. A., Curtin, D., and Safi, I. (1981). Ammonia volatilization as influenced by calcium carbonate particle size and iron oxides. *Soil Sci. Soc. Am. J.* **45,** 338–341.

Ryden, J. C. (1981). N₂O exchange between grassland soil and the atmosphere. *Nature (London)* **292**, 235–236.

Ryden, J. C., and Lund, L. J. (1980). Nature and extent of directly measured denitrification losses from some irrigated vegetable crop production units. *Soil Sci. Soc. Am. J.* **44**, 505–511.

Ryden, J. C., Lund, L. J., and Focht, D. D. (1978). Direct in-field measurement of nitrous oxide flux from soils. *Soil Sci. Soc. Am. J.* **42**, 731–737.

Ryden, J. C., Lund, L. J., Letey, J., and Focht, D. D. (1979). Direct measurement denitrification loss from soils. II. Development and application of field methods. *Soil Sci. Soc. Am. J.* **43**, 110–118.

Ryzhova, I. M. (1979). Effect of nitrate concentration on the rate of soil denitrification. *Sov. Soil Sci. (Engl. Transl.)* **11**, 168–171.

Satoh, T. (1977). Light-activated, inhibited and independent denitrification by a denitrifying phototrophic bacterium. *Arch. Microbiol.* **115**, 293–298.

Sawada, E., Satoh, T., and Kitamura, H. (1978). Purification and properties of a dissimilatory nitrite reductase of a denitrifying phototrophic bacterium. *Plant Cell Physiol.* **19**, 1339–1531.

Scott, B. C., and Laulainen, N. S. (1979). On the concentration of sulphate in precipitation. *J. Appl. Meteorol.* **18**, 138–147.

Scott, D. B., and Scott, C. A. (1978). Nitrate-dependent nitrogenase activity in *Azospirillum* spp. under low oxygen tensions. *In* "International Symposium on the Limitations and Potentials for Biological Nitrogen Fixation in the Tropics" (J. Dobereiner, ed.), pp. 350–351. Plenum, New York.

Sherlock, R. R., and Goh, K. M. (1983). Initial emission of nitrous oxide from sheep urine applied to pasture soil. *Soil Biol. Biochem.* **15**, 615–617.

Sherlock, R. R., and Goh, K. M. (1984). Dynamics of ammonia volatilization from simulated urine patches and aqueous urea applied to pasture. I. Field experiments. *Fert. Res.* **5**, 181–195.

Sherlock, R. R., and Goh, K. M. (1985a). Dynamics of ammonia volatilization from simulated urine patches and aqueous urea applied to pasture. II. Theoretical derivation of a simplified model. *Fert. Res.* **6**, 3–22.

Sherlock, R. R., and Goh, K. M. (1985b). Dynamics of ammonia volatilization from simulated urine patches and aqueous urea applied to pasture. III. Field verification of a simplified model. *Fert. Res.* **6**, 23–36.

Simpson, D. M. H., and Melsted, S. W. (1962). Gaseous ammonia losses from urea solutions applied as foliar spray to various grass sods. *Soil Sci. Soc. Am. Proc.* **26**, 186–189.

Simpson, J. R., and Steele, K. W. (1983). Gaseous nitrogen exchanges in grazed pastures. *In* "Gaseous Loss of Nitrogen from Plant–Soil Systems" (J. R. Freney and J. R. Simpson, eds.), pp. 215–236. Martinus Nijhoff/Dr. W. Junk, The Hague.

Smith, C. J., and Chalk, P. M. (1979). Mineralization of nitrite fixed by soil organic matter. *Soil Biol. Biochem.* **11**, 515–519.

Smith, C. J., and Chalk, P. M. (1980a). Gaseous nitrogen evolution during nitrification of ammonia fertilizer and nitrite transformations in soils. *Soil Sci. Soc. Am. J.* **44**, 277–282.

Smith, C. J., and Chalk, P. M. (1980b). Fixation and loss of nitrogen during transformations of nitrite in soil. *Soil Sci. Soc. Am. J.* **44**, 288–291.

Smith, C. J., Brandon, M., and Patrick, W. H. (1982). Nitrous oxide emission following urea-N fertilization of wetland rice. *Soil Sci. Plant Nutr.* **28**, 161–171.

Smith, D. H., and Clark, F. E. (1960). Volatile losses of nitrogen from acid or neutral soils or solutions containing nitrite and ammonium ions. *Soil Sci.* **90**, 86–92.

Smith, K. A. (1977). Soil aeration. *Soil Sci.* **123**, 284–291.

Smith, K. A. (1980). A model of the extent of anaerobic zones in aggregated soils, and its potential application to estimates of denitrification. *J. Soil Sci.* **31**, 263–277.

Smith, M. S., and Tiedje, J. M. (1979a). Phases of denitrification following oxygen depletion in soil. *Soil Biol. Biochem.* **11**, 261–267.

Smith, M. S., and Tiedje, J. M. (1979b). The effect of roots on soil denitrification. *Soil Sci. Soc. Am. J.* **43**, 951–955.

Smith, M. S., and Zimmerman, K. (1981). Nitrous oxide production by non-denitrifying soil nitrate reducers. *Soil Sci. Soc. Am. J.* **45**, 865–871.

Sorensen, J. (1978). Occurrence of nitric and nitrous oxides in a coastal marine sediment. *Appl. Environ. Microbiol.* **36**, 809–813.

Stanford, G., Vander Pol, R. A., and Dzienia, S. (1975). Denitrification rates in relation to total and extractable soil carbon. *Soil Sci. Soc. Am. Proc.* **39**, 284–289.

Starr, J. L., and Parlange, J. V. (1975). Non-linear denitrification kinetics with continuous flow in soil columns. *Soil Sci. Soc. Am. Proc.* **39**, 875–880.

Steen, W. C., and Stojanovic, B. J. (1971). Nitric oxide volatilization from a calcareous soil and model aqueous systems. *Soil Sci. Soc. Am. Proc.* **35**, 277–282.

Stefanson, R. C. (1972a). Soil denitrification in sealed soil–plant systems. I. Effects of plants, soil water content and soil organic matter content. *Plant Soil* **33**, 113–127.

Stefanson, R. C. (1972b). Soil denitrification in sealed soil–plant systems. II. Effect of soil water content and form of applied nitrogen. *Plant Soil* **37**, 129–140.

Stefanson, R. C. (1972c). Effect of plant growth and form of nitrogen fertilizer on denitrification from four South Australian soils. *Aust. J. Soil Res.* **10**, 183–195.

Stefanson, R. C. (1976). Denitrification from nitrogen fertilizer placed at various depths in the soil–plant system. *Soil Sci.* **121**, 353–363.

Stevens, R. K., Dzubay, T. G., Russworm, G., and Rickel, D. (1978). Sampling and analysis of atmospheric sulphate and related species. *Atmos. Environ.* **12**, 56–68.

Stevenson, F. J., and Swaby, R. J. (1964). Nitrosation of soil organic matter. I. Nature of gases evolved during nitrous acid treatment of lignins and humic substances. *Soil Sci. Soc. Am. Proc.* **28**, 773–778.

Stevenson, F. J., Harrison, R. M., Wetselaar, R., and Leeper, R. A. (1970). Nitrosation of soil organic matter. III. Nature of gases produced by reaction of nitrite with lignins, humic substances, and phenolic constituents under neutral and slightly acidic conditions. *Soil Sci. Soc. Am. Proc.* **34**, 430–435.

Stouthamer, A. H. (1976). Biochemistry and genetics of nitrate reductase in bacteria. *Adv. Microb. Physiol.* **14**, 315–375.

Stutte, C. A., and Weiland, R. T. (1978). Gaseous nitrogen loss and transpiration of several crop and weed species. *Crop Sci.* **18**, 887–889.

Stutte, C. A., Weiland, R. T., and Blem, A. R. (1979). Gaseous nitrogen loss from soybean foliage. *Agron. J.* **71**, 95–97.

Swank, W. T., and Caskey, W. H. (1982). Nitrate depletion in a second-order mountain stream. *J. Environ. Qual.* **11**, 581–584.

Taylor, G. S., Baker, M. B., and Charlson, R. J. (1983). Heterogeneous interactions of the C, N, and S cycles in the atmosphere: The role of aerosols and clouds. *In* "The Major Biogeochemical Cycles and Their Interactions" (B. Bolin and R. B. Cook, eds.), pp. 115–141. Wiley, New York.

Terman, G. L. (1979). Volatilization losses of nitrogen as ammonia from surface-applied fertilizers, organic amendments, and crop residues. *Adv. Agron.* **31**, 189–223.

Terman, G. L., Parr, J. F., and Allen, S. E. (1968). Recovery of nitrogen by corn from solid fertilizers and solutions. *J. Agric. Food Chem.* **16**, 685–690.

Terry, R. E., and Tate, R. L. (1980). Denitrification as a pathway for nitrate removal from organic soils. *Soil Sci.* **129,** 162–166.

Terry, R. E., Nelson, D. W., Sommers, L. E., and Meyer, G. J. (1978). Ammonia volatilization from wastewater sludge applied to soil. *J. Water Pollut. Control Fed.* **50,** 2657–2665.

Terry, R. E., Tate, R. L., and Duxbury, J. M. (1981). Nitrous oxide emissions from drained, cultivated organic soils of South Florida. *J. Air Pollut. Control Assoc.* **31,** 1173–1176.

Thauer, R. K., Jungermann, K., and Decker, K. (1977). Energy conservation in chemotrophic anaerobic bacteria. *Bacteriol. Rev.* **41,** 100–180.

Tiedje, J. M., Firestone, R. B., Firestone, M. K., Betlach, M. R., Kaspar, H. F., and Sorensen, J. (1981). Use of nitrogen-13 in studies of denitrification. *In* "Recent Developments in Biological and Chemical Research with Short-Lived Isotopes" (R. A. Krohn and J. W. Root, eds.), pp. 295–317. Am. Chem. Soc., Washington, D.C.

Tripathi, O. N. (1958). Role of energy-rich materials as conservators in presence of nitrogenous fertilizers in Sagar soil. *Proc.—Indian Acad. Sci., Sect. A* **27A,** 215–220.

Tunney, H. (1980). Agricultural wastes as fertilizers. *In* "Handbook of Organic Waste Conversion" (M. W. M. Bewick, ed.), pp. 1–39. Van Nostrand-Reinhold, Princeton, New Jersey.

Vallis, I., Harper, L. A., Catchpoole, V. R., and Weier, K. L. (1982). Volatilization of ammonia from urine patches in a subtropical pasture. *Aust. J. Agric. Res.* **33,** 97–107.

Van Cleemput, O., and Baert, L. (1984). Nitrite: A key compound in N loss process under acid conditions? *Plant Soil* **76,** 233–241.

Van Cleemput, O., and Patrick, W. H. (1974). Nitrate and nitrite reduction in flooded gamma-irradiated soil under controlled pH and redox potential conditions. *Soil Biol. Biochem.* **6,** 85–88.

Vanderholm, D. H. (1975). Nutrient losses from livestock waste during storage, treatment and handling. *In* "Managing Livestock Wastes," pp. 282–285. Am. Soc. Agric. Eng., St. Joseph, Michigan.

Verma, R. N. S., and Sarkar, M. C. (1974). Some soil properties affecting loss of nitrogen from urea due to ammonia volatilization. *J. Indian Soc. Soil Sci.* **22,** 80–83.

Verstraete, W. (1981). Nitrification. *In* "Terrestrial Nitrogen Cycles: Processes, Ecosystem Strategies and Management Impacts" (F. E. Clark and T. Rosswall, eds.), pp. 303–314. Ecological Bulletins, Stockholm.

Vine, H. (1962). Some measurements of release and fixation of nitrogen in soil of natural structure. *Plant Soil* **17,** 109–130.

Vlek, P. L. G., and Craswell, E. T. (1979). Effect of nitrogen source and management on ammonia volatilization losses from flooded rice–soil systems. *Soil Sci. Soc. Am. J.* **43,** 352–358.

Vlek, P. L. G., and Craswell, E. T. (1981). Ammonia volatilization from flooded soils. *Fert. Res.* **2,** 227–245.

Vlek, P. L. G., and Stumpe, J. M. (1978). Effect of solution chemistry and environmental conditions on ammonia volatilization losses from aqueous systems. *Soil Sci. Soc. Am. J.* **42,** 416–421.

Volk, G. M. (1959). Volatile loss of ammonia following surface application of urea to turf or bare soils. *Agron. J.* **51,** 746–749.

Volk, G. M. (1966). Efficiency of urea as affected by method of application, soil moisture, and lime. *Agron. J.* **58,** 249–252.

Volz, M. G., and Starr, J. L. (1977). Nitrate dissimilation and population dynamics of denitrifying bacteria during short term continuous flow. *Soil Sci. Soc. Am. J.* **41,** 891–896.

Volz, M. G., Ardakani, M. S., Schulz, R. K., Stolzy, L. H., and McLaren, A. D. (1976). Soil nitrate loss during irrigation: Enhancement by plant roots. *Agron. J.* **68,** 621–627.

Wahhab, A., and Uddin, F. (1954). Loss of nitrogen through reaction of ammonium and nitrite ions. *Soil Sci.* **78,** 119–126.

Wahhab, A., Randhawa, M. S., and Alam, S. Q. (1956). Loss of ammonia from ammonium sulphate under different conditions when applied to soils. *Soil Sci.* **84,** 294–295.

Walter, H. M., Keeney, D. R., and Fillery, I. R. (1979). Inhibition of nitrification by acetylene. *Soil Sci. Soc. Am. J.* **43,** 195–196.

Wang, W. C., Yung, Y. L., Lacis, A. A., Mo, T., and Hansen, J. E. (1976). Greenhouse effects due to manmade perturbations of trace gases. *Science* **194,** 685–690.

Warembourg, F. R., and Billies, G. (1979). Estimating carbon transfers in the plant rhizosphere. *In* "The Soil–Root Interface" (J. L. Harley and R. Scotts Russell, eds.), pp. 183–196. Academic Press, New York.

Watkins, S. H., Strand, R. F., De Bell, D. S., and Esch, J. (1972). Factors influencing ammonia losses from urea applied to northwestern forest soils. *Soil Sci. Soc. Am. Proc.* **36,** 354–357.

Weast, R. C., ed. (1977). "Handbook of Chemistry and Physics." CRC Press, Cleveland, Ohio.

Weiland, R. T., and Stutte, C. A. (1979). Pyrochemiluminescent differentiation of oxidized and reduced N forms evolved from plant foliage. *Crop Sci.* **19,** 545–547.

Weiland, R. T., and Stutte, C. A. (1980). Concomitant determination of foliar nitrogen loss, net carbon dioxide uptake, and transpiration. *Plant Physiol.* **65,** 403–406.

Weiland, R. T., Stutte, C. A., and Talbert, R. E. (1979). Foliar nitrogen loss and CO_2 equilibrium as influenced by three soybean (*Glycine max*) postemergence herbicides. *Weed Sci.* **27,** 545–548.

Weiss, R. F. (1981). The temporal and spatial distribution of tropospheric nitrous oxide. *JGR, J. Geophys. Res.* **86,** 7185–7195.

Wetselaar, R., and Farquhar, G. D. (1980). Nitrogen losses from tops of plants. *Adv. Agron.* **33,** 263–302.

Wetselaar, R., Passioura, J. B., and Singh, B. R. (1972). Consequences of banding nitrogen fertilizers in soil. I. Effect of nitrification. *Plant Soil* **36,** 159–175.

Wijler, J., and Delwiche, C. C. (1954). Investigations on the denitrification process in soil. *Plant Soil* **5,** 155–169.

Woldendorp, J. W. (1963). The influence of living plants on denitrification. *Meded. Landbouwhogesch. Wageningen* **63,** 1–100.

Yordy, D. M., and Ruoff, K. L. (1981). Dissimilatory nitrate reduction to ammonia. *In* "Denitrification, Nitrification and Atmospheric Nitrous Oxide" (C. C. Delwiche, ed.), pp. 171–190. Wiley, New York.

Yoshida, T., and Alexander, M. (1970). Nitrous oxide formation by *Nitrosomonas europaea* and heterotrophic microorganisms. *Soil Sci. Soc. Am. Proc.* **34,** 880–882.

Yoshida, T., and Alexander, M. (1971). Hydroxylamine oxidation by *Nitrosomonas europaea*. *Soil Sci.* **111,** 307–312.

Yoshinari, T., Hynes, R., and Knowles, R. (1977). Acetylene inhibition of nitrous oxide reduction and measurement of denitrification and nitrogen fixation in soil. *Soil Biol. Biochem.* **9,** 177–183.

Zablotowicz, R. M., and Focht, D. D. (1979). Denitrification and aerobic, nitrate-dependent acetylene reduction in cowpea *Rhizobium*. *J. Gen. Microbiol.* **111**, 445–448.

Zablotowicz, R. M., Eskew, D. L., and Focht, D. D. (1978). Denitrification in *Rhizobium*. *Can. J. Microbiol.* **24**, 757–760.

Zahniser, M., and Howard, C. J. (1979). The reaction of HO_2 with O_3. *J. Chem. Phys.* **73**, 1620–1626.

Chapter 6

Uptake and Assimilation of Mineral Nitrogen by Plants

R. J. HAYNES

I. INTRODUCTION

Nitrogen is generally viewed as a nutrient element that is central to plant growth because of its role in substances such as proteins, chlorophyll, nucleic acids, and nucleotides and nucleosides. In these, and other organic materials, N exists in a chemically reduced state and commonly constitutes 1.5 to 5% of the dry weight of plants.

The predominant form of N available to plants is NO_3^- since under most soil conditions NH_4^+-N is rapidly nitrified to NO_3^--N (see Chapter 3). The utilization of NO_3^- by higher plants involves several processes, including uptake, storage, translocation, reduction, and incorporation of N into organic forms. Some species reduce considerable quantities of NO_3^- in their roots whereas others translocate most of it to the leaves where it is reduced (Pate, 1980; Guerrero et al., 1981).

Ammonium is the major form of N available to plants under conditions that are unfavorable for the nitrification process to proceed (e.g., poor aeration and/or soil acidity). Ammonium cannot accumulate in cells to any great extent without damage to the plant and it is normally converted to amino acids or amides in the root and translocated to the tops in these organic forms.

In this chapter, the processes involved in the uptake and assimilation of mineral N by plants, and the factors affecting them, are discussed. Storage and transportation of nitrogenous solutes around the plant are reviewed. Mechanisms involved in the preference of plant species for NH_4^+

or NO_3^- nutrition, their adaptation to a fluctuating or localized supply of N, and their response to an under- and oversupply of N are also discussed.

II. PROCESSES OF UPTAKE

Higher plants generally acquire the bulk of their nutrient elements from the environment surrounding the root. Nutrients destined for use by the shoot must therefore first move through the root tissues before entering the xylem and being translocated to the shoots (Luttge and Higinbotham, 1979).

The generally accepted model of radial transport of ions across a root from the external solution to the xylem stream is shown schematically in Fig. 1. Ions diffuse into cell walls of the epidermal cells and active ion uptake may occur at the plasmalemma (the outer cell membrane) of these cells. The ions may then be transported across the cortex, endodermis, and pericycle in the symplast. The symplast represents a continuous system of protoplasm (cell contents) in which the protoplasm of adjacent cells is linked by intercellular bridges (plasmadesmata). Ions may also move passively into the continuum of nonliving cell wall material (apoplast or free space) of the cortex cells and then be absorbed across the plasmalemma of the cortical and endodermal cells, thus entering the symplast. Hydrophobic bands of suberin deposited in the radial walls of the endodermal cells (Casparian strips) essentially restrict apoplastic move-

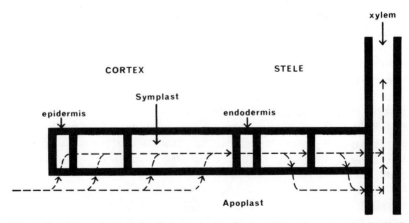

Fig. 1. Possible pathways of radial movement of solutes from the external soil solution through the root to the xylem. [Redrawn from Haynes (1980).]

ment from the free space of the cortex to the free space of the stele. Solutes entering the vascular tissue of roots possessing an intact endodermis must therefore do so by first being absorbed across the plasmalemma of the epidermal, cortical, or endodermal cells and then moving through symplast.

Absorption of ions across the plasmalemma of root cells is generally accepted to be an active process that often overcomes an unfavorable electrochemical gradient through the expenditure of energy. Active uptake is probably accomplished by "carriers," which are visualized as protein units in the plasmalemma and which associate with the ions from the apoplast and then discharge them into the symplast.

Uptake processes show a diminishing returns relationship to increasing ionic concentration in the external solution and the response can be fitted by the Michaelis–Menten equation (see Clarkson, 1974). The important parameters of this equation are K_m (the external concentration sufficient for a half maximum rate of uptake) and V_{max} (the maximum rate of uptake).

A. Ammonium Uptake

1. Description of Ammonium Uptake

The time-dependent uptake of NH_4^+ by plants can be characterized as two phases (Fig. 2). The initial phase (not inhibited by low temperatures or metabolic inhibitors such as KCN) is thought to represent passive exchange–adsorption phenomena in the negatively charged free space of roots. Such a process may be important when plants are transferred from nutrient solutions devoid of NH_4^+ to those containing ambient NH_4^+, but in the situation of a plant growing in the soil, it is likely to be unimportant (Nye and Tinker, 1977).

The second phase of uptake (Fig. 2) is sensitive to both low temperatures and metabolic inhibitors and represents active absorption of NH_4^+. The kinetics of active concentration-dependent NH_4^+ uptake have been studied in a number of plant species and it generally appears to have a multiphasic pattern (i.e., has more than two phases) (Joseph *et al.*, 1975; Hassan and van Hai, 1976; Dogar and van Hai, 1977; Nissen *et al.*, 1980). Three phases in the uptake of NH_4^+ by young soybean plants are shown in Fig. 3; the three phases of uptake coincide with regions of deficiency, luxury consumption, and toxicity as evidenced from the yield response of the plants. Thus, in terms of the efficiency of NH_4^+ uptake, characterization of the first phase of uptake is most important. In any case, the concentration of NH_4^+-N in soil solution is usually only in the region of 10

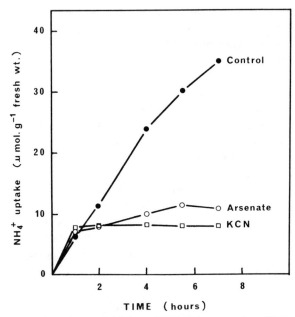

Fig. 2. Influence of respiratory inhibitors (KCN and arsenate) on NH_4^+ uptake by rice seedlings. [Data from Sasakawa and Yamamoto (1978).]

to 50 μM (Novoa and Loomis, 1981). The K_m and V_{max} for NH_4^+ uptake differ greatly among species. The K_m values are generally in the range 10 to 70 μM (Van den Honert and Hooymans, 1955; Tromp, 1962; Lycklama, 1963; Fried et al., 1965.

2. Mechanisms of Ammonium Uptake

On the whole, the literature indicates a similarity between NH_4^+ uptake and the uptake of other monovalent cations, particularly K^+ (Berlier et al., 1969; Epstein, 1972; Hassan and van Hai, 1976). Ammonium uptake appears in some cases to be competitive with that of K^+, indicating that a common uptake system may exist (Epstein, 1972). There is very limited information on the mechanisms by which NH_4^+ is absorbed by plant roots and, indeed, uncertainty surrounds the mechanism by which K^+ is absorbed. Evidence suggests that K^+ uptake either is directly linked to an ATPase that acts as an electrogenic H^+/K^+ pump or is mediated by specific carriers and occurs with simultaneous cotransport of protons maintained by a membrane-bound ATPase (Hodges, 1976; Poole, 1978; Lin, 1979; Clarkson and Hanson, 1980; Spanswick, 1981).

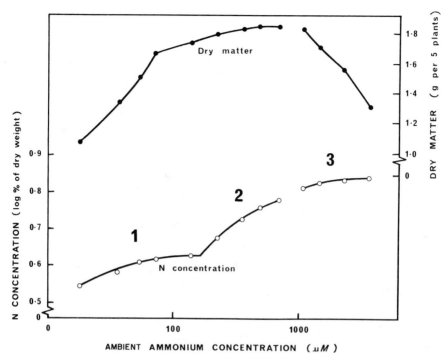

Fig. 3. Effect of external NH_4^+ concentration on total dry matter and N concentration in soybean plant tops after 20 days. [Data from Nissen *et al.* (1980).]

Not all experimental results indicate a similarity between NH_4^+ and K^+ uptake (Rufty *et al.*, 1982a; Zsoldos and Haunold, 1982). Zsoldos and Haunold (1982), for instance, observed a differential effect of low pH on the uptake of K^+ and NH_4^+ by young rice plants. Under reducing conditions Mengel *et al.* (1976) found that the uptake of NH_4^+ by rice plants was unaffected by ambient K^+ concentration. It is possible that at high pH, and under reducing conditions, NH_4^+ may be absorbed by plant roots mainly as NH_3 (Mengel *et al.*, 1976) since the neutral NH_3 molecule is thought to readily diffuse across cell membranes (Heber *et al.*, 1974; Moore, 1974).

B. Nitrate Uptake

1. Description of Nitrate Uptake

Nitrate uptake by NO_3^--depleted plants has been shown to exhibit an initial lag phase, followed by an accelerated exponential increase; with

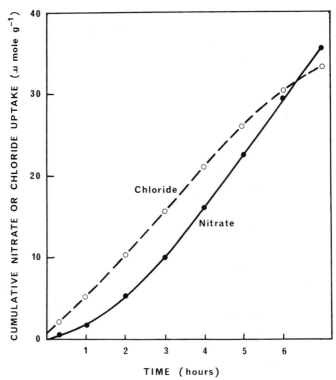

Fig. 4. Comparison of time-dependent uptake of Cl^- and NO_3^- (from 0.5 mM solutions) by corn seedlings. [Data from Jackson *et al.* (1973).]

time, the curve approaches linearity (see Fig. 4). This apparent lag period has been observed in many plant species (Huffaker and Rains, 1978; Jackson, 1978) and is in contrast to that of other ions (e.g., Cl; Fig. 4), which generally exhibit approximately linear rates during the initial stages of uptake. The accelerated rate appears to be inducible and is dependent on a critical internal NO_3^- concentration in a way similar to an inducible enzyme's response to its substrate (Jackson, 1978). The system is not inducible by other anions such as sulfate, chloride, chlorate, borate, or molybdate at concentrations commonly employed in standard nutrient solutions.

The concentration-dependent kinetics of NO_3^- uptake do not appear to have been studied in any detail although the K_m for the first phase of uptake generally falls in the range 10 to 100 μM (Van den Honert and Hooymans, 1955; Lycklama, 1963; Rao and Rains, 1976; Van de Dijk *et al.*, 1982; Youngdahl *et al.*, 1982) with V_{max} values ranging from 2.5 to 10 mol gm^{-1} (fresh weight of plant) hr^{-1}. In general the concentration of NO_3^--N in soil solution is low: 1000 μM or less with 100 μM not uncommon (Novoa and Loomis, 1981).

2. Mechanisms of Nitrate Uptake

Uptake of NO_3^- by plants is an energy-requiring process and it is restricted by inhibitors of RNA and protein synthesis (Jackson et al., 1973; Tomkins et al., 1978) as well as inhibitors of respiratory and oxidative phosphorylation (Rao and Rains, 1976).

Huffaker and Rains (1978) have suggested that a protein located at the plasmalemma of plant cells in some way mediates NO_3^- uptake. The activity of the protein would be enhanced by the substrate (NO_3^-) and the overall process of NO_3^- uptake would thus be accelerated as the endogenous NO_3^- level increased. However, as yet, the actual mechanisms of NO_3^- uptake is purely speculative. It is generally thought that NO_3^- transport is linked to a membrane-bound ATPase (Huffaker and Rains, 1978; Poole, 1978), although mechanisms of anion absorption based on a transmembrane pH gradient have been proposed (e.g., Kirby, 1981).

Although NO_3^- uptake by NO_3^--depleted plants appears to be inducible by a critical internal NO_3^- concentration, increasing concentrations of NO_3^- in roots do generally result in a marked decrease in the uptake of ambient NO_3^- (Jackson et al., 1976a). This suggests the existence of a feedback control mechanism for active NO_3^- uptake. Indeed, in both Hordeum vulgare and Daucus carota roots, net NO_3^- influx appears to be subject to feedback control by the intercellular NO_3^- plus Cl^- concentration (Cram, 1973; Smith, 1973).

It is noted here that the net uptake of NO_3^- by plants is the difference between active influx of NO_3^- across the plasmalemma and passive efflux of NO_3^- down a diffusion gradient (Morgan et al., 1973; Jackson et al., 1976a). In nutrient solution experiments, passive efflux has sometimes been shown to be a significant component of the net uptake rate (Morgan et al., 1973; Ashley et al., 1975).

C. Factors Influencing Uptake

1. Repression of Nitrate Uptake by Ammonium

Although uptake rates of NH_4^+ are normally unaffected by the presence or absence of NO_3^- in nutrient solution (Mengel and Viro, 1978; Munns and Jackson, 1978; Youngdahl et al., 1982), ambient NH_4^+ has been shown to restrict net NO_3^- uptake by numerous plant species (Minotti et al., 1969; Rao and Rains, 1976; Sahulka, 1977; Buczek, 1979; MacKown et

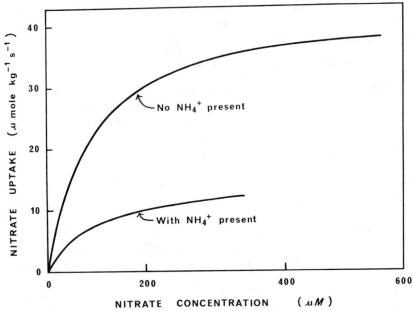

Fig. 5. Uptake of NO_3^- by rice seedlings as a function of concentration in the presence or absence of ambient NH_4^+. [Redrawn from Youngdahl *et al.* (1982).]

al., 1982; Youngdahl *et al.*, 1982). Such an effect is clearly illustrated in Fig. 5. Nonetheless, there have been some examples where NH_4^+ had little or no effect on NO_3^- uptake (Schrader *et al.*, 1972; Neyra and Hageman, 1975; Edwards and Barber, 1976a).

In the majority of cases, NH_4^+ appears to have an inhibitory effect on NO_3^- uptake that is independent of any such effect on NO_3^- reductase enzyme activity (e.g., Rao and Rains, 1976; MacKown *et al.*, 1982). Indeed, Rao and Rains (1976) showed that ambient NH_4^+ reduced NO_3^- uptake in short-term experiments without having any detectable effect on nitrate reductase activity.

Some workers believe that it is the endogenous level of cytoplasmic NH_4^+ in root tissue that inhibits the NO_3^- uptake mechanism (Jackson *et al.*, 1976b; Jackson, 1978) while others believe that end product regulation of NO_3^- uptake is exerted by amino acids accumulated in the roots during NH_4^+ nutrition (Heimer and Filner, 1971; Doddema and Otten, 1979). Evidence for both modes of action has been presented (e.g., Jackson, 1978; Doddema and Otten, 1979) and further research will be required to characterize the exact mechanism(s) involved.

2. Rhizosphere pH

Active uptake of cations at the plasmalemma of roots involves active excretion of OH^- or HCO_3^- ions (Nye, 1981). These processes are essential to maintain electroneutrality since no net charge can cross the soil–root boundary.

With NH_4^+ nutrition, plants absorb cations in excess of anions (N being the element often absorbed in the largest amounts) so that plant growth results in net efflux of H^+ ions into the rhizosphere. In nutrient solutions or sand culture, pH values may fall as low as 2.8 with NH_4^+ nutrition (Maynard and Barker, 1969). Similarly, when soil is supplied with NH_4^+-rather than NO_3^--N, the growth of field- or container-grown plants has been shown to cause a decrease in soil pH close to the root surface (Miller *et al.*, 1970; Riley and Barber, 1971; Smiley, 1974).

In contrast, when NO_3^- is the major form of N supplied, plants absorb an excess of anions and there is a net efflux of HCO_3^- or OH^- ions resulting in an increase in rhizosphere pH (Pierre *et al.*, 1970; Bagshaw *et al.*, 1972; Pierre and Banwart, 1973; Smiley, 1974).

The contrasting effects of the uptake of NH_4^+ or NO_3^- by plants on the pH of leachates from sand cultures is clearly shown in Table I. The effects of NH_4^+ and NO_3^- nutrition on rhizosphere pH have rather detrimental effects on the absorption of the respective ions. For example, in nutrient solution, maximum absorption of NH_4^+ by plants occurs at a pH of around 7 to 8 while that for NO_3^- occurs at around pH 4 to 5 (Sheat *et al.*, 1959; Rao and Rains, 1976). Thus, the pH for maximum absorption of the respective ions and the pH at which the rhizosphere tends to, following their absorption, are at opposite ends of the pH scale. It is interesting to note that the toxic effect of NH_4^+ nutrition is at least partially due to the rhizosphere acidification associated with NH_4^+ uptake (see Section V,D).

Table I

Effect of Nitrogen Source on the pH of Leachate from Sand Culture after Growth of Corn,[a] Cucumber, and Onion Plants[b]

Treatment	Corn	Cucumber	Onion
NH_4^+	2.8	3.9	3.8
NO_3^-	6.9	7.1	7.2

[a] Growth period was: corn, 14 days; cucumber, 10 days; and onion, 31 days.
[b] Data from Maynard and Barker (1969).

3. Interactions among Ions

Competition among ions during the uptake process and enhancement of one ion by another are common physiological occurrences during accumulation of ions by plants (Epstein, 1972). Unfortunately, although much is known regarding the effects of NH_4^+ and NO_3^- nutrition on the uptake of other ions by plants, less is known regarding the effects of the other ions on NH_4^+ and NO_3^- uptake.

a. Ammonium. Generally, NH_4^+, Ca^{2+}, Mg^{2+}, and K^+ appear to compete with each other during ion accumulation by plants (Haynes and Goh, 1978; Reisenauer, 1978). Commonly NH_4^+ uptake results in large reductions in K^+ uptake and to a lesser extent *vice versa* (Ajayi *et al.*, 1970; Moraghan and Porter, 1975). However, Rufty *et al.* (1982a) observed that although increasing concentrations of ambient NH_4^+ progressively inhibited K^+ uptake, increasing concentrations K^+ had no effect on NH_4^+ uptake. Some exceptions where low concentrations of ambient NH_4^+ have increased or had no effect on cation uptake have also been reported (Blair *et al.*, 1970; Rayar and van Hai, 1977; Reisenauer, 1978).

The decreased cation uptake accompanying increasing ambient NH_4^+ levels is usually discussed in terms of competition at the plasmalemma either with NH_4^+ ions *per se* or with H^+ ions excreted during active NH_4^+ uptake (Cox and Reisenauer, 1973; Haynes and Goh, 1978). High concentrations of H^+ ions in the free space of roots could also inhibit movement of other cations within the free space (Haynes, 1980).

In comparison with NO_3^- nutrition, NH_4^+ nutrition generally results in increased uptake of phosphate and sulfate (Blair *et al.*, 1970). In the soil situation, the lowering of rhizosphere pH, caused by NH_4^+ nutrition, increases the ratio of $H_2PO_4^-$ to HPO_4^{2-} ions (Soon and Miller, 1977). The $H_2PO_4^-$ ion is absorbed several times faster than $H_2PO_4^{2-}$ and, in addition, HPO_4^{2-} salts have a tendency to precipitate at the root–soil surface (Miller *et al.*, 1970).

b. Nitrate. Nitrate nutrition generally has the opposite effect to ammonium; that is, nitrate stimulates cation uptake and inhibits anion uptake (Haynes and Goh, 1978). Although NO_3^- uptake is apparently a rather specific process and not greatly affected by the presence of other anions (e.g., Cl^-, Br^-, and SO_4^{2-}), in the ambient medium (Rao and Rains, 1976), NO_3^- uptake does generally result in a decreased uptake of other anions (Blair *et al.*, 1970). An inverse relationship between NO_3^- and Cl^- accumulation in plants has been observed by several workers (Weigel *et al.*, 1973; Hiatt and Leggett, 1974; Kafkafi *et al.*, 1982). The possibility that NO_3^- uptake is subject to feedback control by intercellular NO_3^- plus Cl^-

concentration has previously been noted. Additions of high levels of NO_3^- can partially alleviate the detrimental effect of soil salinity on plant growth by reducing Cl^- uptake by the plant (Kafkafi et al., 1982).

The phenomenon of increased cation uptake with increasing ambient NO_3^- concentration is well documented (Jackson and Williams, 1968; Cox and Reisenauer, 1973; Jackson et al., 1974). The uptake of both monovalent and and divalent cations is stimulated and it may be that the rise in rhizosphere pH, as a consequence of rapid NO_3^- uptake, produces favorable conditions for cation uptake (Maas, 1969). An association between the onset of the accelerated phase of NO_3^- uptake and the enhancement of K^+ uptake was observed by Jackson et al. (1974). Conversely, increasing the supply of Ca^{2+} and K^+ in nutrient solution generally accelerates the rate of NO_3^- uptake (Minotti et al., 1968, 1969).

c. *Perspective*. Since N is normally taken up by plants in greater quantities than most other elements (although the uptake of K and Ca is often of a similar order) it would seem likely that effects of N sources on uptake of other ions would be considerably greater than the reverse situation. Nonetheless, in situations where plants are relying principally on NH_4^+ as a source of N, the supply of K^+ could possibly have an inhibitory influence on N uptake.

4. Supply of Photosynthates

Since both NH_4^+ and NO_3^- uptake appear to be active processes, it is likely that carbohydrate (energy) supply will influence both processes. Indeed, Sasakawa and Yamamoto (1978) found that removal of the endosperm from *Oryza* seedlings suppressed both NH_4^+ and NO_3^- uptake (Fig. 6) while exogenously supplied sucrose restored uptake. Low light intensity greatly reduces uptake of both forms of N (Ta and Ohira, 1981).

Several workers have demonstrated the strong requirement by the root for a continual energy supply in order to sustain the NO_3^- uptake system (Minotti and Jackson, 1970; Jackson et al., 1973; Pearson and Steer, 1977). Removal of endosperm from *Zea* seedlings, for example, inhibited NO_3^- uptake more than that of Cl^- (Jackson et al., 1973). The diurnal rhythm of NO_3^- uptake by plants, which is greatest around midday (Pearson and Steer, 1977; Clement et al., 1978), closely coincides with translocation of photosynthates away from leaves (Pearson and Steer, 1977).

The uptake of NH_4^+ by plants also shows a wide diurnal variation (Van Egmond, 1978) that can be disturbed by providing continuous light or by supplying glucose to the nutrient medium during darkness. The decline in NH_4^+ uptake in darkness is apparently due to the depletion of carbohydrate reserves in roots that are used for the assimilation of NH_4^+ (Reise-

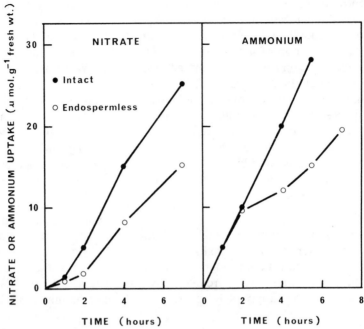

Fig. 6. Effect of endosperm removal on the uptake of ambient NH_4^+ and NO_3^- (0.5 mM; pH 6.5) by rice seedlings grown in solution culture. [Data from Sasakawa and Yamamoto (1978).]

nauer, 1978). It seems that the supply of C skeletons for the immediate assimilation and detoxification of absorbed NH_4^+ may limit NH_4^+ uptake.

Such an effect is less important during NO_3^- absorption, since NO_3^- is not as toxic as NH_4^+ and may be accumulated in root vacuoles or translocated to the shoots. Thus, Michael *et al.* (1970) showed that a reduction of the supply of carbohydrates to the roots, by ringing *Phaseolus vulgaris* plants to remove the phloem, resulted in reduced uptake of N; the uptake of NH_4^+ was reduced markedly more than that of NO_3^-.

5. Temperature

It is well known that NO_3^- uptake is much more hindered by low temperatures than is the uptake of NH_4^+ (Williams and Vlamis, 1962; Zsoldos, 1971; Frota and Tucker, 1972; Sasakawa and Yamamoto, 1978; Clarkson and Warner, 1979). Frota and Tucker (1972), for example, found that the absorption of NH_4^+ by *Lactuca sativa* was greater than that for NO_3^- at 8°C and reached a maximum of 25°C in the rooting medium. Nitrate uptake usually becomes greater than NH_4^+ at around 23°C and increases up to 35°C (Lycklama, 1963; Frota and Tucker, 1972).

Grasmanis and Nicholas (1971) found that apple trees (*Pyrus malus*) growing outside in nutrient solutions receiving both NH_4^+- and NO_3^--N in a ratio of 1 : 7 absorbed both ions in nearly equivalent amounts, except for the winter period when NH_4^+ uptake was in excess of that of NO_3^-. This difference was attributed to the relatively lower rate of uptake of NO_3^- at the lower winter temperatures.

6. Mycorrhizal Associations

Three major types of symbiotic associations between fungi and higher plants are grouped together as mycorrhizal associations. These are the ectotrophic, vascular arbuscular (VA), and ericaceous mycorrhizae. Ectotrophic mycorrhizas are mainly restricted to a few tree families and are characterized by a fungal sheath that completely envelopes the tips of lateral roots and grows between the cortical cells. The fungal sheath markedly influences the amounts and types of compounds reaching the roots and mycorrhizal infection also increases the longevity of rootlets (Smith, 1980). Ectotrophic mycorrhiza can also produce mycelial strands that may penetrate the soil and litter for up to 12 cm (Skinner and Bowen, 1974).

In the VA and ericaceous mycorrhiza, there is extensive fungal growth both between and within root cortical cells, but no sheath develops. An extensive loose network of hyphae extends into the soil to a distance of up to 8 cm (Read and Stribley, 1975; Rhodes and Gerdemann, 1975).

The extensive growth of all three types of mycorrhizal fungi in the soil, outside the root, enables plants to use soil nutrients more effectively through the exploration of a greater volume of soil than would otherwise be achieved. This is particularly important when considering the nutrition of plants for immobile nutrients such as phosphorus (Nye and Tinker, 1977). However, mycorrhizal associations can also be important factors influencing N nutrition of plants, particularly when the relatively immobile NH_4^+ rather than the mobile NO_3^- is the major source of plant-available N (Bowen and Smith, 1981; Smith, 1980).

In culture, ectomycorrhizal and ericoid mycorrhizal fungi generally appear to prefer NH_4^+- to NO_3^--N (Melin and Nilsson, 1952; Lundeberg, 1970; Stribley and Read, 1974). Furthermore, ericaceous mycorrhizas have consistently been shown to increase N uptake by ericaceous plants such as *Vacinnium* spp. (Read and Stribley, 1973; Stribley and Read, 1974, 1976). These plants grow in soils in which nitrification rates are low and NH_4^+ is the major form of mineral N available. This is also true of

Pinus spp., which have also sometimes been observed to have increased N concentrations as a result of ectotrophic infection (Smith, 1980).

It is the VA mycorrhizae that might be expected to be most important in respect to the utilization of NO_3^- since plants associated with them are common in soils with high rates of nitrification. However, the high mobility of NO_3^- in soils probably means that mycorrhizal associations have little influence on NO_3^- uptake. Increases in N content of plants bearing VA mycorrhiza are occasionally observed (e.g., Holevas, 1966; Possingham and Groot Obbink, 1971). In general, however, additions of NO_3^--N to the soil decrease VA mycorrhizal infection (Kruckelmann, 1975; Chambers *et al.*, 1980; Azcon *et al.*, 1982) apparently by reducing the infectivity of mycorrhizal propagules (spores or infected root fragments) (Mosse and Phillips, 1971; Chambers *et al.*, 1980). Applications of nitrogenous fertilizers also tend to depress the development of ectotrophic mycorrhizae associated with some forest tree species (Redhead, 1980).

There is increasing interest in the possible role of mycorrhizae as agents for the use of simple organic nitrogenous compounds by plants. Indeed, in culture, the ericaceous mycorrhizal fungi of *Vaccinium macrocarpon* can utilize glutamic acid and alanine as effectively as NH_4^+ (Pearson and Read, 1975). It cannot, however, effectively use N in complex organic compounds such as humic acid (Stribley and Read, 1980). Absorption and translocation of amino acids has been demonstrated for both ectotrophic (Melin and Nilsson, 1952) and ericaceous (Stribley and Read, 1975, 1980) mycorrhizal associations. Results presented in Table II demonstrate that

Table II

Effect of Several Nitrogen Sources on Shoot Dry Weight of *Vaccinium macrocarpon* in Mycorrhizal and Nonmycorrhizal Sterile Treatments[a]

Nitrogen source[b]	Mycorrhizal	Nonmycorrhizal
No nitrogen	6.0	8.3
Ammonium	43.4	41.4
Glycine	43.9	11.6
Alanine	47.6	22.5
Aspartic acid	29.5	7.0
Glutamic acid	39.4	8.2
Glutamine	44.6	18.2

[a] Data from Stribley and Read (1980).
[b] Plants grown in sand culture for 8 weeks with nitrogen supplied at a concentration of 20.5 mg N liter^{-1}.

mycorrhizal plants of *V. macrocarpon* were able to use all amino acids tested, except aspartic acid, as readily as they used NH_4^+, while nonmycorrhizal plants could not.

Thus, through the penetration of organic matter, mycorrhizal fungi may be able to compete successfully with other soil microorganisms for simple organic nitrogenous compounds, with a consequent benefit to the host plant (Bowen and Smith, 1981).

D. Foliar Absorption

So far in this discussion, the absorption of N by roots has been considered. However, in agricultural systems, the application of foliar-applied nutrients is becoming increasingly important. Furthermore, it is well known that plant foliage can absorb nitrogenous gases, particularly NH_3.

1. Foliar-Applied Nitrogen

Urea is probably the most widely used nutrient spray material in crop production. It is applied singly or in combination with many formulated mixtures. Other nitrogenous compounds such as ammonium nitrate and potassium nitrate are also used to a limited extent. The role and effectiveness of foliar applications of nitrogenous substances in crop production are discussed in Chapter 7.

The discussion below concentrates on mechanisms and pathways of foliar absorption.

a. Cuticular penetration. The structure of the terrestrial plant leaf is adapted to lessen water loss, hence it is not designed to absorb water and nutrients. Indeed, the external and internal surfaces of aerial plant parts are covered with a fatty lipoidal layer known as the cuticle, which retards outward water movement. Much evidence points to the leaf cuticle being the major barrier to foliar penetration of water-soluble material (Hull *et al.*, 1975; Leece, 1976). Substances may penetrate the cuticle directly or through apparently thinner, more permeable areas of the cuticle associated with guard cells of the stomata and trichomes (unicellular or multicellular projections of the epidermis) (Schonherr and Bukovac, 1970; Hull *et al.*, 1975).

Because of the presence of air within the substomatal cavity, stomatal penetration is not a common pathway of entry for aqueous solutions with a surface tension similar to that of water (Green and Bukovac, 1974). However, the use of certain surfactants can lower the surface tension sufficiently for stomatal penetration (Green and Bukovac, 1974; Weinbaum and Neumann, 1977). The use of such surfactants can markedly increase the uptake and use of foliar-applied N by the plant (Weinbaum and Neumann, 1977), possibly because of the relative ease with which solutes can traverse the thin cuticle within the substomatal cavity.

While diffusion is assumed to be the major driving force for the movement of substances through the cuticle, urea penetrates the cuticular membrane with a velocity higher than that which would be expected by simple diffusion (Yamada *et al.*, 1965). Indeed, the extent of urea penetration through the cuticle exceeds that of most ions by 10- to 20-fold and is independent of concentration. This increased permeability for urea also favors foliar penetration of ions such as iron, phosphate, and magnesium when they are applied together with the urea (Yamada *et al.*, 1966). The exact mechanism of the action of urea on the cuticle is unclear, although Yamada *et al.* (1965) concluded that its effect is based on a loosening of the structure by changing ester, ether, and diether bonds between macromolecules of cutin.

2. Subcuticular Movement

The movement of foliar-applied nutrients through the leaf following cuticular penetration is still not completely understood (Haynes and Goh, 1977; Kannan, 1980).

Initial movement through the cell wall, immediately below the cuticle, may be facilitated by areas of wall growth and repair. Such areas of wall stress are thought to result from diurnal fluctuations in leaf turgidity and solutes appear to diffuse more readily through these areas than through the intact cell wall (Lyon and Meuller, 1974; Hull *et al.*, 1975).

Following movement through the cell walls of the epidermis, solutes may follow a number of alternative pathways through the leaf to the vascular tissue (Fig. 7).

Solutes could move passively in the apoplast through the walls of leaf cells and then be actively loaded into the phloem. Alternatively, they may be actively transported across the plasmalemma into leaf cells, where they could be transported in the symplast and then be actively loaded into the phloem. Assimilation and transformation of solutes within the symplast may occur (e.g., urea hydrolysis and ammonia assimilation) and then organic molecules such as amino acids may be used within the leaf or transported in the symplast to the phloem. At some point during inward movement, foliar-applied solutes or their assimilates within the symplast could also leak back into the apoplast.

The relative importance of the above pathways is not known, although it is known that a considerable portion of foliar-applied urea is rapidly metabolized in leaves (Dilley and Walker, 1961a,b; Shim *et al.*, 1973) so that a symplastic pathway is indicated.

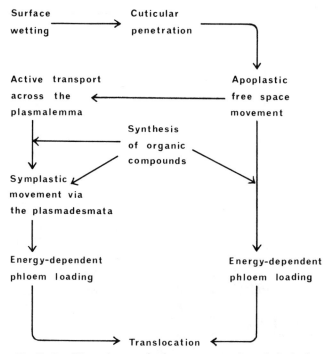

Fig. 7. Possible pathways of solute movement through the leaf.

3. Foliar Absorption of Gases

In artificial atmosphere, it has been demonstrated that plant foliage can absorb NO_2 (Tingey, 1968; Hill and Chamberlain, 1976; Yoneyama et al., 1980) and NO (Hill, 1971). The extent of such absorption under natural and/or agricultural conditions is unknown.

Other workers have demonstrated that NH_3 can be absorbed from the air by plants (Hutchinson et al., 1972; Porter et al., 1972) although plants are also known to release small amounts of NH_3 to the atmosphere (Martin and Ross, 1968; Farquhar et al., 1979).

Absorption of NH_3 (originating from volatilization at the soil surface) by the plant cover above has been measured under field conditions (Denmead et al., 1976, 1978). Denmead et al. (1976), for example, found considerable losses of NH_3 from grazed pastures (13 gm N ha^{-1} hr^{-1}), but losses from ungrazed pastures were comparatively small (2 gm N ha^{-1} hr^{-1}). Measurements within the canopy of the ungrazed pasture (Fig. 8) indicated a large production of NH_3 near the ground and almost complete absorption of it by the plant cover.

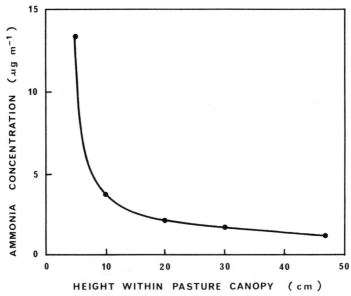

Fig. 8. Ammonia concentrations at various heights within an ungrazed ryegrass–subterranean clover pasture. [Redrawn from Denmead *et al.* (1976). Reprinted with permission from Pergamon Press.]

The role of foliar absorption of nitrogenous gases in terrestrial N cycles requires further study. So too does the mechanism of absorption by plant foliage. It is generally assumed to occur primarily by diffusion into stomata and thence into the intercellular spaces of leaves. However, calculations by Denmead *et al.* (1976, 1978) indicated that NH_3 absorption by plants appeared to be too large in magnitude to be accounted for by stomatal uptake alone.

III. PROCESSES OF ASSIMILATION

A. Reduction of Nitrate to Nitrite

Many detailed reviews dealing with different aspects of NO_3^- reduction exist (Hewitt, 1975; Aparicio and Maldonado, 1978; Notton and Hewitt, 1978; Oaks, 1978; Beevers and Hageman, 1980; Guerrero *et al.*, 1981) and the process is only outlined below.

The first step in the assimilatory reduction of NO_3^- in higher plants is

catalyzed by the enzyme complex nitrate reductase. The enzyme catalyzes the reduction of NO_3^- to NO_2^- by reduced pyridine nucleotides:

$$NO_3^- + NAD(P)H + H^+ \overset{2e^-}{\rightleftharpoons} NO_2^- + NAD(P) + H_2O$$

The enzyme can be divided into three subclasses (Guerrero *et al.*,
1981). EC 1.6.6.1 is specific for NADH (reduced nicotinamide dinucleotide) and is present in most higher plants, EC 1.6.6.2 can use either
NADH or NADPH (reduced nicotinamide dinucleotide phosphate) with
the same effectiveness and is most prevalent in green algae, while nitrate
reductases of molds EC 1.6.6.3 are NADPH specific. A simultaneous
presence of two different reducing enzymes, NADH- and NAD(P)H-de-
pendent, has been demonstrated in soybean leaves (Campbell, 1976) and
young rice seedlings (Shen *et al.*, 1976).

The nitrate reductase enzyme complex has a high molecular weight
varying from 220,000 to 600,000 depending on the organisms in which it
occurs (Notton and Hewitt, 1978). It contains several prosthetic groups
(Notton and Hewitt, 1978; Guerrero *et al.*, 1981) such as FAD (flavin
adenine dinucleotide), cytochrome b_{557}, and molybdenum, which seem to
be ubiquitous constituents of this enzyme.

The nitrate reductase complex can be visualized as being made up of
two moieties that participate jointly and sequentially in the transfer of
electrons from NAD(P)H to NO_3^- (see Fig. 9). The first diaphorase

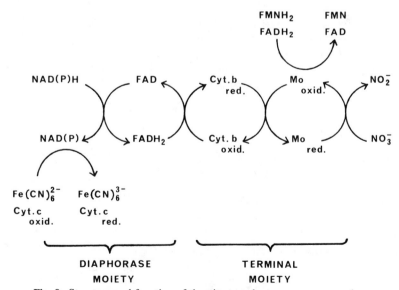

Fig. 9. Structure and function of the nitrate reductase enzyme complex.

(NAD(P)H-activating) moiety results in the reduction by NAD(P)H of a variety of electron acceptors (e.g., ferricyanide or cytochrome c). The participation of hydrogen acceptor FAD in the diaphorase moiety is well known (Losada and Guerrero, 1979; Beevers and Hageman, 1980).

The second, terminal (NO_3^--activating) moiety of the complex is capable of reducing NO_3^- to NO_2^- using either reduced violgen dyes or reduced flavins ($FADH_2$ (reduced flavin adenine dinucleotide) and $FMNH_2$ (reduced flavin adenine mononucleotide)). The participation of Mo in the terminal moiety is well established (Hewitt and Notton, 1980) and it is speculated that the change in oxidation state during NO_3^- reduction may be the Mo(VI)/Mo(IV) couple (Guerrero et al., 1981). The active participation of cytochrome b_{557} in electron flow is also well known (Beevers and Hageman, 1980) and its site of action appears to be between FAD and Mo.

Thus, as illustrated in Fig. 9, the pathway of electrons from NAD(P)H to nitrate through nitrate reductase can be depicted as

$$NAD(P)H \longrightarrow (FAD \longrightarrow cyt.\ b_{557} \longrightarrow Mo) \longrightarrow NO_3^-$$

The intracellular localization of nitrate reductase in green tissues is still unclear since, although it is frequently referred to as a cytoplastic enzyme, it has in some cases been associated, at least in part, with the chloroplast (Losada and Guerrero, 1979). A variety of shuttle systems have been proposed to explain how reducing power is transferred from the chloroplast to cytoplasmic NAD(P)H for nitrate reduction (Krause and Heber, 1976).

Nitrate reductase exists in the cells of roots as well as those of shoots and the capacity of root tissue to assimilate NO_3^- appears to be related to its carbohydrate content (Minotti and Jackson, 1970). The mechanism by which reductive energy is directed from the carbohydrate to a physiological reductant is, however, unknown (Hewitt, 1975; Guerrero et al., 1981).

B. Reduction of Nitrite to Ammonium

Several workers have compiled detailed reviews of the enzymatic reduction of nitrite to ammonium (e.g., Vennesland and Guerrero, 1979; Beevers and Hageman, 1980; Guerrero et al., 1981) and the major features are outlined below.

The enzyme responsible for the reduction of nitrite to ammonium in photosynthetic cells is ferredoxin-nitrite reductase (EC 1.7.7.1). The reaction, which involves the transfer of six electrons, is shown below:

$$NO_2^- + GFd_{red} + 8H^+ \xrightarrow{6e^-} NH_4^+ + GFd_{ox} + 2H_2O$$

The ferredoxin-nitrite reductases from several sources can also catalyze the reduction of hydroxylamine to ammonium, but at a lower rate than that for nitrite. Nitrite reduction in nonphotosynthetic organisms (e.g., fungi) is carried out by a nitrite reductase (NAD(P)H) (EC 1.6.6.4).

Ferredoxin-nitrite reductase appears to be composed of a single polypeptide chain of about 600 amino acid residues and has a relatively small molecular weight of approximately 62,000 (Hucklesby et al., 1978). It appears to contain two prosthetic groups, a tetranuclear 4Fe–4S iron–sulfur center and a heme-containing protein known as sirohem (Vega et al., 1980).

The pathway of electron flow from reduced ferredoxin to nitrite, via nitrite reductase, can be depicted as follows (Vega et al., 1980; Guerrero et al., 1981):

$$Fd_{red} \longrightarrow (4Fe–4S) \longrightarrow sirohem \longrightarrow NO_2^-$$

Sirohem is thought to be the substrate binding site and the nitrite is reduced to ammonium without the accumulation of free nitrogenous compounds of intermediate redox state (Vega and Kamin, 1977; Vega et al., 1980).

In the leaves of higher plants, nitrite reductase is localized in the chloroplast (Hewitt, 1975; Guerrero et al., 1981) but, in addition, plastids containing nitrite reductase occur in the root (Dalling et al., 1973; Emes and Fowler, 1979). The enzyme from roots, like that from leaves, can use ferredoxin but not nicotinamide or flavin nucleotides as electron donors. Nevertheless, there are no reports of ferredoxin in extracts from roots and the source of reductant for nitrite reduction in roots has not yet been established (Guerrero et al., 1981).

C. Urea Hydrolysis

Urea is normally formed in plants as a result of the hydrolysis of arginine to ornithine catalyzed by the enzyme arginase (EC 3.5.3.1) (Thompson, 1980):

$$arginine + H_2O \rightleftharpoons ornithine + urea$$

Urea is not known to be incorporated *in toto* into organic molecules and it is metabolized in higher plants to NH_3 and CO_2 by the urease enzyme (EC 3.5.1.5) as follows:

$$O=C\begin{matrix} NH_2 \\ \\ NH_2 \end{matrix} + H_2O \rightleftharpoons CO_2 + 2NH_3$$

The NH_3 is then assimilated as described in the next section. As already noted, foliar-applied urea can sometimes be a significant source of N for crop plants.

Since two C—N bonds are broken during urea hydrolysis, the reaction is the net result of two component reactions; however, the exact nature of these is not clear (Reithel, 1971). The urease enzyme has a molecular weight of about 480,000 and is composed of six identical subunits (Bailey and Boulter, 1969). Nickel appears to be an active site component of urease from some plants (Polacco, 1977; Gordon *et al.*, 1978).

The urease enzyme is widespread among plant species (Thompson, 1980) and has been found in leaves, roots, and bark of plants with actively growing tissues possessing greater activity than senescing ones (Shim *et al.*, 1973). Urease has generally been found to be a substrate-inducible enzyme induced by additions of urea to tissues (Cook and Sehgal, 1970; Shim *et al.*, 1973; Polacco, 1976). This fact makes foliar applications of urea a viable commercial proposition. The increase in urease activity following foliar applications of urea is apparently due to *de novo* synthesis of the enzyme (Matsumoto *et al.*, 1968; Shim *et al.*, 1973).

D. Ammonia Assimilation

Ammonia assimilation has a central role in plant N metabolism since, in addition to NH_4^+ being absorbed directly by roots, NH_4^+ is also the product of NO_3^- and urea assimilation as well as molecular nitrogen fixation. Since amino acids, either free or protein-bound, are the predominant form of organic N, the major product of assimilation is usually considered to be amino N. Several workers have reviewed the processes of ammonia assimilation in detail (Fowler and Barker, 1978; Lea and Miflin, 1979; Miflin and Lea, 1980).

The major pathway of ammonia assimilation in higher plants is through the glutamate synthase cycle (Fig. 10), which involves two reactions oper-

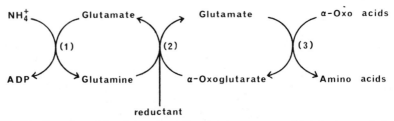

Fig. 10. Operation of the glutamate synthase cycle. Enzymes: (1) glutamine synthetase, (2) glutamate synthase, and (3) transaminase.

ating in series and catalyzed by two enzymes: glutamine synthetase and glutamate synthase. A key characteristic of this route is its cyclical nature in which glutamate acts as both the acceptor and product of ammonia assimilation.

1. Glutamine Synthetase

Initial incorporation of NH_3 into the amide position of glutamine is catalyzed by the enzyme glutamine synthetase (EC 6.3.1.2):

$$\text{glutamate} + NH_3 + ATP \rightleftharpoons \text{glutamine} + ADP + P_i$$

Two isoforms of glutamine synthetase are known to exist in the leaves of many plant species (Guiz et al., 1979; Mann et al., 1979a,b; Hirel and Gadal, 1980; Nishimura et al., 1982); one, named GS_1, is located in the cytosol and the other, named GS_2, is located in the chloroplast. Research suggests that GS_1 may be present in leaves, roots, and seeds while GS_2 is restricted to green tissues (Hirel et al., 1982).

The enzyme has a molecular weight in the range of 350,000 to 400,000 and consists of eight apparently identical subunits (McParland et al., 1976; Stewart et al., 1980). It exhibits an absolute requirement for divalent cations (Mg^{2+} and to a lesser degree Mn^{2+} and CO^{2+}) (O'Neal and Joy, 1974). Its reaction mechanism probably involves binding of substrates in an ordered sequence (Stewart et al., 1980).

2. Glutamate Synthase

In the presence of a reducing source, glutamate synthase catalyzes the transfer of the amide group of glutamine to α-oxoglutarate resulting in the formation of the amino acid glutamate:

$$\text{glutamine} + \alpha\text{-oxoglutarate} + Fd_{red} \text{ (or NAD(P)H)} \rightleftharpoons 2\,\text{glutamate} + Fd_{ox} \text{ (or NAD(P))}$$

Three forms of glutamate synthase are known to exist in plants: the NADH-dependent (EC 1.4.1.14) and NADPH-dependent (EC 1.4.1.13) forms are found in nonphotosynthetic leaf and root tissues (Fowler et al., 1974; Suzuki et al., 1982) while the ferredoxin-dependent (EC 1.4.7.1) enzyme is found in photosynthetic tissues (Stewart and Rhodes, 1978; Suzuki and Gadal, 1982) and also in roots (Miflin and Lea, 1975; Suzuki et al., 1982). The enzyme from both roots and leaves is located in the plastids (Emes and Fowler, 1979; Suzuki et al., 1981; Suzuki and Gadal, 1982).

Most reports indicate the enzyme is a single polypeptide chain with a molecular weight in the range 140,000 to 180,000 (Wallsgrove et al., 1977; Matoh et al., 1980; Tamura et al., 1980). Very little is known of the reaction mechanism of the enzyme (Stewart et al., 1980), although it may be sequential with respect to α-oxoglutarate and glutamine, the former binding first.

3. Further Transformation

Incorporation of NH_4^+ into an amino acid is then followed by transamination reactions in which the amino group ($-NH_2$) is transferred to another metabolite thus forming other amino acids or amino compounds. Transamination reactions are catalyzed by enzymes known as amino transferases, which catalyze the transfer of the amino group of an amino acid to the keto group of a keto acid according to the general formula shown below:

$$R_1-\underset{\underset{NH_2}{|}}{\overset{\overset{H}{|}}{C}}-COOH + R_2-\overset{\overset{O}{\|}}{C}-COOH \underset{\text{transferase}}{\rightleftharpoons} R_1-\overset{\overset{O}{\|}}{C}-COOH + R_2-\underset{\underset{NH_2}{|}}{\overset{\overset{H}{|}}{C}}-COOH$$

Amino acids are considered to be the "building blocks" of proteins since they are assembled in specific sequences of up to 50 to 100 monomer units to form different proteins (Larsen, 1980). Proteins serve as structural units of cytoplasm and membranes, as carriers in transport functions, and as enzyme catalysts that determine the pattern and rate of chemical reactions in plant cells. Not all amino acids are combined into proteins and 20 to 40 different ones are found in the free state in various plant species. The proportions of different amino acids in both free and combined states are characteristic of plant species and sometimes cultivars within species.

Amino acids are also synthesized within the plant into a variety of complex nitrogenous compounds involved in plant growth and metabolism (e.g., chlorophyll, growth regulators, alkaloids, purine and pyrimidine bases, and their nucleoside and nucleotide derivatives and nucleic acids).

E. Detoxification of Ammonia

The accumulation of NH_4^+ in plant tissues has particularly deleterious effects on plant growth and metabolism (see Section V,D). In addition to the glutamate synthase cycle there are other potential routes of ammonia assimilation that might operate at high levels of tissue NH_4^+ (Givan, 1979). Reactions in this category involve enzymes with a relatively low affinity for NH_4^+.

The enzyme glutamate dehydrogenase catalyzes the combination of α-oxoglutarate with ammonia to yield glutamic acid:

$$NH_3 + \alpha\text{-oxoglutarate} + NAD(P)H \rightleftharpoons \text{L-glutamate} + NAD(P)^+$$

The reaction is known to be an important pathway of NH_3 assimilation in green algae (Miflin and Lea, 1980) and may be of minor significance in higher plants (Lea and Miflin, 1974; Lewis and Probyn, 1978). The high K_m for NH_4^+ (upwards of 5 mM) means that this enzyme can probably only assimilate ammonia when intracellular ammonia concentrations are unusually high. A role for glutamate dehydrogenase at high ammonia levels is suggested by the finding that considerable increases in the overall level of the enzyme often take place in the roots and leaves of plants grown in high concentrations of NH_4^+ (Shepard and Thurman, 1973; Barash et al., 1973, 1975; Rhodes et al., 1976; Small et al., 1977; Taylor and Havill, 1981). Since roots are the major site of assimilation of absorbed NH_4^+, there is a greater increase in glutamate dehydrogenase activity in roots than shoots when plants are supplied with high levels of NH_4^+ (Taylor and Havill, 1981).

Givan (1979) has suggested that at high levels of tissue NH_4^+, the enzyme asparagine synthetase could also become a primary assimilating enzyme, although there is no direct evidence that this is so (Skokut et al., 1978; Givan, 1979). This enzyme normally transfers the amide N from glutamine to aspartate, thereby producing asparagine (Rognes, 1975). Nonetheless, asparagine synthetase can also react with NH_4^+ directly although its K_m for NH_4^+ is at least an order of magnitude higher than its K_m for glutamine (Givan, 1979).

F. Reassimilation of Ammonia

Besides the primary assimilation of NH_3, it is known that there are several reactions in plant tissues that release NH_3 from the organic form, particularly from glycine (during photorespiration), or the breakdown of asparagine via a transaminase enzyme (Lloyd and Joy, 1978) or via asparaginase (Miflin and Lea, 1980). Much research has centered on the first process because of its potential magnitude.

Photorespiration occurs in the mitochondria of most temperate plants simultaneously with the fixation of CO_2 by photosynthesis. It involves the oxidation of intermediates of the photosynthetic process, resulting in a return of CO_2 to the atmosphere. The principal reaction is shown below:

$$2 \text{ glycine} + 2H_2O \rightleftharpoons \text{serine} + CO_2 + NH_3 + 2H^+ + 2e^-$$

Keys et al. (1978) and others have calculated that under normal growth conditions, the flux of NH_3 produced through photorespiration could be an order of magnitude greater than that fixed during primary ammonia assimilation. Thus ammonia reassimilation in leaves is likely to be quantitatively more important than primary assimilation and if it failed to take place then the plant would quickly be depleted of all its organic N.

The combined action of glutamine synthetase and glutamate synthase plays a major role in the reassimilation of NH_3 in plants (Keys *et al.*, 1978; Woo *et al.*, 1982). The NH_3 released from glycine in the mitochondria is thought to be refixed by cytoplasmic glutamine synthetase into glutamine. Glutamine and α-oxoglutarate then move into the chloroplast, where, in the presence of reduced ferredoxin and glutamate synthase, two molecules of glutamate are formed. One of these molecules of glutamate becomes the acceptor for another ammonia molecule (Fig. 10).

It is now generally recognized that gaseous losses of N into the atmosphere can occur from growing plants (Wetselaar and Farquhar, 1980; see chapter 5). These include losses of NH_3 particularly during leaf senescence (Martin and Ross, 1968; Crasswell and Martin, 1975; Farquhar *et al.*, 1979), which could well be linked to a loss in efficiency of NH_3 reassimilation during the photorespiratory cycle.

G. Sites of Nitrogen Assimilation

Under most conditions, N is available in the soil predominantly as NO_3^--N, which is absorbed directly by plant roots and may be translocated unchanged or may first be reduced and metabolized in the roots to a range of amino acids and amides. Should significant quantities of NH_4^+ be present in soils, plants also directly absorb this form of N.

Nevertheless, in contrast to NO_3^- nutrition, during NH_4^+ nutrition virtually all the NH_4^+ absorbed by plant roots is assimilated rapidly in the roots and translocated as organic compounds. Kinetic studies of $^{15}NH_4^+$-fed plants (Arima and Kumazawa, 1976) suggest that the synthesis of glutamine occurs at or near the plasmalemma of the root cells. Amino acids and amines are the major forms of N transported to the shoot from NH_4^+-fed plants (Muhammad and Kumazawa, 1974).

Nitrite reductase and the enzymes of the glutamate synthase cycle are present in both tops and roots of plants and, as discussed in the next section, the reduction of NO_3^- to NO_2^- is generally the rate-controlling step of N assimilation. Thus, it is the relative activity of the nitrate reductase enzyme in the roots and shoots that determines the pattern of N assimilation within NO_3^--fed plants. Three broad groups of plants can be recognized in relation to the predominant site of NO_3^- reduction.

The xylem of many species of woody plants contains all their N in organic form (Pate, 1971, 1973). Nitrate reduction is assumed to take place almost entirely in the roots of these species. The lack of nitrate reductase activity has been demonstrated in plant leaves of several species (Routley, 1972). In such plants, nitrate reduction in the roots must be of paramount importance. These include many species of the family Ericaceae such as lowbush blueberry (*Vaccinium angustifolium*), cranberry (*Vaccinium macrocarpon*), and many species of *Rhododendron* (Routley, 1972; Dirr, 1974).

In contrast, there are plants in which little NO_3^- reduction takes place in the root system, since their xylem sap contains 95 to 99% of its nitrogen as free NO_3^-, and nitrate reductase activity cannot be detected or it is at very low levels in the roots. Examples are *Xanthium pennsylvanicum* and *X. borago* (Pate, 1973).

The majority of species examined fall into the third category and exhibit a pattern intermediate between the two extremes with both root and shoot tissues having appreciable nitrate reductase activity and xylem sap containing both free NO_3^- and organic N. These plants, when supplied with continuous NO_3^- at a concentration just maintaining a maximum growth rate, are able to maintain active nitrate reductases in both their root and shoot systems (Pate, 1973). The proportion of NO_3^- reduced in the roots not only differs between species (e.g., Olday *et al.*, 1976b) but also between cultivars and hybrids of one species (Olday *et al.*, 1976a; Jackson, 1978).

To some extent, the contribution of root and shoot tissue to the overall NO_3^- assimilation process may depend on the external NO_3^- concentration. An increase in the concentration of NO_3^- in the external medium can greatly increase nitrate reductase activity in shoots with a concomitant decrease in activity in roots (Wallace and Pate, 1967), so that there is an increase in the proportion of NO_3^--N to reduced N in the xylem exudate (Olday *et al.*, 1976a). However, for some plants such as *Phaseolus vulgaris* and *Lycopersicon esculentum*, the proportion of NO_3^- translocated is apparently unaffected by the external NO_3^- concentration (Lorenz, 1976; Thomas *et al.*, 1979).

Despite the above discussion, it is noted that considerable uncertainty surrounds the use of the relative proportions of NO_3^- and organic N in the xylem as an indicator of the relative extent of NO_3^- reduction in the roots (Rufty *et al.*, 1982b) because of circulation of organic N within plants (see Section IV,C).

The capacity of the roots to reduce NO_3^- appears to be particularly intense during the early stages of plant development (Oaks *et al.*, 1972; Wallace, 1975; Oaks, 1978); as the plant matures, the activity in the root tends to decrease and the leaves tend to reduce an increasing proportion of absorbed NO_3^-. The nitrate reductase enzyme appears to be unstable in mature regions of the root (Oaks *et al.*, 1972) due to an increase in activity of an inactivating system with age (Wallace, 1975).

H. Regulation of Nitrogen Assimilation

In general, it seems that if ammonium is absorbed by plants or generated within the plant, then it is assimilated (Miflin and Lea, 1980). Ammonium is extremely toxic if it accumulates in plant tissues (see Section V,D) and plants generally lack any mechanism to deal with its accumulation other than assimilation. This is probably because most land plants evolved under conditions of N limitation and where NO_3^- was the major form of available soil N so that there was little selection pressure to evolve such regulation of NH_4^+ accumulation.

Thus the controls that do operate on N metabolism tend to ensure that NH_4^+ is not generated internally under conditions such that it cannot be assimilated (Miflin and Lea, 1980). For example, NO_3^- reduction in the leaves only occurs in the light and in the presence of CO_2 (conditions necessary for carbohydrate synthesis) (Rathnam, 1978; Tishner and Hutterman, 1978; Sherrard et al., 1979) and in the roots when there is sufficient carbohydrate supply to produce the energy required for reduction and at the same time C skeletons for assimilation (Minotti and Jackson, 1970; Hallmark and Huffaker, 1978).

The level of nitrite reductase in different cells and tissues is usually much higher than that of nitrate reductase and accordingly accumulation of NO_2^- seldom occurs (Guerrero et al., 1981). Thus, it is the reduction of NO_3^- to NO_2^- rather than the further reduction of NO_2^- to ammonia that is the overall rate-controlling step in nitrate reduction (Hewitt et al., 1978; Guerrero et al., 1981). Factors regulating nitrate reductase activity hence exert a regulatory effect on the supply of reduced N to the plant. Nitrate reductase activities are known to fluctuate in response to changes in environmental conditions such as light, temperature, pH, CO_2 and O_2 tensions, water potential, nitrogen source, and other factors (see Guerrero et al., 1981). Such fluctuations usually also influence the capacity of the plant to assimilate NO_3^-.

1. Regulation of Nitrate Reduction

The regulation of nitrate reductase activity in higher plants is very complex and not fully understood because many interrelationships exist among regulatory factors. Indeed, its regulation appears to differ from species to species as well as in different plant parts. Several authors have reviewed this topic in some detail (Hewitt et al., 1978; Oaks, 1978; Lee, 1980; Srivastava, 1980; Guerrero et al., 1981; Schrader and Thomas, 1981).

The amount of active nitrate reductase can be considered as a function of (1) the controlled synthesis of the active enzyme and (2) further changes in the activity state of preexisting enzyme (inactivation, (re)activation, or degradation). It is often difficult to determine through which of these two methods the many control mechanisms of nitrate reductase activity work.

a. Regulation of enzyme synthesis. In higher plants nitrate reductase is generally regarded as a substrate-inducible enzyme; the presence of NO_3^- is thought to induce the *de novo* synthesis of nitrate reductase (Zielke and Filner, 1971; Hewitt *et al.*, 1976; Srivastava, 1980). Nonetheless, its requirement is not absolute and considerable enzyme levels are sometimes found in some plants in the absence of NO_3^- (Guerrero *et al.*, 1981).

The mechanism by which NO_3^- enhances the level of activity of nitrate reductase is not well defined. Much of the work dealing with the induction of nitrate reductase by NO_3^-, while showing that protein synthesis is necessary (e.g., Sluiters-Scholten, 1973; Jones *et al.*, 1978), does not distinguish between *de novo* synthesis of the enzyme and assembly of preexisting polypeptides. Indeed, the nitrate reductase enzyme of higher plants appears to be a constitutive enzyme made up of several components (Hewitt *et al.*, 1978) so that the "induction" of nitrate reductase by NO_3^- may involve an assembly of preexisting enzyme components and not strictly *de novo* synthesis.

Light also appears to be involved in the induction process in photosynthetic tissues (Hewitt, 1975; Vennesland and Guerrero, 1979) although its requirement is not absolute (Roth-Bejerano and Lips, 1973; Muller and Grafe, 1978). Light-promoted synthesis of ATP in photosynthetic tissues may have a positive effect on general protein synthesis and hence the synthesis of the nitrate reductase enzyme (Guerrero *et al.*, 1981). Alternatively, light may indirectly effect induction by increasing levels of NO_3^- in the leaf cytoplasm. This could be achieved either by increasing movement of NO_3^- into the leaves (Beevers *et al.*, 1965) or stimulating the movement of NO_3^- across membranes from a vacuolar "storage" pool to a cytoplasmic "metabolic" pool (Jones and Sheard, 1975; Aslam *et al.*, 1976). The importance of these two pools of NO_3^- is discussed in subsection III.H.1.c. dealing with substrate availability.

In some plants, NH_4^+, or rather some products of NH_4^+ metabolism (certain amino acids), can play a crucial role in the regulation of the synthesis of active nitrate reductase (Guerrero *et al.*, 1981). Thus, NH_4^+ and some amino acids have been shown to prevent the NO_3^--promoted increase in nitrate reductase activity in a variety of plant tissues (e.g., Frith, 1972; Behrend and Mateles, 1975; Radin, 1975; Oaks *et al.*, 1977; MacKown *et al.*, 1982).

Such a mechanism would tend to ensure that accumulation of NH_4^+ in tissues does not occur. Nevertheless, effects of NH_4^+ are rather variable (Oaks, 1978) and no effect or an enhancement of nitrate reductase activity in the presence of ambient NH_4^+ has been reported (Oaks et al., 1977; Sahulka, 1977; Buczek, 1979).

b. Regulation of enzyme activity. The occurrence of rapid variations in nitrate reductase activity in response to changes in particular environmental factors cannot be ascribed to long-term changes in the amount of enzyme synthesized. Such variations are attributable to the actions of inactivating proteins that appear to be of two types: one is a nitrate reductase-degrading protein and the other is a nitrate reductase-specific binding protein.

The presence of proteases that preferentially inactivate and degrade nitrate reductase has been demonstrated particularly in root tissue (Wallace, 1978; Yamaya et al., 1980). The lower levels of nitrate reductase that are found in older parts of roots have been related to a higher activity of such inactivating proteins (Wallace, 1978; Oaks, 1978).

Other types of proteins that cause inactivation of nitrate reductase through binding to the enzyme protein (Yamaya et al., 1980), and do not appear to cause its degradation, have been found in leaf tissue, seedlings, and cell cultures of many different plant species (Kadam et al., 1974; Jolly and Tolbert, 1978; Sherrard et al., 1979; Yamaya et al., 1980). Such inactivating proteins are thought to be involved in the reversible enzyme activity changes that occur in response to light–dark transitions (Jolly and Tolbert, 1978; Tishner and Hutterman, 1978; Sherrard et al., 1979).

c. Substrate availability. Once NO_3^- enters the plant, across the plasmalemma of root cells, it may enter storage organelles (e.g., vacuoles) in the root or be reduced in the root. Alternatively, NO_3^- may be translocated to other plant parts (e.g., leaves) where similarly it may enter storage organelles or be reduced. Hence, the processes of storage and translocation can play an important role in controlling substrate (NO_3^-) availability in the cytoplasm of cells and thus nitrate reductase activity. The situation is further complicated since some evidence suggests that NO_3^- reduction in the shoot may, in some way, regulate NO_3^- uptake by roots and also its translocation from root to shoot (Benzioni et al., 1971; Pate, 1980; Kirkby and Armstrong, 1980).

The existence of two different pools of NO_3^- in plant tissues, namely, a storage of nonmetabolic pool located in vacuoles and an active metabolic pool located in the cytoplasm, is widely accepted (Ferrari et al., 1973; Aslam et al., 1976; Shaner and Boyer, 1976a; Hageman, 1978; Oaks, 1978). It is the metabolic pool that regulates nitrate reductase activity. For example, both Shaner and Boyer (1976a) and Udayakumar *et al.* (1981) showed that the NO_3^- flux into leaves is more important in the regulation of nitrate reductase activity than is the total leaf NO_3^- content. Although the storage pool of NO_3^- contributed largely to total leaf NO_3^- content, nitrate reductase activity was apparently controlled by the flux of NO_3^- entering the metabolic pool via the transpiration stream.

The relationship between the storage and metabolic pools and the factors affecting transfer could obviously be of great importance in the efficient use of NO_3^- by plants. Several workers have found that the storage pool of NO_3^- in plant roots cannot readily supply the cytoplasmic metabolic pool for sustained nitrate reductase activity or xylem loading (Martin, 1973; Aslam and Oaks, 1975; Jackson, 1978). Movement out of the storage pool is, however, indicated when reduction exceeds uptake (e.g., Jackson *et al.*, 1976b; Huffaker and Rains, 1978). The size of the metabolic pool can be increased by addition of glucose (Aslam and Oaks, 1975) or light (Aslam *et al.*, 1976) and possibly other factors such as changes in the hormone balance (Knypl, 1978). These factors probably act through affecting the permeability of cell membranes (Guerrero *et al.*, 1981).

d. Other factors. Provision of reductant NAD(P)H for nitrate reductase is also a key regulatory factor. Light may indirectly stimulate nitrate reductase by provision of reductant via photosynthesis (Klepper *et al.*, 1971; Nicholas *et al.*, 1976; Vennesland and Guerrero, 1979). In the presence of light, CO_2 may enhance NO_3^- reduction (Huffaker and Rains, 1978; Stulen and Lanting, 1978; Aslam *et al.*, 1979) presumably through increasing the amount of fixed C available to provide reductant (Schrader, 1978). Nonetheless, increasing CO_2 concentrations can decrease leaf nitrate reductase activity probably by decreasing stomatal opening and hence transpiration flow and nitrate flux to the leaves (Meeker *et al.*, 1974; Neyra and Hageman, 1976; Huffaker and Rains, 1978).

Temperature also influences nitrate reductase activity. Increasing temperatures may increase nitrate reductase activity; Benzioni and Heimer (1977) found that raising the temperature from 20 to 45°C caused a three- to fourfold increase in *in vivo* enzyme activity. Higher temperatures (above a certain optimum) inhibit nitrate reductase activity although the magnitude of the inactivation varies according to species (Kauffman *et al.*, 1971; Onweunne *et al.*, 1971; Alofe *et al.*, 1973; Amos and Scholl, 1977). Inactivation of nitrate reductase activity is not generally as drastic at temperatures lower than optimum than at those higher than optimum (Srivastava, 1980).

Water stress can also decrease nitrate reductase activity and decreases in water potential below -44 to -200 kPa have been shown to cause decreases in enzyme activity in barley and maize (Huffaker *et al.*, 1970; Shaner and Boyer, 1976b). Water stress may decrease nitrate reductase activity by causing a decline in NO_3^- flux to leaf tissues or by the inhibition of protein synthesis (Srivastava, 1980).

The availability of nutrient elements may also influence nitrate reductase activity; in particular the trace elements sulfur and molybdenum are important. An adequate sulfur supply is essential to maintain enzyme activity since the nitrate reductase enzyme appears to have a requirement for an active sulfhydryl site-SH associated with the bonding and function of NAD(P)H (Schrader *et al.*, 1968). Sulfur deficiency in maize seedlings was shown by Friedrich and Schrader (1978) to result in the loss of nitrate reductase activity more quickly than other enzymes involved in NO_3^- assimilation. Since Mo is a constituent of the enzyme and must be present for synthesis of active nitrate reductase (Hewitt and Notton, 1980), Mo deficiency can also limit NO_3^- reduction (Cantliffe *et al.*, 1974; Hewitt *et al.*, 1978).

IV. TRANSPORTATION OF NITROGENOUS SUBSTANCES

The basic scheme for N uptake, reduction, and movement in higher plants is shown in Fig. 11. As discussed previously, absorbed NH_4^+ is quickly assimilated to amino acids in roots. A very small amount of NH_4^+ may be translocated in the xylem. Absorbed NO_3^- may be stored in roots, reduced, and synthesized into amino acids in root tissues or transported across the root and deposited in the xylem for movement toward the shoots. The amino acids synthesized in roots may similarly be stored there or transported to the shoots. During xylem transport, portions of both NO_3^- and amino acids can be absorbed in the stem and petiole cells while the bulk moves into the leaves where further storage or reduction of NO_3^- can occur.

Eventually, amino acids from any of the storage regions may be deposited in the phloem for translocation to younger regions of the shoot, reproductive organs, or back to the roots. The amino acids are used for the synthesis of proteins and other complex nitrogenous compounds involved in plant growth and metabolism. Under conditions of N stress, the breakdown and remobilization of protein N may occur.

A. Xylem Transport

Solutes absorbed by the roots or synthesized in the roots are usually transported to the shoots initially in the xylem; solutes move upward through the xylem vessels in the transpiration stream.

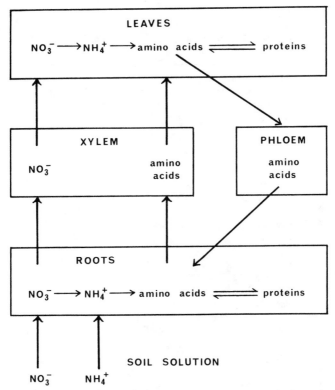

Fig. 11. Basic scheme of nitrogen uptake and reduction and protein formation in higher plants.

Although the xylem is the principal pathway for long-distance transport of nitrogenous solutes from roots to transpiring organs (Pate, 1973, 1975, 1980; Hill-Cottingham and Lloyd-Jones, 1978), there are some exceptions. For example, Martin (1971) found that upward transport of ^{15}N in *Phaseolus vulgaris* occurred in both the xylem and phloem, while Joy and Antcliff (1966) found that readily soluble organic nitrogenous compounds may be directly transported upward in the phloem of *Beta vulgaris*.

1. Composition of Xylem Sap

The xylem sap often contains both NO_3^- and organic nitrogenous compounds (e.g., see Table III). Total concentrations of N in the xylem sap are usually in the range 0.01 to 0.21% (w/v); the C : N ratio is low, in the range 1.5 to 6, indicating the relative importance of nitrogenous compounds in xylem transport.

Reduced N in roots is incorporated into a limited number of amino

Table III

Comparison of the Composition of Xylem and Phloem Sap from Stalks of *Brassica Oleracea*[a]

Form of nitrogen (μg N ml^{-1})	Xylem sap	Phloem sap
Nitrate	117.7	T[b]
Aspartic acid	13.1	29.1
Asparagine	T	T
Glutamine	317.6	T
Threonine	T	34.6
Serine	T	70.2
Glutamic acid	73.3	243.1
Proline	0	319.6
Glycine	T	15.7
Alanine	5.5	48.8
Valine	T	39.5
Isoleucine	T	21.0
Leucine	T	32.5
Tyrosine	0	T
Phenylalanine	0	T
γ-Aminobutyric acid	17.1	T
Lysine	T	9.3
Histidine	T	9.3
Arginine	T	99.8
Total	544.3	1014.2
C : N ratio	2 : 1	22 : 1

[a] Data from Pate (1973). Reprinted with permission from Pergamon Press.
[b] T = Trace amount.

acids, amides, and other solutes for transport to the shoot; each plant species has a characteristic spectrum of these compounds (Pate, 1973, 1976; Hill-Cottingham and Lloyd-Jones, 1978). The amides, glutamine and asparagine and amino acids closely related to them, are the commonest constituents of xylem exudate (Pate, 1976). In a few species alkaloids, ureides, or certain nonprotein amino acids may play an important role (Pate, 1976; Thomas *et al.*, 1979). The ureides can account for up to 90% of the total N transported from the nodules of certain tropical legumes (Pate, 1980).

The ratio of organic N to NO_3^--N in xylem sap obviously varies widely since, as already discussed (Section III), species differ greatly in their ability to reduce NO_3^- in their roots. Species with particularly active nitrate reductase activity in their roots (e.g., *Lupinus, Rhaphanus,* and *Pisum*) have very high levels of organic N relative to free NO_3^- in the xylem sap, while in species with relatively weak nitrate reductase activity in their roots (e.g., *Xanthium, Stellaria, Gossypium,* and *Cucumis*) over 95% of xylem N may consist of free NO_3^- (Pate, 1973, 1980). However, when plants from this latter group are supplied with NH_4^+-N or urea N, then large amounts of amide are synthesized in the roots and are transported in the xylem (Wallace and Pate, 1967). Most species fall into a broad intermediate category in which both free NO_3^- and organic forms of N are present in xylem, with the probability that more NO_3^- than organic N will be present when high levels of fertilizer NO_3^- are applied (Pate, 1973, 1980).

A combination of absorbed cations (mainly K^+, Ca^{2+}, and Mg^{2+}) act as counterions for NO_3^- during its upward translocation in the xylem (Kirkby and Knight, 1977; Breteler and Ten Cate, 1978; Frost *et al.,* 1978; Kirkby *et al.,* 1981).

The ability of stem tissues, mediated by xylem parenchyma cells (McNeil *et al.,* 1979), to abstract solutes from the ascending xylem stream tends to modify the composition of the xylem stream (Pate, 1976, 1980). Certain nitrogenous solutes in the xylem are more readily retrieved by stems than others: arginine is most effectively absorbed, asparagine, glutamine, valine, and serine are absorbed with moderate effectiveness, and aspartic and glutamic acid are retrieved ineffectively (Pate, 1980).

The quantity and form of N transported in the xylem sap are known to change during plant development. Levels of N generally increase during the early stages of development and remain high until after flowering, whereupon they decline markedly (Hofstra, 1964). Nitrate reductase activity of roots appears to be greatest in young plants and as the plant matures a greater proportion of absorbed NO_3^- is translocated in that form (Oaks, 1978).

Quantities of N transported in the xylem exudate are known to follow a circadian rhythm; maximum exudation rate and N concentration in exudate occur at or near noon while the minima of these parameters occur around midnight (van Die, 1959; Pate, 1971).

2. Xylem Loading

Active loading of substances into the xylem is generally considered to occur across the plasmalemma of xylem parenchyma cells (Lauchli, 1976; Bowling, 1981) and probably involves specific protein carriers. The vacuoles of the xylem parenchyma cells may act as sinks for solutes that have been transported radially across the root (Lauchli, 1976).

a. Nitrate. The process of xylem loading appears to be more sensitive to alterations in aerobic metabolism, temperature, and possibly protein synthesis than is the uptake process since anaerobic conditions, low temperatures, and metabolic inhibitors decrease the percentage of absorbed NO_3^- that is translocated (Ezta and Jackson, 1975).

Nitrate nutrition itself apparently influences the rate of water movement in the xylem. Addition of NO_3^- to the growth medium has been shown by several workers to stimulate exudation rates from the xylem of decapitated plants (Ashcroft *et al.*, 1972; Cooil, 1974; Ezta and Jackson, 1975). The mechanism by which xylem exudation is stimulated is unknown, but exudation is known to be a function of both the osmotic pressure difference between the xylem sap and the root bathing solution and the hydraulic conductivity of the root tissue. Thus, NO_3^- nutrition must induce the deposition of larger amounts of osmoticum into the xylem and/or it must enhance the hydraulic conductivity of the root tissue (Ezta and Jackson, 1975).

b. Organic nitrogenous compounds. Selective loading of organic nitrogenous compounds is indicated by the fact that the relative proportion of the reduced N constituents in the xylem sap differs from that in the root tissue (Pate, 1971; Edgar and Draper, 1974).

Little is known concerning the mechanisms of xylem loading of organic nitrogenous compounds at the xylem parenchyma cells. Nevertheless, the transport of amino acids across the plasmalemma of plant cells is known to be pH-, concentration-, and energy-dependent and carrier-mediated (Soldal and Nissen, 1978; McDaniel *et al.*, 1981, 1982). Amino acids are thought to be cotransported across the plasmalemma together with H^+ ions (Etherton and Rubinstein, 1978; Novacky *et al.*, 1978; Kinraide and Etherton, 1982). The driving force for the transport is an electrochemical potential difference in H^+ that is maintained by the active (ATP-consuming) extrusion of H^+ across the plasmalemma (Hodges, 1976; Dupont *et al.*, 1981; Kinraide and Etherton, 1982).

B. Phloem Transport

1. Composition of Phloem Sap

In contrast to the xylem sap, nitrogenous solutes in the phloem are virtually all organic (Pate, 1976, 1980; Table III). Nitrate is either absent or present in trace amounts (Pate, 1975). As in xylem exudates, frequently the glutamine/glutamate and asparagine/aspartate fractions are both high (Hall and Baker, 1972). However, the types of nitrogenous solutes found in phloem sap vary from species to species. For example, in *Brassica oleracea*, proline is a major component (Table III).

At the time of senescence, amides (Garner and Peel, 1971; van Die and Tammes, 1975) and peptides (Duke *et al.*, 1978) may appear in the phloem sap. These are thought to be products of protein hydrolysis (Schrader and Thomas, 1981).

The amount of N transported in phloem can vary with plant age and stage of development. For example, the buildup of nitrogenous substances in the phloem at leaf senescence is well documented for deciduous woody species in autumn when N bound in foliage is withdrawn into the trunk (Ziegler, 1975). Furthermore, during growth of reproductive organs, much N from storage pools in leaves can be mobilized and transported in the phloem to these organs (Schrader and Thomas, 1981).

It is interesting to note here that the major inorganic ion in the phloem sap is invariably K^+ (Geiger, 1975) and it is cycled efficiently from the downward phloem stream to the ascending xylem stream (Pate, 1975). Thus, the highly mobile K^+ can be circulated within the plant, acting as one of the counterions for the upward transport of NO_3^- in the xylem and then being translocated downward again in the phloem (e.g., Kirkby *et al.*, 1981).

2. Phloem Loading

Phloem loading is the process by which major translocated substances are selectively and actively delivered to sieve tubes at the source region prior to translocation. It is a selective process since the composition of a sieve tube sap does not correspond to the composition of solutes in the leaf blade nor in the tissue surrounding the sieve tubes (Geiger, 1975). Absorption of solutes into the phloem is thought to occur principally through the extensive network of small-diameter minor veins that offers a large surface area within the leaf close to the production of assimilates.

The mechanism of transport of amino compounds across the plasmalemma of plant cells was discussed in relation to xylem loading. The active loading of solutes into phloem is thus thought to be mediated by specific carriers and occurs with simultaneous cotransport of protons (Giaquinta, 1976, 1977; Servaites *et al.*, 1979). The proton gradient is maintained by a membrane-bound ATPase.

Loading of the phloem does not necessarily occur from the leaf or stem symplasm. Indeed, the existence of transfer cells (modified parenchyma cells of vascular strands) in leaf veins, particularly those bridging sieve and xylem elements, apparently facilitates the rapid transfer and loading of solutes from the xylem into the phloem (Pate, 1975, 1980). Thus, solutes can be transported from roots to the transpiring leaves in the xylem stream and then almost immediately be transported to other parts of the plant in the phloem.

C. Circulation, Storage, and Remobilization of Nitrogen

1. Circulation of Nitrogen

As outlined in previous sections, the N required for growth of shoots is transported from the roots via the xylem as a mixture of NO_3^- and organic N, the proportions varying with species and conditions. Mature (vigorously transpiring) leaves receive most of the xylem-borne N (Pate, 1980). Developing leaves, vegetative apices, and fruiting bodies are major sinks for N for amino acid and protein synthesis, but since they have relatively small surface areas, their requirements cannot be met by transpirationally (xylem) derived N. They must depend on amino acids retranslocated from the mature leaves via the phloem.

In plants that translocate a large proportion of their absorbed N in the xylem stream as NO_3^-, a substantial quantity of the organic N retranslocated in the phloem is derived from NO_3^--N that has been recently reduced in the leaves (e.g., Pate *et al.*, 1975). Xylem-borne organic nitrogenous substances are known to differ with respect to the rapidity of their transfer to the phloem in the leaves and also with respect to the extent of metabolism prior or during transfer. Asparagine, for example, is rapidly transferred to the phloem with little or no conversion to other compounds. In contrast, aspartate and glutamate tend to be metabolized before their C and N is transferred to the phloem and the transfer occurs more slowly (Sharkey and Pate, 1975; Pate, 1976; McNeil *et al.*, 1979; Urquhart and Joy, 1982).

Nitrogenous compounds can also be retranslocated from the leaves back to the roots and any excess not required for root growth can be reloaded into the xylem stream. It was calculated by Simpson *et al.* (1982) that of the total increment of N in wheat seedlings after 22 days growth, between 79 to 100% of the absorbed N was cycled in the plant (root → shoot → root → shoot).

Some workers have attempted to estimate the proportion of absorbed NO_3^- that is reduced in the roots by measurement of the ratio of reduced N to NO_3^--N in the xylem sap. Such measurements can, however, give large overestimates of the extent of reduction (Rufty *et al.*, 1982b) because of the circulation of organic N within plants.

2. Storage and Remobilization of Nitrogen

The existence of storage and metabolic pools of NO_3^- in plant tissue has already been discussed. The vacuoles apparently represent the major repository for the storage pool of NO_3^- (Granstedt and Huffaker, 1982). The proportion of NO_3^- found in the vacuoles at a given time is known to be affected by several factors. Accumulation of NO_3^- is a function of the rate of NO_3^- uptake from soil solution and the rate of reduction (Aslam *et al.*, 1979); more NO_3^- accumulates in darkness than in light (Aslam *et al.*, 1979). Light also stimulates the release of NO_3^- from storage pools into metabolic pools (Aslam *et al.*, 1976).

Some plant species (e.g., members of the families Chenopodiaceae, Compositae, Amaranthaceae, Solanaceae, and Cruciferae) tend to store large quantities of NO_3^-, sometimes in amounts up to 5% of tissue dry weight (Schrader, 1978). In some plants, release of storage NO_3^- from vacuoles can supply enough N to sustain growth for several days (Novoa and Loomis, 1981). In species of salt bush (members of the Chenopodiaceae), uptake of NO_3^- is particularly active during periods of rainfall, and the NO_3^- that builds up in the plant at these times is slowly utilized during periods of drought (Pate, 1980).

Other species store soluble N predominantly in organic form. Herbaceous plants accumulate considerable quantities of soluble N in fleshy tissues of stems and storage organs while woody species tend to utilize the bark and ray parenchyma of trunk and root for storage. Different plants store N in different forms; ureides are the principal forms in stems of *Acer*, asparagine is the major form in *Lupinus* and *Trifolium*, glutamine in *Lycopersicon*, and citrulline in *Alnus* (Pate, 1980).

When the soluble N reserves of plants are inadequate to sustain the N demand of the plant, it is then that breakdown of insoluble leaf protein occurs. In species that have a high demand for N during fruiting (e.g., the protein-rich grain crops), phloem loading from mobilization of soluble N, and breakdown products of insoluble N reserves in the leaves, is of particular importance. In plants with an adequate supply of N, protein losses are high only in older leaves, but in plants growing under a low N supply, mature leaves lose protein more or less linearly with age (Novoa and Loomis, 1981). The decline in N content of the leaves and stems of wheat plants accompanying an increase in the quantity of grain N is illustrated in Fig. 12. With extreme N deficiency, protein levels of very young leaves may be reduced. Under low N availability virtually all of the N in wheat grain is derived from remobilization, whereas under conditions where N absorption from the soil is still possible during grain development remobilization may account for less than 50% of grain N (Evans *et al.*, 1975).

Remobilization of N is a major determinant of the overall N use efficiency of the whole crop (Huffaker and Rains, 1978; Novoa and Loomis, 1981). It represents an adaptation for high efficiency in the use of scarce supplies of N. For example, remobilization is so efficient in oats that a plant can acquire enough N during the vegetative phase for the entire life cycle (Leopold, 1961). Nonetheless, wheat varieties have been shown to differ in their ability to remobilize N (McNeal *et al.*, 1971) so that it may

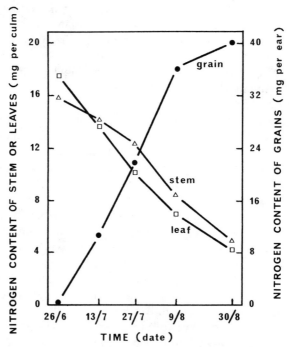

Fig. 12. Quantity of nitrogen in the leaves, stems, and grains of the main culm of the postfloral period of the growth of wheat. [Data from Spiertz (1977).]

be possible to select crop species that are highly efficient in remobilization and thus utilize mineral N from the soil more efficiently.

V. ECOLOGICAL AND PHYSIOLOGICAL ASPECTS OF NITROGEN NUTRITION

There is increasing interest in attempting to relate the ecological distribution of plant species to their physiological characteristics. Nitrogen, the element taken up in greatest quantities by plants, must be of major significance in plant ecology because of its involvement in plant growth and metabolism. This is well illustrated by the fact that when N fertilizers are applied to natural ecosystems marked changes in species composition and abundance occur (e.g., Thurston, 1969).

Studies of the distribution of higher plants have shown that species possess biochemical and physiological attributes that enable them to survive and exploit their characteristic ecological niches (e.g., Grime and Hunt, 1975; Osborne and Whittington, 1981). Among these attributes are adaptations for the uptake and assimilation of the major form of N (NH_4^+ or NO_3) present in their soil environment and the ability to efficiently use a low, fluctuating, and often localized supply of N (Lee and Stewart, 1978; Taylor *et al.*, 1982).

A. Preference for Ammonium or Nitrate

In most arable and calcareous soils, the predominant form of mineralized N is NO_3^- and this ion is therefore generally considered to be the major form of N used by most higher plants. However, in acid soils nitrification is often, although not always, inhibited (Ellenberg, 1977; Havill *et al.*, 1974; Osborne and Whittington, 1981). This had led to the assumption that NH_4^+ rather than NO_3^- is the predominant form of available N absorbed by plants in acid soils. This suggestion is strongly supported by experimental observations that some calcifuge (acid-loving) species prefer, or at least tolerate, a N supply predominantly in the form of NH_4^+ (Ingestad, 1970, 1971, 1973, 1976, 1979; Gigon and Rorison, 1972; Krajina *et al.*, 1973; Wiltshire, 1973; Havill *et al.*, 1974).

Several workers have shown that forest tree species, originating from acidic soil environments, generally prefer NH_4^+-N while those from more fertile soils of higher pH tend to prefer NO_3^--N (Krajina *et al.*, 1973; Nelson and Selby, 1974). Krajina *et al.* (1973), for example, found that of four tree species studied, two preferred NO_3^- (*Pseudotsuga menziesii* (Mirb.) Franco and *Thuya plicata* Donn.), one preferred either NH_4^+ or a combination of NH_4^+ plus NO_3^- (*Tsuga heterophylla* (Raf.) Sang.), and one had a preference for NH_4^+ (*Pinus contorta* Doug.). In nature, *P. menziesii* and *T. plicata* are found in soils where nitrification takes place, while *P. contorta* and *T. heterophylla* generally grow in acid soils where nitrification does not actively occur (Krajina, 1969).

Very similar results have been found for grassland plants (Gigon and Rorison, 1972; Wiltshire, 1973). One of the most studied groups of calcifuge species are those belonging to the family Ericaceae. These include species of the genera *Azalea, Rhododendron, Erica,* and *Vaccinium* and in general they have a distinct preference for NH_4^+ rather than NO_3^--N (Townsend, 1966; Routley, 1972; Havill *et al.*, 1974).

For the majority of plant species, however, it seems that a mixture of NO_3^- plus some NH_4^+ produces greatest growth; the optimum ratio probably differs for different species and may change with plant age (Michael *et al.*, 1970; Ta and Ohira, 1981). Yield increases by adding small amounts of NH_4^+ to all NO_3^- systems have been observed for many plant species

(Joiner and Knoop, 1969; Van den Driessche, 1971; Cox and Reisenauer, 1973; Green *et al.,* 1973; Reisenauer *et al.,* 1982; Edwards and Horton, 1982; Precheur and Maynard, 1983). Such results are not unexpected since in most fertile soils where nitrification occurs, the major form of N is NO_3^- though small quantities of NH_4^+ are usually present.

B. Reasons for Preferences

The mechanisms by which plants exhibit a preference for NH_4^+ or NO_3^- are unclear and probably differ among various groups of plants. Some of the suggested mechanisms are discussed below.

1. Nitrate Reductase Activity

The lack of the ability to effectively reduce large amounts of NO_3^--N is the most obvious reason for a preference for NH_4^+-N. As noted in Section III,G, many members of the family Ericaceae possess very low levels of nitrate reductase activity in their leaves even when they are supplied with NO_3^--N (Townsend and Blatt, 1966; Townsend, 1970; Routley, 1972; Dirr, 1974; Havill *et al.,* 1974; Lee and Stewart, 1978). Nonetheless, significant nitrate reductase activity has been observed in their roots (Townsend, 1970; Dirr, 1974). It seems that in many ericaceous plants, the roots are the main site of NO_3^- assimilation and NO_3^- cannot be reduced efficiently in the leaves. When the root nitrate reductase system becomes saturated with NO_3^-, the leaves cannot act as a major sink for the assimilation of the surplus NO_3^-, hence these plants cannot make efficient use of the NO_3^- as their major source of N.

Nonetheless, a restricted capacity to utilize and reduce NO_3^- is by no means a general characteristic of calcifuge species (Havill *et al.,* 1974; Stewart *et al.,* 1974; Osborne and Whittington, 1981). For example, results presented in Table IV show that although two calcifuge *Vaccinium* species belonging to the family Ericaceae possessed virtually no capacity to reduce NO_3^- in their leaves, this was not true for two other calcifuge plants. The calcicole species all possessed significant leaf nitrate reductase activity. The possession of significant nitrate reductase activity by some calcifuge species is not altogether surprising since several workers have observed significant flushes of NO_3^- in soils even in the pH range 3.5 to 4.0 (Davy and Taylor, 1974; Taylor *et al.,* 1982).

2. Lime-Induced Chlorosis

Plants that generally prefer NH_4^+ as a source of N often exhibit the physiological disorder lime-induced chlorosis when grown on NO_3^- alone (Cain, 1952, 1954; Colgrove and Roberts, 1956; Ingestad, 1973; Nelson

Table IV

Nitrate Reductase Activity in Plants from Acidic and Calcareous Soil Sites[a]

Plant species	Mean *in vivo* nitrate reductase activity[b]	
	Field assay	72 hr after nitrate addition
Calcifuge species		
Vaccinium myrtillus	<0.10	<0.10
Vaccinium vitis-idaea	<0.10	<0.10
Deschampsia flexuosa	0.62	2.80
Molonia caerulea	0.52	1.50
Calcicole species		
Koeleria cristata	0.90	5.50
Teucrium scorodonia	1.30	2.61
Minuartia verna	1.90	4.09
Poterium sanguisorba	1.11	3.80

[a] Data from Havill *et al.* (1974).

[b] *In vivo* nitrate reductase activity determined in leaves of plants in undisturbed field sites and in field sites 72 hr after the addition of 106 gm $NaNO_3$ m^{-2}.

and Selby, 1974). A vast literature deals with lime-induced chlorosis in calcifuge species and a variety of explanations have been proposed. Although this chlorosis is of an Fe deficiency type, high total Fe contents in leaves are sometimes recorded (Cain, 1952, 1954).

Nitrate nutrition is associated with high cation accumulation (Cain, 1954; Ingestad, 1973, 1976, 1979), which in turn results in the production of balancing organic anions in the plant once the NO_3^- is reduced (Dijkshoorn, 1973; Raven and Smith, 1976). Symptoms of iron chlorosis in many species, under a wide range of conditions, have been associated with high organic acid anion concentrations in plants (e.g., Su and Miller, 1961; Wallace, 1971; Nelson and Selby, 1974). It is thought that organic acids produced during NO_3^- nutrition may bind with Fe in both roots and leaves, thereby interfering with the functions of Fe in the plant (Wallace, 1971; Nelson and Selby, 1974).

3. Aluminum and Manganese Toxicities

It is interesting to note that under acidic soil conditions, where NH_4^+ predominates, high concentrations of potentially toxic metals such as Al and Mn also exist in soil solution (Foy and Fleming, 1978). Thus one might expect that there would be interactions between NH_4^+ and NO_3^- nutrition and Al and Mn toxicities.

Indeed, in general, ambient NH_4^+ inhibits the plant uptake of Mn and Al while ambient NO_3^- enhances their uptake (Rorison, 1980; McGrath and Rorison, 1982). Such results are not unexpected since, as discussed in Section II,C, NH_4 nutrition generally inhibits cation uptake while NO_3^- nutrition enhances it. Thus, in acid soils NH_4^+ nutrition may, to some extent, be involved in the tolerance of plants to high levels of Al and/or Mn. By effectively absorbing NH_4^+ from soil solution, calcifuge plants may, in fact, be simultaneously minimizing their uptake of Al^{3+} and Mn^{2+} ions and thus decreasing the likelihood of Al and Mn toxicities.

Nonetheless, Foy and Fleming (1978) associated Al tolerance in wheat varieties to their ability to use NO_3^- efficiently in the presence of NH_4^+ and therefore to raise the pH of the rhizosphere and precipitate and detoxify the Al.

4. Ammonium Toxicity

The most obvious reason for the preference of most plants for NO_3^- rather than NH_4^+ is the very toxic effects that high levels of ambient NH_4^+ have on the majority of plant species. There are, however, large differences among species in their tolerances to the NH_4^+ ion (Haynes and Goh, 1978; Barker and Mills, 1980).

In general, NH_4^+ toxicity is characterized by an immediate restriction in plant growth, chlorosis of leaves, marginal necrosis, necrotic spots, stem lesions, and finally plant death (Maynard et al., 1966, 1968; Barker et al., 1966a,b; Maynard and Barker, 1969). Toxicity of NH_4^+ is also characterized by greatly restricted root growth with the production of short, thick, less branched, and darkly colored roots (Maynard and Barker, 1969; Cox and Reisenauer, 1973; Warncke and Barber, 1973). Warncke and Barber (1973) also observed that root growth was restricted more than top growth so that the shoot:root ratio increased.

It seems that most plants have evolved under conditions of a very low NH_4^+ supply and where NO_3^- is the predominant form of available N. Thus these plants are not adapted for the use of NH_4^+ as their major or sole source of N. The mechanisms by which NH_4^+ exhibits its toxic effects on plant growth are discussed in Section V,D.

5. Combination of Nitrate Plus Ammonium

As noted previously, most plant species produce highest dry matter yields under a N regime of high NO_3^- plus low NH_4^+. Although addition of NH_4^+ to culture solution characteristically depresses the intake of NO_3^- (see Section III,G), at sufficiently low levels of NH_4^+, the depression of

NO_3^- uptake is less than the rapid uptake of NH_4^+ and as a result the total intake of N, plant protein content, and growth rate increase. For example, maximum yields of ryegrass were obtained by Reisenauer *et al.* (1982) from supplying low levels of NH_4^+ (36 μM) with adequate NO_3^- (72 μM; see Fig. 13). Higher levels of NH_4^+ produced toxic reactions and reduced growth rates. Other workers have observed higher plant protein contents from plants provided with NO_3^- and NH_4^+ rather than either alone (Weissman, 1964).

The exact reason for the superiority of NO_3^- plus NH_4^+ over NO_3^- alone is not clear. It has, however, often been ascribed to the reduced energy requirement in using NH_4^+, instead of NO_3^-, in protein synthesis (Cox and Reisenauer, 1973; Middleton and Smith, 1979). It may also be due to the inability of the NO_3^- reducing system to supply the plant with maximum usable levels of reduced N (Reisenauer *et al.*, 1982). Certainly the quantities of reduced N in the xylem sap are generally higher in plants fed NO_3^- plus NH_4^+ than NO_3^- alone (Pate, 1971, 1973).

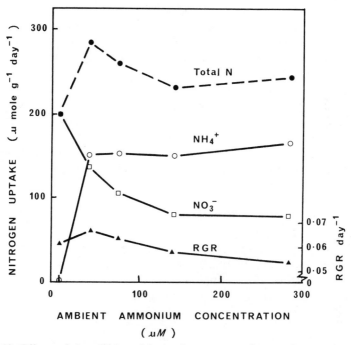

Fig. 13. Effects of the addition of increasing amounts of ammonium to an optimum supply of nitrate (72 μM) on the uptake of nitrate, ammonium, and total mineral nitrogen and on the relative growth rate (R.G.R.) of ryegrass plants. [Redrawn from Reisenauer *et al.* (1982).]

C. Responses to a Limiting Supply of Nitrogen

1. Nutrient Status of Natural Habitat

Several workers have attempted to relate short-term kinetic analysis of NO_3^- uptake (e.g., K_m and V_{max} values) with the adaptation of plant species to nutrient-rich and nutrient-poor habitats (Huffaker and Rains, 1978; Van de Dijk *et al.*, 1982). Such research has met with limited success.

For example, Huffaker and Rains (1978) found that of three grass species studied, *Bromus mollis* had the lowest K_m value for NO_3^- uptake (15 μM), *Avena fatua* had a slightly higher K_m value, while *Lolium multiflorum* had a K_m for NO_3^- uptake of 30 μM. In agreement with such K_m values, under natural conditions *B. mollis* occupies low-fertility soils (low NO_3^-), *A. fatua* is intermediate, and *L. multiflorum* requires relatively fertile conditions to compete successfully with a mixture of other grasses.

Nonetheless, Van de Dijk *et al.* (1982) found no significant differences in K_m values for NO_3^- uptake (range 10 to 20 μM) between five grassland species characteristic of soils of widely different nutrient status. On the other hand, the values of V_{max} for NO_3^- uptake by the species were related to the nutrient status of their natural habitats. The plant *Urtica dioica*, characteristic of extremely nutrient-rich habitats, had the highest V_{max} value.

Other research (Van de Dijk, 1981), however, showed that when plants were grown at very low NO_3^- concentrations ($\leq 1 \ \mu M$), the uptake capacity of *U. dioica* was less than that of other species with lower calculated V_{max} values for NO_3^-. Thus measurements of K_m and V_{max} under arbitrarily chosen conditions (e.g., from 0 to 50 μM NO_3^-) do not always yield complete information, particularly if one is considering growth at very low NO_3^- concentrations.

2. Fluctuating Supply of Nitrogen and Enzyme Activity

In many natural ecosystems the rate of N mineralization shows a distinct seasonal trend with peaks in availability generally occurring in spring and autumn (Williams, 1969; Davy and Taylor, 1974; Gupta and Rorison, 1975; Taylor *et al.*, 1982). Such seasonal flushes in soil N availability result in transient high concentrations of mineral N followed by periods of comparatively low supply.

Plants take advantage of the transient high levels of mineral N since the activity of N-assimilating enzymes in plants shows seasonal trends similar to those of mineral N. Taylor *et al.* (1982), for instance, observed a pronounced seasonal trend in soil mineral N availability (Fig. 14) and a similar seasonal pattern was evident in the levels of N-assimilating en-

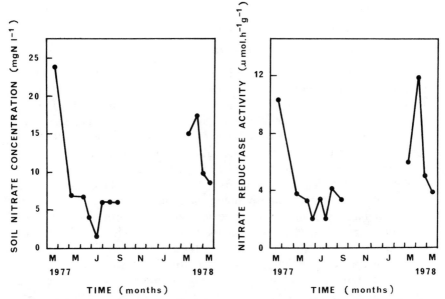

Fig. 14. Seasonal changes in soil nitrate concentrations and *in vivo* nitrate reductase activity in shoot tissue of *Deschampsia flexuosa*. [Data from Taylor *et al.* (1982).]

zymes such as nitrate reductase (Fig. 14) and to a lesser extent glutamine synthetase. Havill *et al.* (1977) showed that several grassland species retained their maximum potential to assimilate NO_3^- (maximum *in vivo* nitrate reductase activity in response to a nonlimiting supply of NO_3^-) throughout the growing season. Such results are not surprising since it is well known that the nitrate reductase enzyme is substrate-inducible (see Section III,G) and maintenance of its activity is dependent on a continued supply of NO_3^-.

The activity of the NH_4^+-assimilating enzymes (glutamine synthetase and glutamate synthase) does not change greatly depending on whether plants are fed NH_4^+ or NO_3^- (Lee and Stewart, 1978; Taylor and Havill, 1981), presumably because they are utilized whichever source of N is initially absorbed. As noted previously, when plants are supplied with large quantities of NH_4^+-N, there is often a significant increase in levels of glutamate dehydrogenase, indicating that this enzyme is probably involved in the assimilation of NH_4^+ when it is supplied at high rates (Givan, 1979; Taylor and Havill, 1981).

It therefore seems that quantitative differences in enzyme levels (as influenced by control of their synthesis/degradation and/or activation/

inactivation) do constitute a method by which plant species adapt to a fluctuating supply of N.

3. Localized Supply of Nitrogen

Soils are known to possess both lateral and vertical spatial variability and large spatial differences in NO_3^- concentrations occur in soils (Biggar, 1978; Greenwood, 1978). Thus, when the supply of mineral N is low, the plant must be able to adapt to extract NO_3^- (or NH_4^+) from localized zones within the rooting zone. The mobility of NO_3^-, however, means that such localized zones will be characteristically transient in nature.

Much of the research concerned with the response of plants to a localized supply of N has originated from agronomic studies related to the effects of the placement and banding of fertilizer N. Such research has shown that a crop plant can continue normal growth with less than 15% of its roots exposed to NO_3^- (Burns, 1980), and second, a proliferation of roots occurs in the NO_3^--rich soil zone (Drew, 1975).

In split root experiments, up to three-fold increases in the rate of N uptake per unit root weight or length have been observed, depending on the rate of growth, as a decreasing proportion of roots is exposed to NO_3^- (Drew et al., 1973; Drew, 1975; Drew and Saker, 1975; Frith and Nichols, 1975) or NH_4^+ (Drew, 1975). Nevertheless, there appears to be an absolute limit to the extent to which inflow from a restricted root volume can occur so that this mechanism alone is not sufficient to maintain N uptake and plant growth when there are substantial reductions in the proportions of roots exposed to N (e.g., Drew and Saker, 1975; Edwards and Barber, 1976b).

However, when NO_3^- is restricted to a small proportion of the roots, plants also compensate by increasing lateral root development and growth in the NO_3^--rich zone (Hacket, 1972; Drew, 1975; Drew and Saker, 1975; Robinson and Rorison, 1983; see Table V), often at the expense of other roots (Drew et al., 1973). There have also been many quantitative observations of root proliferation in soil layers enriched with nitrogenous fertilizers (e.g., in fertilizer bands) (Duncan and Ohlrogge, 1958; Wiersum, 1958; Wilkinson and Ohlrogge, 1961). Such morphological adaptations may often take a week or more to complete, even under ideal conditions, and the growth rate of the shoots may be depressed in the intervening period if physiological compensation is insufficient (Drew and Saker, 1975).

Some plants appear to respond to a localized supply of N by increasing root growth in regions of the root system not supplied with N as well as in the localized region (Robinson and Rorison, 1983). Such a response may be of practical significance since continued survival of the plant may well

Table V

Weight (mg) of Portions of Roots after Different Zones (A, B, and C) along a Single Seminal Axis of an Intact Barley Plant Were Exposed for Extended Periods to Contrasting Concentrations of NO_3^-, either Low (L) or High (H)[a]

Root zone	Treatment[b]			
	HHH	LHL	LLL	HLH
Zone A (0–6 cm)	16	3	6	28
Zone B (6–10 cm)	19	29	5	3
Zone C (>10 cm)	15	7	15	29

[a] Data from Drew et al. (1973).

[b] Concentrations of NO_3^- used were L = 0.01 mM and H = 0.10 mM. Four treatments were imposed in which zones A, B, and C of the seminal root axis were subjected to four treatments: HHH, all high NO_3^-; LHL, high in the middle zone only; LLL, all low NO_3^-; and HLH, low in the middle zone only.

depend on the growth of roots other than those in the N-rich zone of the soil. This is because zones of depletion soon occur around roots currently absorbing NO_3^- (Nye and Tinker, 1977) since NO_3^- is highly mobile in soils and its concentration in soil solution is not buffered against depletion (i.e., NO_3^- is not adsorbed by most soils).

D. Responses to an Oversupply of Nitrogen

In contrast to many natural environments where the supply of mineral N is scarce, in many agricultural soils nitrogenous fertilizers are added to facilitate maximum growth of the fast-growing agricultural crop plants. This sometimes results in a temporary oversupply of mineral N especially where fast-acting fertilizers are applied in a single rather than in split applications.

In general, the tolerance of plants to an ample supply of NH_4^+ is low, whereas the tolerance for NO_3^- is high. Toxic reactions occur when NH_4^+ accumulates in plants and its translocation to shoots is especially deleterious (e.g., Puritch and Barker, 1967). In contrast, plants will accumulate NO_3^- and transport it through the plant with few toxic effects (Mills and Jones, 1979).

The response of plants to an oversupply of mineral N is discussed below.

1. Ammonium Toxicity

Phytotoxic levels of NH_4^+ do not usually occur in fertile soils since the nitrification process generally occurs rather rapidly (Chapter 3). Conditions that lead to a predominance of NH_4^+-N are heavy applications of ammoniacal fertilizers followed by cool spring soil conditions that inhibit nitrification. Ammonium toxicity can also occur when fertilizers such as urea, anhydrous ammonia, or diammonium phosphates are applied as band applications (Court et al., 1964; Bennett and Adams, 1970).

Researchers, using a variety of plants, have suggested many different reasons for the toxic effects of NH_4^+. Indeed, it seems that NH_4^+ has several toxic effects on plants (Haynes and Goh, 1978; Goyal et al., 1982a,b; Reisenauer et al., 1982); the importance of these probably varies depending on particular experimental conditions encountered and the species involved. The major mechanisms by which NH_4^+ is thought to inhibit plant growth are outlined below.

a. Rhizosphere pH. As discussed in Section II,C, NH_4^+ nutrition results in acidification of the plant's rhizosphere. Media acidification associated with NH_4^+ absorption has been shown to be toxic to many crop plants such as peas, beans, maize, tomatoes, and asparagus (Maynard et al., 1966; Maynard and Barker, 1969). Toxicity can often be alleviated by control of media pH at or near neutrality using $CaCO_3$ as a buffer (Barker et al., 1966a,b; Barker and Maynard, 1972; Precheur and Maynard, 1983). Acidity control appears to result in greater incorporation of absorbed NH_4^+ into amino acids, amides, and ethanol-soluble N by the root tissue (Barker et al., 1966a) and therefore the limitation of NH_4^+ transport to the shoots. The reduction of NH_4^+ levels in the shoots of NH_4^+-fed plants subjected to acidity control is illustrated in Fig. 15. The mechanism by which control of the pH of the rooting medium encourages assimilation of NH_4^+ in the roots is unknown. It is, however, an important phenomenon since once NH_4^+ ions reach the shoots, the biochemistry and physiology of the plant are greatly disrupted.

It is interesting to reflect that NO_3^- nutrition results in an increase in rhizosphere pH. This may explain why ambient concentrations of NH_4^+, in excess of those required to induce toxicity symptoms, can be maintained when NO_3^- supplied part of the total ambient N (McElhannon and Mills, 1977; Goyal et al., 1982a). Goyal et al. (1982a) found that NO_3^- equivalent to 10% or more of the NH_4^+ concentration alleviated the inhibitory effects of NH_4^+ on the growth of radish. Although the pH of nutrient solutions in the study was regulated near neutrality, NH_4^+ nutrition would still have decreased rhizosphere pH.

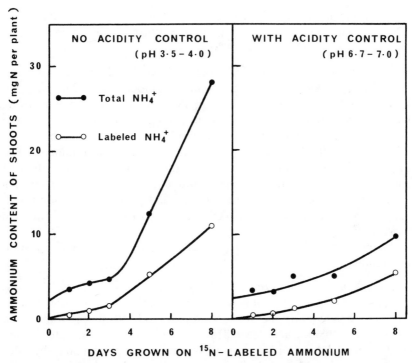

Fig. 15. Influence of root environment acidity on accumulation patterns of free ammonium in shoots of bean plants. Plants were grown with ammonium, with or without acidity control, for 2 days prior to being supplied with ^{15}N-labeled ammonium. [Data from Barker *et al.* (1966a).]

b. Cation deficiencies. As discussed in Section II,C, NH_4^+ nutrition generally results in the inhibition of uptake of cations such as K^+ and Ca^{2+} by plants. This is attributable to ionic competition during cation uptake either with NH_4^+ ions *per se* or with H^+ ions excreted during active NH_4^+ uptake.

Indeed, deficiencies of K and Ca in plants are among the most commonly cited phenomena associated with NH_4^+ toxicity (Adams, 1966; Maynard *et al.*, 1968; Ajayi *et al.*, 1970; Hoff *et al.*, 1974). Ajayi *et al.* (1970), for example, corrected symptoms of NH_4^+ toxicity by adding excessive amounts of ambient K, while Ca deficiency was suggested as a mechanism of $(NH_4)_2HPO_4$ injury to plants (Adams, 1966). Similarly, blossom-end rot of tomatoes (a physiological disease associated with Ca deficiency and enhanced by plant water stress) is increased by NH_4^+ nutrition, in comparison with that of NO_3^- (Willcox *et al.*, 1973; Pill *et al.*, 1978; Pill and Lambeth, 1980). Ammonium nutrition also decreases fruit Ca concentrations (Pill *et al.*, 1978).

c. Plant water stress. Several workers have noted wilting as a symptom of NH_4^+ toxicity (e.g., Stuart and Haddock, 1968; Maynard and Barker, 1969; McElhannon and Mills, 1977). Others have demonstrated that NH_4^+ nutrition generally results in decreases in plant water uptake, xylem exudation, and leaf water potential (Quebedeaux and Ozbun, 1973; Pill and Lambeth, 1977, 1980; Pill *et al.*, 1978). Pill and Lambeth (1980) associated blossom-end rot in tomatoes under NH_4^+ nutrition with a reduced transpiration rate resulting in reduced water flux into fruit. The reduction in leaf xylem pressure potential and plant transpiration rate in NH_4^+-fed, in comparison with NO_3^--fed, tomato plants is illustrated in Table VI.

The mechanism by which NH_4^+ enhances plant water stress appears to be unknown.

d. Carbohydrate metabolism. Yield reductions caused by excessive supplies of NH_4^+ can to some extent be attributed to metabolic disturbances associated with the detoxification of NH_4^+ within the root (Reisenauer *et al.*, 1982). Such detoxification results in an immediate demand for carbon skeletons with the plant, which are supplied mainly by intermediates in glycolysis and the tricarboxylic acid cycle (Givan, 1979). Supply of α-ketoglutarate to higher plant tissues, for example, can greatly

Table VI

Influence of Nitrogen Form and Solution pH on Leaf Xylem Pressure Potential and Transpiration Rate of Tomatoes[a]

Form of nitrogen and solution pH	Leaf xylem pressure potential[b] (kPa)	Transpiration rate[c] (gm dm^{-2}/3 hr)
Nitrate		
4.0	−404	2.097
6.5	−389	2.277
Ammonium		
4.0	−565	1.277
6.5	−504	1.349

[a] Data from Pill and Lambeth (1977).
[b] Measurements made after 4 weeks growth.
[c] Mean values for three consecutive weekly measurements over 3-hr periods.

reduce internal NH_4^+ concentrations and can thus alleviate NH_4^+ toxicity (Matsumoto et al., 1971).

There is considerable evidence that intracellular levels of tricarboxylic acids (e.g., malic or oxalic acids) decline markedly and there is an immediate sharp rise in cellular amide (e.g., glutamine) levels when plants are subjected to high levels of exogenous NH_4^+ (Kirkby, 1968; Michael et al., 1970; Reisenauer, 1978). These observations suggest that during periods of NH_4^+ assimilation a very high level of demand is placed on carbohydrates and on carbohydrate-degrading reactions that provide the necessary substrate for amide formation.

Most research indicates that NH_4^+-fed plants at least maintain or increase their respiration rates (Willis and Yemm, 1955; Berner, 1971; Wakiuchi et al., 1971; Ikeda et al., 1974; Goyal et al., 1982b). An increase in respiration has been explained as a cellular response to detoxify NH_4^+ by rapidly turning over the carbon skeletons needed for NH_4^+ assimilation (Givan, 1979).

The increased respiration rate and high demand on storage carbohydrates during the growth of plants supplied with high levels of exogenous NH_4^+ are likely to be particularly damaging when the plant begins to translocate NH_4^+ to the shoots. This is because a major effect of NH_4^+ accumulation in leaves is the inhibition of photosynthesis and therefore production of carbohydrates (Goyal et al., 1982b). Ammonium ions restrict photosynthesis through uncoupling of noncyclic photophosphorylation in isolated chloroplasts (Ohmura, 1958; Gibbs and Calo, 1959; Krogman et al., 1959; Losada and Arnon, 1963) although its exact mechanism of action is unknown (Losada et al., 1973).

Indeed, the symptoms of NH_4^+ toxicity, yellowing and necrosis, indicate a disruption of the structure and integrity of chloroplasts. Puritch and Barker (1967) showed that the chloroplasts of NH_4^+-toxic leaves were severely disrupted and that the tissue had impaired photosynthetic capabilities. Goyal et al. (1982b) observed that the chlorophyll content of leaves of plants fed NH_4^+ as a source of N decreased rapidly and continuously. Nonetheless, at less than toxic levels, NH_4^+ nutrition tends to have a positive influence on the development of chloroplasts (Golvano et al., 1982).

2. Excess of Nitrate

Plants can tolerate high levels of soil NO_3^- and concentrations of NO_3^- may rise to several percent (dry weight basis) in the plant before phytotoxicity is apparent (Maynard and Barker, 1971). Thus, the plant adapts to a high level of NO_3^- supply by storing NO_3^- in vacuoles of plant tissue for reduction and use at a later date. Nonetheless, an excess supply of NO_3^-

can be toxic to plants although the exact mechanism of toxicity appears to be unknown (Mills and Jones, 1979).

The accumulation of NO_3^- in plants has received much study (see Chapter 7) because high concentrations of NO_3^- in food can be toxic to animals ingesting it. This applies to both livestock and man, particularly infants. Studies have shown that heavy applications of NO_3^- fertilizers can lead to the accumulation of large amounts of NO_3^- in plants (Maynard et al., 1976; Lorenz, 1978; Mills and Jones, 1979). Accumulation of NO_3^- is, nevertheless, also influenced by factors other than soil NO_3^- levels such as light, temperature, and moisture stress (Lorenz, 1978).

To some extent, an oversupply of NO_3^- has an effect similar to that of ammonium, that is, it results in a depletion of the plant's supply of storage carbohydrates during the assimilation of NH_4^+, following NO_3^- reduction (Michael et al., 1970). Thus excess NO_3^- nutrition tends to increase the demand on the plant's carbohydrate metabolism.

Accumulation of large amounts of NO_3^- in plant tissues would also upset the cation–anion balance of the plant. Normally, plants absorb more anions than cations when NO_3^- is the major form of N available. The ratio of cations to anions absorbed is usually in the range 1 : 2 to 7 : 10 (Nye, 1981). Nitrate normally constitutes more than one-half the total equivalent anions absorbed but this anion is rapidly reduced to organic components within the plant. Thus, within the plant there is almost always an excess of free cations over anions and ionic balance is maintained by the plant synthesizing organic acid anions such as malate and citrate (Raven and Smith, 1976; Haynes and Goh, 1978). In a situation where NO_3^- accumulates in plants at high levels (e.g., 1 to 5%), it is not clear how the plant maintains ionic balance.

The accumulation of NO_3^- in plant tissues might also upset the process of osmoregulation. The major osmotic components of nonhallophytes are potassium salts of organic acids and sugars (Helleburst, 1976) and the ion content of the plant is regulated by redistribution of ions in different organs and in different cell compartments. High concentrations of vacuolar and cytoplasmic NO_3^- may well cause considerable alterations to osmoregulatory mechanisms. A major osmotic component might well become potassium nitrate.

VI. CONCLUSIONS

The mechanisms of absorption of both NH_4^+ and NO_3^- by plant roots are active energy-requiring processes and therefore influenced by the supply of photosynthates and energy from shoot to root. Other factors also influence the uptake of NH_4^+ and NO_3^-, including temperature, ambient pH, and competition among other cations and anions during active uptake. Ectotrophic and ericaceous mycorrhizal associations may increase the uptake of N by plants in natural environments in which NH_4^+ is the major form of available N, while there is some evidence that such mycorrhizae can also facilitate the use of simple organic nitrogenous compounds by plants.

Plants may absorb a limited quantity of N through their foliage. Foliar sprays of nitrogenous solutes (e.g., urea) are used to a limited extent in crop production. Plant foliage is also known to be capable of absorbing nitrogenous gases such as NO, N_2O, and NH_3 from the atmosphere. The absorption of NH_3 in this way may be of quantitative importance in some terrestrial N cycles.

Nitrate is the major form of N absorbed by most plants. The utilization of NO_3^- by plants involves a series of process including storage, transport, reduction, and assimilation. The first step in the reduction of NO_3^- is mediated by the enzyme complex nitrate reductase, which catalyzes the reduction of NO_3^- to NO_2^-. This enzyme is present in both the roots and shoots of most species; some species reduce considerable amounts of NO_3^- in their roots while others reduce most of it in the shoots. In plant tissues, levels of the enzyme nitrite reductase, which catalyzes the reduction of NO_2^- to NH_4^+, are usually much higher than those of nitrate reductase so that NO_2^- seldom accumulates in tissues.

The major pathway of ammonia assimilation is mediated through the combined action of two enzymes, glutamine synthetase and glutamate synthase, which operate in series and result in the formation of one molecule of glutamate. This pathway occurs in both the roots and shoots of plants. Since accumulation of NH_4^+ in plant tissues is extremely deleterious to plant growth and metabolism, NH_4^+, once actively absorbed, is rapidly assimilated in the root tissues and translocated to the shoots in the form of amides and amino acids.

Incorporation of ammonia into amino acids may be followed by transamination reactions in which the amino groups can be transferred to other metabolites thus forming other amino acids or amino compounds. Amino acids are assembled in specific sequences to form different proteins.

Nitrogenous substances are transported from the roots to the shoots in the xylem. Most of the xylem-borne N (NO_3^- plus organic compounds) is received by the vigorously transpiring, fully expanded leaves. Much of this N is then reexported as amino compounds, via the phloem, to developing leaves, vegetative apices, fruiting structures, and actively growing roots.

Plants vary greatly in their ability to absorb and utilize NH_4^+ and NO_3^- as sources of N. Most plants appear to prefer a N regime consisting of high NO_3^- and low NH_4^+. Some calcifuge species that grow naturally in acid soils where little nitrification occurs (e.g., some conifers, grasses, and members of the Ericaceae) have adapted to use NH_4^+ in preference to NO_3^-.

Since most of the N-assimilating enzymes are substrate-inducible, plants can take advantage of a fluctuating supply of mineral N. Plants can also adapt to a localized supply of N by increasing the rate of N uptake per unit root weight or length in the N-rich zone and also by increased lateral root growth in that zone.

In situations of high NH_4^+ supply, NH_4^+ toxicity may inhibit plant growth through a variety of mechanisms. These include lowering of rhizosphere pH, cation deficiencies, plant water stress, and interruptions to carbohydrate metabolism. Although plants can tolerate high levels of soil NO_3^-, and NO_3^- can accumulate in plant tissues to levels of several percent, very high levels of NO_3^- nutrition can be toxic to plants. The exact mechanisms involved in toxicity are unclear.

REFERENCES

Adams, F. (1966). Calcium deficiency as a causal agent of ammonium phosphate injury to cotton seedlings. *Soil Sci. Soc. Am. Proc.* **30,** 485–488.

Ajayi, O., Maynard, D. N., and Barker, A. V. (1970). The effects of potassium on ammonium nutrition of tomato (*Lycopersicon esculentum* Mill.). *Agron. J.* **62,** 818–821.

Alofe, C. O., Schrader, L. E., and Smith, R. R. (1973). Influence of high day and variable night temperatures on nitrate reductase activity of young corn (*Zea mays* L.) plants. *Crop Sci.* **13,** 625–629.

Amos, J. A., and Scholl, R. L. (1977). Effect of temperature on leaf nitrate reductase, glutamine synthetase, and NADH-glutamate dehydrogenase of juvenile maize genotypes. *Crop Sci.* **17,** 445–448.

Aparicio, P. J., and Maldonado, J. M. (1978). Regulation of nitrate assimilation in photosynthetic organisms. *In* "Nitrogen Assimilation of Plants" (E. J. Hewitt and C. V. Cutting, eds.), pp. 207–215. Academic Press, New York.

Arima, Y., and Kumazawa, K. (1976). A kinetic study of amide and amino acid synthesis in roots of rice seedlings supplied with ^{15}N-labelled ammonium. 2. Physiological significance of glutamine in nitrogen absorption and assimilation. *J. Sci. Soil Manure, Jpn.* **46,** 355–361.

Ashcroft, R. T., Wallace, A., and Abou-Zamzam, A. M. (1972). Nitrogen pretreatments vs nitrate treatments after detopping on xylem exudation in tobacco. *Plant Soil* **36,** 407–416.

Ashley, D. A., Jackson, W. A., and Volk, R. J. (1975). Nitrate uptake and assimilation by wheat seedlings during initial exposure to nitrate. *Plant Physiol.* **55,** 1102–1106.

Aslam, M., and Oaks, A. (1975). Effect of glucose on the induction of nitrate reductase in corn roots. *Plant Physiol.* **56,** 634–639.

Aslam, M., Oaks, A., and Huffaker, R. C. (1976). Effect of light and glucose on the induction of nitrate reductase and glucose on the induction of nitrate reductase and on the distribution of nitrate in etiolated barley leaves. *Plant Physiol.* **58**, 588–591.

Aslam, M., Huffaker, R. C., Rains, D. W., and Rao, K. P. (1979). Influence of light and ambient carbon dioxide concentration on nitrate assimilation by intact barley seedlings. *Plant Physiol.* **63**, 1205–1209.

Azcon, R., Gomez-Ortega, M., and Barea, J. M. (1982). Comparative effects of foliar- or soil-applied nitrate on vesicular-arbuscular mycorrhizal infection in maize. *New Phytol.* **92**, 553–559.

Bagshaw, R., Vaidyanathan, L. V., and Nye, P. H. (1972). The supply of nutrient ions by diffusion to plant roots in soil. VI. Effects of onion plant roots on pH and phosphate desorption characteristics in a sandy soil. *Plant Soil* **37**, 627–639.

Bailey, C. J., and Boulter, D. (1969). The subunit structure of Jack-bean urease. *Biochem. J.* **113**, 669–677.

Barash, I., Sadon, T., and Mor, H. (1973). Induction of a specific isoenzyme of glutamate dehydrogenase by ammonia in oat leaves. *Nature (London)* **244**, 150–151.

Barash, I., Mor, H., and Sadon, T. (1975). Evidence for ammonium-dependent *de novo* synthesis of glutamate dehydrogenase in detached oat leaves. *Plant Physiol.* **56**, 856–858.

Barker, A. V., and Maynard, D. N. (1972). Cation and nitrate accumulation in pea and cucumber plants as influenced by nitrogen nutrition. *J. Am. Soc. Hortic. Sci.* **97**, 27–30.

Barker, A. V., and Mills, H. A. (1980). Ammonium and nitrate nutrition of horticultural crops. *Hortic. Rev.* **2**, 395–423.

Barker, A. V., Volk, R. J., and Jackson, W. A. (1966a). Root environment acidity as a regulatory factor in ammonium assimilation by the bean plant. *Plant Physiol.* **41**, 1193–1199.

Barker, A. V., Volk, R. J., and Jackson, W. A. (1966b). Growth and nitrogen distribution patterns in bean plants (*Phaseolus vulgaris* L.) subjected to ammonium nutrition. I. Effect of carbonates and acidity control. *Soil Sci. Soc. Am. Proc.* **30**, 228–232.

Beevers, L., and Hageman, R. H. (1980). Nitrate and nitrite reduction. *In* "The Biochemistry of Plants" (B. J. Miflin, ed.), Vol. 5, pp. 115–167. Academic Press, New York.

Beevers, L., Schrader, L. E., Flesher, D., and Hageman, R. H. (1965). The role of light and nitrate in the induction of nitrate reductase in radish cotyledons and maize seedlings. *Plant Physiol.* **40**, 691–698.

Behrend, J., and Mateles, R. I. (1975). Nitrogen metabolism in plant cell suspension cultures. I. Effect of amino acids on growth. *Plant Physiol.* **56**, 584–589.

Bennet, A. C., and Adams, F. (1970). Concentrations of NH_3 (aq.) required for incipient toxicity to seedlings. *Soil Sci. Soc. Am. Proc.* **34**, 259–263.

Benzioni, A., and Heimer, Y. M. (1977). Temperature effect on nitrate reduction activity *in vivo*. *Plant Sci. Lett.* **9**, 225–231.

Benzioni, A., Vaadia, Y., and Lips, H. (1971). Nitrate uptake by roots as regulated by nitrate reductase products of the shoot. *Physiol. Plant.* **24**, 288–290.

Berlier, Y., Guiraud, G., and Sauvair, Y. (1969). Etude avec l'azote 15 de l'absorption et du métabolisme de l'ammonium fourni a concentration croissante a des racines excisées de mais. *Agrochimica* **13**, 250–260.

Berner, E. (1971). Studies of the nitrogen metabolism of barley leaves. II. The effect of nitrate and ammonium on respiration and photosynthesis. *Physiol. Plant., Suppl.* **6**, 46–56.

Biggar, J. W. (1978). Spatial variability of nitrogen in soils. *In* "Nitrogen in the Environment" (D. R. Nielsen and J. G. MacDonald, eds.), Vol. 1, pp. 201–211. Academic Press, New York.

Blair, G. J., Miller, M. H., and Mitchell, W. A. (1970). Nitrate and ammonium as sources of nitrogen for corn and their influence on the intake of other ions. *Agron. J.* **62**, 530–532.

Bowen, G. D., and Smith, S. E. (1981). The effects of mycorrhizas on nitrogen uptake by plants. *In* "Terrestrial Nitrogen Cycles: Processes, Ecosystem Strategies and Management Impacts" (F. E. Clark and T. Rosswall, eds.), pp. 405–426. Ecological Bulletins, Stockholm.

Bowling, D. J. F. (1981). Release of ions to the xylem in roots. *Physiol. Plant.* **53**, 392–397.

Breteler, H., and Ten Cate, C. H. H. (1978). Ionic balance of root–shoot nitrate transfer in dwarf bean. *Physiol. Plant.* **42**, 53–56.

Buczek, J. (1979). Ammonium and potassium effect on nitrate assimilation in cucumber seedlings. *Acta Soc. Bot. Pol.* **48**, 157–169.

Burns, I. G. (1980). Influence of the spatial distribution of nitrate on the uptake of N by plants: A review and a model for rooting depth. *J. Soil Sci.* **31**, 155–173.

Cain, J. C. (1952). A comparison of ammonium and nitrate nitrogen for blueberries. *Proc. Am. Soc. Hortic. Sci.* **59**, 161–166.

Cain, J. C. (1954). Blueberry chlorosis in relation to leaf pH and mineral composition. *Proc. Am. Soc. Hortic. Sci.* **64**, 61–70.

Campbell, W. H. (1976). Separation of soybean leaf nitrate reductases by affinity chromatography. *Plant Sci. Lett.* **7**, 239–247.

Cantliffe, D. J., MacDonald, G. E., and Peck, N. H. (1974). Reduction in nitrate accumulation by molybdenum in spinach grown at low pH. *Commun. Soil Sci. Plant Anal.* **5**, 273–282.

Chambers, C. A., Smith, S. E., and Smith, F. A. (1980). Effects of ammonium and nitrate ions on mycorrhizal infection, nodulation and growth of *Trifolium subterraneum*. *New Phytol.* **85**, 47–62.

Clarkson, D. T. (1974). "Ion Transport and Cell Structure in Plants." McGraw-Hill, New York.

Clarkson, D. T., and Hanson, J. B. (1980). The mineral nutrition of higher plants. *Annu. Rev. Plant Physiol.* **31**, 239–298.

Clarkson, D. T., and Warner, A. J. (1979). Relationships between root temperature and the transport of ammonium and nitrate ions by Italian and perennial ryegrass (*Lolium multiflorum* and *Lolium perenne*). *Plant Physiol.* **64**, 557–561.

Clement, C. R., Hopper, M. J., Jones, L. H. P., and Leafe, E. L. (1978). The uptake of nitrate by *Lolium perenne* from flowing nutrient solution. II. Effect of light, defoliation and relationship to CO_2 flux. *J. Exp. Bot.* **29**, 1173–1183.

Colgrove, M. S., and Roberts, A. N. (1956). Growth of the azalea as influenced by ammonium and nitrate nitrogen. *Proc. Am. Soc. Hortic. Sci.* **68**, 522–536.

Cooil, B. (1974). Accumulation and radial transport of ions from potassium salts by cucumber roots. *Plant Physiol.* **53**, 158–163.

Cook, J. A., and Sehgal, P. (1970). Studies on cotyledons *in vitro:* Factors regulating urease. *J. Exp. Bot.* **21**, 672–676.

Court, M. N., Stephen, R. C., and Waid, J. S. (1964). Toxicity as a cause of the inefficiency of urea as a fertilizer. II. Experimental. *J. Soil Sci.* **15**, 49–65.

Cox, W. J., and Reisenauer, H. M. (1973). Growth and ion uptake by wheat supplied with nitrogen as nitrate, or ammonium, or both. *Plant Soil* **38**, 363–380.

Cram, W. J. (1973). Effects of Cl^- and HCO_3^- and malate and fluxes and CO_2 fixation in carrot and barley root cells. *J. Exp. Bot.* **25**, 253–268.

Crasswell, E. T., and Martin, A. E. (1975). Isotopic studies of the nitrogen balance in a cracking clay. I. Recovery of added nitrogen from soil and wheat in a glasshouse and gas lysimeter. *Aust. J. Soil Res.* **13**, 43–52.

Dalling, M. J., Hucklesby, D. P., and Hageman, R. H. (1973). A comparison of nitrite reductase enzymes from green leaves, scutella and roots of corn (*Zea mays* L.). *Plant Physiol.* **51**, 481–484.

Davy, A. J., and Taylor, K. (1974). Seasonal patterns of nitrogen availability in the Chiltern hills. *J. Ecol.* **62**, 793–807.

Denmead, O. T., Freney, J. R., and Simpson, J. R. (1976). A closed ammonia cycle within a plant canopy. *Soil Biol. Biochem.* **8**, 161–164.

Denmead, O. T., Nulsen, R., and Thurtell, G. W. (1978). Ammonia exchange over a corn crop. *Soil Sci. Soc. Am. J.* **42**, 840–842.

Dijkshoorn, W. (1973). Organic acids and their role in ion uptake. *In* "Chemistry and Biochemistry of Herbage" (G. W. Butler and R. W. Bailey, eds.), Vol. 2, pp. 161–188. Academic Press, New York.

Dilley, D. R., and Walker, D. R. (1961a). Assimilation of ^{14}C, ^{15}N-labelled urea by excited apple and peach leaves. *Plant Physiol.* **36**, 757–761.

Dilley, D. R., and Walker, D. R. (1961b). Urease activity of peach and apple leaves. *Proc. Am. Soc. Hortic. Sci.* **77**, 121–134.

Dirr, M. A. (1974). Nitrogen form and growth and nitrate reductase activity of the cranberry. *Hort Science* **9**, 347–348.

Doddema, H., and Otten, H. (1979). Uptake of nitrate by mutants of *Arabidopsis thaliana,* disturbed in uptake or reduction of nitrate. III. Regulation. *Physiol. Plant.* **45**, 339–346.

Dogar, M. A., and van Hai, T. (1977). Multiphasic uptake of ammonium by intact rice roots and its relationship with growth. *Z. Pflanzenphysiol.* **84**, 25–35.

Drew, M. C. (1975). Comparison of the effects of a localized supply of phosphate, nitrate, ammonium and potassium on the growth of the seminal root system, and the shoot, in barley. *New Phytol.* **75**, 479–490.

Drew, M. C., and Saker, L. R. (1975). Nutrient supply and the growth of the seminal root system of barley. II. Localized, compensatory increases in lateral root growth and rates of nitrate uptake when nitrate supply is restricted to only part of the root system. *J. Exp. Bot.* **26**, 79–90.

Drew, M. C., Saker, L. R., and Ashley, T. W. (1973). Nutrient supply and the growth of the seminal root system in barley. I. The effect of nitrate concentration on the growth of axes and laterals. *J. Exp. Bot.* **24**, 1189–1202.

Duke, S. H., Schrader, L. E., Miller, M. G., and Niece, R. L. (1978). Low temperature effects on soybean (*Glycine max* (L.) Merr. cv. Wells) free amino acid pools during germination. *Plant Physiol.* **62**, 642–647.

Duncan, W. G., and Ohlrogge, A. J. (1958). Principles of nutrient uptake from fertilizer bands. II. Root development in the band. *Agron. J.* **50**, 605–608.

Dupont, F. M., Burke, L. L., and Spanswick, R. M. (1981). Characterization of a partially purified adenosine triphosphate from a corn root plasma membrane fraction. *Plant Physiol.* **67**, 59–63.

Edgar, K. F., and Draper, S. R. (1974). Amino acids in *Hordeum distichon* and *Panicum miliaceum* grown in sand culture. *Phytochemistry* **13**, 325–327.

Edwards, J. H., and Barber, S. A. (1976a). Nitrogen uptake characteristics of corn roots at low N concentration as influenced by plant age. *Agron. J.* **68**, 17–19.

Edwards, J. H., and Barber, S. A. (1976b). Nitrogen flux into corn roots as influenced by shoot requirement. *Agron. J.* **68**, 471–473.

Edwards, J. H., and Horton, B. D. (1982). Interaction of peach seedlings to $NO_3^- : NH_4^+$ ratios in nutrient solutions. *J. Am. Soc. Hortic. Sci.* **107**, 142–147.

Ellenberg, H. (1977). Stickstoff als Standortfaktor, insbesondere fur mitteleuropaische, Pflanzengesellschaften. *Oecol. Plant.* **12**, 1–22.

Emes, M. J., and Fowler, M. W. (1979). The intracellular location of the enzymes of nitrate assimilation in the apices of seedlings pea roots. *Planta* **194**, 249–253.

Epstein, E. (1972). "Mineral Nutrition of Plants: Principles and Perspectives." Wiley, New York.

Etherton, B., and Rubinstein, B. (1978). Evidence for amino acid-H^+ co-transport in oat coleoptiles. *Plant Physiol.* **61**, 933–937.

Evans, L. T., Wardlaw, J. F., and Fisher, R. A. (1975). The physiological basis of crop yield. *In* "Crop Physiology" (L. T. Evans, ed.), pp. 327–355. Cambridge Univ. Press, London and New York.

Ezta, F. N., and Jackson, W. A. (1975). Nitrate translocation by detopped corn seedlings. *Plant Physiol.* **56**, 148–156.

Farquhar, G. D., Wetselaar, R., and Firth, P. M. (1979). Ammonia volatilization from senescing leaves of maize. *Science* **203**, 1257–1258.

Ferrari, T. E., Yoder, O. C., and Filner, P. (1973). Anaerobic nitrite production by plant cells and tissues: Evidence for two nitrate pools. *Plant Physiol.*, **51**, 423–431.

Fowler, M. W., and Barker, R. D. J. (1978). Assimilation of ammonium in non-chlorophyllous tissue. *In* "Nitrogen Assimilation of Plants" (E. J. Hewitt and C. V. Cutting, eds.), pp. 489–500. Academic Press, New York.

Fowler, M. W., Jessup, W., and Sarkissian, G. S. (1974). Glutamate synthase type activity in higher plants. *FEBS Lett.* **46**, 340–342.

Foy, C. D., and Fleming, A. L. (1978). The physiology of plant tolerance to excess available aluminum and manganese in acid soils. *In* "Crop Tolerance to Suboptimal Land Conditions" (G. A. Jung, ed.), pp. 301–328. Am. Soc. Agron., Madison, Wisconsin.

Fried, M., Zsoldos, F., Vose, P. B., and Shatokhin, I. L. (1965). Characterising the NO_3^- and NH_4^+ uptake process in rice roots by use of ^{15}N-labelled NH_4NO_3. *Physiol. Plant.* **18**, 313–320.

Friedrich, J. W., and Schrader, L. E. (1978). Sulphur deprivation and nutrient metabolism in maize seedlings. *Plant Physiol.* **61**, 900–903.

Frith, G. J. T. (1972). Effect of ammonium nutrition on the activity of nitrate reductase in the roots of apple seedlings. *Plant Cell Physiol.* **13**, 1085–1090.

Frith, G. J. T., and Nichols, D. G. (1975). Nitrogen uptake by apple seedlings as affected by light, and nutrient stress in part of the root system. *Physiol. Plant.* **34**, 129–133.

Frost, W. B., Blevins, D. G., and Barnett, N. M. (1978). Cation pretreatment effects on nitrate uptake, xylem exudate, and malate levels in wheat seedlings. *Plant Physiol.* **61**, 323–326.

Frota, J. N. E., and Tucker, T. C. (1972). Temperature influence on ammonium and nitrate absorption by lettuce. *Soil Sci. Soc. Am. Proc.* **36**, 97–100.

Garner, D. C. J., and Peel, A. J. (1971). Metabolism and transport of ^{14}C-labelled glutamic and aspartic acids in the phloem of willow. *Phytochemistry* **10**, 2385–2387.

Geiger, D. R. (1975). Phloem loading. *In* "Encyclopedia of Plant Physiology, New Series" (M. H. Zimmermann and J. A. Milburn, eds.), Vol. 1, pp. 395–431. Springer-Verlag, Berlin and New York.

Giaquinta, R. T. (1976). Evidence for phloem loading from the apoplast. *Plant Physiol.* **57**, 872–875.

Giaquinta, R. T. (1977). Possible role of pH gradient and membrane ATPase in the loading of sucrose into the sieve tubes. *Nature (London)* **267**, 369–370.

Gibbs, M., and Calo, N. (1959). Factors affecting light induced fixation of carbon dioxide by isolated spinach chloroplasts. *Plant Physiol.* **34**, 318–323.

Gigon, A., and Rorison, I. H. (1972). The response of some ecologically distinct plant species to nitrate- and ammonium-nitrogen. *J. Ecol.* **60**, 93–102.

Givan, G. V. (1979). Metabolic detoxification of ammonia in tissues of higher plants. *Phytochemistry* **18**, 375–382.

Golvano, M. P., Felipe, M. R., and Cintas, A. M. (1982). Influence of nitrogen sources on chloroplast development in wheat seedlings. *Physiol. Plant.* **56**, 353–360.

Gordon, W. R., Schwemmer, S. S., and Hillman, W. S. (1978). Nickel and the metabolism of urea by *Lemna paucicostata* Hegelm. 6746. *Planta* **140**, 265–268.

Goyal, S. S., Lorenz, O. A., and Huffaker, R. C. (1982a). Inhibitory effects of ammoniacal nitrogen on growth of radish plants. I. Characterization of toxic effects of NH_4^+ on growth and its alleviation by NO_3^-. *J. Am. Soc. Hortic. Sci.* **107**, 125–129.

Goyal, S. S., Huffaker, R. C., and Lorenz, O. A. (1982b). Inhibitory effects of ammoniacal nitrogen on growth of radish plants. II. Investigation on the possible causes of ammonium toxicity to radish plants and its reversal by nitrate. *J. Am. Soc. Hortic. Sci.* **107**, 130–135.

Granstedt, R. C., and Huffaker, R. C. (1982). Identification of the leaf vacuole as a major nitrate storage pool. *Plant Physiol.* **70**, 410–413.

Grasmanis, V. O., and Nicholas, D. J. D. (1971). Annual uptake and distribution of N^{15}-labelled ammonia and nitrate in young Jonathan/MM 104 apple trees grown in solution cultures. *Plant Soil* **35**, 95–112.

Green, D. W., and Bukovac, M. J. (1974). Stomatal penetration: Effect of surfactants and role in foliar absorption. *Am. J. Bot.* **61**, 100–106.

Green, J. L., Holley, W. D., and Thaden, B. (1973). Effects of the $NH_4^+ : NO_3^-$ ratio, chloride, N-Serve and simazine on carnation flower production and plant growth. *Proc. Fla. State Hortic. Soc.* **86**, 383–388.

Greenwood, D. J. (1978). Influence of spatial variability in soil on microbial activity, crop growth, and agronomic research. *In* "Nitrogen in the Environment" (D. R. Nielsen and J. G. MacDonald, eds.), Vol. 1, pp. 213–222. Academic Press, New York.

Grime, J. P., and Hunt, R. (1975). Relative growth rate: Its range and adaptive significance in a local flora. *J. Ecol.* **64**, 975–988.

Guerrero, M. G., Vega, J. M., and Losada, M. (1981). The assimilatory nitrate-reducing system and its regulation. *Annu. Rev. Plant Physiol.* **32**, 169–204.

Guiz, C., Hirel, B., Shedlofsky, G., and Gadal, P. (1979). Occurrence and influence of light on the relative proportions of two glutamine synthetases in rice leaves. *Plant Sci. Lett.* **15**, 271–277.

Gupta, P. L., and Rorison, I. H. (1975). Seasonal differences in the availability of nutrients down a podzolic profile. *J. Ecol.* **63**, 521–534.

Hacket, C. (1972). A method of applying nutrients locally to roots under controlled conditions and some morphological effects of locally applied nitrate on the branching of wheat roots. *Aust. J. Biol. Sci.* **25**, 1169–1180.

Hageman, R. H. (1978). Integration of nitrogen assimilation in relation to yield. *In* "Nitrogen Assimilation of Plants" (E. J. Hewitt and C. V. Cutting, eds.), pp. 591–611. Academic Press, New York.

Hall, S. M., and Baker, D. A. (1972). The chemical composition of *Ricinus* phloem exudate. *Planta* **106**, 131–140.

Hallmark, W. B., and Huffaker, R. C. (1978). The influence of ambient nitrate, temperature, and light on nitrate assimilation in Sudangrass seedlings. *Physiol. Plant.* **44**, 147–152.

Hassan, M. M., and van Hai, T. (1976). Ammonium and potassium uptake from citrus roots. *Physiol. Plant.* **36**, 20–22.

Havill, D. C., Lee, J. A., and Stewart, G. R. (1974). Nitrate utilization by species from acidic and calcareous soil. *New Phytol.* **73**, 1221–1231.

Havill, D. C., Lee, J. A., and De-Felice, J. (1977). Some factors limiting nitrate utilization in acidic and calcareous grasslands. *New Phytol.* **78**, 649–659.

Haynes, R. J. (1980). Ion exchange properties of roots and ionic interactions within the root apoplasm: Their role in ion accumulation by plants. *Bot. Rev.* **46**, 75–99.

Haynes, R. J., and Goh, K. M. (1977). Review on physiological pathways of foliar absorption. *Sci. Hortic. (Amsterdam)* **7**, 291–302.

Haynes, R. J., and Goh, K. M. (1978). Ammonium and nitrate nutrition of plants. *Biol. Rev. Cambridge Philos. Soc.* **53**, 465–510.

Heber, U., Kirk, M. R., Gimmler, H., and Schafer, G. (1974). Uptake and reduction of glycerate by isolated chloroplasts. *Planta* **120**, 32–46.

Heimer, Y. M., and Filner, P. (1971). Regulation of the nitrate assimilation pathway in cultured tobacco cells. *Biochim. Biophys. Acta* **230**, 352–367.

Helleburst, J. A. (1976). Osmoregulation. *Annu. Rev. Plant Physiol.* **27**, 485–505.

Hewitt, E. J. (1975). Assimilatory nitrate–nitrite reduction. *Annu. Rev. Plant Physiol.* **26**, 73–100.

Hewitt, E. J., and Notton, B. A. (1980). Nitrate reductase systems in eukaryotic and prokaryotic organisms. *In* "Molybdenum and Molybdenum-Containing Enzymes" (M. Coughland, ed.), pp. 273–325. Pergamon, Oxford.

Hewitt, E. J., Hucklesby, D. P., and Notton, B. A. (1976). Nitrate metabolism. *In* "Plant Biochemistry" (J. Bonner and J. E. Varner, eds.), 3rd ed., pp. 633–681. Academic Press, New York.

Hewitt, E. J., Hucklesby, D. P., Mann, A. F., Notton, B. A., and Rucklidge, G. J. (1978). Regulation of nitrate assimilation in plants. *In* "Nitrogen Assimilation of Plants" (E. J. Hewitt and C. V. Cutting, eds.), pp. 255–287. Academic Press, New York.

Hiatt, A. J., and Leggett, J. E. (1974). Ionic interactions and antagonism in plants. *In* "The Plant Root and Its Environment" (E. W. Carson, ed.), pp. 101–134. Univ. of Virginia Press, Charlottesville.

Hill, A. C. (1971). Vegetation: A sink for atmospheric pollutants. *J. Air Pollut. Control Assoc.* **21**, 341–346.

Hill, C. A., and Chamberlain, E. M. (1976). The removal of water-soluble gases from the atmosphere by vegetation. *ERDA Symp. Ser.* **38**, 153–170.

Hill-Cottingham, D. G., and Lloyd-Jones, C. P. (1978). Translocation of nitrogenous compounds in plants. *In* "Nitrogen Assimilation of Plants" (E. J. Hewitt and C. V. Cutting, eds.), pp. 397–405. Academic Press, New York.

Hirel, B., and Gadal, P. (1980). Glutamine synthetase in rice: A comparative study of the enzymes from roots and leaves. *Plant Physiol.* **66**, 619–623.

Hirel, B., Perrot-Rechenmann, C., Suzuki, A., Vidal, J., and Gadal, P. (1982). Glutamine synthetase in spinach leaves. Immunological studies and immunocytochemical localization. *Plant Physiol.* **69**, 983–987.

Hodges, T. K. (1976). ATPases associated with membranes of plant cells. *In* "Encyclopedia of Plant Physiology, New Series" (U. Luttge and M. G. Pitman, eds.), Vol. 2, Part A, pp. 260–283. Springer-Verlag, Berlin and New York.

Hoff, J. E., Wilcox, G. E., and Jones, C. M. (1974). The effect of nitrate and ammonium nitrogen on the free amino acid composition of tomato plants and tomato fruits. *J. Am. Soc. Hortic. Sci.* **99**, 27–30.

Hofstra, J. J. (1964). Amino acids in the bleeding sap of fruiting tomato plants. *Acta Bot. Neerl.* **13**, 148–158.

Holevas, C. D. (1966). The effect of vesicular-arbuscular mycorrhiza on the uptake of soil phosphorus by strawberry (*Fragaria* sp. var. Cambridge Favourite). *J. Hortic. Sci.* **41**, 57–64.

Hucklesby, D. P., Cammack, R., and Hewitt, E. J. (1978). Properties and mechanisms of nitrite reductase. *In* "Nitrogen Assimilation of Plants" (E. J. Hewitt and C. V. Cutting, eds.), pp. 245–254. Academic Press, New York.

Huffaker, R. C., and Rains, D. W. (1978). Factors influencing nitrate acquisition by plants; assimilation and fate of reduced nitrogen. *In* "Nitrogen in the Environment" (D. R. Nielsen and J. G. MacDonald, eds.), Vol. 2, pp. 1–43. Academic Press, New York.

Huffaker, R. C., Radin, T., Kleinkopf, G. E., and Cox, E. L. (1970). Effects of mild water stress on enzymes of nitrate assimilation and of the carboxylative phase of photosynthesis in barley. *Crop Sci.* **10**, 471–474.

Hull, H. M., Morton, H. L., and Wharrie, J. R. (1975). Environmental influences on cuticle development and resultant foliar penetration. *Bot. Rev.* **41**, 421–452.

Hutchinson, G. L., Millington, R. J., and Peters, D. B. (1972). Atmospheric ammonia: Absorption by plant leaves. *Science* **175**, 771–772.

Ikeda, M., Yamada, Y., and Harada, T. (1974). Glucose metabolism in detached leaves of tomato plants grown with ammonium and nitrate as nitrogen sources. *Soil Sci. Plant Nutr.* **20**, 185–194.

Ingestad, T. (1970). A definition of optimum nutrient requirements in birch seedlings. I. *Physiol. Plant.* **23**, 1127–1138.

Ingestad, T. (1971). A definition of optimum nutrient requirements in birch seedlings. II. *Physiol. Plant.* **24**, 118–125.

Ingestad, T. (1973). Mineral nutrient requirements of *Vaccinium vitis idaea* and *V. myrtillus*. *Physiol. Plant.* **29**, 239–246.

Ingestad, T. (1976). Nitrogen and cation nutrition of three ecologically different plant species. *Physiol. Plant.* **38**, 29–34.

Ingestad, T. (1979). Mineral nutrient requirements of *Pinus silvestris* and *Picea abies* seedlings. *Physiol. Plant.* **45**, 373–380.

Jackson, W. A. (1978). Nitrate acquisition and assimilation by higher plants: Processes in the root system. *In* "Nitrogen in the Environment" (D. R. Nielsen and J. D. MacDonald, eds.), Vol. 2, pp. 45–88. Academic Press, New York.

Jackson, W. A., and Williams, D. C. (1968). Nitrate-stimulated uptake and transport of strontium and other cations. *Soil Sci. Soc. Am. Proc.* **32**, 689–704.

Jackson, W. A., Flesher, D., and Hageman, R. H. (1973). Nitrate uptake by dark-grown corn seedlings: Some characteristics of apparent induction. *Plant Physiol.* **51**, 120–127.

Jackson, W. A., Johnson, R. E., and Volk, R. J. (1974). Nitrate uptake patterns in wheat seedlings as influenced by nitrate and ammonium. *Physiol. Plant.* **32**, 108–114.

Jackson, W. A., Kwik, K. D., Volk, R. J., and Butz, R. G. (1976a). Nitrate influx and efflux by intact wheat seedlings: Effects of prior nitrate nutrition. *Planta* **132**, 149–156.

Jackson, W. A., Kwik, K. D., and Volk, R. J. (1976b). Nitrate uptake during recovery from nitrogen deficiency. *Physiol. Plant.* **36**, 174–181.

Joiner, J. N., and Knoop, W. E. (1969). Effect of ratios of NH_4^+ to NO_3^- and levels of N and K on chemical content of *Chrysanthemum morifolium* "Bright Golden Ann." *Proc. Fla. State Hortic. Soc.* **82**, 403–407.

Jolly, S. O., and Tolbert, N. E. (1978). NADH-nitrate reductase inhibitor from soybean leaves. *Plant Physiol.* **62**, 197–203.

Jones, R. W., and Sheard, R. W. (1975). Phytochrome, nitrate movement and induction of nitrate reductase in etiolated pea terminal buds. *Plant Physiol.* **55**, 954–959.

Jones, R. W., Abbott, A. J., Hewitt, E., Best, G. R., and Watson, E. F. (1978). Nitrate reductase activity in Paul's scarlet rose suspension cultures and the differential role of nitrate and molybdenum in induction. *Planta* **141**, 183–189.

Joseph, R. A., van Hai, T., and Lambert, J. (1975). Multi-phasic uptake of ammonium by soybean roots. *Physiol. Plant.* **34**, 321–325.

Joy, K. W., and Antcliff, A. J. (1966). Translocation of amino acids in sugar beet. *Nature (London)* **211**, 210–211.

Kadam, S. S., Gandhi, A. P., Sawhney, S. K., and Naik, M. S. (1974). Inhibitor of nitrate reductase in the roots of rice seedlings and its effect on the enzyme activity in the presence of NADH. *Biochim. Biophys. Acta* **350**, 162–170.

Kafkafi, U., Valoras, N., and Letey, J. (1982). Chloride interaction with nitrate and phosphate nutrition in tomato (*Lycopersicon esculentum* L.). *J. Plant Nutr.* **5**, 1369–1385.

Kannan, S. (1980). Mechanisms of foliar uptake of plant nutrients: Accomplishments and prospects. *J. Plant Nutr.* **2**, 717–735.

Kauffman, J. E., Beard, J. B., and Penner, D. (1971). The influence of temperature on nitrate reductase activity of *Agrostis palustris* and *Cynodon dactylon*. *Physiol. Plant.* **25**, 378–381.

Keys, A. J., Bird, I. F., Cornelius, M. J., Lea, P. J., Wallsgrove, R. M., and Miflin, B. J. (1978). Photorespiratory nitrogen cycle. *Nature (London)* **275**, 741–743.

Kinraide, T. B., and Etherton, B. (1982). Energy coupling in H^+-amino acid cotransport. ATP dependence of the spontaneous electrical repolarization of the cell membranes of oat coleoptiles. *Plant Physiol.* **69**, 648–652.

Kirkby, E. A. (1968). Influence of ammonium and nitrate nutrition on the cation–anion balance and nitrogen and carbohydrate metabolism of white mustard plants grown in dilute nutrient solutions. *Soil Sci.* **105**, 133–141.

Kirkby, E. A. (1981). Plant growth in relation to nitrogen supply. *In* "Terrestrial Nitrogen Cycles: Processes, Ecosystem Strategies and Management Impacts" (F. E. Clark and T. Rosswall, eds.), pp. 249–267. Ecological Bulletins, Stockholm.

Kirkby, E. A., and Armstrong, M. J. (1980). Nitrate uptake by roots as regulated by nitrate assimilation in the shoot of castor oil plants. *Plant Physiol.* **65**, 286–290.

Kirkby, E. A., and Knight, A. H. (1977). Influence of the level of nitrate nutrition on ion uptake assimilation, organic acid accumulation, and cation–anion balance in whole tomato plants. *Plant Physiol.* **60**, 349–353.

Kirkby, E. A., Armstrong, M. J., and Leggett, J. E. (1981). Potassium recirculation in tomato plants in relation to potassium supply. *J. Plant Nutr.* **3**, 955–966.

Klepper, L., Flesher, D., and Hageman, R. H. (1971). Generation of reduced nicotinamide adenine dinucleotide for nitrate reduction in green leaves. *Plant Physiol.* **48**, 580–590.

Knypl, J. S. (1978). Hormonal control of nitrate assimilation: Do phytohormones and phytochrome control the activity of nitrate reductase? *In* "Nitrogen Assimilation of Plants" (E. J. Hewitt and C. V. Cutting, eds.), pp. 541–556. Academic Press, New York.

Krajina, V. J. (1969). Ecology of forest trees in British Columbia. *Ecol. West. North Am.* **2**, 1–146.

Krajina, V. J., Madoc-Jones, S., and Mellor, G. (1973). Ammonium and nitrate in the nitrogen economy of some conifers growing in Douglas-fir communities of the Pacific northwest of America. *Soil Biol. Biochem.* **5**, 143–147.

Krause, G. H., and Heber, U. (1976). Energetics of intact chloroplasts. *In* "The Intact Chloroplast" (J. Barber, ed.), pp. 171–214. Elsevier, Amsterdam.

Krogman, D. W., Jagerdorf, A. T., and Avron, M. (1959). Uncouplers of spinach chloroplast photosynthetic phosphorylation. *Plant Physiol.* **34**, 272–277.

Kruckelmann, H. W. (1975). Effects of fertilizers, soil, soil tillage and plant species on the frequency of *Endogone* chlamidophores and mycorrhizal infection in arable soils. *In* "Endomycorrhizas" (F. E. Sanders, B. Mosse, and P. B. Tinker, eds.), pp. 512–525. Academic Press, New York.

Larsen, P. O. (1980). Physical and chemical properties of amino acids. *In* "The Biochemistry of Plants" (B. J. Miflin, ed.), Vol. 5, pp. 225–271. Academic Press, New York.

Lauchli, A. (1976). Symplasmic transport and ion release to the xylem, *In* "Transport and Transfer Processes in Plants" (I. F. Wardlaw and J. B. Passioura, eds.), pp. 101–112. Academic Press, New York.

Lea, P. J., and Miflin, B. J. (1974). Alternative route for nitrogen assimilation in higher plants. *Nature (London)* **251**, 614–616.

Lea, P. J., and Miflin, B. J. (1979). Photosynthetic ammonia assimilation. *In* "Encyclopedia of Plant Physiology, New Series" (M. Gibbs and E. Latzko, eds.), Vol. 6, pp. 445–456. Springer-Verlag, Berlin and New York.

Lee, J. A., and Stewart, G. R. (1978). Ecological aspects of nitrogen assimilation. *Adv. Bot. Res.* **6**, 1–43.

Lee, R. B. (1980). Sources of reductant for nitrate assimilation in non-photosynthetic tissue: A review. *Plant, Cell Environ.* **3**, 65–90.

Leece, D. R. (1976). Composition and ultrastructure of leaf cuticles from different trees relative to differential foliar absorption. *Aust. J. Plant Physiol.* **3**, 833–847.

Leopold, A. C. (1961). Senescence in plant development. *Science* **134**, 1727–1732.

Lewis, O. A. M., and Probyn, T. A. (1978). [15]N incorporation and glutamine synthetase inhibition studies of nitrogen assimilation in leaves of the nitrophile, *Datura stramonium* L. *New Phytol.* **81**, 519–526.,

Lin, W. (1979). Potassium and phosphate uptake in corn roots. Further evidence for an electrogenic H^+/K^+ exchanger and on OH^-/Pi antiporter. *Plant Physiol.* **63**, 952–955.

Lloyd, N. D. H., and Joy, K. W. (1978). 2-Hydroxysuccinamide acid: A product of asparagine metabolism in plants. *Biochem. Biophys. Res. Commun.* **81**, 186–192.

Lorenz, H. (1976). Nitrate, ammonium and amino acids in the bleeding sap of tomato plants in relation to form and concentration of nitrogen in the medium. *Plant Soil* **45**, 169–175.

Lorenz, O. A. (1978). Potential nitrate levels in edible plant parts. *In* "Nitrogen in the Environment" (D. R. Nielsen and J. G. MacDonald, eds.), Vol. 2, pp. 201–219. Academic Press, New York.

Losada, M., and Arnon, D. I. (1963). Selective inhibitors of photosynthesis. *In* "Metabolic Inhibitors" (R. H. Hochester and J. H. Quastel, eds.), Vol. 2, pp. 559–573. Academic Press, New York.

Losada, M., and Guerrero, M. G. (1979). The photosynthetic reduction of nitrate and its regulation. *In* "Photosynthesis in Relation to Model Systems" (J. Barber, ed.), pp. 365–408. Elsevier, Amsterdam.

Losada, M., Herrera, J., Maldonano, J. M., and Paneque, A. (1973). Mechanism of nitrate reductase reversible inactivation by ammonia in *Chlamydomonas. Plant Sci. Lett.* **1,** 31–37.

Lundeberg, G. (1970). Utilisation of various nitrogen sources, in particular bound soil nitrogen by mycorrhizal fungi. *Stud. For. Suec.* **79,** 1.

Luttge, U., and Higinbotham, N. (1979). "Transport in Plants." Springer-Verlag, Berlin and New York.

Lycklama, J. C. (1963). The absorption of ammonium and nitrate by perennial ryegrass. *Acta Bot. Neerl.* **12,** 361–423.

Lyon, N. C., and Mueller, W. C. (1974). A freeze-etch study of plant cell walls for ectodesmata. *Can. J. Bot.* **52,** 2023–2036.

Maas, E. V. (1969). Calcium uptake by excised maize roots and interactions with alkali cations. *Plant Physiol.* **44,** 985–989.

McDaniel, C. N., Lyons, R. A., and Blackman, M. S. (1981). Amino acid transport in suspension cultured plant cells. IV. Biphasic saturatable uptake kinetics of L-leucine in isolates of six *Nicotiana tabacum* plants. *Plant Sci. Lett.* **23,** 17–23.

McDaniel, C. N., Holterman, R. K., Bone, R. F., and Wozniak, P. M. (1982). Amino acid transport in suspension-cultured plant cells. III. Common carrier system for the uptake of L-arginine, L-aspartic acid, L-histidine, L-leucine, and L-phenylalanine. *Plant Physiol.* **69,** 246–249.

McElhannon, W. S., and Mills, H. A. (1977). The influence of N concentration and NO_3/NH_4 ratio on the growth of lima and snap beans and southern pea seedlings. *Commun. Soil Sci. Plant Anal.* **8,** 677–687.

McGrath, S. P., and Rorison, I. H. (1982). The influence of nitrogen source on the tolerance of *Holcus lanatus* L. and *Bromus erectus* Huds. to manganese. *New Phytol.* **91,** 443–452.

MacKown, C. T., Jackson, W. A., and Volk, R. J. (1982). Restricted nitrate influx and reduction in corn seedlings exposed to ammonium. *Plant Physiol.* **69,** 353–359.

McNeal, F. N., Berg, M. A., Brown, P. L., and McGuire, C. F. (1971). Productivity and quality response of five spring wheat genotypes, *Triticum aestivum* L., to nitrogen fertilizer. *Agron. J.* **63,** 908–910.

McNeil, D. L., Atkins, C. A., and Pate, J. S. (1979). Uptake and utilization of xylem-borne amino compounds by shoot organs of a legume. *Plant Physiol.* **63,** 1076–1081.

McParland, R. H., Guevara, J. G., Becker, R. R., and Evans, H. J. (1976). The purification and properties of the glutamine synthetase from the cytosol of soya-bean root nodules. *Biochem. J.* **153,** 597–606.

Mann, A. F., Fentem, P. A., and Stewart, G. R. (1979a). Identification of two forms of glutamine synthetase in barley (*Hordeum vulgare*). *Biochem. Biophys. Res. Commun.* **88,** 515–521.

Mann, A. F., Fentem, P. A., and Stewart, G. R. (1979b). Tissue localization of barley (*Hordeum vulgare*) glutamine synthetase isoenzymes. *FEBS Lett.* **110,** 265–267.

Martin, A. E., and Ross, P. J. (1968). A nitrogen-balance study using labelled fertilizer in a gas lysimeter. *Plant Soil* **28,** 182–186.

Martin, P. (1971). Pathways of upward translocation of nitrogen in kidney bean plants after uptake by the root. *Z. Pflanzenphysiol.* **64,** 206–222.

Martin, P. (1973). Nitratstickstoff in Buschbohnenblaettern unter dem Gesichtspunkt der Kompartimentierung der Zellen. *Z. Pflanzenphysiol.* **70,** 158–162.

Matoh, T. Suzuki, F., and Ida, S. (1980). Corn leaf glutamate synthase: Purification and properties of the enzyme. *Plant Cell Physiol.* **20,** 1329–1340.

Matsumoto, H., Hasegawa, Y., Kobayashi, M., and Takahashi, E. (1968). Inducible formation of urease in *Canavalia ensiformis*. *Physiol. Plant.* **21**, 872–881.

Matsumoto, H., Wakiuchi, N., and Takahashi, E. (1971). Changes of some mitochondrial enzyme activities of cucumber leaves during ammonium toxicity. *Physiol. Plant.* **25**, 353–357.

Maynard, D. N., and Barker, A. V. (1969). Studies on the tolerance of plants to ammonium nutrition. *J. Am. Soc. Hortic. Sci.* **94**, 235–239.

Maynard, D. N., and Barker, A. V. (1971). Critical nitrate levels for leaf lettuce, radish, and spinach plants. *Commun. Soil Sci. Plant Anal.* **2**, 461–470.

Maynard, D. N., Barker, A. V., and Lachman, W. H. (1966). Ammonium-induced stem and leaf lesions of tomato plants. *Proc. Am. Soc. Hortic. Sci.* **88**, 516–520.

Maynard, D. N., Barker, A. V., and Lachman, W. H. (1968). Influence of potassium on the utilization of ammonium by tomato plants. *J. Am. Soc. Hortic. Sci.* **92**, 537–542.

Maynard, D. N., Barker, A. V., Minotti, P. L., and Peck, N. H. (1976). Nitrate accumulation in vegetables. *Adv. Agron.* **28**, 71–118.

Meeker, G. B., Purvis, A. C., Neyra, C. A., and Hageman, R. H. (1974). Uptake and accumulation of nitrate as a major factor in the regulation of nitrate reductase activity of corn (*Zea mays* L.) leaves: Effects of high ambient CO_2 and malate. *In* "Mechanisms of Regulation of Plant Growth" (R. L. Bieleski, A. R. Ferguson, and M. M. Cresswell, eds.), Bull. No. 12, pp. 49–58. Royal Society of New Zealand, Wellington.

Melin, E., and Nilsson, H. (1952). Transport of labelled nitrogen from an ammonium source to pine seedlings through mycorrhizal mycelium. *Sven. Bot. Tidskr.* **46**, 281–285.

Mengel, K., and Viro, M. (1978). The significance of plant energy status for the uptake and incorporation of NH_4^+-nitrogen by young rice plants. *Soil Sci. Plant Nutr. (Tokyo)* **24**, 407–416.

Mengel, K., Viro, M., and Hehl, G. (1976). Effect of potassium on uptake and incorporation of ammonium-nitrogen by rice plants. *Plant Soil* **43**, 479–486.

Michael, G., Martin, P., and Owissia, I. (1970). Uptake of ammonium and nitrate-N from labelled ammonium nitrate in relation to carbohydrate supply of the roots. *In* "Nitrogen Nutrition of the Plant" (E. A. Kirkby, ed.), pp. 22–29. Waverly Press, Leeds.

Middleton, K. R., and Smith, G. S. (1979). A comparison of ammoniacal and nitrate nutrition of perennial ryegrass through a thermodynamic model. *Plant Soil* **53**, 487–504.

Miflin, B. J., and Lea, P. J. (1975). Glutamine and asparagine as nitrogen donors for reductant-dependent glutamate synthesis in pea roots. *Biochem. J.* **149**, 403–409.

Miflin, B. J., and Lea, P. J. (1980). Regulation of nitrate assimilation in plants. *In* "The Biochemistry of Plants" (B. J. Miflin, ed.), Vol. 5, pp. 169–202. Academic Press, New York.

Miller, M. H., Mamaril, C. P., and Blair, G. J. (1970). Ammonium effects on phosphorus absorption through pH changes and phosphorus precipitation at the soil–root interface. *Agron. J.* **62**, 524–527.

Mills, H. A., and Jones, J. B. (1979). Nutrient deficiencies and toxicities in plants: Nitrogen. *J. Plant Nutr.* **1**, 101–122.

Minotti, P. L., and Jackson, W. A. (1970). Nitrate reduction in the roots and shoots of wheat seedlings. *Planta* **95**, 36–44.

Minotti, P. L., Williams, D. C., and Jackson, W. A. (1968). Nitrate uptake and reduction as affected by calcium and potassium. *Soil Sci. Soc. Am. Proc.* **32**, 692–698.

Minotti, P. L., Williams, D. C., and Jackson, W. A. (1969). Nitrate uptake in wheat as influenced by ammonium and other cations. *Crop Sci.* **9,** 9–14.

Moore, D. P. (1974). Physiological effects of pH on roots. *In* "The Plant Root and Its Environment" (E. W. Carson, ed.), pp. 135–151. Virginia Univ. Press, Charlottesville.

Moraghan, J. T., and Porter, O. A. (1975). Maize growth as affected by root temperature and form of nitrogen. *Plant Soil* **43,** 479–486.

Morgan, M. A., Volk, R. J., and Jackson, W. A. (1973). Simultaneous influx and efflux of nitrate during uptake by perennial ryegrass. *Plant Physiol.* **51,** 267–272.

Mosse, B., and Phillips, J. M. (1971). The influence of phosphate and other nutrients on the development of vesicular-arbuscular mycorrhiza in culture. *J. Gen. Microbiol.* **69,** 157–166.

Muhammad, S., and Kumazawa, K. (1974). Assimilation and transport of nitrogen in rice. I. ^{15}N-labelled ammonium nitrogen. *Plant Cell Physiol.* **15,** 747–758.

Muller, A. J., and Grafe, R. (1978). Isolation and characterization of cell lines of *Nicotiana tabacum* lacking nitrate reductase. *Mol. Gen. Genet.* **161,** 67–76.

Munns, D. N., and Jackson, W. A. (1978). Nitrate and ammonium uptake by rooted cuttings of sweet potato. *Agron. J.* **70,** 312–316.

Nelson, L. E., and Selby, R. (1974). The effect of nitrogen sources and iron levels on the growth and composition of sitka spruce, and scots pine. *Plant Soil* **41,** 573–588.

Neyra, C. A., and Hageman, R. H. (1975). Nitrate uptake and induction of nitrate reductase in excised corn roots. *Plant Physiol.* **56,** 692–695.

Neyra, C. A., and Hageman, R. H. (1976). Relationships between carbon dioxide, malate, and nitrate accumulation and reduction in corn (*Zea mays* L.) seedlings. *Plant Physiol.* **58,** 726–730.

Nicholas, J. C., Harper, J. E., and Hageman, R. H. (1976). Nitrate reductase activity in soybeans (*Glycine max* L.). II. Energy limitations. *Plant Physiol.* **58,** 736–739.

Nishimura, M., Bhusawana, P., Strzalka, K., and Akazawa, T. (1982). Developmental formation of glutamine synthetase in greening pumpkin cotyledons and its subcellular localization. *Plant Physiol.* **70,** 353–356.

Nissen, P., Fageria, N. K., Rayar, A. J., Hassan, M. M., and van Hai, T. (1980). Multiphasic accumulation of nutrients by plants. *Physiol. Plant.* **49,** 222–240.

Notton, B. A., and Hewitt, E. J. (1978). Structure and properties of higher plant nitrate reductase, especially *Spinacea oleracea*. *In* "Nitrogen Assimilation of Plants" (E. J. Hewitt and C. V. Cuttings, eds.), pp. 227–244. Academic Press, New York.

Novacky, A., Fisher, E., Ullrich-Eberius, C. I., Luttge, U., and Ullrich, W. (1978). Membrane potential changes during transport of glycine as a neutral amino acid and nitrate in *Lemna gibba* G 1. *FEBS Lett.* **88,** 264–267.

Novoa, R., and Loomis, R. S. (1981). Nitrogen and plant production. *Plant Soil* **58,** 177–204.

Nye, P. H. (1981). Changes of pH across the rhizosphere induced by roots. *Plant Soil* **61,** 7–26.

Nye, P. H., and Tinker, P. B. (1977). "Solute Movement in the Soil–Root Systems." Blackwell, Oxford.

Oaks, A. (1978). Nitrate reductase in roots and its regulation. *In* "Nitrogen Assimilation of Plants" (E. J. Hewitt and C. V. Cutting, eds.), pp. 217–226. Academic Press, New York.

Oaks, A., Wallace, W., and Stevens, D. (1972). Synthesis and turnover of nitrate reductase in corn roots. *Plant Physiol.* **50,** 649–654.

Oaks, A., Aslam, M., and Boesel, I. (1977). Ammonium and amino acids as regulators of nitrate reductase in corn roots. *Plant Physiol.* **59,** 391–394.

Ohmura, T. (1958). Photo-phosphorylation by chloroplasts. *J. Biochem. (Tokyo)* **45**, 319–331.

Olday, F. C., Barker, A. V., and Maynard, D. N. (1976a). A physiological basis for different patterns of nitrate accumulation in two spinach cultivars. *J. Am. Soc. Hortic. Sci.* **101**, 217–219.

Olday, F. C., Barker, A. V., and Maynard, D. N. (1976b). A physiological basis for different patterns of nitrate accumulation in cucumber and pea. *J. Am. Soc. Hortic. Sci.* **101**, 219–221.

O'Neal, D., and Joy, K. W. (1974). Glutamine synthetase of pea leaves. *Plant Physiol.* **54**, 773–779.

Onweunne, I. C., Laude, H. M., and Huffaker, R. C. (1971). Nitrate reductase activity in relation to heat stress in barley seedlings. *Crop Sci.* **11**, 195–200.

Osborne, B. A., and Whittington, W. J. (1981). Variation in nitrate reductase activity between *Agrostic* species and ecotypes. *New Phytol.* **89**, 581–590.

Pate, J. S. (1971). Movement of nitrogenous solutes in plants. *In* "Nitrogen-15 in Soil–Plant Studies," pp. 165–187. IAEA, Vienna.

Pate, J. S. (1973). Uptake, assimilation and transport of nitrogen compounds by plants. *Soil Biol. Biochem.* **5**, 109–119.

Pate, J. S. (1975). Exchange of solutes between phloem and xylem and circulation in the whole plant. *In* "Encyclopedia of Plant Physiology, New Series" (M. H. Zimmermann and J. A. Milburn, eds.), Vol. 1, pp. 451–473. Springer-Verlag, Berlin and New York.

Pate, J. S. (1976). Nutrients and metabolites of fluids recovered from xylem and phloem: Significance in relation to long distance transport in plants. *In* "Transport and Transfer Processes in Plants" (I. F. Wardlaw and J. B. Passioura, eds.), pp. 253–345. CSIRO, Canberra, Australia.

Pate, J. S. (1980). Transport and partitioning of nitrogenous solutes. *Annu. Rev. Plant Physiol.* **31**, 313–340.

Pate, J. S., Sharkey, P. J., and Lewis, O. A. (1975). Xylem to phloem transfer of solutes in fruiting shoots of a legume, studied by a phloem bleeding technique. *Planta* **122**, 11–26.

Pearson, C. T., and Steer, B. T. (1977). Daily changes in nitrate uptake and metabolism in *Capsicum annuum*. *Planta* **137**, 107–112.

Pearson, V., and Read, D. J. (1975). The physiology of the mycorrhizal endophyte of *Calluna vulgaris*. *Trans. Br. Mycol. Soc.* **64**, 1–7.

Pierre, W. H., and Banwart, W. L. (1973). Excess-base and excess-base/nitrogen ratio of various crop species and parts of plants. *Agron. J.* **65**, 91–96.

Pierre, W. H., Meisinger, J., and Birchett, J. R. (1970). Cation–anion balance in crops as a factor in determining the effect of nitrogen fertilizers. *Agron. J.* **62**, 106–112.

Pill, W. G., and Lambeth, B. H. (1977). Effects of ammonium and nitrate nutrition with and without pH adjustment on tomato growth, ion composition, and water relations. *J. Am. Soc. Hortic. Sci.* **102**, 78–81.

Pill, W. G., and Lambeth, B. H. (1980). Effects of soil water regime and nitrogen form on blossom-end rot, yield, water relations, and elemental composition of tomato. *J. Am. Soc. Hortic. Sci.* **105**, 730–734.

Pill, W. G., Lambeth, V. N., and Hinckley, T. M. (1978). Effects of nitrogen form and level on ion concentration, water stress, and blossom-end rot incidence in tomato. *J. Am. Soc. Hortic. Sci.* **103**, 265–268.

Polacco, J. C. (1976). Nitrogen metabolism in soybean tissue culture. I. Assimilation of urea. *Plant Physiol.* **58**, 350–357.

Polacco, J. C. (1977). Is nickel a universal component of plant ureases? *Plant Sci. Lett.* **10,** 249–255.

Poole, R. J. (1978). Energy coupling for membrane transport. *Annu. Rev. Plant Physiol.* **29,** 437–460.

Porter, L. K., Viets, F. G., and Hutchinson, G. L. (1972). Air containing nitrogen-15 ammonia foliar absorption by corn seedlings. *Science* **175,** 759–761.

Possingham, J. V., and Groot Obbink, J. (1971). Endotrophic mycorrhiza and the nutrition of grape vines. *Vitis* **10,** 120–130.

Precheur, R. J., and Maynard, D. N. (1983). Growth of asparagus transplants as influenced by nitrogen form and lime. *J. Am. Soc. Hortic. Sci.* **108,** 169–172.

Puritch, G. S., and Barker, A. V. (1967). Structure and function of tomato leaf chloroplasts during ammonium toxicity. *Plant Physiol.* **42,** 1229–1238.

Quebedeaux, B., and Ozbun, J. L. (1973). Effects of ammonium nutrition on water stress, water uptake and root pressure in *Lycopersicon esculentum* Mill. *Plant Physiol.* **52,** 677–679.

Radin, J. W. (1975). Differential regulation of nitrate reductase induction in roots and shoots of cotton plants. *Plant Physiol.* **55,** 178–182.

Rao, K. P., and Rains, D. W. (1976). Nitrate absorption by barley. I. Kinetics. *Plant Physiol.* **57,** 55–58.

Rathnam, C. K. M. (1978). Malate and dihydroxyacetone phosphate-dependent nitrate reduction in spinach leaf protoplasts. *Plant Physiol.* **62,** 220–223.

Raven, J. A., and Smith, F. A. (1976). Nitrogen assimilation and transport in vascular land plants in relation to intracellular pH regulation. *New Phytol.* **76,** 415–431.

Rayar, A. J., and van Hai, T. (1977). Effect of ammonium on uptake of phosphorus, potassium, calcium and magnesium by intact soybean plants. *Plant Soil* **48,** 81–87.

Read, D. J., and Stribley, D. P. (1973). Effect of mycorrhizal infection on nitrogen and phosphorus nutrition of ericaceous plants. *Nature (London)* **244,** 81–82.

Read, D. J., and Stribley, D. P. (1975). Some mycological aspects of the biology of mycorrhiza in the Ericaceae. *In* "Endomycorrhizas" (F. E. Sanders, B. Mosse, and P. B. Tinker, eds.), pp. 105–117. Academic Press, New York.

Redhead, J. F. (1980). Mycorrhiza in natural vegetation. *In* "Tropical Mycorrhiza Research" (P. Mikola, ed.), pp. 127–142. Oxford Univ. Press (Clarendon), London and New York.

Reisenauer, H. M. (1978). Absorption and utilization of ammonium nitrogen by plants. *In* "Nitrogen in the Environment" (D. R. Nielsen and J. G. MacDonald, eds.), Vol. 2, pp. 157–189. Academic Press, New York.

Reisenauer, H. M., Clement, C. R., and Jones, L. H. P. (1982). Comparative efficacy of ammonium and nitrate for grasses. *Proc. Int. Plant Nutr. Colloq., 9th, 1982,* Vol. 2, pp. 539–544.

Reithel, F. J. (1971). Ureases. *In* "The Enzymes" (P. D. Boyer, ed.), 3rd ed., Vol. 4, pp. 1–21. Academic Press, New York.

Rhodes, D., Rendon, G. A., and Stewart, G. R. (1976). The regulation of ammonia assimilating enzymes in *Lemna minor* L. *Planta* **129,** 203–210.

Rhodes, L. H., and Gerdemann, J. W. (1975). Phosphate uptake zones of mycorrhizal and non-mycorrhizal onions. *New Phytol.* **75,** 555–561.

Riley, D., and Barber, S. A. (1971). Effect of ammonium fertilization on phosphorus uptake as related to root-induced pH changes at the root–soil interface. *Soil Sci. Soc. Am. Proc.* **35,** 301–306.

Robinson, D., and Rorison, I. H. (1983). A comparison of the responses of *Lolium perenne*

L., *Holcus lanatus* L., and *Deschampsia flexuosa* (L). Trin. to a localized supply of nitrogen. *New Phytol.* **94,** 263–273.

Rognes, S. E. (1975). Glutamine-dependent asparagine synthetase from *Lupinus luteus*. *Phytochemistry* **14,** 1975–1982.

Rorison, I. H. (1980). The effects of soil acidity on nutrient availability and plant response. *In* "Effects of Acid Precipitation on Terrestrial Ecosystems" (T. C. Hutchinson and M. Havas, eds.), pp. 283–304. Plenum, New York.

Roth-Bejerano, N., and Lips, H. (1973). Induction of nitrate reductase in leaves of barley in the dark. *New Phytol.* **72,** 253–257.

Routley, D. G. (1972). Nitrate reductase in leaves of Ericaceae. *HortScience* **7,** 84–87.

Rufty, T. W., Jackson, W. A., and Roper, D. C. (1982a). Inhibition of nitrate assimilation in roots in the presence of ammonium: The moderating influence of potassium. *J. Exp. Bot.* **33,** 1122–1137.

Rufty, T. W., Volk, R. J., McClure, P. R., Israel, D. W., and Raper, C. D. (1982b). Relative content of NO_3^- and reduced N in xylem exudate as an indicator of root reduction of concurrently absorbed $^{15}NO_3^-$. *Plant Physiol.* **69,** 166–170.

Sahulka, J. (1977). The effect of some ammonium salts on nitrate reductase level, on *in vivo* nitrate reduction and on nitrate content in excised *Pisum sativum*. *Biol. Plant.* **19,** 113–128.

Sasakawa, H., and Yamamoto, Y. (1978). Comparison of the uptake of nitrate and ammonium by rice seedlings. Influences of light, temperature, oxygen concentration, exogenous sucrose and metabolic inhibitors. *Plant Physiol.* **62,** 665–669.

Schonherr, J., and Bukovac, M. J. (1970). Preferential polar pathways in the cuticle and their relationship to ectodesmata. *Planta* **92,** 189–201.

Schrader, L. E. (1978). Uptake, accumulation, assimilation, and transport of nitrogen in higher plants. *In* "Nitrogen in the Environment" (D. R. Nielsen and J. G. MacDonald, eds.), Vol. 2, pp. 101–141. Academic Press, New York.

Schrader, L. E., and Thomas, R. J. (1981). Nitrate uptake, reduction and transport in the whole plant. *In* "Nitrogen and Carbon Metabolism" (J. D. Bewley, ed.), pp. 49–93. Martinus Nijhoff/Dr. W. Junk, The Hague.

Schrader, L. E., Ritenour, G. L., Bilrich, G. L., and Hageman, R. H. (1968). Some characteristics of nitrate reductase from higher plants. *Plant Physiol.* **43,** 930–940.

Schrader, L. E., Domska, D., Jung, P. E., and Patterson, L. A. (1972). Uptake and assimilation of ammonium-N and nitrate-N and their influence on the growth of corn (*Zea mays* L.). *Agron. J.* **64,** 690–695.

Servaites, J. C., Schrader, L. E., and Jung, D. M. (1979). Energy-dependent loading of amino acids and sucrose into the phloem of soybean. *Plant Physiol.* **64,** 546–550.

Shaner, D. L., and Boyer, J. S. (1976a). Nitrate reductase activity in maize (*Zea mays* L.) leaves. I. Regulation by nitrate flux. *Plant Physiol.* **58,** 499–504.

Shaner, D. L., and Boyer, J. S. (1976b). Nitrate reductase activity in maize (*Zea mays* L.) leaves. II. Regulation by nitrate flux at low leaf water potential. *Plant Physiol.* **58,** 505–509.

Sharkey, P. J., and Pate, J. S. (1975). Selectivity in xylem to phloem transfer of amino acids in fruiting shoots of white lupin (*Lupinus albus* L.). *Planta* **127,** 251–262.

Sheat, D. E. G., Fletcher, B. H., and Street, H. E. (1959). Studies on the growth of excised roots. VIII. The growth of excised tomato roots supplied with various inorganic sources of nitrogen. *New Phytol.* **58,** 128–141.

Shen, T. C., Funkhouser, E. A., and Guerrero, M. G. (1976). NADH- and NAD(P)H-nitrate reductases in rice seedlings. *Plant Physiol.* **58,** 292–294.

Shepard, D. V., and Thurman, D. A. (1973). Effect of nitrogen sources upon the activity of L-glutamate dehydrogenase of *Lemna gibba*. *Phytochemistry* **12**, 1937–1946.

Sherrard, J. H., Kennedy, J. A., and Dalling, M. J. (1979). *In vitro* stability of nitrate reductase from wheat leaves. *Plant Physiol.* **64**, 640–645.

Shim, K., Splittstoesser, W. E., and Titus, J. S. (1973). Changes in urease activity in apple trees as related to urea applications. *Physiol. Plant.* **28**, 327–334.

Simpson, R. J., Lambers, H., and Dalling, M. J. (1982). Translocation of nitrogen in a vegetative wheat plant (*Triticum aestivum*). *Physiol. Plant.* **56**, 12–17.

Skinner, M. F., and Bowen, G. D. (1974). The uptake and translocation of phosphate by mycelial strands of pine mycorrhizas. *Soil Biol. Biochem.* **6**, 53–56.

Skokut, T. A., Wolk, C. P., Thomas, J., Meeks, J. C., Shaffer, P. W., and Chien, W. S. (1978). Initial organic products of assimilation of [^{13}N] ammonium and (^{13}N) nitrate by tobacco cells cultured on different sources of nitrogen. *Plant Physiol.* **62**, 299–304.

Sluiters-Scholten, C. M. T. H. (1973). Effect of chloramphenicol and cycloheximide on the induction of nitrate reductase and nitrite reductase in bean leaves. *Planta* **113**, 229–240.

Small, J. G. C., Onraet, A., Grierson, D. S., and Reynolds, G. (1977). Studies on insect-free growth, development and nitrate assimilating enzymes of *Drosera aliciae*. Hamet. *New Phytol.* **79**, 127–133.

Smiley, R. W. (1974). Rhizosphere pH as influenced by plant, soil and nitrogen fertilizers. *Soil Sci. Soc. Am. Proc.* **38**, 795–799.

Smith, F. A. (1973). The internal control of nitrate uptake into excised barley roots with differing salt contents. *New Phytol.* **72**, 769–782.

Smith, S. E. (1980). Mycorrhizas of autotrophic higher plants. *Biol. Rev. Cambridge Philos. Soc.* **55**, 475–510.

Soldal, T., and Nissen, P. (1978). Multiphasic uptake of amino acids by barley roots. *Physiol. Plant.* **43**, 181–188.

Soon, Y. K., and Miller, M. H. (1977). Changes in rhizosphere due to ammonium and nitrate fertilization and phosphorus uptake by corn seedlings (*Zea mays* L.). *Soil Sci. Soc. Am. Proc.* **41**, 77–80.

Spanswick, R. M. (1981). Electrogenic ion pumps. *Annu. Rev. Plant Physiol.* **32**, 267–289.

Spiertz, J. H. J. (1977). The influence of temperature and light intensity on grain growth in relation to the carbohydrate and nitrogen economy of the wheat plant. *Neth. J. Agric. Sci.* **25**, 182–197.

Srivastava, H. S. (1980). Regulation of nitrate reductase activity in higher plants. *Phytochemistry* **19**, 723–733.

Stewart, G. R., and Rhodes, D. (1978). Nitrogen metabolism of halophytes. III. Enzymes of ammonia assimilation. *New Phytol.* **80**, 307–316.

Stewart, G. R., Lee, J. A., Orebamjo, T. O., and Havill, D. C. (1974). Ecological aspects of nitrogen metabolism. *In* "Mechanisms of Regulation of Plant Growth" (R. L. Bieleski, A. R. Ferguson, and M. M. Creswell, eds.), Bull. No. 12, pp. 41–47. Royal Society of New Zealand, Wellington.

Stewart, G. R., Mann, A. F., and Fentem, P. A. (1980). Enzymes of glutamate formation: Glutamate dehydrogenase, glutamine synthase and glutamate synthase. *In* "The Biochemistry of Plants" (B. J. Miflin, eds.), Vol. 5, pp. 271–327. Academic Press, New York.

Stribley, D. P., and Read, D. J. (1974). The biology of mycorrhiza in the Ericaceae. IV. The effect of mycorrhizal infection on uptake of ^{15}N from labelled soil by *Vaccinium macrocarpon* Ait. *New Phytol.* **73**, 1149–1163.

Stribley, D. P., and Read, D. J. (1975). Some nutritional aspects of the biology of ericaceous mycorrhizae. *In* "Endomycorrhizas" (F. E. Sanders, F. E. Mosse, and P. B. Tinker, eds.), pp. 195–207. Academic Press, New York.

Stribley, D. P., and Read, D. J. (1976). The biology of mycorrhiza int he Ericaceae. VI. The effects of mycorrhizal infection and concentration of ammonium nitrogen on growth of cranberry (*Vaccinium macrocarpon* Ait.) in sand culture. *New Phytol.* **77,** 63–72).

Stribley, D. P., and Read, D. J. (1980). The biology of mycorrhiza in the Ericaceae. VII. The relationship between mycorrhizal infection and the capacity to utilize simple and complex organic sources. *New Phytol.* **86,** 365–371.

Stuart, D. M., and Haddock, J. L. (1968). Inhibition of water uptake in sugar beet roots by ammonia. *Plant Physiol.* **43,** 345–350.

Stulen, I., and Lanting, L. (1978). Influence of CO_2 reduction on nitrate in corn seedlings. *Physiol. Plant.* **42,** 283–286.

Su, L. T., and Miller, G. W. (1961). Chlorosis in higher plants as related to organic acid content. *Plant Physiol.* **36,** 415–420.

Suzuki, A., and Gadal, P. (1982). Glutamate synthase from rice leaves. *Plant Physiol.* **69,** 848–852.

Suzuki, A., Gadal, P., and Oaks, A. (1981). Intracellular distribution of enzymes associated with nitrogen assimilation in roots. *Planta* **151,** 457–461.

Suzuki, A., Vidal, J., and Gadal, P. (1982). Glutamate synthase isoforms in rice. Immunological studies of enzymes in green leaf, etiolated leaf and root tissue. *Plant Physiol.* **70,** 827–832.

Ta, T. C., and Ohira, K. (1981). Effects of various environmental and medium conditions on the response of Indica and Japonica rice plants to ammonium and nitrate nitrogen. *Soil Sci. Plant Nutr. (Tokyo)* **27,** 347–355.

Tamura, G., Oto, M., Hirasawa, M., and Aketagawa, J. (1980). Isolation and partial characterization of homogeneous glutamate synthase from *Spinacia oleracea*. *Plant Sci. Lett.* **19,** 209–215.

Taylor, A. A., and Havill, D. C. (1981). The effect of inorganic nitrogen on the major enzymes of ammonia assimilation in grassland plants. *New Phytol.* **87,** 53–62.

Taylor, A. A., De-Felice, J., and Havill, D. C. (1982). Seasonal variation in nitrogen availability and utilization in an acidic and calcareous soil. *New Phytol.* **92,** 141–152.

Thomas, R. J., Feller, U., and Erismann, K. J. (1979). The effect of different inorganic nitrogen sources and plant age on the composition of bleeding sap of *Phaseolus vulgaris* (L.). *New Phytol.* **82,** 657–669.

Thompson, J. F. (1980). Arginine synthesis, proline synthesis, and related processes. *In* "The Biochemistry of Plants" (B. J. Miflin, ed.), Vol. 5, pp. 375–402. Academic Press, New York.

Thurston, J. M. (1969). The effect of liming and fertilisers on the botanical composition of permanent grassland and on yield of hay. *In* "Ecological Aspects of the Mineral Nutrition of Plants" (I. H. Rorison, ed.), pp. 3–10. Blackwell, Oxford.

Tingey, D. T. (1968). Foliar absorption of nitrogen dioxide. M.A. Thesis, University of Utah, Salt Lake City.

Tishner, T., and Hutterman, A. (1978). Light-mediated activation of nitrate reductase in synchronous *Chlorella*. *Plant Physiol.* **62,** 284–286.

Tomkins, G. A., Jackson, W. A., and Volk, R. J. (1978). Accelerated nitrate uptake in wheat seedlings: Effects of ammonium and nitrite pretreatments and of 6-methylpurine and puromycin. *Physiol. Plant.* **43,** 166–171.

Townsend, L. R. (1966). Effect of nitrate and ammonium nitrogen on the growth of the lowbush blueberry. *Can. J. Plant Sci.* **46,** 209–210.

Townsend, L. R. (1970). Effect of form of N and pH on nitrate reductase activity in lowbush blueberry leaves and roots. *Can. J. Plant Sci.* **50,** 603–605.

Townsend, L. R., and Blatt, C. R. (1966). Lowbush blueberry: Evidence for the absence of a nitrate reducing system. *Plant Soil* **25,** 456–460.

Tromp, J. (1962). Interactions in the absorption of ammonium, potassium, and sodium ions by wheat roots. *Acta Bot. Neerl.* **11,** 147–192.

Udayakumar, M., Devendra, R., Reddy, V. S., and Sastry, K. S. K. (1981). Nitrate availability under low irradiance and its effect on nitrate reductase activity. *New Phytol.* **88,** 289–297.

Urquhart, A. A., and Joy, K. W. (1982). Transport, metabolism, and redistribution of xylem-borne amino acids in developing pea shoots. *Plant Physiol.* **69,** 1226–1232.

Van de Dijk, S. J. (1981). Differences in nitrate uptake of species from habitats rich or poor in nitrogen when grown at low nitrate concentrations, using a new growth technique. *Plant Soil* **62,** 265–278.

Van de Dijk, S. J., Lanting, L., Lambers, H., Posthumus, F., Stulen, I., and Hofstra, R. (1982). Kinetics of nitrate uptake by different species from nutrient-rich and nutrient-poor habitats as affected by the nutrient supply. *Physiol. Plant.* **55,** 103–110.

Van den Driessche, R. (1971). Response of conifer seedlings to nitrate and ammonium sources of nitrogen. *Plant Soil* **34,** 421–439.

Van den Honert, T. H., and Hooymans, J. J. (1955). On the absorption of nitrate by maize in water culture. *Acta Bot. Neerl.* **4,** 376–384.

van Die, J. (1959). Diurnal rhythm in the amino acid content of xylem exudate from tomato plants bleeding under constant environmental conditions. *Proc. K. Ned. Akad. Wet., Ser. C* **62,** 50–58.

van Die, J., and Tammes, P. M. L. (1975). Phloem exudation from monocotyledonous axes. *In* "Encyclopedia of Plant Physiology, New Series" (M. H. Zimmermann and J. A. Milburn, eds.), Vol. 1, pp. 196–222. Springer-Verlag, Berlin and New York.

Van Egmond, F. (1978). Nitrogen nutritional aspects of the ionic balance of plants. *In* "Nitrogen in the Environment" (D. R. Nielsen and J. G. MacDonald, eds.), Vol. 2, pp. 171–189. Academic Press, New York.

Vega, J. M., and Kamin, H. (1977). Spinach nitrite reductase. Purification and properties of a siroheme-containing iron–sulphur enzyme. *J. Biol. Chem.* **252,** 896–909.

Vega, J. M., Cardenas, J., and Losada, M. (1980). Ferredoxin-nitrite reductase. *In* "Methods in Enzymology" (A, S. Pietro, ed.), Vol. 69, pp. 255–270.

Vennesland, B., and Guerrero, M. G. (1979). Reduction of nitrate and nitrite. *In* "Encyclopedia of Plant Physiology, New Series" (M. Gibbs and E. Latzko, eds.), Vol. 6, pp. 425–444. Springer-Verlag, Berlin and New York.

Wakiuchi, N., Matsumoto, H., and Takahashi, E. (1971). Changes of some enzyme activities of cucumber during ammonium toxicity. *Physiol. Plant.* **24,** 248–253.

Wallace, A. (1971). The competitive chelation hypothesis of lime-induced chlorosis. *In* "Regulation of the Micro-Nutrient Status of Plants by Chelating Agents and other Factors" (A. Wallace, ed.), pp. 230–239. A. Wallace, Los Angeles, California.

Wallace, W. (1975). Effects of a nitrate reductase inactivating enzyme and NAD(P)H on the nitrate reductase from higher plants and *Neurospora*. *Biochim. Biophys. Acta* **377,** 239–250.

Wallace, W. (1978). Comparison of a nitrate reductase-inactivating enzyme from the maize root with a protease from yeast which inactivates tryptophan synthase. *Biochim. Biophys. Acta* **524,** 418–427.

Wallace, W., and Pate, J. S. (1967). Nitrate assimilation in higher plants with special reference to cocklebur (*Xanthium pennsylvanicum* Wallr.). *Ann. Bot. (London)* [N.S.] **31**, 213–228.

Wallsgrove, R. M., Harel, E., Lea, P. J., and Miflin, B. J. (1977). Studies on glutamate synthase form leaves of higher plants. *J. Exp. Bot.* **28**, 588–596.

Warncke, D. D., and Barber, S. A. (1973). Ammonium and nitrate uptake by corn (*Zea mays* L.) as influenced by nitrogen concentration and NH_4^+/NO_3 ratio. *Agron. J.* **65**, 950–953.

Weigel, R. D., Schillinger, J. A., McCaw, B. A., Gauch, H. G., and Hsiao, E. (1973). Nutrient-NO_3^- levels and the accumulation of chloride in leaves of snap beans and roots of soybeans. *Crop Sci.* **13**, 411–412.

Weinbaum, S. A., and Neumann, P. M. (1977). Uptake and metabolism of ^{15}N-labelled potassium nitrate by french prune (*Prunus domestica* L.) leaves and the effects of two surfactants. *J. Am. Soc. Hortic. Sci.* **102**, 601–604.

Weissman, G. S. (1964). Effects of ammonium and nitrate nutrition on protein level and exudate composition. *Plant Physiol.* **39**, 947–952.

Wetselaar, R., and Farquhar, G. D. (1980). Nitrogen losses from tops of plants. *Adv. Agron.* **33**, 263–302.

Wiersum, L. K. (1958). Density of root branching as affected by substrate and separate ions. *Acta Bot. Neerl.* **7**, 174–109.

Wilkinson, S. R., and Ohlrogge, A. J. (1961). Fertilizer nutrient uptake as related to root development in the fertilizer band: Influence of nitrogen and phosphorus fertilizer on endogenous auxin content of soybean roots. *Trans. Int. Congr. Soil Sci., 7th, 1960*, Vol. 3, pp. 234–242.

Willcox, G. E., Hoff, J. E., and Jones, C. M. (1973). Ammonium reduction of calcium and magnesium content of tomato and sweet corn leaf tissue and influence of blossom-end rot of tomato fruit. *J. Am. Soc. Hortic. Sci.* **98**, 86–89.

Williams, D. E., and Vlamis, J. (1962). Differential cation and anion absorption as affected by climate. *Plant Physiol.* **37**, 198–202.

Williams, J. T. (1969). Mineral nitrogen in British grassland soils. I. Seasonal patterns and simple models. *Oecol. Plant.* **4**, 307–320.

Willis, A. J., and Yemm, E. W. (1955). The respiration of barley plants. VIII. Nitrogen assimilation and the respiration of the root system. *New Phytol.* **54**, 163–181.

Wiltshire, G. H. (1973). Responses of grasses to nitrogen source. *J. Appl. Ecol.* **10**, 429–435.

Woo, K. C., Morot-Gaudry, J. F., Summons, R. E., and Osmond, B. C. (1982). Evidence for the glutamine synthetase glutamate synthase pathway during the photorespiratory nitrogen cycle in spinach leaves. *Plant Physiol.* **70**, 1514–1517.

Yamada, Y., Wittwer, S. H., and Bukovac, M. J. (1965). Penetration of ions through isolated cuticles. *Plant Physiol.* **38**, 28–32.

Yamada, Y., Rasmussen, H. P., Bukovac, M. J., and Wittwer, S. H. (1966). Binding sites for inorganic ions and urea on isolated cuticular membrane surfaces. *Am. J. Bot.* **53**, 170–172.

Yamaya, T., Oaks, A., and Boesel, I. L. (1980). Characteristics of nitrate reductase-inactivating proteins obtained from corn roots and rice cultures. *Plant Physiol.* **65**, 141–145.

Yoneyama, T., Hashimoto, A., and Toksuka, T. (1980). Absorption of atmospheric NO_2 by plants and soils. IV. Two routes of nitrogen uptake by plants from atmospheric NO_2: Direct incorporation into aerial plant parts and uptake by roots after absorption into the soil. *Soil Sci. Plant Nutr. (Tokyo)* **26**, 1–7.

Youngdahl, L. J., Pacheco, R., Street, J. J., and Vlek, P. L. G. (1982). The kinetics of ammonium and nitrate uptake by young rice plants. *Plant Soil* **69,** 225–232.

Ziegler, H. (1975). Nature of transported substances. *In* "Encyclopedia of Plant Physiology, New Series" (M. H. Zimmerman and J. A. Milburn, eds.), Vol. 1, pp. 59–100. Springer-Verlag, Berlin and New York.

Zielke, H. R., and Filner, P. (1971). Synthesis and turnover of nitrate reductase induced by nitrate in cultured tobacco cells. *J. Biol. Chem.* **246,** 1772–1779.

Zsoldos, F. (1971). Ammonium and nitrate ion uptake by plants. *In* "Nitrogen-15 in Soil–Plant Studies," pp. 81–101. IAEA, Vienna.

Zsoldos, F., and Haunold, E. (1982). Influence of 2,4-*D* and low pH on potassium, ammonium and nitrate uptake by rice roots. *Physiol. Plant.* **54,** 63–68.

Chapter 7

Nitrogen and Agronomic Practice

K. M. GOH and R. J. HAYNES

I. INTRODUCTION

Nitrogen is a major essential nutrient element and is required by plants in substantial quantities. It is a constituent of all proteins, of many metabolic intermediates involved in synthesis and energy transfer, and of nucleic acids. When supplies of soil water are adequate, N is most commonly the key limiting factor for crop production. Thus, on the average, considerably more N than any other element is supplied to crops as fertilizer and is removed from agricultural lands in harvested crops (Olson and Kurtz, 1982).

The major aim of farmers in applying N fertilizers is to obtain an increase in plant yield with a consequent economic return from the additional expense of applying fertilizers. Excessive N applications are not only undesirable from an economic viewpoint since environmental and crop quality problems associated with excessive fertilizer N use are widely recognized (National Research Council, 1978; Keeney 1982a). Accurate fertilizer N recommendations are therefore important for cost-efficient and environmentally sound agricultural production.

There is, however, no well accepted method for soil testing for available soil N. This is because, in contrast to most essential nutrients, there is no mechanism for long-term storage of plant-available N in soils, and approximately 97 to 99% of soil N is present in organic forms that are unavailable to plants. Some of this N slowly becomes available through microbial decomposition of soil organic matter with the release of mineral forms of N. The mineralization rate in the field is affected by many cultural and environmental factors that make it difficult to predict plant requirement for fertilizer N in a given situation.

This chapter discusses recent findings related to practical agronomic aspects of the use of fertilizer N. These include the pattern of N uptake by crops, the responses of crop plants to applied N, factors affecting such responses, methods used to assess soil N availability and therefore fertilizer requirements of crops, and the dependence of modern agricultural systems on a supply of N either as fertilizer N or through symbiotic N_2 fixation.

II. NITROGEN IN PLANT PRODUCTION

A. Nitrogen Uptake and Content of Plants

1. Time and Rate of Uptake

Typically, N uptake by field crops involves a period of very slow accumulation followed by a rapid nearly linear rate of accumulation that coincides with rapid plant growth (Tinker, 1978; Pearson and Muirhead, 1984). For field crops, the rate of N uptake can be extremely rapid (3–5 kg N ha^{-1} day^{-1}) during the rapid growth phase (Viets, 1965; Tinker, 1978; Remy and Viaux, 1982). Olson and Kurtz (1982) calculated approximate rates of N uptake for nonirrigated wheat (*Triticum aestivum* L.) and irrigated corn (*Zea mays* L.) of approximately 4 kg N ha^{-1} day^{-1} during their rapid phases of uptake. A near linear rate of uptake does not imply that N concentrations in the plant remain constant. In fact, N concentrations in the young plant are initially high and characteristically decline as the plant ages and accumulates dry matter (Tinker, 1978; Pearson and Muirhead, 1984).

Figure 1 demonstrates a slow rate of dry matter accumulation followed by a linear phase of rapid growth for a corn crop. Early-season accumulation of N was more rapid than that of dry matter but the rate of N accumulation decreased late in the season and continued at a decreased rate until maturity. Such a pattern suggests that there was initially a large supply of mineral N in the soil since the time of maximum N accumulation in relation to corn development depends primarily on available N supply once the crop enters the phase of rapid growth (Russelle *et al.*, 1981). Maximum N accumulation rates can be delayed by delayed applications of fertilizer N (Jordan *et al.*, 1950; Russelle *et al.*, 1981, 1983).

In other crops, too, the pattern of N uptake is greatly influenced by N

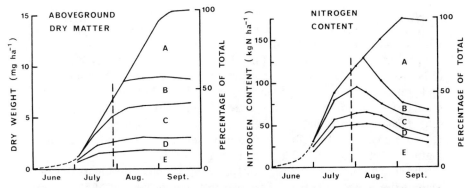

Fig. 1. Aboveground dry matter and nitrogen accumulation and their distribution in various plant parts during the growth of a corn crop. (A) Grain; (B) cob, husks, silks, shank, and ear shoots; (C) stalk and tassel; (D) leaf sheaths; (E) leaves. [Data from Hanway (1962a,b). Reproduced from *Agron. J.* **54**, pp. 218 and 223 by permission of the American Society of Agronomy.]

availability. Results presented in Fig. 2 show that when N was limiting sugar beet (*Beta vulgaris* L.; unfertilized and 125 kg N ha^{-1} treatments) growth and N uptake slowed greatly after July, which coincided with the initiation of the rapid phase of growth.

Ultimately, however, the response of many grain crops to applied N depends as much, or more, on N redistribution within the plant as on

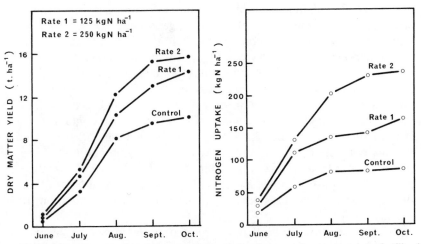

Fig. 2. Total dry matter and nitrogen accumulation by sugar beet grown in unfertilized soil and at two rates of applied nitrogen. [Data from Last and Draycott (1975a,b).]

uptake during the rapid phase of growth. As discussed in Chapter 6, the pathway of absorbed N is not unidirectional from roots to tops but involves cycling of a variable amount of amino acid N from tops to roots in the phloem. As well as the cycling of recently acquired N, large amounts of N previously accumulated in vegetative plant parts can be remobilized and translocated to the developing grain. Indeed, translocation during grain development often greatly affects grain yield and grain protein content (Pearson and Muirhead, 1984). For example, during anthesis wheat can contain about 80% of the N that is present in its aboveground parts at maturity (Daling *et al.,* 1976; Austin *et al.,* 1977).

The substantial redistribution of N in the corn plant during its development is illustrated in Fig. 1. There was little net movement of N from one plant part to another until grain formation began; then N was transported from all other plant parts to the grain. Transport from the cob, husk, and shank appeared to precede that from the leaves. At maturity, the grain contained about two-thirds of the total N in the plant and about half of that had resulted from net transport from other aboveground plant parts. Rapid changes in distribution of N within plants and consequent changes in N concentrations in plant parts indicate the difficulty of determining if plants are receiving an optimum supply of N through N analysis of plant parts.

2. Amount and Distribution of Accumulated Nitrogen

The quantities of N present in different crops vary greatly with species, cultivar, and the environment in which they were produced. While variations in yield explain part of this variation, differences in N accumulation and storage characteristics of different genotypes are also important. Some typical values for N content of different crop plants and N distribution in harvested and nonharvested plant parts are shown in Table I.

Because of their capacity to fix atmospheric N_2 in root nodules, leguminous forage legumes contain larger amounts of N in harvested tissue than do forage grasses, cereal grain crops, or other arable crops. Grain legumes such as soybeans (*Glycine max* (L.) Merr.) also contain relatively large amounts of N in their harvested crop. In the case of these legumes a varying portion of their N content comes from symbiotic N fixation and the remainder from uptake of soil mineral N. In the case of grass/legume pastures, the pasture relies almost entirely on biological N_2 fixation and the N content of the grass originates primarily from N_2 fixation by the legume component.

It is evident from Table I that large variations in N content exist between harvested and nonharvested plant parts. For example, while wheat straw contains a small amount of N compared with the grain, in sugar beet

Table I

Nitrogen Concentrations in Plant Parts Associated with Deficiency, Sufficiency, and Excess in Some Crop Plants[a]

Crop	Plant part	Form of N measured	Deficient	Low	Sufficient	Excessive
Wheat (spring) (*Triticum aestivum* L.)	Total aboveground plant at head emergence from the boot	Total N	<1.25	1.25–1.75	1.75–3.0	>3.0
Rice (*Oryza sativa* L.)	Most recent fully extended leaf at maximum tillering	Total N	<2.4	2.4–2.8	2.8–3.6	>3.6
Corn (*Zea mays* L.)	Ear leaf at silk	Total N	<2.25	2.25–2.75	2.75–3.5	>3.5
Sugar beet (*Beta vulgaris* L.)	Petioles of recently matured leaves	NO_3^--N	<0.1	0.1–0.2	0.2–0.3	>0.3
Cabbage (*Brassica oleracea* L.)	Midrib of wrapper leaf	NO_3^--N	<0.5	0.5–0.7	0.7–0.9	>0.9
Lettuce (heading) (*Lactuca sativa* L.)	Midrib of wrapper leaf	NO_3^--N	<0.4	0.4–0.6	0.6–0.8	>0.8
Potatoes (early season) (*Solanum tuberosum* L.)	Petiole of 4th leaf from growing tip	NO_3^--N	<0.8	0.8–1.0	1.0–1.2	>1.2
Potatoes (late season)	Petiole of 4th leaf from growing tip	NO_3^--N	<0.3	0.3–0.4	0.4–0.5	>0.5

[a] Ranges compiled from Geraldson *et al.* (1973) and Olson and Kurtz (1982).

the tops contain as much N as the harvested roots. Removal of sugar beet tops from the field therefore increases the fertilizer requirement of the following crop appreciably. For sugarcane (*Saccharum* spp.) a considerable amount of N is held in stalks and trash and can be lost when the cane field is burned to facilitate stalk harvest. Crop improvement, through plant breeding, can affect plant N content. For example, the improved rice (*Oryza sativa* L.) variety IR8, which has a higher harvest index than the tall variety Peta, contains a considerably higher proportion of its aboveground N in the grain.

The amount of N absorbed by plants includes that in both tops and roots. Few reliable data are available on the N content of plant roots largely because of the difficulty in obtaining representative samples. Root growth under field conditions can vary greatly with season, soil type, soil management practices, and other environmental factors (Russell, 1977; Cannell, 1982). As a general rule, total N in roots at harvest is thought to be approximately half that in the aboveground forage component (Olson and Kurtz, 1982).

The quantity of N removed during harvest is an important agronomic parameter since it is often a major determinant of the fertilizer N requirement of a particular crop. In other words, fertilizer recommendations should take account of the major debits (e.g., crop removals and leaching and gaseous losses) and credits (e.g., residual NO_3^- in the soil profile and mineralization of soil organic matter and organic residues) of soil N. In practice, the major determinant of the appropriate rate of fertilizer N is often the crop and its probable yield since the N requirement of modern high-yielding crops is usually large in comparison with variations in the N-supplying capacity of soils of an area (Olson and Kurtz, 1982).

3. Deficiency, Sufficiency, and Excess

a. Visual symptoms. Nitrogen deficiency symptoms are most closely associated with restricted chlorophyll synthesis. Thus, N deficiency is characterized by a reduction in plant growth and a general loss of green color progressing to yellow coloration particularly in the older leaves. It occurs first in older leaves since N is translocated to developing portions of the plant when the plant is in an N-stressed condition. As the severity of the deficiency increases the entire plant turns yellow, the older leaves turn brown and die, the growth ceases. In fruiting plants, N deficiency is also expressed as a failure of fruit to set, cessation of fruit development, and ultimately fruit drop (Mills and Jones, 1979).

When the supply of soil N is adequate, plant growth is normally vigorous and plants are dark green in color. Most plants continue to grow well with an excess supply of N and exhibit a deep green color. Vegetative growth is, however, often stimulated to the detriment of flower or fruit set and development and fruit quality may be adversely affected. Plants may become very succulent and easily subjected to diseases and insect attack and cold hardiness may be reduced (Mills and Jones, 1979).

b. Plant analysis. The nutritional status of plants is often determined by measuring the concentration of total N or an extractable fraction (e.g., NO_3^--N) in plant tissues. An assumption of such a method is that the chemical composition of the plant reflects its nutrient supply in relation to growth. Such an assumption is not always valid since the composition of a plant is the result of the interaction of nutrient supply and plant growth. Any factor that limits growth (e.g., light, temperature, moisture, some nutrient) may cause N to accumulate in the plant. Sulfur deficiency, for example, can result in high levels of NO_3^--N accumulating in plant tissues since protein synthesis is inhibited (Terman *et al.*, 1976; Goh and Kee, 1978). High concentrations of NO_3^- can also accumulate in plant tissues when plant growth is limited by drought, while plants damaged by pests and diseases can show higher N concentrations than adjacent healthy plants (Olson and Kurtz, 1982). In contrast, very good growth conditions or supply of a growth-limiting factor such as another nutrient or water can result in high growth rates and consequent dilution of N in plant tissues.

The schematic relationship between plant yield and N concentration in plant tissue is shown in Fig. 3. Very low concentrations of N indicate acute deficiency and yield increases rapidly with small increases in N concentrations in the deficient zone. Concentration values increase as nutrient supply is increased until luxury uptake and finally toxicity occurs. The "optimum" or "critical" level is the N concentration in a plant sample below which growth rate, yield, or quality significantly declines. In reality, an optimum concentration or critical level is not usually distinguishable because of the multitude of factors interacting to influence concentration and yield but an optimum or sufficient range can usually be defined. Plant N concentrations associated with deficiency, sufficiency, and excess in some agricultural crops are presented in Table II.

The time of sampling and particular plant part sampled are defined in Table II since both factors have very great effects on measured tissue N concentrations. Under similar conditions, tissue N concentrations can also differ among cultivars of the same species (e.g., Chavalier and Schrader, 1977; Anderson *et al.*, 1984). As noted previously, N concentrations are at a maximum during early growth and as the season progresses and plant dry matter increases N concentrations in the whole plant and in vegetative plant parts typically decline (Jones and Eck, 1973; Munson and Nelson, 1973; Ward *et al.*, 1973). For example, the concentration of

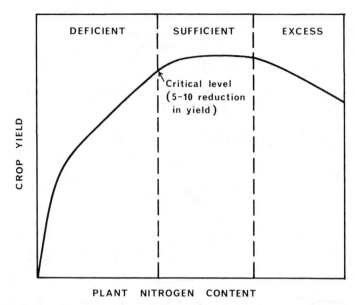

Fig. 3. Schematic representation of the relationship between crop yield and nitrogen content of plant tissues.

NO_3^- in the petiole of potato plant (*Solanum tuberosum* L.) leaves decreases greatly as the season progresses and consequently the concentration below which deficiency is considered to occur is 8000, 6000, and 3000 ppm NO_3^--N in early, mid-, and late-season, respectively (Geraldson *et al.*, 1973).

Various plant parts usually possess significantly different N concentrations. Changes in N concentration with time in different parts of corn plants are shown in Fig. 4. Dry matter and total N accumulation in these plants were shown in Fig. 1. The decline in total N content was greater for leaf sheaths and stalks than for leaves. Leaf sheath and stalks contained considerably higher concentrations of NO_3^- than leaves. Furthermore, nitrogen was not evenly distributed within leaves, and Jones (1970) found that N concentration in midribs of corn leaves was less than that in the blade and margin and older leaves had lower N contents than younger leaves. Water-soluble N, such as NO_3^-, tends to be higher in stems and conductive tissues of leaves than in the leaf blade (Mills and Jones, 1979).

Although total N concentrations in plant parts can be used as a reasonable reliable diagnostic index of the nutritional status of plants, analysis of total N is time-consuming and costly. In contrast, laboratory analysis of NO_3^- from plant tissues is considerably more rapid and suitable as a rou-

Table II

Approximate Total Nitrogen Content of Various Crop Plants and Its Distribution in Different Plant Parts[a]

Crop	Plant part	Yield (kg ha^{-1})	Nitrogen content (kg N ha^{-1})
Wheat	Grain	5,400	100
(*Triticum aestivum* L.)	Straw	6,000	45
Rice IR8	Panicle	9,100	116
(*Oryza sativa* L.)	Straw	8,000	48
Rice Peta	Panicle	6,400	69
	Straw	11,900	74
Corn	Grain	10,000	150
(*Zea mays* L.)	Stover	9,000	80
Soybeans	Grain	2,800	180
(*Glycine max* (L.) Merr.)	Straw	5,400	75
Sugar beet	Roots	68,000	140
(*Beta vulgaris* L.)	Tops	36,000	145
Sugarcane	Stalks	112,000	180
(*Saccharum* spp.)	Tops and trash	50,000	225
Onions	Bulbs	50,000	110
(*Allium cepa* L.)	Tops	na[b]	35
Grass–legume pasture	Tops	8,740	236
Perennial ryegrass	Roots	8,707	144
(*Lolium perenne* L.)			
White clover	Tops	3,305	122
(*Trifolium repens* L.)	Roots	3,712	137

[a] Data compiled from Gregg (1976), Lorenz and Maynard (1980), and Olson and Kurtz (1982).

[b] na = Not available.

tine test. Indeed, in some parts of the world, kits are available for field estimation of the NO_3^- content of xylem sap using diphenylamine sulfuric acid reagent and a color chart. In some regions of the Federal Republic of Germany such a method is used in conjunction with residual NO_3^- in the soil profile to determine N requirements of wheat (Becker and Aufhammer, 1982; Sauderbeck and Timmermann, 1983).

B. Nature of Plant Responses to Applied Nitrogen

Yield responses of plants due to N fertilizer additions may occur as dry matter yield, protein yield, quality improvement, or other features. Some such changes are incidental while others are direct effects.

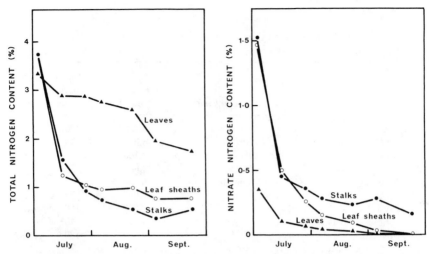

Fig. 4. Concentrations of total N and nitrate N in leaves, leaf sheaths, and stalks of corn plants at different times during the growing season. [Data from Hanway (1962c). Reproduced from *Agron. J.* **54** p. 223 by permission of the American Society of Agronomy.]

It can often be difficult to establish a relationship between maximum yield and N requirement by field experiments (Needham, 1982) since five, or more, rates of applied N are often required and yield potentials are strongly influenced by the supply of N and water from the soil and the position and activity of plant roots. Thus, factors that affect the supply of N from soil reserves, such as previous cropping history, winter rainfall (which facilitates leaching), kind of soil, and plant and other climatic factors, are very important (Cooke, 1982). Nonetheless, the general nature of crop responses to applied N is discussed below.

1. Pattern of Yield Response to Applied Nitrogen

The simplest response of plants to applied N, when N is the major growth-limiting factor, is that dry matter yield increases with increasing rates of N up to a maximum and then either stays constant or declines with further rates of N. Such a yield response for corn is shown in Fig. 5. Total uptake of N by the crop increased up to the maximum yield so that maximum fertilizer uptake efficiency (uptake of ^{15}N fertilizer as a percentage of applied N) was achieved at the same fertilizer rate as was required for maximum yield. Indeed, fertilizer uptake efficiency is normally relatively constant with increasing rates of N up to the level at which maximum yield is first obtained; further fertilizer additions decrease uptake efficiency (Broadbent and Mikkelsen, 1968; Westerman *et al.*, 1972;

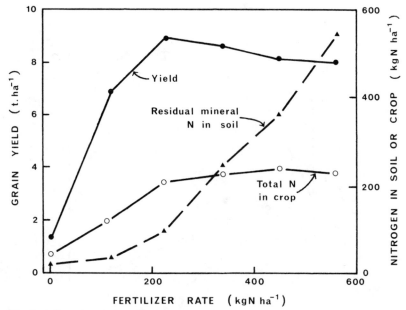

Fig. 5. Effect of rate of applied N on grain yield and total N uptake by a corn crop and on residual mineral N remaining in the soil after harvest. [Redrawn from Broadbent and Carlton (1978).]

Broadbent and Carlton, 1978). Thus, the potential for excess NO_3^- in the soil profile rises sharply above the fertilizer rate required to give maximum yield (Fig. 5). The magnitude of the positive response to applied N is likely to be primarily dependent on the size of the available and potentially available pool of N in the soil and the demand for N by the crop as determined by its potential dry matter production (Greenwood *et al.*, 1980; Olson and Kurtz, 1982).

Responses to applied N do not necessarily follow a pattern similar to that shown in Fig. 5. For example, in highly fertile soils with an abundant N supply, applications of N can have no effect or even decrease crop yields. If some factor other than N (e.g., soil moisture or another nutrient) is limiting growth then even if the supply of soil N is low, applications of fertilizer N are not likely to have a significant effect on plant growth or yields.

Plant growth is affected by many environmental factors and these can greatly modify the response of plants to applied N. Environmental factors also influence the availability (mineralization–immobilization) and losses of N (leaching and gaseous losses) from the soil. Figure 6 shows yields of

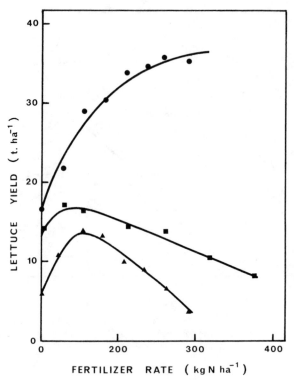

Fig. 6. Effect of rate of applied N on lettuce yields on one soil in different years. [Data from Greenwood *et al.* (1974). Reprinted with permission from Cambridge University Press.]

lettuce (*Lactuca sativa* L.) grown in adjacent sites on the same soil in different years. In one season, the response was positive, while in the others yields were first increased and then decreased even though the N status of the soil was similar in each year. The differences were attributed, at least in part, to the extent to which rain and irrigation moved NO_3^- down the soil profile and when this occurred (Greenwood *et al.*, 1974).

2. *Mathematical Description of Yield Responses*

As already noted, the response of a plant to applied N is an integration of many complex and interacting mechanisms, many of which are not measured or understood in most field experiments. In many experiments there are only sufficient measurements to justify the fitting of the data in a relatively simple mathematical expression (Wood, 1980). From an agronomic viewpoint, the most worthwhile aspect of describing yield responses mathematically is that it allows summarizing data from a series of experiments into a series of fitted curves describing crop responses, thereby facilitating comparisons between sites. In addition, various parameters of interest can be derived from the curves and used to compare plant responses.

A large variety of methods are used to treat response data mathematically (Greenwood *et al.*, 1971; Middleton and Smith, 1978; Wood, 1980; Karlovsky, 1982). Some common response curves are illustrated in Figs. 7 and 8, where the *x* and *y* axes represent rate of applied N and yield, respectively.

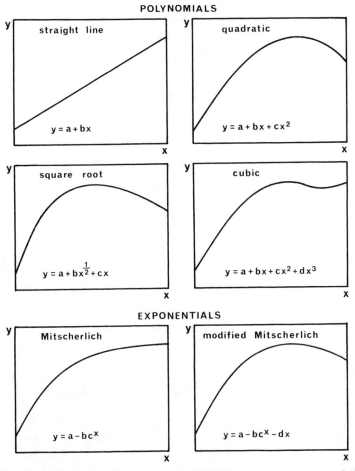

Fig. 7. Examples of polynomial and exponential split-line response curves of plants to applied fertilizer. [Redrawn from Wood (1980).]

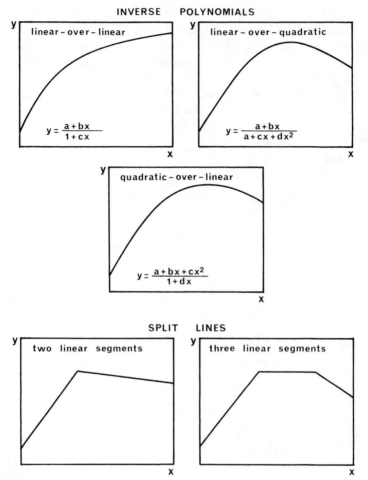

Fig. 8. Examples of inverse polynomial and split-line response curves of plants to applied fertilizer. [Redrawn from Wood (1980).]

The linear response is the simplest and most useful function for describing crop response. The quadratic caters for a curved response with yield rising to a maximum and then declining. Exponential functions describe yield increases toward a theoretical maximum with diminishing returns as more fertilizer is added and are sometimes referred to as Mitscherlich curves. Inverse polynomial curves can also be used (Nelder, 1966; Greenwood *et al.*, 1971), while the split-line or two intersecting straight-line models have been used by Boyd *et al.* (1976) for cereal responses to N.

Different cultivars of one crop grown in the same soil can have different response patterns. Thus, Balko and Russell (1980) found that of ten corn genotypes studied, one showed no response to applied N, five genotypes exhibited significant linear responses, and four genotypes showed significant quadratic responses. When N fertilizer experiments are carried out over a wide range of locations and conditions no one response model is likely to fit all the experimental results. Sparrow (1979), for example, found that of eight models tested against N response of spring barley in 83 experiments, the inverse quadratic, linear over linear, and two intersecting straight lines represented the yield–fertilizer relationship well although no one model fitted best at every site.

The modeling of yield response to N can become quite complex. Greenwood *et al.* (1980), for instance, described the response curves of 21 crops to applied N in terms of two parameters; one represented the beneficial component of the response (which increases yields) and the other the detrimental component (which suppresses yields at high N rates), i.e.,

$$\frac{a}{y} = \frac{1}{(1 - (N_s + N_f))/\alpha N}\left[1 + \frac{a}{BN(N_s + N_f)}\right]$$

where y = crop yield; a = theoretical maximum yield when the adverse effects of fertilizer are negligible; N_s = total amount of available N in the soil before fertilizer application; N_f = the amount of fertilizer N applied; αN = minimum amount of N required to prevent crop growth; and BN = coefficient of crop response to N. Greenwood *et al.* (1980) observed that the beneficial component was largely determined by potential demand for N by the crop and the detrimental component varied considerably with weather and was generally large for root crops.

Often, several models fit equally well to a set of data, and in general the mathematical function chosen should be as simple as possible while adequately representing the data (Wood, 1980). Indeed, models are tools for interpretation of experimental results and are less important than the thought and care that go into the design and performance of the experiments.

Conventional fertilizer trials are normally conducted with simple designs such as factorial arranged in randomized blocks. The use of the relatively simple continuous function (systemic) designs can result in large savings in trial site areas compared to conventional factorial designs (e.g., Munns and Fox, 1977). More complex (incomplete) designs can also be used in which fewer experimental units are required than with complete factorials (Box and Hunter, 1957; Littel and Mott, 1974; Dougherty *et al.*, 1979). Multiple regression analysis provides the basis for making inferences on effects of treatment levels not actually included as long as these levels fall within the range of extreme treatments. Computer-generated three-dimensional response surfaces can be drawn using multiple regression analysis of factorial or incomplete factorial designs. Such a response surface for the positive interaction of N and S applications on vegetative yield of rape (*Brassica napus* L.) is shown in Fig. 9.

3. Effect of Nitrogen on Yield Components

Thus far, the overall yield response to applied N has been outlined. In reality, N often increases harvestable yield through its effect on various yield components.

In small-grain crops leaf area is usually increased by N applications (Langer and Liew, 1973; Spiertz and Ellen, 1978). This is due to an increased number of tillers, and therefore leaves (Chandler, 1969; Pearman *et al.*, 1977), and also to increased leaf size (Pearman *et al.*, 1977). Longevity of leaves is also extended (Langer and Liew, 1973; Thomas *et al.*, 1978). Thus, the potential photosynthetic capacity of the crop is raised and the rate of photosynthesis can be increased (Pearman *et al.*, 1977; Spiertz and van der Haar, 1978). Increased respiration rates have also

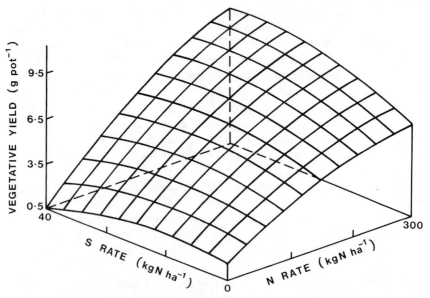

Fig. 9. Computer-drawn three-dimensional response surface of the effects of applied N and S on vegetative yield (stems plus leaves) of rape. [Redrawn from Janzen and Bettany (1984). Reproduced from *Soil Sci. Soc. Am. J.* **48,** p. 102 by permission of the Soil Science Society of America.]

been reported in response to N applications (Spiertz and van der Haar, 1978). The overall effect of N applications is therefore to increase the source capacity of the grain plant; the sink capacity is determined by the number and size of grains and their rates of growth.

In wheat, grain yield follows the relationship

$$\text{grain yield (ha}^{-1}) = \text{ears (ha}^{-1}) \times \text{spikelets (ear}^{-1}) \times \text{grain (spikelet}^{-1})$$
$$\times \text{weight (grain}^{-1})$$

Nitrogen applications typically increase tiller production and tiller survival so that more tillers are available to form ears (Pushman and Bingham, 1976; Dougherty et al., 1979; Langer, 1979). Thus, N-induced yield increases are generally the result of increases in ear density (Remy and Viaux, 1982; Sturm and Effland, 1982). Nitrogen generally promotes the rate of floret development, the number of fertile florets (Langer and Hanif, 1973; Thomas et al., 1978), and the number of grains set (Langer and Liew, 1973; Thomas et al., 1978). Applications of N can increase (Langer and Liew, 1973; Pearman et al., 1977; Spiertz and van der Haar, 1978; Spiertz and Ellen, 1978) or decrease grain weights or have no effect (Pushman and Bingham, 1976), depending on competition among individual grains for assimilates and N (Langer, 1979).

Compensatory phenomena occur among yield components so that optimization of yields by a simple combination of optimum yield components is difficult to achieve or predict (Evans and Wardlaw, 1976). For example, where ear number is little affected by N applications an increase in ear weight up to 70% over control can be induced by N applications (Gasser and Iordanou, 1976; Spratt and Gasser, 1970). Such an increase can be the result of an increase in the number of spikelets per ear (Holmes, 1973; Langer and Liew, 1973; Pearman et al., 1977).

Nitrogen applications are not always beneficial to yields of small grains. Applied N tends to increase the height of the crop (McNeal et al., 1971) and elongation of basal stem internodes enhances the danger of lodging with consequent yield losses. The use of growth regulators, such as chlormequat, to reduce stem growth in small grains is one of the best examples of their use in crop production, although dwarf varieties have also been developed (Harper, 1983). In general, the application of chlormequat reduces stem length and lodging, increases yields, and also results in the optimum rate of N application being increased (Dilz et al., 1982; Remy and Viaux, 1982), although this is not always the case (Needham, 1982).

In some corn genotypes, applied N can influence morphology (ear number), which in turn affects utilization of accumulated N. In semiprolific corn genotypes, increasing N rates greatly increase the number of plants with two ears (Balko and Russell, 1980; Kamprath et al., 1982; Anderson *et al.,* 1984). Kamprath *et al.* (1982) found that ear number per plant of two semiprolific genotypes of corn was linearly correlated with N concentration of corn plants at silking. Two-eared plants accumulated more ear and total dry weight and total N and also partitioned more of their dry matter and N to the ears than did one-eared plants (Anderson *et al.,* 1984).

For pasture grasses, N applications generally enhance the activity of the shoot apex, resulting in more primordia and unemerged leaves per tiller, and the number of emerging leaves per tiller per week is increased (Wilman *et al.,* 1977; Wilman and Mohamad, 1980, 1981). The number and rate of emergence of tillers are also increased as long as defoliation is reasonably frequent (Dobson *et al.* 1978; Wilman and Mohamad, 1980). The rate of extension of leaf blades can be doubled by applied N resulting in quicker recovery after defoliation and the mature size of individual blades can be doubled without a reduction in the number of leaves in the canopy provided defoliation is not too infrequent (Wilman *et al.,* 1977; Wilman and Mohamad, 1980). The larger leaves, along with the taller canopy, provide a greater surface area for photosynthesis and make the crop more accessible to grazing (Wilman and Wright, 1983).

C. Effect of Applied Nitrogen on Crop Quality

The predominant positive impact of fertilizer N on crop quality is in its enhancement of the total N content of crops. However, excess N tends to result in the accumulation of NO_3^- and sometimes organic acids (e.g., oxalate) in plant parts. Both can be toxic to man and livestock if ingested in large amounts. By stimulating luxuriant new growth, excess N can improve appearance but lessen the resistance of harvested vegetative parts to mechanical damage and for fruits and vegetables it can cause a reduction in red color, a loss of flavor, and a decline in keeping quality (Mills and Jones, 1979).

1. Protein Content

In grain and forage crops the major effect of fertilizer N on crop quality is its enhancement of protein content (Olson and Kurtz, 1982). In grain crops, increasing rates of N above those that give maximum yield often further increase the protein content of the grain (Johnson *et al.,* 1973; Benzian and Lane, 1979, 1981). The grain protein content typically increases linearly with N rate up to a maximum irrespective of the optimum N rate for yields (Benzian and Lane, 1981; Benzian *et al.,* 1983). Thus, when high rates of N depress crop growth and yield there can be a negative relationship between grain yield and grain protein content (Evans and Wardlaw, 1976; Novoa and Loomis, 1981).

Delayed applications of N are often more effective in maximizing protein yields than early treatments, which tend to stimulate vegetative growth (Olson and Kurtz, 1982). Closely related to protein content of wheat grain is baking quality (loaf volume) of flour produced from the grain. Increasing rates of applied N, particularly when applied late, often increase protein yield and loaf volume of flour from good-quality wheats (Feyter and Cossens, 1977; Dougherty et al., 1978, 1979).

Fertilizer N can also have detrimental effects on grain crop quality. In small grains excessive N fertilization can cause lodging, delayed maturity, and shriveled kernels with abnormally high protein content (Olson and Kurtz, 1982; Graham et al., 1983). In wheat, excessive N applications can lead to increased susceptibility to mildew and rust diseases and consequently reduced baking quality (Olson and Kurtz, 1982). Excessive N resulting in high protein contents also decreases the malting quality of barley and the quality of soft wheats used for pastry flour (Olson and Kurtz, 1982).

Increasing rates of applied N can not only increase the protein content of grain crops but also change the relative content of amino acids within the proteins (Table III). Increases in cereal protein content of corn, wheat, rice, rye (*Secale cereale* L.), and sorghum (*Sorghum bicolor* (L.) Moench) usually result in increases in the prolamine protein fraction (zeins) that dilutes the albumins and globulins, which are rich in essential amino acids such as lysine (Keeney, 1970; Juliano, 1972; Eppendorfer, 1975; Rendiz and Jimenez, 1978). Lysine is the most limiting essential amino acid in cereal grains and a reduction in its relative content represents a lowering in nutritional quality per unit protein for human and other monogastric animal consumption (Eppendorfer, 1975; Olson and Kurtz, 1982). The greater protein and amino acid production caused by N fertilization normally more than balances the relative decrease in protein quality (Eppendorfer, 1975). Any such changes in protein fractions do not, in any case, affect feed quality for ruminant animals, which have the capacity to synthesize their own essential amino acids.

Numerous investigators have reported that increasing levels of applied N increase the total N and protein content of the herbage of temperate grasses (e.g., Starbursk and Heide, 1974; Wilman et al., 1976; Wilman and Wright, 1983). Concentrations of water-soluble carbohydrates are, however, lowered because of their increased demand for N assimilation (Nowakowski, 1962; Wilson and Flynn, 1979; Wilman, 1980). Generally, the effect on herbage digestibility is either slight or absent (Wilman et al., 1976; Wilson, 1982).

2. Nitrate Accumulation

The accumulation of NO_3^- in plant parts is a natural phenomenon that occurs when the uptake of NO_3^- by the roots exceeds its reduction and

Table III

Effect of Application of Fertilizer Nitrogen on Amino Acid and Crude Protein Content of Corn Grain[a]

Amino acid composition	Control (no N added)	Fertilized treatment ($360 \ kg \ N \ ha^{-1}$)	Percentage increase
Lysine	0.24	0.26	8
Glycine	0.32	0.36	13
Arginine	0.37	0.42	14
Aspartic acid	0.51	0.62	15
Threonine	0.28	0.34	21
Histidine	0.21	0.27	29
Methionine	0.18	0.24	33
Valine	0.34	0.46	35
Isoleucine	0.25	0.34	36
Alanine	0.52	0.71	37
Serine	0.35	0.48	37
Tyrosine	0.25	0.35	40
Phenylalanine	0.32	0.46	44
Glutamic acid	1.26	1.83	45
Leucine	0.79	1.22	54
Proline	0.67	1.06	58
Crude protein	7.24	9.93	34

[a] Data from Rendiz and Jimenez (1978).

subsequent assimilation within the plant. The degree of accumulation is controlled by the genetic potential of the plant and is modified by environmental factors, fertilizer management, and crop production practices (Maynard *et al.,* 1976; Maynard and Barker, 1979).

Nitrate accumulation is of great interest to human and animal nutritionists since, as noted in Chapter 4, the microbial reduction of NO_3^- to NO_2^- in the alimentary tract can result in NO_2^- toxicity, particularly in ruminant livestock (e.g., cattle) and human infants. The most significant feature of acute toxicity of NO_2^- is methemoglobinemia in which hemoglobin is oxidized to methemoglobin with the result that oxygen transport in the blood is reduced and anoxia and death can occur. Most cases of human infant methemoglobinemia have been the result of ingestion of drinking water rather than of vegetables with a high NO_3^- content (National Research Council, 1972), but there are numerous reports of ruminants dying of NO_2^- intoxication after they ingested plants containing high amounts of NO_3^- (Walters and Walker, 1978). Reduced milk production in cows, increased abortion in cattle, and vitamin A deficiency have also been implicated in animals feeding on silage or pasture forage containing excessive amounts of NO_3^- (i.e., $>0.2\%$ NO_3^--N) (Walters and Walker, 1978; Wolff and Wasserman, 1972).

It has also been suggested that the presence of NO_2^- and secondary amines together in the alimentary tract could lead to the formation of carcinogenic nitrosamine compounds (Maynard et al., 1976; Walters and Walker, 1978). In addition to potential health hazards, high NO_3^- in canned vegetables may adversely affect storage life since the NO_3^- can act as an oxidizing agent resulting in the removal of tin from cans (Maynard et al., 1976).

Differences in plant species and cultivars, and changes in the environment, can greatly affect NO_3^- accumulation in plant foliage and these factors are considered below along with the effects of fertilizer applications.

a. Plant effects. Nitrate accumulation varies among plant species and cultivars, plant part, and age of plant. Species of grasses, for example, differ widely in the amounts of NO_3^- they accumulate (Deinum and Sibma, 1980), while cultivars of perennial ryegrass (Lolium perenne L.) can differ significantly with respect to NO_3^- accumulation (Butler et al., 1962; Vose and Breese, 1964). Plants belonging to the families Ameranthaceae, Chenopodiaceae, Cruciferae, Compositae, Gramineae, and Solanaceae have a tendency to accumulate NO_3^- (Wright and Davison, 1964). Annual weeds and grasses and cereal crops are more likely than perennial grasses and legumes to accumulate NO_3^-. Forage crops such as beet (Beta vulgaris L.) tops, rape (Brassica vulgaris L.), and oat (Avena sativa L.) hay generally have high NO_3^- contents (Viets and Hageman, 1971). Leafy vegetables characteristically accumulate more NO_3^- than other plants (Maynard, 1978).

The location and activity of the nitrate reductase enzyme complex are particularly important with regard to NO_3^- accumulation in plants. Plants that reduce most of the absorbed NO_3^- in roots obviously accumulate considerably less NO_3^- in aboveground parts than those that reduce NO_3^- predominantly in leaves (Olday et al., 1976b). Differential nitrate assimilation appears to account for differences in NO_3^- accumulation in spinach (Spinacia oleracea L.) cultivars. Savoy-leaf spinach types accumulate considerably more NO_3^- than smooth-leaf types; the latter types possess a higher nitrate reductase activity and greater dry matter content (Maynard and Barker, 1974, 1979; Barker et al., 1974; Olday et al., 1976a). The best opportunity for excess NO_3^- uptake, and maximum accumulation, occurs in early-maturing vegetables and these can contain considerably more NO_3^- than late-maturing cultivars (Kowal and Barker, 1981).

In general, NO_3^- levels are highest in petioles and stems, intermediate in leaves and to a lesser extent roots, and lowest in reproductive parts (Pimpini et al., 1970; Barker and Maynard, 1972; Maynard and Barker, 1972; Lorenz, 1978). Thus, the highest concentrations of NO_3^- in vegetable foods occur when leaves, petioles, or stem constitute the edible portions. Nitrate often tends to accumulate in the older portions of the plant, where nitrate reductase activity tends to be depressed (Barker and Maynard, 1972; Darwinkel, 1976; Maynard et al., 1976). For example, in cabbage (*Brassica oleracea* L.) NO_3^- tends to accumulate in older, mature wrapper leaves with lower levels occurring in younger tissue or the head portions (Kowal and Barker, 1981).

b. Environmental effects. Many environmental factors influence plant NO_3^- concentrations through their effect on nitrate reductase activity and nitrate uptake. In general, low light intensities, high temperatures, and moisture stress tend to lower nitrate reductase activity and enhance NO_3^- accumulation (Maynard et al., 1976; Maynard and Barker, 1979; Blanc et al., 1979; Deinum and Sibma, 1980).

Nitrate concentrations are greater when plants are exposed to low light intensities or to short photoperiods (Knipmeyer et al., 1962; Cantliffe, 1972a). Thus, Minotti and Stankey (1973) found highest concentrations of NO_3^- in beet plants between 4 and 8 A.M. and lowest concentrations about 4 P.M. In vegetables, time of harvest could therefore be an important consideration in terms of their NO_3^- content. Effects of temperature on NO_3^- accumulation are not clearcut (Maynard and Barker, 1979) but increasing temperatures generally tend to favor NO_3^- accumulation (Cantliffe, 1972b; Deinum and Sibma, 1980), especially in combination with low light intensity (Hoff and Wilcox, 1970).

Many cases of high NO_3^- accumulation in herbage and subsequent toxicity to animals have been related to sudden drought (Wright and Davison, 1964; Deinum and Sibma, 1980). Such accumulation probably results because water stress decreases both nitrate reductase activity and photosynthesis, and therefore NO_3^- assimilation, prior to the time that NO_3^- absorption from the soil is depressed (Maynard et al., 1976; Maynard and Barker, 1979).

c. Fertilizer effects. The amount, source, timing, and method of application all govern the effects of N fertilizers on NO_3^- accumulation (Maynard et al., 1976). For many plant species a direct relationship between N fertilizer and plant NO_3^- accumulation has been established (see Lorenz, 1978). Indeed, N fertilizers are generally recognized as the most significant factors affecting NO_3^- accumulation in many vegetables (Maga et al., 1976; Venter, 1979; Kowal and Barker, 1981; Goh and Ali, 1983) and grasses (Deinum and Sibma, 1980). Peck *et al.* (1971) reported that the addition of 56 kg N ha^{-1} increased the NO_3^- content of beets by 100%. In grasses, Deinum and Sibma (1980) found that NO_3^- accumulation rarely occurs with an N supply below 400 kg N ha^{-1} since under such conditions most of the absorbed NO_3^- is reduced in the roots. Similarly, Baker and Tucker (1971) found that in forage crops (e.g., wheat, oats, rye, and barley), nitrate concentrations rarely exceeded potentially toxic levels to animals (>0.2%) when 90 kg N ha^{-1} or less was applied.

Nitrate is the major form of N absorbed by plants regardless of the source of applied N since NH_4^+ is rapidly nitrified in most agricultural soils. Hence, within limits, the form of applied N has little effect on NO_3^- accumulation (e.g., Crawford *et al.*, 1961; Peck *et al.*, 1971; Barker, 1975). To some extent, the longer the plant is in contact with NO_3^- the greater will be the tendency to accumulate NO_3^- (Maynard *et al.*, 1976). Thus, urea or NH_4^+ fertilizers may result in less accumulation of plant NO_3^- than NO_3^- fertilizers when side-dressed to a growing crop (Barker *et al.*, 1971; Peck *et al.*, 1971). Experimentally, NO_3^- accumulation in spinach radish and beet root can be almost eliminated by the use of $(NH_4)_2SO_4$ plus nitrification inhibitor (nitrapyrin), but NH_4^+ toxicity is a problem that limits yields (Mills *et al.*, 1976; Vityakon, 1979). The use of slow-release fertilizers can also sometimes significantly lower NO_3^- accumulation in plants (Schuphan *et al.*, 1967; Siegel and Vogt, 1975).

Most studies show that as much NO_3^- is accumulated in plants from organic manures as from inorganic fertilizers if adequate time is allowed for mineralization (Peck *et al.*, 1971; Barker, 1975). Organic manures that mineralize slowly lead to less NO_3^- accumulation in vegetables than materials that mineralize more rapidly and therefore release more NO_3^- (Barker, 1975).

Applications of fertilizers other than N can also influence NO_3^- accumulation. Molybdenum is a component of the nitrate reductase enzyme (Chapter 6) and applications of Mo to Mo-deficient soils can decrease NO_3^- contents of plants (Cantliffe *et al.*, 1974; Hildebrandt, 1976). Deficiencies of Mn and Cu also appear to favor NO_3^- accumulation (Maynard *et al.*, 1976; Hildebrandt, 1976), probably by restricting plant growth. Applications of P have little effect on NO_3^- accumulation but K applications usually tend to stimulate the uptake and accumulation of NO_3^- (Maynard *et al.*, 1976; Breimer, 1982).

d. Methods of reducing nitrate accumulation. It is evident from the preceding discussion that some species of plants naturally accumulate NO_3^- in their leaves. Although concentrations of plant NO_3^- are partially related to the fertilizer N applied, if crops are fertilized for maximum yields, NO_3^- accumulation appears to be a natural and unavoidable phenomenon (Maynard *et al.*, 1976). Side-dressing with NH_4^+ rather than NO_3^- fertilizers or the use of slow-release fertilizers or nitrification inhibitors may reduce NO_3^- accumulation to some extent (Sahrawat and Keeney, 1984). It seems that if NO_3^- accumulation is considered a problem, then selection and development of cultivars with a low capacity for NO_3^- accumulation will be a feasible and desirable practice (Mills and Jones, 1979).

3. Organic Acid Accumulation

Since much of the relatively large amount of N that plants require often enters the roots in the form of NO_3^-, more anions than cations are usually absorbed by plants (Nye, 1981). However, reduction of NO_3^- (and SO_4^{2-}) in plant tissues results in a surplus of inorganic cations over inorganic anions within the plant and the surplus is balanced by synthesis of organic acid anions (Dijkshoorn, 1973; Raven and Smith, 1976; Franceschi and Horner, 1980). Thus increasing levels of NO_3^- nutrition stimulate the synthesis of organic acid anions in plants (Blevins *et al.*, 1974; Kirkby and Knight, 1977; Bailley-Fenech *et al.*, 1979) and physiological studies have shown that organic acid contents are higher in NO_3^-- than in NH_4^+-fed plants (e.g., Kirkby, 1968; Breteler and Smit, 1974; Nelson and Selby, 1974). Organic acids that have been identified include malic, citric, oxalic, fumaric, succinic, malonic, quinic, and polyuronic acids. Malic acid is dominant in a large proportion of plants including ryegrass (Dijkshoorn, 1973). Oxalic acid is the dominant acid in plants belonging to the Chenopodiaceae, Polygonaceae, and Portulaceae, and some tropical grasses (e.g., *Setaria*) (Fassett, 1973; Zindler-Frank, 1976).

Like NO_3^- accumulation, the accumulation of oxalate in plant tissue is of interest to both human and animal nutritionists. The consumption of food containing high amounts of oxalate has been reported to lead to the formation of calcium oxalate crystals in various tissues (e.g., kidney stones) and Ca deficiency (hypocalcemia) in humans and animals (R. J. Jones *et al.*, 1970; Jurkowska, 1971; Fassett, 1973). Severe poisoning leading to mortality of livestock has been reported after grazing the tropical grass *Setaria* spp., which has a characteristically high oxalate content (R. J. Jones *et al.*, 1970; Schenk *et al.*, 1982). Severe poisoning of humans following ingestion of large quantities of the leaves of certain oxalate-accumulating plants (e.g., rhubarb) has been reported although oxalate may not be the only toxic substance in the leaves (Fassett, 1973).

Plants that accumulate large amounts of oxalate are usually those that accumulate high levels of cations, particularly K^+ and/or Na^+ in their tissue (Osmond, 1963, 1967; Smith, 1972). Fertilizer K increases oxalate accumulation by increasing the cation excess in plant tissue (Smith, 1972, 1978). Similarly, applications of fertilizer N result in increased uptake of NO_3^- and accompanying cations and the subsequent reduction of NO_3^- causes a larger cation excess and increased oxalate accumulation (Osmond, 1963; Joy, 1964; Van Tuil, 1965, 1970; Smith, 1972; Roughan et al., 1976).

There seems to be no obvious practical method of greatly reducing organic acid levels in plants through fertilizer use, since in general maximum yields are associated with a specified organic acid content (de Wit et al., 1963; Dijkshoorn, 1973). Experimentally, the use of NH_4^+ fertilizers in conjunction with a nitrification inhibitor has resulted in a reduction in oxalate accumulation in some vegetable crops (Vityakon, 1979; Sahrawat and Keeney, 1984).

III. FACTORS AFFECTING CROP RESPONSES TO NITROGEN

A major obstacle in the development of reliable methods for predicting plant nitrogen requirements is the difficulty of identifying and quantifying factors that consistently affect N responses. Considerable variability occurs for the same crop between years within a site or between sites within a year (Cooke, 1982; Keeney, 1982a; Tinker and Widdowson, 1982). Factors affecting responses of plants to applied N have been reviewed extensively elsewhere (Whitehead, 1970; Ingestad, 1977; Cooke, 1982; Keeney, 1982a; Olson and Kurtz, 1982) and only the most important factors are outlined below.

The impact of such factors can, however, be rather complex. For example, on infertile soils a supply of adequate N can induce deficiencies of other nutrients such as P, K, and S. Obviously, when some nutrient other than N is limiting growth, then a response to applied N is unlikely to occur.

A. Form, Time, and Method of Nitrogen Application

1. Form of Applied Nitrogen

Many physiological studies utilizing liquid culture methods have shown that species differ widely in response to NH_4^+ and NO_3^- forms of N (see Chapter 6). Nonetheless, most agronomic research has shown that the two forms of fertilizer N are virtually interchangeable (Tinker, 1978). The major reason for this is that under most well drained soil conditions NH_4^+ is transformed to NO_3^- within a period of days (Chapter 3). The choice of

N fertilizer form is therefore normally an economic rather than an agronomic decision.

When differences in plant response to NH_4^+ and NO_3^- fertilizers occur they can usually be explained by simple physical effects such as volatilization of NH_3 or leaching losses of NO_3^- rather than by physiological effects (Tinker, 1978).

Because of their acidifying effects, it is important to avoid the repeated use of NH_4^+ fertilizers (e.g., ammonium sulfate) on soils of low buffer capacity (Russell, 1968). During nitrification two protons are added to the soil per NO_3^- anion accumulated (see Chapter 3). Urea has half the acidifying effect of NH_4^+ fertilizers per unit of applied N since during urea hydrolysis one proton is consumed per NH_4^+ ion formed. The magnitude of plant uptake is, however, important since for each NO_3^- ion absorbed by plant roots, one OH^- or HCO_3^- ion is excreted (see Chapter 6).

Soil acidification caused by the use of NH_4^+ fertilizers is relatively common (Boawn et al., 1960; Pierre et al., 1971; Helyar, 1976; Schnug and Fink, 1982). It can, to some extent, be beneficial since lowering of soil pH can alleviate deficiencies of trace elements such as Mn, Zn, and B by increasing their solubility (Schnug and Fink, 1982) and therefore minimize the need for supplementary additions of trace element fertilizers.

2. Time of Application

To maximize the use of fertilizer N by crops, N should ideally be applied as it is required by the crop plants. This requires that N be applied in a number of split applications, but each application increases production costs and, furthermore, the use of field equipment is often limited to early in the season. In addition, the effects of timing of fertilizer applications on crop yields are often small compared to those for rate of application (Needham, 1982). For row crops and grasslands, however, split applications consisting of one or two side-dressings are often used (Keeney, 1982a). As discussed in the following sections, with irrigated crops there is scope for supplying N, as required, in the irrigation water. The use of slow-release fertilizers is also a possibility (see Section V,B) but these require further development before their widespread use becomes practicable.

For cereal crops, the most effective time for N application is often related to the latest time compatible with the period of rapid N uptake by the plant (e.g., Olson et al., 1964; Miller et al., 1975; Dougherty et al., 1979). Such an application time reduces the opportunities for N losses through leaching, runoff, NH_3 volatilization, and denitrification. Furthermore, applied N is available throughout the period of grain formation without being used earlier for production of unnecessary vegetative growth (Olson and Kurtz, 1982), so that shorter plants with higher grain : stover ratios are produced. Results shown in Table IV illustrate a slight decrease in foliage dry matter and N content of corn and a concomitant increase in grain yield and N content when N was side-dressed rather than applied at planting.

Research with wheat (Ellen and Spiertz, 1975, 1980), corn (Olson *et al.,* 1964; Stanley and Rhoades, 1977; Bigeriego *et al.,* 1979; Russelle *et al.,* 1981), barley (Widdowson *et al.,* 1961; Bowerman and Harris, 1974; Easson, 1984), and rice (Beachell *et al.,* 1972; De Datta *et al.,* 1972) has generally confirmed that application of fertilizer N early in the season rather than at or before planting enhances total N uptake and utilization for grain protein provided sufficient rainfall or irrigation water is applied to carry fertilizer N to the root system. When delayed too long, N applications can result in decreases in yields and fertilizer N recovery by crops (International Atomic Energy Agency, 1970; Jung *et al.,* 9172; Easson, 1984).

Table IV

Effect of Time of Nitrogen Application on Dry Matter and Nitrogen Accumulation in the Aboveground Portion of a Corn Crop[a,b]

Treatment[b]	Dry matter (kg ha^{-1})		
	Foliage	Cob	Grain
Control	7,699	1,322	9,680
At planting	9,000	1,485	10,635
Side-dressed	8,730	1,478	11,120
LSD $P \leq 0.05$	710	103	866

Treatment	Nitrogen content (kg N ha^{-1})		
	Foliage	Cob	Grain
Control	45.2	5.2	114.8
At planting	66.8	6.9	149.9
Side-dressed	59.7	6.7	156.1
LSD $P \leq 0.05$	5.5	0.7	13.5

[a] Data from Bigeriego *et al.* (1979). Reproduced from *Soil Sci. Soc. Am. J.* **43**, p. 529 by permission of the Soil Science Society of America.

[b] Nitrogen was supplied at a rate of 112 kg N ha^{-1} either at planting or side-dressed when plants were 45 cm tall.

Delayed applications of N are not desirable for all crops. For sugar beet, for example, an early supply of N is required for development of an extensive leaf surface area and therefore maximum photosynthetic capacity. In contrast, high levels of available N late in the season promote continued vegetative growth at the expense of sugar storage in roots (Alexander *et al.,* 1954; Carter and Traveller, 1981).

In cool climates, application of fertilizer N in the fall for the next growing season can be a common practice. In general, N should not be applied in the fall until the soil temperatures fall below 10°C to avoid rapid nitrification of ammoniacal N (Keeney, 1982a). Heavy winter rainfall could then leach the NO_3^- from the potential crop rooting zone. Warm, wet spring conditions could still result in extensive losses of N through leaching and/or denitrification even after a cold winter. Thus, year-to-year climatic differences are likely to greatly influence the effectiveness of fall-applied N. Results have therefore ranged from comparable yields from spring and fall applications (Beauchamp, 1977; Hendrickson *et al.,* 1978b) to large yield advantages with spring applications (Stevenson and Baldwin, 1969; Warren *et al.,* 1975; Ellen and Spiertz, 1980). Applying N (e.g., urea) in concentrated amounts at discrete points in the soil, in narrow bands, or in large pellets can increase plant uptake of fall-applied N, since nitrification is inhibited due to high pH and high NH_3 concentrations in the fertilizer zone following urea hydrolysis (Nyborg and Malhi, 1979).

Generally, in dry areas, fall dressings can be as effective as spring dressings (e.g., Bullen and Lessells, 1957).

3. Method of Application

a. Soil application. There are four major methods of application: (1) surface broadcast, (2) incorporation by cultivating equipment, (3) injection of liquids or gases through knives, and (4) distribution in soluble form with irrigation water. The effects of various application methods vary with soil and climatic conditions and the particular form of N applied. Whatever method of application is used, care must be taken to ensure that appreciable amounts of fertilizer N do not contact the seed since this can have a severe negative effect on germination particularly when urea or NH_4^+ fertilizers are used (Olson and Kurtz, 1982; Harmsen, 1984).

When NH_4^+-forming fertilizers are applied to the surfaces of dry soils of reasonably high pH, losses of N through NH_3 volatilization are likely to be a problem (see Chapter 5). Volatilization of NH_3 is a particular problem following applications of urea, although immediate incorporation after application can greatly reduce losses (Overrein and Moe, 1967; Fenn and Kissel, 1976). Volatilization losses can also occur following injection of liquids or gases (e.g., aqua and anhydrous NH_3) if the soil is too dry or too wet or injection is not at a sufficient depth (Blue and Eno, 1954; Parr and Papendick, 1966).

Yield responses to applied N and fertilizer N recovery can sometimes be improved if fertilizer is banded in the root zone rather than broadcast on the soil surface (e.g., Daigger and Sander, 1976; Toews and Soper, 1978; Tomar and Soper, 1981; Reinertsen et al., 1984). To some extent, such increased fertilizer use efficiency appears to be due to less immobilization of applied N (Knapp, 1979; Tomar and Soper, 1981). As shown in Table V, Tomar and Soper (1981) found that banded applications tended to increase grain yield and N uptake by barley and that the positive effect of banding was greatest when straw residues were added to the soil to stimulate immobilization of fertilizer and soil mineral N.

Deep placement of N fertilizers in the reduced zone of the soil profile of paddy rice can increase plant recovery of applied N by minimizing loss of

Table V

Effect of Placement of Applied Nitrogen and Straw Applications on Grain Yield and Nitrogen Accumulation in Aboveground Portion of a Barley Crop[a,b]

	Grain yield (kg ha^{-1})		
Treatment[c]	No straw	Straw surface applied	Straw incorporated
Control	2128ab	1535a	1659a
Surface broadcast	3629def	3218cde	2612bc
Banded	3965ef	4322f	3007cd
	Nitrogen content (kg N ha^{-1})		
Treatment	No straw	Straw surface applied	Straw incorporated
Control	78a	62a	61ab
Surface broadcast	130de	105cd	98bc
Banded	134e	143e	105cd

[a] Data from Tomar and Soper (1981). Reproduced from *Agron. J.* **73**, p. 993 by permission of the American Society of Agronomy.

[b] Means followed by the same letter do not differ significantly at $P \leq 0.05$ by Duncan's Multiple Range Test.

[c] Nitrogen was applied at a rate of 100 kg N ha^{-1} either broadcast or in a band 2.5 cm wide and 10 cm deep. Ground oat straw (0.45% N) was applied at a rate of 5000 kg N ha^{-1} either thoroughly mixed into the surface 10 cm of soil or placed uniformly on the soil surface.

N through volatilization and denitrification (Mikkelson *et al.*, 1978; Prasad and Rao, 1978).

Application of N through irrigation systems (fertigation) allows the opportunity to supply N throughout the season in accord with crop requirements. Variations in soil permeability and microrelief or excessive winds can result in nonuniform distribution of N when it is injected into flood or sprinkler irrigation systems, respectively. Injection of N into center-pivot or lateral-move sprinklers is routinely practiced for corn in some regions of North America (Rehm and Wiese, 1975; Wesley, 1979). The technique is thought to be particularly suited to maintain N nutrition with minimum leaching losses for sandy soils (Rehm and Wiese, 1975; Watts and Martin, 1981; Gascho *et al.*, 1984).

Injection of fertilizer into trickle irrigation systems is particularly effective since the fertilizer is deliver directly to the active root zone, i.e., the wetted soil volume (Haynes, 1985). In comparison with broadcast fertilizer applications, the quantity of applied N required to give maximum yield can sometimes be halved if it is injected through the trickle system (Miller *et al.*, 1976; Smith *et al.*, 1979).

b. Foliar applications. With foliar treatments, N can be applied directly to the crop whenever desired. Urea is the preferred form of N used in foliar treatments because of its rapid penetration into leaves (see Chapter 6). Foliar applications are most appropriate for crops that are irrigated by sprinkler systems or sprayed frequently (e.g., with pesticides). Foliar sprays of urea are used commonly for high-value truck crops and tree fruit crops (W. W. Jones *et al.*, 1970; Keeney, 1982a). Leaf-burn can be a problem where the spray solution has a greater than 1 to 3% N content or the biuret content (a phytotoxic contaminant of urea formed during its manufacture) of fertilizer urea is high (Jones and Embleton, 1965; Viets, 1965). Indeed, many of the negative responses to foliar N treatments have been attributed to leaf injury (e.g., Parker and Boswell, 1980; Poole *et al.*, 1983a). Foliar injury may be reduced by spraying in the early morning or evening (Poole *et al.*, 1983b).

At present, foliar sprays of N are not widely used for grain crops although some interesting experimental results have been reported. For example, for wheat, foliar sprays of N show potential for grain yield increases when foliar applications are made prior to the heading stage and for increases in percentage grain protein when applied after anthesis (Finney *et al.*, 1957; Sadaphal and Das, 1966; Pushman and Bingham, 1976; Altman *et al.*, 1983). Cultivars, however, differ in the amount of foliar-applied N recovered in the grain (Altman *et al.*, 1983).

Similarly, for brown rice, foliar fertilization at or after panicle initiation can increase yields and grain N (Thom *et al.*, 1981). With corn, results have been less encouraging, with yields unchanged or decreased by foliar treatments, although increases in grain N have been observed (Harder *et al.*, 1982; Olson and Kurtz, 1982; Below *et al.*, 1984). Some investigators have also observed significant increases in soybean yields and N content following foliar treatments during pod-fill (Garcia and Hanway, 1976; Terman, 1977; Syverud *et al.*, 1980; Vasilas *et al.*, 1980) although other investigators have observed little effect (Parker and Boswell, 1980; Poole *et al.*, 1983b).

B. Nitrogen-Supplying Capacity of Soils

Both the amount of residual mineral N in the root zone before planting and the amount of soil organic N mineralized during the growing season can greatly affect the response of plants to applied N under field conditions.

1. Residual Mineral Nitrogen

Many investigators have shown that residual mineral N in the soil profile can be an important source of N for plants and should be accounted for when making fertilizer recommendations (e.g., Carson, 1975; Giles *et al.*, 1975; Carter *et al.*, 1976; Stanford *et al.*, 1977; Ris *et al.*, 1981). An appreciable effect of the quantity of residual NO_3^- in fallow soils before planting on the subsequent response of wheat to applied N was demonstrated by Olson *et al.* (1976) (Fig. 10). When residual soil NO_3^- levels were less than 45 kg N ha^{-1}, wheat grain yield increased sharply with fertilizer N applied up to 67 kg N ha^{-1}. In contrast, when residual NO_3^- was greater than 135 kg N ha^{-1}, grain yields were depressed by additions of fertilizer N. Nevertheless, heavy rainfall in the spring can result in leaching of NO_3^- below the root zone so that under some conditions (e.g., irrigated sandy soils) residual mineral N present before planting may have little or no effect on subsequent fertilizer N response.

The position of any residual available N in the soil profile in relation to the available water supply and consequent root activity can, however, also influence plant response. For example, in dryland farming, root activity generally increases at lower depths in the profile throughout the season as water becomes depleted from the upper soil horizons. Thus, NO_3^- present in the deeper portions of the soil profile can be taken up relatively late in the season, resulting in enhanced grain protein content in both wheat and corn (Gass *et al.*, 1971; Smika and Grabouski, 1976).

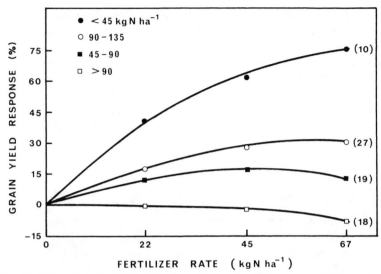

Fig. 10. Effect of residual NO_3^--N in the soil on grain yield response of fallow wheat to fertilizer N. [Redrawn from Olson *et al.* (1976). Reproduced from *Agron. J.* **68**, p. 770 by permission of the American Society of Agronomy.]

Upward water movement through the soil profile due to evapotranspiration at the soil surface can also result in movement of NO_3^- from the subsoil into the rooting zone (Wetselaar, 1961; Vlek *et al.*, 1981).

2. Mineralization of Soil Organic Nitrogen

In some localities, organic matter is still the major source of N for crop production. The supply of N is often so large in soils first brought under cultivation that fertilizers are not needed. Stephen (1982), for example, noted that N applications increased wheat yields in the second and third years but not in the first year after cultivation of pastures. As cropping is continued on the same site, the amount of readily mineralizable N becomes depleted and outside sources of N are required (e.g., fertilizers or organic wastes). The soil no longer serves primarily as a source of N but rather as a reservoir where biological mineralization–immobilization results in mineral N being increased at some times and decreased at others. A flush of mineralization in cropped soils in spring usually occurs (Hart *et al.*, 1979; Tinker, 1978).

Many attempts have been made to estimate the amount of potentially mineralizable N in soils (see Section IV,B) but in practice some subjective assessment is usually made. In many regions of the United States, fertilizer recommendations take into account previous cropping history and, in particular, previous use of leguminous crops or applications of organic wastes (Keeney, 1982b). Crop residues from previous crops are an important source of nitrogen (Power, 1981). In the United Kingdom, a subjective soil N index is used that is based on previous cropping history. There are three classes; the lowest represents soils that are under continuous arable cropping and the highest includes soils just out of alfalfa, grass/legume pastures, or pastures on which organic wastes have been spread (Needham, 1982).

C. Available Soil Water

In general, a positive crop yield response occurs for the interaction between applied fertilizer N and applied irrigation (e.g., Olson *et al.*, 1964; Viets, 1967; Spratt and Gasser, 1970; Pushman and Bingham, 1976; Singh *et al.*, 1979; Balasubramanian and Chari, 1983). Singh *et al.* (1979), for example, did not find a response for unirrigated wheat beyond an N application of 80 kg ha^{-1}, whereas the response to N was linear up to 120 kg N ha^{-1} on an irrigated plot. Such an interaction is not surprising since when plant growth and yield are limited by available moisture the N requirement is relatively low. If irrigation water is applied and growth of a crop is greatly increased N requirement also increased. Furthermore, protein synthesis is typically greatly reduced by water stress (Hsiao, 1973; Hsiao and Acevedo, 1974) and the activities of some enzymes involved in N metabolism are decreased although others are increased (Todd, 1972).

Irrigation frequency can be an important factor since the longer the plant remains water-stressed, the more likely that available moisture rather than N limits growth (Balasubramanian and Chari, 1983; Eck, 1984). For example, Eck (1984) withheld irrigation from a corn crop for 2 or 4 weeks during vegetative growth (Fig. 11). Applied N slightly increased grain yield under water stress and greatly increased yields under full irrigation. At the lower N rates, N deficiency limited yield to the extent that water stress had only a small effect, but with adequate or excessive N, water stress was the major limiting factor.

Water use efficiency (crop yield per unit of water used) is facilitated by a balanced nutrient supply (Olson *et al.*, 1964; Arnon, 1975). A plant that is growing slowly because of nutrient deficiency or toxicity uses a similar amount of water as a nutritionally balanced plant yet produces a considerably lower yield. Thus, field and laboratory experiments have shown that fertilizer N additions can increase water use efficiency of various plant species (Viets, 1962; Ramig and Rhoades, 1963; Olson *et al.*, 1964;

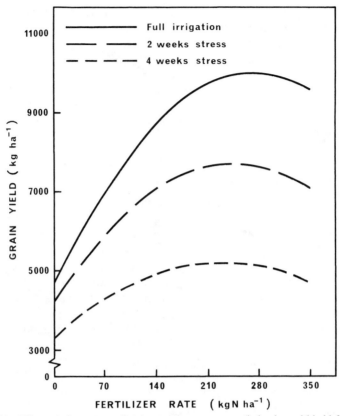

Fig. 11. Effect of nitrogen applications and water stress (irrigation withheld for 2 or 4 weeks during vegetative growth) on corn grain yields. [Redrawn from Eck (1984). Reproduced from *Agron. J.* **76,** p. 424 by permission of the American Society of Agronomy.]

Fowler and Hageman, 1978). Aktan (1976) found that water use efficiency of winter wheat and barley was decreased markedly by either a deficiency or an excess of N.

High N applications have in some cases been shown to improve drought resistance of various plant species (Brown, 1972; Tesha and Kumar, 1978; Tesha and Eck, 1983), possibly by increasing their photosynthetic efficiency (Tesha and Eck, 1983).

D. Tillage Method

Cultivation results in soil disturbance, increases in soil porosity and aeration, and exposure of less accessible organic substrates to mineralization by the soil biomass (see Chapter 2). Cultivation therefore generally accelerates net mineralization of soil organic N. In contrast, no-tillage normally results in the accumulation of organic matter in the surface soil (Blevens et al., 1977, 1983; Fleige and Baeumer, 1974; Douglas and Goss, 1982). Indeed, although many advantages of no-tillage arise from the presence of a mulch of dead vegetation, such residues, which often have a high C : N ratio, tend to stimulate immobilization of mineral N in the surface soil (Elliott et al., 1981; Kitur et al., 1984). Thus, untilled soils can contain considerably less NO_3^- than cultivated ones, particularly during autumn (Dowdell and Cannell, 1975; Dowdell et al., 1983). Additional causes of lower levels of NO_3^- in untilled soils may include increased NO_3^- leaching due to improved pore continuity (Goss et al., 1978; McMahon and Thomas, 1976) and greater denitrification losses of N_2O and N_2 associated with higher soil moisture content and a source of readily available C in the soil (Burford et al., 1981; Rice and Smith, 1982; Aulakh et al., 1984).

It is, therefore, not surprising that long-term experiments with corn (Moschler et al., 1972; Bandel et al., 1975; Moschler and Martens, 1975; Blevens et al., 1977; Legg et al., 1979; Kitur et al., 1984), wheat (Dowdell and Crees, 1980; Smith and Howard, 1980; Vaidyanathan and Leitch, 1980; Fredrickson et al., 1982; Ellis et al., 1982), and barley (Hodgson et al., 1977; Smith and Howard, 1980) have shown that without added N or at low N rates, N uptake and yields are usually lower for untilled than for tilled soils, but with adequate or high N rates equal or higher yields can be achieved without tillage. This is illustrated in Table VI, which shows that at the lower rate of N (252 kg N ha^{-1}), ^{15}N fertilizer recovery in corn grain and stover was less for untilled than for tilled soils, while the amount of ^{15}N remaining in the soil (immobilized) at the end of the experiment was greater without tillage.

E. Genetic Effects on Crop Response

Differences between species and genotypes of plants in their capacity to absorb, translocate, and utilize soil and fertilizer N are well known (Chandler, 1970; Clark, 1983; Mengel, 1983). For instance, grain protein content obtainable from different crops follows the order wheat > rice = corn > sorghum. Currently available varieties of grain sorghum do not yield as well as corn under identical N fertility conditions even though sorghum takes up more N from the soil (Olson and Kurtz, 1982). The reason is that grain sorghum translocates much less of its N from vegetative tissue to grain, and consequently the stover of sorghum contains approximately 50% more total N than does that of corn (Perry and Olson, 1975).

Table VI

Nitrogen Balance following Three Years of Application of ^{15}N-Labeled Fertilizer (f) at Two Rates to No-Till (NT) and Conventional Till (CT) Corn[a,b]

Tillage	Total nitrogen (f) applied	Fertilizer N recovered in grain				Nitrogen (f) in stover 1982	Total nitrogen (f) in soil	Nitrogen (f) not accounted for (%)
		1980	1981	1982	Total			
NT	252	20a	10a	18a	57a	13a	107b	29a
CT	252	38b	33b	28b	100b	17a	70a	26a
NT	504	49b	50c	145c	145c	36b	199c	25a
CT	504	43b	49c	141c	141c	37b	188c	27a

[a] Values within a column followed by the same letter are not significantly different ($P \leq 0.05$) by t-tests using least-square means.
[b] Data from Kitur et al. (1984). Reproduced from Agron. J. **76**, p. 241 by permission of the American Society of Agronomy.

Several factors are responsible for genotypic differences in responses (Clark, 1983; Mengel, 1983) and these can be divided into two major categories: (1) plant growth rate and morphology and (2) capacity for uptake and metabolism.

1. Plant Growth Rate and Morphology

The rate of plant growth is an obvious factor affecting N requirement. High-yielding new crop cultivars are characterized by rapid growth rates resulting in high production of plant material per unit of land area or time. Such production can only be attained when it is matched by an adequate supply of N and other essential nutrients. Thus, in soils low in N, high-yielding cultivars respond readily to additions of fertilizer N, whereas low-yielding cultivars with a lower growth rate may have their N requirement met adequately by the natural N supply of a fertile soil.

High production of stalks and leaves is associated with the high grain potential of high-yielding inbred corn lines and therefore these lines require high rates of fertilizer N (and other nutrients) to obtain maximum yield (Agboola, 1972; Mengel, 1983). In the case of modern high-yielding wheat and rice cultivars low straw production is a typical feature. The small, erect leaves enhance maximum utilization of solar energy and plants are planted much more densely than older cultivars. These dense plantings produce grain yields in the range of 8 to 10 tons ha^{-1}, which contain protein N equivalent to 160–200 kg N ha^{-1}. High fertilizer rates are therefore required since most soils are not capable of providing such quantities of N. Underdeveloped cultivars that only yield 2–3 tons ha^{-1} require 40–60 kg N ha^{-1} for grain production and hence respond only marginally to fertilizer applications (Chandler, 1970).

2. Capacity for Uptake and Metabolism

Genotypes of various crop plants differ greatly in their capacity to produce dry matter at a given level of N supply and in the amount of dry matter produced per unit of N absorbed (N efficiency ratio) (e.g., O'Sullivan et al., 1974; Chavalier and Schrader, 1977; Reed and Hageman, 1980). There is not necessarily a close relationship among genotypes between the capacity to produce high dry matter and N efficiency ratios (Chavalier and Schrader, 1977).

Genotypes differ not only in their capacity to absorb N but also to translocate and partition N within the plant (Clark, 1983). For grain crops this is important since remobilization of N from stalks and leaves is an important source of grain N. For genotypes of corn (Beauchamp et al., 1976; Chavalier and Schader, 1977; Rodriguez, 1977), wheat (Johnson et al., 1967; Halloran and Lee, 1979; Fjell et al., 1984), sorghum (de Franca,

1981), and soybeans (*Glycine max* L.) (Jeppson *et al.*, 1978; Zeiher *et al.*, 1982) there is no close relationship between the proportion of total plant N in grain at harvest and the total amount of N accumulated in plants. However, prolific and semiprolific corn genotypes generally partition more of the total plant N into grain production than do nonprolific ones (Anderson *et al.*, 1984). Similarly, high-protein genotypes of wheat generally translocate N to grain from other vegetative plant parts more readily and completely than do normal cultivars (Johnson *et al.*, 1967; Lal *et al.*, 1978; Fjell *et al.*, 1984). Some high-protein genotypes of wheat also require continued assimilation of N by leaves during grain development (Mikesell and Paulsen, 1971; McNeal *et al.*, 1972). Large differences among wheat cultivars in the proportions of aboveground N distributed to the head and also in the amount of head N distributed to the grain at harvest are shown in Table VII.

F. Leguminous Crops

Leguminous crops obtain much of their N through symbiotic N_2 fixation by *Rhizobium* bacteria present in nodules on their root systems. Leguminous plants include grain crops such as soybeans, peanuts (*Arachis hypogaea* L.), beans (*Phaseolus vulgaris* L., *Vigna unguiculata* (L.) Walp., and *Vicia fabia* L.), and peas (*Pisum sativum* L.) and forage crops such as alfalfa and related *Medicago* species and clovers (*Trifolium* spp.), which are grown in monoculture or in mixed culture with grasses. Much research has centered on whether a significant positive response can be obtained by applying fertilizer N to such crops.

Table VII

Nitrogen Distribution in Aboveground Parts of Five Wheat Cultivars[a]

Wheat cultivar	Percentage distribution		
	Head N	Grain N	Grain N
	Total plant N	Total plant N	Head N
Argentine IX	54.8	51.6	94.1
Petit Rojo	52.5	48.0	91.3
Gatcher	60.4	47.9	79.4
Olympic	35.9	26.9	75.1
Timgalen	38.9	25.0	64.6

[a] Data from Halloran and Lee (1979).

It is well established that large amounts of applied N reduce root hair infection (Munns, 1968; Dazzo and Brill, 1978), nodule number (Dart and Mercer, 1965), nodule mass (Summerfield *et al.,* 1977), and the total amount of N_2 fixed (Allos and Bartholomew, 1959). The degree of inhibition varies with the form of the N compound (Dart and Wildon, 1970), species and cultivar of crop (Allos and Bartholomew, 1959; Gibson, 1974), and strain of *Rhizobium* (Pate and Dart, 1961), as well as environmental and nutritional conditions (Gibson, 1974; Pankhurst, 1978). When mineral N is dissipated from the root environment the N_2-fixing capacity of the *Rhizobium* bacteria is resumed. In general, as fertilizer N rates are increased, legumes fix less atmospheric N_2 and use proportionately more mineral N from the soil while yields are often not greatly affected (e.g., Oghoghorie and Pate, 1971; Ham and Caldwell, 1978; Westermann *et al.,* 1981; Rennie *et al.,* 1982). Nevertheless, when applied early, low rates of fertilizer N often produce beneficial effects on plant development and subsequent nodule formation and function depending on plant species and cultivar, *Rhizobium* strain and environmental factors (Dart and Wildon, 1970; Gibson, 1974; Agboola, 1978; Mahon and Child, 1979; Huxley, 1980; Westermann *et al.,* 1981; Eaglesham *et al.,* 1983). Peanuts (Chesney, 1975) and particularly beans (*P. vulgaris*) (Robinson *et al.,* 1974; Wilkes and Scarisbrick, 1974; Peck and MacDonald, 1984) often show a positive response to early N fertilization. In general, an early supply of mineral N may result in rapid initial plant growth and leaf area development so that after such N has been depleted, the plant is capable of N_2 fixation at higher than normal rates (Mahon and Child, 1979).

There is also some evidence of response to applied N in later phases of growth of grain legumes, particularly soybeans. Positive responses have been associated with decreases in amounts of N_2 fixed during late growth stages (during seed development) due to nodule senescence (Weil and Ohlrogge, 1972; Lawn and Brun, 1974b) or competition between nodules and seeds for assimilates (Lawn and Brun, 1974a; Bhangoo and Albritton, 1976). Such responses do not, however, appear to be widespread (Welch *et al.,* 1973).

In grass–legume pastures, applications of fertilizer N generally favor the grass component (Haynes, 1981) since N_2 fixation by the legume is depressed and, further, pasture legumes are generally weaker competitors for mineral N than are grasses (Vallis, 1978). Nonetheless, the grass component of grass–legume pastures is often N-deficient (Henzell, 1981) and strategic use of fertilizer N as a management tool to improve pasture production can be practiced. Generally, relatively small amounts of N ($25–50$ kg N ha^{-1}) are applied in early spring or autumn when legume growth is characteristically weak (Ball and Field, 1982; O'Connor, 1982; Steele *et al.,* 1982).

G. Disease and Pest Incidence

The interaction between N nutrition and disease and pest incidence is one that can have important agronomic implications. The plant response to applied N can be modified by diseases and pests and, conversely, N applications may exert secondary, often unpredictable, effects on crop yield through their effects on growth, survival, and virulence of disease and pest organisms, and for host tolerance to diseases and pests (Henis, 1976; Jones, 1976; Kiraly, 1976).

Disease incidence and severity caused by root-infecting fungi are often increased by applications of N fertilizers, although cases of no effect or decreases have also been reported (Henis, 1976). For small-grain cereals the incidence of take-all (*Gaeumannomyces graminis*), eyespot (*Cercosporella herpotrichoides*), and foot rot (*Fusarium roseum* var. *culmorum*) diseases tends to be increased by fertilizer N applications (Lemaire and Jouan, 1976; Lynch, 1983) as does the incidence of stem and root rot of maize caused by *Fusarium spp.* (Krüger, 1976). To some extent, root diseases appear to differ in their response to NH_4^+ and NO_3^- forms of fertilizer (Huber *et al.,* 1968; Smiley and Cooke, 1973; Henis, 1976). This may partially be attributable to a pH effect (Smiley, 1975) with NH_4^+ acting to initially reduce pH and NO_3^- acting to increase it following their uptake by plant roots (see Chapter 6). Ammonia is also very toxic to survival propagules of fungi present in soils (Lyda, 1981) so that applications of anhydrous NH_3 have been reported to reduce the incidence of some diseases such as cotton root rot (*Phymatotrichum omnivorum*) in infested soils (Rush *et al.,* 1979).

Susceptibility of plants to leaf diseases caused by obligate parasites tends to be increased by applications of N fertilizers (Jenkyn, 1976; Krüger, 1976; Temiz, 1976). By increasing growth, N gives a denser crop and therefore a more humid environment for pathogenic infection (Kiraly, 1976), although the number of successful infections, colony growth rate, and spore production in some pathogens are also increased (Jenkyn and Griffiths, 1976; Kiraly, 1976). Facultative parasites, particularly those that cause necrotic leaf spots, cannot grow in the absence of dead tissues so that high N rates can reduce their incidence by stimulating healthy fresh green growth (Kiraly, 1976). High rates of N can exacerbate the symptoms of some viruses (e.g., yellow dwarf virus of onions) but mask the effect of others (e.g., sugar beet yellows). In general, high N levels increase both the susceptibility of plants to virus infection and plant growth and therefore multiplication of the virus within the plant (Martin, 1976).

In many pest–crop situations, N fertilizers increase the number of pests by increasing yields and the amount and quality of food available. Nevertheless, the increased growth induced by N applications means that the intensity of attack per unit of plant weight is initially reduced and also the capacity of plants to compensate for lost parts is improved (Jones, 1976).

The presence of diseases can suppress the response of plants to applied N. Soil-borne diseases generally restrict root growth and destroy conducting tissue in roots and therefore interfere with uptake and translocation of water and mineral nutrients. The effectiveness of N fertilizers is therefore reduced (Remy and Viaux, 1982), although plants infected with root diseases, such as take-all, may still respond to high rates of N (Needham, 1982). Leaf diseases directly reduce photosynthesis and therefore yield. Thus, the use of fungicides on wheat can slightly increase the rate of response up to the optimum yield so that applied N is used more efficiently (Needham, 1982; Tinker and Widdowson, 1982).

IV. ASSESSMENT OF SOIL NITROGEN AVAILABILITY

There is no generally accepted method of soil testing for available N. This is a reflection of the fact that 97–99% of the N in soils is present in organic forms that are not directly available to plants until after mineralization has occurred (see Chapter 2). The amount of N mineralized depends on temperature, moisture, and other environmental factors and is therefore difficult to predict. In addition, the major form of mineral N (NO_3^-) is subject to losses via leaching and denitrification and conversion back to organic forms by microbial immobilization.

Crop yields, and hence their N requirements, are also affected by numerous soil, climate, and management variables, many of which are very difficult, if not impossible, to predict in advance (e.g., yearly weather patterns). Furthermore, other management factors such as the current economics of fertilizer use also need be considered. Thus, recommendations for fertilizer N applications always involve some degree of judgment rather than simple interpretations of soil test results.

Two major types of soil tests for N exist. First, residual mineral N in the soil profile can be routinely measured in spring and fertilizer recommendations can then be modified depending on the amount of available N already present. A second and/or complementary approach is to obtain an estimate of the amount of "potentially mineralizable" N present in the soil. This is achieved experimentally with a large number of diverse incubation methods and chemical extractants collectively known as N availability indices.

A. Residual Mineral Nitrogen

As discussed in Section III, the quantity of residual mineral N in the rooting zone before planting can significantly influence crop responses to applied N. This is particularly so in climates where extensive leaching and/or denitrification do not occur to any great extent during the growing season. Residual NO_3^- in the soil profile usually originates from previous N fertilization (Herron *et al.*, 1971; Soper *et al.*, 1971; Onken and Sunderman, 1972; Ludwick *et al.*, 1976; Jolley and Pierre, 1977) or mineralization of soil organic N, especially in crop-fallow systems (Carter *et al.*, 1974, 1975; Oyanedal and Rodriguez, 1977). Although soil NO_3^- is often used as a measure of residual mineral N, appreciable amounts of NH_4^+ can also be present in soils in early spring (Needham, 1982). It is, therefore, advisable to measure both NH_4^+ and NO_3^- rather than simply assuming that amounts of NH_4^+ present will be negligible.

Critical variables in estimating residual NO_3^- in the soil profile include (1) depth of profile sampled, (2) time of sampling, and (3) number of individual samples taken from a given field.

The effective rooting depth of a crop determines the depth of sampling required to adequately assess the quantity of available residual NO_3^- in the soil profile (Soper *et al.*, 1971; Onken and Sunderman, 1972; Ludwick *et al.*, 1977). Crops such as wheat, maize, and sugar beet can utilize NO_3^- down to depths of 150 cm (Herron *et al.*, 1971; Anderson *et al.*, 1972; Daigger and Sander, 1976), although rooting depth depends on many factors including soil type, presence of impeding soil layers, and distribution of nutrients and water in the soil profile (Russell, 1977; Cannell, 1982). In practice, recommendations for sampling depth vary for different crops in different areas but are usually in the range of 60 to 180 cm for cereals (Keeney, 1982b; Stanford, 1982; Sauderbeck and Timmermann, 1983).

Sampling must be carried out each season shortly before spring planting or very early in the season to reflect the available pool of NO_3^- in the soil. Sampling in autumn, preceding the next year's crop, is unwise since leaching losses of NO_3^- are likely to occur during winter or, in contrast, accumulation of NO_3^- could also occur through mineralization (Ludwick *et al.*, 1977). Hence, for autumn-sown crops (e.g., winter wheat), profile NO_3^- is generally sampled in the early spring before rapid growth begins (Ludecke, 1974; Becker and Aufhammer, 1982).

Spatial variability of NO_3^- in soils is recognized as a major problem in sampling fields (see Chapter 4) and applications of fertilizer N tend to increase variability (Biggar, 1978). Minimum sampling requirements to give a satisfactory estimate of profile NO_3^-, in relation to spatial variability, will differ among soils, climates, and management systems and need to be evaluated locally. This is an area that requires further research. Keeney (1982b) recommends that at least 20 core samples be taken from each field to a minimum depth of 60 cm.

Residual mineral N in the soil profile is used in fertilizer N recommendations in many of the western states of the United States (Keeney, 1982b) as well as in the Federal Republic of Germany (Becker and Aufhammer, 1982; Sauderbeck and Timmermann, 1983), the Netherlands, and Belgium (Ris *et al.*, 1981; Dilz *et al.*, 1982).

B. Potentially Mineralizable Nitrogen

Nitrogen availability indices are a measure of the potential of a soil to supply N to plants. Since mineralization is affected by many environmental and cultural factors (Chapter 2), such indices can only be expected to measure an amount of N that is roughly proportional to the amount mineralized under field conditions. The indices can be subdivided into biological methods (aerobic and anaerobic incubations) and chemical methods.

1. Biological Indices

a. Short-term incubations. These methods involve incubation of soils for 1–6 weeks under aerobic or anaerobic (waterlogged) conditions. Mineral N is generally measured before and after incubation. Biological methods are usually considered to be unsuitable for routine soil testing since they consume both space (they require constant temperature cabinets) and time (they require a minimum of 1 week for incubation). The many different procedures that have been proposed have been reviewed by several investigators (Bremner, 1965; Dahnke and Vasey, 1973; Campbell, 1978; Keeney, 1982b; Stanford, 1982; Sahrawat, 1983).

Control of water content during aerobic incubation is a major problem for soils having a wide range of water-holding capacities since moisture content has a major effect on mineralization rate. Reasonably reliable results can, however, be obtained when a constant level of water is added to soil–sand mixtures (Keeney and Bremner, 1967). Greenhouse studies have generally shown a close relationship between N released on aerobic incubation and N uptake by plants (Baerug *et al.*, 1973; Gasser and Kalembasa, 1976; Geist, 1977; Smith *et al.*, 1977; Stalk and Clapp, 1980).

Anaerobic incubation methods have several advantages over aerobic procedures: (1) problems associated with estimation and maintenance of optimum moisture content are avoided, (2) only NH_4^+ need be measured since nitrification does not occur under anaerobic conditions, (3) more N is mineralized in a given period under anaerobic conditions, and (4) higher temperatures can be maintained during incubation (therefore mineralization rates are higher) since the temperature optimum for nitrification need not be considered (Keeney, 1982b). Thus, shorter incubation times (e.g., 1 week) can be used but initial NO_3^- concentrations must be determined separately if they are to be utilized in N recommendations.

Correlations between N released on anaerobic incubation (6–14 days at 30–40°C) and N uptake by plants in greenhouse studies have generally been close (Osborne and Storrier, 1976; Geist, 1977; Powers, 1980; Stalk and Clapp, 1980) and similar to those for aerobic indices (Keeney and Bremer, 1966a; Cornforth, 1968; Stalk and Clapp, 1980).

As discussed in Chapter 2, the microbial biomass in the soil acts as a dynamic source and sink of mineral N (Carter and Rennie, 1984) and the biomass can contribute substantial amounts to the pool of mineral N in soils (Jenkinson and Ladd, 1981). Air drying kills much of the biomass and short-term incubation results in remineralization of dead biomass N (Marumoto, 1984; Marumoto et al., 1982). Thus, it is likely that many of the short-term incubation methods outlined above (aerobic or anaerobic) give a relative estimate of the size of the pool of biomass N in soils. Estimation of the size of the pool of soil biomass N using the chloroform fumigation technique (Jenkinson and Powlson, 1976; Shen et al., 1984) may give similar relative results.

b. Long-term incubations. Stanford and co-workers (Stanford and Smith, 1972; Stanford et al., 1974; Stanford, 1977) have developed a long-term incubation procedure to define the mineralizable (labile) pool of N in soils. The procedure involves estimation of the N mineralized for a soil sample over an extended period (up to 30 weeks) with the mineral N removed at regular intervals during the incubation. The incubation is carried out at a temperature of 35°C and at a soil moisture tension of -60 to -70 kPa. The N mineralization potential (N_0) is estimated from the cumulative amount of N mineralized (N_t) based on the assumption that N mineralized obeys first-order kinetics ($-dN/dt = KN$), i.e., log ($N_0 - N_t$) = log $N_0 - K_t/2.303$. The parameters K and N_0 can be considered as definitive soil characteristics upon which quantitative estimates of N mineralization can be based (Stanford, 1977, 1982).

Results of short-term incubations do not necessarily reflect the long-term N-supplying capacity of a soil since Stanford and Smith (1972) and Stanford et al. (1974) found that results of the first incubation period (1–2 weeks) were virtually meaningless with regard to N_0. This suggests that the initial rate of mineralization is greatly affected by crop residues, sample preparation, and other factors and the true rate of N mineralization is established only after such effects have been overcome.

The disadvantage of determining the long-term mineralization capacity of soils is that it is laborious, time-consuming, and expensive.

2. Chemical Indices

It seems impossible to devise a chemical extraction procedure that simulates the action of microorganisms in releasing plant-available forms of soil N. Many chemical indices of soil N availability have, nevertheless, been proposed since they are more rapid, precise, and convenient than biological incubation procedures. The numerous extractants that have proposed have been reviewed in detail by Bremner (1965), Dahnke and Vasey (1973), Campbell (1978), Keeney (1982b), Stanford (1982), and Sahrawat (1983). They can be divided into three broad groups: (1) weak extractants (e.g., hot water), (2) intermediate extractants (e.g., alkaline permanganate), and (3) strong extractants (e.g., 6 N H_2SO_4). A range of such extractants is shown in Table VIII.

Mild extractants of NH_4^+ or total N such as hot water (Keeney and Bremner, 1966a; Lathwell *et al.*, 1972; Gasser and Kalembasa, 1976; Osborne and Storrier, 1976), hot 0.01 M $CaCl_2$ (Smith and Stanford, 1971; Stanford and Smith, 1976; Stanford, 1977), cold 0.01 M $NaHCO_3$ (Mac-Lean, 1964; Smith, 1966), and hot 1 M or 2 M KCl (Øien and Selmer-Olsen, 1980; Whitehead, 1981) have given results that correlated closely with biological availability indices or with greenhouse tests of plant uptake of N. Total N extracted with hot 0.01 M $CaCL_2$, cold 0.01 M $NaHCO_3$, or UV absorbance of the extracts at 260 nm was correlated with N uptake by maize under field conditions (Fox and Piekielek, 1978a,b). A recent development is the use of the electro-ultrafiltration (EUF) technique, which extracts different fractions of N from soil at differing voltages and temperatures (Németh, 1979; Németh *et al.*, 1979; Kutscha-Lissberg and Prillinger, 1982).

The relatively large amounts of NH_4^+ or total N extracted by intermediate extractants such as alkaline permanganate often show inconsistent and generally poor relationships with biological indices of N availability (Keeney and Bremner, 1966a; Jenkinson, 1968; Stanford and Legg, 1968; Cornforth and Walmsley, 1971; Singh and Brar, 1973; Osborne and Storrier, 1976; Stanford, 1978) except in studies involving a narrow range of soils (Cornforth, 1968; Herlihy, 1972).

Strong extractants remove a substantial proportion of total soil N so that the amounts of N extracted by hot 6 M HCl (Keeney and Bremner, 1966b), boiling 4.5 M NaOH (Geist and Hazard, 1975), or $K_2Cr_2O_7$—H_2SO_4 oxidation (Sahrawat, 1982; Stanford, 1982) are closely correlated with total N content of soils. These methods remove considerably more N than could conceivably be mineralized in the short term and

Table VIII

A Range of Chemical Nitrogen Availability Indices

Extractant	Temperature	Time (hr)	Form of N measured	Reference[a]
Mild extractants				
Water	100°C	1	Total N	Keeney and Bremner (1966a)
0.01 M CaCl$_2$	100°C	64	Total or NH$_4^+$-N	Stanford (1968)
0.01 M CaCl$_2$	121°C	1	NH$_4^+$-N	Stanford and Demar (1969)
0.01 M NaHCO$_3$	Room	0.25	Total N or UV absorbance	Fox and Piekielek (1978a,b)
1 M KCl	100°C	1	NH$_4^+$- and NO$_3^-$	Whitehead (1981)
2 M KCl	80°C	20	NH$_4^+$- and NO$_3^-$-N	Øien and Selmer-Olsen (1980)
0.01 M CaCl$_2$	121°C	16	Soluble carbohydrate	Smith and Stanford (1971)
0.1 N Ba(OH)$_2$	Room	0.5	Soluble carbohydrate	Gasser and Kalembassa (1976)
Intermediate extractants				
Alkaline KMnO$_4$	100°C	0.25	NH$_4^+$-N	Keeney and Bremner (1966a)
Na$_2$CrO$_4$ + H$_3$PO$_4$	100°C	2	NH$_4^+$-N	Nommik (1976)
Neutral 0.5 N Na$_4$P$_2$O$_7$	100°C	6	NH$_4^+$-N	Stanford (1968)
1 M NaOH	100°C	0.5	NH$_4^+$-N	Stanford and Legg (1968)
1 M NaOH	Room	4.2	NH$_4^+$-N	Cornforth and Walmsley (1971)
1 M HCl	Room	26	NH$_4^+$-N	Cornforth and Walmsley (1971)
1 M H$_2$SO$_4$	Room	1	NH$_4^+$-N	Stanford and Smith (1978)
Strong extractants				
6 N H$_2$SO$_4$	Room	28	NH$_4^+$-N	Gallagher and Bartholomew (1964)
4.5 M NaOH	NaOH distillation		NH$_4^+$-N	Geist and Hazard (1975)
K$_2$Cr$_2$O$_7$—H$_2$SO$_4$	Walkley–Black oxidation		NH$_4^+$-N	Sahrawat (1982)

[a] For a more complete list of workers who used these and other tests see Bremner (1965), Keeney (1982b), and Stanford (1982).

generally are not reliable indices of N availability (Stanford, 1982). Conventional chemical fractionation of soil N (i.e., fractionation of the acid hydrolysate into amino acid N, hexosamine N, and unidentified N) has been found to be of little or no value for testing soil for available N (e.g., Keeney and Bremner, 1966b; Kadirgamathoiyah and MacKenzie, 1970; Osborne, 1977).

3. Applicability of Indices

None of the proposed biological or chemical indices for estimating available N in soils have been implemented for general use. Most indices have shown satisfactory correlations with greenhouse results of N uptake with a variety of crop species. However, correlations between incubation indices and yield of plants are generally higher for greenhouse than for field studies (Gasser and Jephcott, 1964; Robinson, 1968a,b). Michrina *et al.* (1981) showed that five chemical indices of N availability (organic matter content, total N, boiling $CaCl_2$ extractable N, 0.01 M $NaHCO_3$ extractable N, and UV absorbance of $NaHCO_3$ extract) were highly correlated with crop uptake of N in the greenhouse but there was no significant correlation between field results and the five indices. Some investigations found satisfactory correlations between biological indices (e.g., Jenkinson, 1968; Walmsley and Forde, 1976; Quin *et al.*, 1982; Steele *et al.*, 1982) or chemical indices (Joshi *et al.*, 1973; Fox and Piekielek, 1978a,b; Whitehead *et al.*, 1981) and field results. Relationships between indices and field results can, however, vary from year to year. Fox and Piekielek (1978a,b) found several chemical indices that were well correlated with N-supplying capacity of field soils to corn over a two-year period but, following further testing over a four-year period with a wider range of weather and soil conditions, none of the indices were sufficiently accurate to use for routine testing (Fox and Piekielek, 1983). An anaerobic incubation method was no better correlated with field results than were the chemical indices (Fox and Piekielek, 1984).

Under field conditions the major problem is that variations in climatic and cultural conditions affect plant growth (and N uptake) as well as the rate of N mineralization. The most accurate N availability index will give a good indication of the relative sizes of the pools of readily mineralizable N in different soils. It gives no indication of what proportion of that mineralizable pool will be mineralized during the growing season. Interpretations of availability indices therefore must vary from year to year. Adjusting availability indices by estimating the amount of mineralizable N mineralized during the growing season (using mean weekly or monthly soil moisture content and temperature) shows some promise (Stanford, 1978; Whitehead *et al.*, 1981) but does not seem practical for a simple soil test procedure.

In some cases, for example, when comparing tilled versus untilled fields, availability indices can give unrealistic results. In untilled fields there is generally an accumulation of organic matter and N in the surface soil and this is accompanied by increased values for N availability indices as compared to cultivated soils (e.g., Broder *et al.*, 1984). However, N requirements of crops are often higher without tillage than under conventional tillage (see Section III,D) since net immobilization of N often occurs under no tillage (Kitur *et al.*, 1984) while net mineralizaton or balanced organic matter turnover frequently occur under conventional tillage.

Since N availability indices have not been shown to be closely related to field results and few are applicable as routine soil testing procedures (particularly incubation methods), they have not gained wide acceptance. A 1978 survey of approaches to N fertilizer recommendations in the United States showed that only one state used an incubation method and one a chemical method to estimate N availability (Keeney, 1982b).

C. Modeling Approaches

Some investigators have attempted to predict N availability in the field by measuring certain climatic variables (e.g., rainfall, soil moisture content, and soil temperature) and relating these to the rate of mineralization (Campbell *et al.*, 1975; Hart and Goh, 1980). Such an approach, incorporating the effects of soil moisture and temperature on mineralization, has been used in developing a simple model to predict fertilizer N requirements of crops (Stanford, 1973, 1977). The model incorporates the N requirement for an attainable crop yield, N mineralized during cropping, residual mineral N, and efficiency of uptake by the crop:

$$N_f = [N_c - (N_i + N_m)]/E$$

where N_f = amount of fertilizer N required; N_c = nitrogen uptake by a crop associated with an economically optimum yield; N_i = initial amount of mineral N in the soil profile; N_m = estimated quantity of N mineralized in the growing season; and E = the efficiency or fractional recovery of N by the plant.

The amount of N mineralized is based on estimation of the potentially mineralizable N (N_0) by incubation or the amount of NH_4^+ released after autoclaving the soil in 0.01 M $CaCl_2$. These values are then modified by taking into account weekly or monthly variation in temperature and soil moisture content (Stanford, 1977). The above approach has been used with reasonable success to experimentally estimate fertilizer N requirements of sugar beet (Carter *et al.*, 1974, 1975; Stanford, 1977) and wheat

(Oyanedel and Rodriguez, 1977; Prado and Rodriguez, 1978). More recently, Myers (1984) developed a static model, based on a series of empirical relationships incorporating the major components (N_c, N_i, N_m, and E) of the above equation, which predicts the fertilizer N required for maximum and optimum economic yield. Such models may be of value in the future for making fertilizer recommendations.

Dynamic simulation models have also been developed to describe crop responses to applied N (e.g., Scaife, 1974; Barnes *et al.*, 1976; Frissel and Van Veen, 1981; Greenwood *et al.*, 1984). Such models often give reliable predictions but require numerous inputs and therefore the determination of many physical and biological parameters.

A relatively simple dynamic model was developed by Greenwood *et al.* (1984) for calculating day-to-day increments in N uptake and dry matter of crops with different levels of N fertilizers. A flow diagram of the various processes incorporated in the model is shown in Fig. 12. Inputs required by the model are the initial distribution of NO_3^- down the profile, the maximum crop yield, the maximum depth of rooting, and the mineralization rate. Net mineralization of N is assumed to be proportional to the mass of organic N in the soil and to take place at a constant rate irrespective of changes in temperature or soil moisture content. Agreement between predicted and experimental results was generally good (Greenwood *et al.*, 1984).

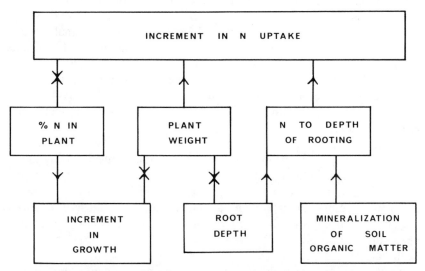

Fig. 12. Flow diagram of the various processes incorporated into a simulation model for calculating day-to-day increments in nitrogen uptake and dry matter of crops grown with different levels of nitrogenous fertilizer. [From Greenwood *et al.* (1984).]

Relatively simple N prediction models, such as that of Greenwood *et al.* (1984), can be incorporated into wider dynamic N simulation models that facilitate the linking of soil N cycling processes, environmental parameters, management factors, and plant yield (Duffy *et al.*, 1975; Hauck and Tanji, 1982; Tanji, 1982).

D. Fertilizer Nitrogen Recommendations

1. Rationale for Predicting Nitrogen Requirements

The capacity of the soil to supply N for crop use is determined by several factors: (1) the amount of residual mineral N present in the potentially active root zone early in the season before crop growth commences; (2) the amount of potentially mineralizable N present in the soil; (3) the proportion of the potentially mineralizable pool of soil N that is mineralized during the growing season; and (4) the amount of residual and mineralized N that is subsequently immobilized or lost from the plant–soil system by leaching or gaseous losses.

Factor (1) can be measured reasonably easily following field sampling, and a relative estimate of factor (2) can be made by a large number of laboratory methods. The amount of mineralizable N actually mineralized can be roughly simulated experimentally using mean weekly or monthly soil temperature and moisture values, but such an approach is not suitable for routine predictive soil testing. In reality, factors (2) and (3) are usually accounted for by using more or less subjective methods that make provision for soil type, productivity, previous cropping history, and additions of organic residues and manures (Keeney, 1982b; Needham, 1982). Losses and transformations of N noted under (4) depend on environmental and cultural factors and are usually corrected for subjectively, often after taking into account the results of field trials carried out in the locality being considered.

For making fertilizer N recommendations, the requirement of the crop for N is obviously important. The fertilizer requirement of a crop under field conditions is influenced by several factors: (1) the requirement of the crop for N as determined by its yield and N content; (2) the effectiveness of a crop to recover available mineral N from the soil profile as affected by stage of growth and root distribution; and (3) the availability of fertilizer N when added to supplement soil N supply as influenced by immobilization, leaching, and gaseous losses.

In practice, for high-yielding agricultural crops, the major determinant of the appropriate rate of fertilizer N is often the particular crop and its yield (Olson and Kurtz, 1982). As noted in Section II,A, crop plants differ greatly in the amount of N they accumulate. The effectiveness of crop recovery of available N (2) and the transformations and losses of applied fertilizer N (3) are affected by many diverse factors and may be more or less accounted for using local knowledge.

2. Practical Recommendations

In arid areas, where summer leaching of NO_3^- is minimal, fertilizer recommendations are often based on (1) expected yields, (2) estimates of the amount of N required for different yield levels, and (3) the amount of residual NO_3^- in the soil (Keeney, 1982b; Olson and Kurtz, 1982). The availability of residual NO_3^- to the crop is normally considered to be equal to that of fertilizer N. Fertilizer is applied to bring the sum of residual mineral N in the soil profile plus added fertilizer N to a predetermined desired value. The desired value can be adapted to individual site conditions (e.g., soil and climatic conditions, expected mineralization rate, and expected yields) and attempts to use the residual mineral N method over a wide range of soil and climatic conditions have generally been unsuccessful (Becker and Aufhammer, 1982; Needham, 1982). In some states of the United States soil organic matter content, either directly or indirectly, is used as a rough indicator of mineralizable N. An example of N recommendations for grain sorghum in Nebraska based on expected attainable yield, residual NO_3^- in the rooting profile, and total N content of the soil is shown in Table IX. In regions where manures and legumes constitute important sources of N, allowances are also made for their contribution (Keeney, 1982b).

In humid regions, where soils are less likely to contain or retain residual NO_3^-, fertilizer recommendations can be made by multiplying the expected yield by a factor that can be adjusted to take into account the economic optimum as well as such factors as expected mineralization rate and the use of manures or legumes (Keeney, 1982b; Olson and Kurtz, 1982). As noted in Section III,B, in the United Kingdom fertilizer recommendations are based on a subjective soil N index that mainly takes account of previous cropping history and manural practice. Recommendations are further modified by other site factors such as soil texture and depth, soil structure, tillage method, and amount of winter rainfall, as well as yield potential of the crop (Needham, 1982).

As Keeney (1982b) concluded, it seems that given natural soil and weather variability, along with on-farm variability, acceptable recommendations can be made with more or less subjective approaches, taking into account expected crop uptake and residual profile NO_3^- where applicable. Although a large amount of time and effort has been spent on developing a laboratory N availability index, none has gained wide acceptance. The

Table IX

Nitrogen Recommendations for Sorghum Crops in Nebraska Based on Residual Nitrate in the Soil Profile to 180 cm at Planting and Soil Organic Matter Content[a]

Soil NO_3^- content (kg N ha^{-1})	Grain sorghum with yield objective in kg N ha^{-1}				
	2,025	4,050	6,075	8,100	10,125
Recommended N rate kg N ha^{-1}					
Soil organic matter content = 1%					
<56	10	56	90	125	170
112	0	0	35	65	110
168	0	0	0	10	56
>168	0	0	0	0	0
Soil organic matter content = 2%					
<56	0	35	65	100	145
112	0	0	10	45	90
168	0	0	0	0	35
>168	0	0	0	0	0
Soil organic matter content = 3%					
<56	0	10	45	80	125
112	0	0	0	20	65
168	0	0	0	0	10
>168	0	0	0	0	0

[a] From Olson and Kurtz (1982). Reproduced from "Nitrogen in Agricultural Soils" (F. J. Stevensen, ed.) Agronomy Mono. **22** p. 593 by permission of the American Society of Agronomy, Madison, Wisconsin.

additional expense of a laboratory index is not apparently worthwhile and it does not seem likely to be so in the foreseeable future.

V. NITROGEN SUPPLY FOR CROPS

Most modern high-yielding agricultural systems depend on the addition of fertilizer N or an efficient *Rhizobium*–legume symbiosis for their N supply. A varying portion of crop N can also be derived from mineralization of soil organic matter. A major reason for the high N requirement of modern agricultural systems is that the harvested crop is fed to animals or people located elsewhere and little if any of the organic wastes from domestic or farm sources are returned to the soil. Sewage, for example, contains large amounts of N and much of this is discharged into rivers, lakes, and oceans (Cooke, 1977; National Research Council, 1978).

The roles of fertilizer N and biologically fixed N_2 in crop production are outlined below.

A. Nitrogenous Fertilizers

Fertilizer N is applied to approximately 11% of the earth's land surface but the amount is determined largely by the type and productivity of the agricultural system to which the fertilizer is applied (Hauck, 1981). Humans are the major beneficiaries of the use of fertilizer N since they consume, either directly or indirectly, the crop plants that grow more productively after N applications. In 1981–1982 world consumption of fertilizer N was 60.4 Tg N (Food and Agriculture Organization, 1982).

1. Requirements

As shown in Section II, amounts of N utilized by different crops differ greatly. For high yields of grain crops such as rice or wheat they must absorb more than 200 kg N ha^{-1} (N content of tops plus roots). In warmer climates, potential yield, and consequently N uptake, is higher because of the longer growing season and the fact that C4 plants, with their efficient photosynthetic pathways, can be grown in place of C3 plants. Thus, for high yields of C4 crops, such as corn or sugarcane, uptake in excess of 300 kg N ha^{-1} may be required (Greenwood, 1982).

The amount of soil organic N mineralized varies but if a surface soil contains 0.08–0.40% total N and 1–3% of this is mineralized in a growing season then 8–120 kg N ha^{-1} would be made available for potential crop use. This amount of N is not sufficient for high yields of many crops and furthermore it is released slowly, whereas the crop usually requires most of its N early in the season during the period of rapid growth. Thus, there is usually a need for application of fertilizer N to attain maximum yield of modern high-yielding crops.

2. Origin and Production

Only a small proportion of the N fertilizer consumed globally is mined from N-containing deposits such as Chile saltpeter ($NaNO_3$). The bulk of commercial N fertilizers is produced synthetically from atmospheric N_2 via NH_3 synthesis. Almost all NH_3 is produced in anhydrous form using the Haber–Bosch process (Pesek et al., 1971; Rankin, 1978), which involves a catalytic reaction between molecular nitrogen (N_2) and hydrogen that is liberated from a hydrocarbon source such as natural gas or water gas:

$$N_2 + 3H_2 \rightleftharpoons 2NH_3$$

The reaction takes place at a pressure of 200–1000 at 500°C. The reaction is exothermic but production of the starting materials requires considerably larger quantities of energy.

Since N reserves are large (the earth's atmosphere consists of almost 80% N) the quantities of fertilizer N produced depend mainly on energy requirements. Production of 1 kg of fertilizer N requires about 90 MJ of energy (Lewis and Tatchell, 1978), which corresponds to the caloric value of 2 kg of oil (White, 1977). However, Greenwood (1982) calculated that if all the N in crops were derived from fertilizer with an efficiency of only 25% it would require 64×10^9 kg fertilizer N yr^{-1} to grow enough food for the global population. The energy required for its manufacture would be equivalent to only about 2% of the world annual fossil fuel consumption. Thus, at least on a global basis, shortages of energy are not a serious constraint to the present-day production and use of N fertilizer.

Anhydrous NH_3 is the most concentrated fertilizer available (82% N) and is the raw material for all other synthetic nitrogenous fertilizers. The liquefied gas, anhydrous NH_3, requires special equipment for storage, transportation, and application and consequently numerous solid or dry-type fertilizers have been developed. With NH_3 and by-products produced by an NH_3 plant it is possible to manufacture urea and NH_4NO_3. With additional raw materials many nitrogenous fertilizers can be produced. Some common nitrogenous fertilizers are listed in Table X. Urea is fast becoming the most important solid fertilizer in world agriculture (Engelstad and Hauck, 1974; Beaton, 1978).

Table X

Chemical Formulae and Nitrogen Content of Some Common Nitrogenous Fertilizers

Fertilizer material	Chemical formula	Nitrogen content (%)
Gaseous ammonia	NH_3	82
Aqua ammonia	NH_3, NH_4OH	24
Ammonium sulfate	$(NH_4)_2SO_4$	21
Ammonium nitrate	NH_4NO_3	35
Lime ammonium nitrate	$NH_4NO_3 + CaCO_3$	26
Ammonium sulfate nitrate	$(NH_4)_2SO_4 \cdot NH_4NO_3$	26
Ammonium chloride	NH_4Cl	26
Urea	$CO(NH_2)_2$	46
Calurea	$CO(NH_2)_2 + NH_4NO_3$	34
Calcium cyanamide	$CaCN_2$	21
Calcium nitrate	$Ca(NO_3)_2$	15
Sodium nitrate (Chile saltpeter)	$Na(NO_3)_2$	16

3. Recovery of Applied Fertilizer Nitrogen

Nitrogen budgets have been constructed for many different agricultural systems utilizing both conventional and ^{15}N tracer techniques (Allison, 1966; Hauck, 1971, 1973; Zamyatina, 1971; Broadbent and Carlton, 1978; Keeney, 1982a; Legg and Meisinger, 1982). The final N balance sheet is the product of many physical, chemical, and biological processes interacting with one another over time. Thus, the fate of applied N varies widely depending on soils, crops, cropping system, and environmental factors. A number of N balances for various cropping systems are shown in Table XI. The data indicate that plant recovery of applied N varies greatly (8–75%) and that N unaccounted for (lost through NO_3^- leaching or gas-

Table XI

Some Estimates of Plant Recovery, Soil Recovery, and Balance of Applied Nitrogen Accounted for and Unaccounted for in Various Crops[a]

Cropping system	Percentage of applied nitrogen				Source
	Plant recovery	Soil recovery	Balance (accounted portion)	Unaccounted losses	
Cropland	50–75	na	60–111	0–40	Hauck (1971)
Oats (podzolic soils)	51–71	8–22	100*	14–35	Zamyatina (1971)
Maize (podzolic soils)	61	19	100*	20	Zamyatina (1971)
Oats (Chernozem)	32	45	100*	23	Zamyatina (1971)
Winter rye (Chernozem)	68	12	100*	20	Zamyatina (1971)
Corn	30–68	33–62	100*	16–25	Broadbent and Carlton (1978)
Wheat (spring)	22–29	34–65	100*	6–16	Feigenbaum et al. (1983)
Wheat (winter)	27–33	47–54	99	1	Olson and Swallow (1984)
Pasture	50–70	na	70–90	25–35	Hauck (1971)
Wheatgrass	27–67	29–49	70–95	5–30	Power and Legg (1984)
Rhodegrass pasture	8–27	16–27	98	51–65	Catchpoole (1975)
Paddy	26–45	na	na	20–50	Hauck (1971)
Paddy	51–61	24–27	78–85	15–22	Reddy and Patrick (1978)

[a] na = Not available; * refers to calculated value only.

eous losses) can be large (up to 65%). Kundler (1970) generalized that ranges of N recovery were: incorporated into soil organic matter, 10–40%; leaching loss, 5–10%; gaseous loss, 10–30%; and crop recovery, 30–70%. Such values are useful generalizations but cannot be applied to specific agricultural systems. They do, however, illustrate that crop recovery of applied N often is in the range of 30 to 70%.

Practices that increase the efficiency of fertilizer N use by crop plants are therefore of great importance from both the agronomic and environmental standpoints. The effects of form, placement, timing, and rate of fertilizer application on N response of crops have already been discussed (Section III), while factors affecting plant uptake, mineralization–immobilization, and losses of N via NO_3^- leaching and gaseous pathways were the subjects of previous chapters. Recent developments aimed at reducing losses of applied N and increasing fertilizer use efficiency by crops include the use of slow-release fertilizers and use of nitrification and urease inhibitors in conjunction with conventional fertilizers.

4. Recent Developments

a. Slow-release fertilizers. An obvious method of increasing fertilizer efficiency is to provide plants with a continuous supply of N at a rate that matches their physiological needs. This concept has led to the development of a wide range of slow-release fertilizers (Hauck and Koshino, 1971; Prasad *et al.,* 1971; Hauck, 1972; Parr, 1973; Maynard and Lorenz, 1979; Oertli, 1980).

Three main methods are commonly used to regulate the rate of N release from nitrogenous fertilizers (Table XII), including (1) use of a low-solubility coating over a soluble fertilizer (e.g., osmocote and sulfur-coated urea), (2) manufacture of compounds of low or limited solubility (e.g., magnesium ammonium phosphate and isobutylidene-diurea), and (3) manufacture of low-solubility compounds that depend on microbial activity for N release (e.g., urea formaldehyde).

The use of slow-release fertilizers has often resulted in improved yield or performance of plants in comparison with plants given conventional fertilizers applied once or as split applications (Maynard and Lorenz, 1979; Oertli, 1980). The difference is thought to be the result of decreased losses and increased uptake of fertilizer N from the slow-release compounds (Hauck, 1981). For example, lysimeter studies with corn fertilized with 240 kg N ha^{-1} showed leaching losses of NO_3^- with sulfur-coated urea (SCU) to be similar to those of unfertilized control plots and one-fourth of those for plots treated with urea (Nnadi, 1975). Greater residual effect of sulfur-coated urea compared to conventional N fertilizers has

Table XII

Some Properties and Nitrogen Release of Some Slow-Release Fertilizers[a]

Method used to control nitrogen release	Fertilizer	Nutrients supplied	Span of effectiveness
Release of soluble fertilizers controlled by a coating	Sulfur-coated urea (SCU)	N, S	Variable—about 1% daily
Release by diffusion or mass flow through cracks or pinholes in coat barrier	Osmocote	N or N, P, K	3 to 4, 8 to 9 months, or 1 to 2 years
Low-solubility characteristics	Magnesium ammonium phosphates (MagAmp)	N, P, Mg	About 100 days for coarse granules
	Isobutylidene-diurea (IBDU)	N	58% in 21 weeks with 1- to 2-mm-diam. particles
Low-solubility characteristics controlled by microbiol activity	Urea-formaldehyde (UF)	N	60% in 6 months for 75% insoluble material with 60% AI

[a] Compiled from Maynard and Lorenz (1979) and Oertli (1980).

also been reported (Oertli, 1975; Cox and Addiscott, 1976; Allen et al., 1978).

Because of their high cost, slow-release fertilizers have gained acceptance primarily for use on high-value horticultural crops (Oertli, 1980; Maynard and Lorenz, 1979), particularly container-grown ornamental plants and turfgrasses. Condensation products of urea and SCU have also received some acceptance in wetland rice-growing areas since their use can reduce losses of N through nitrification–denitrification and NH_3 volatilization (Sanchez et al., 1973; Mikkelsen and De Datta, 1980; Vlek et al., 1979). Slow-release fertilizers may have application for other crops if they are grown on coarse-textured soils subject to heavy leaching (Oertli, 1975; Liegel and Walsh, 1976; Allen et al., 1978). Economics will play a decisive role in whether slow-release fertilizers remain restricted to a few specialty crops or whether they find wide acceptance. Their widespread use is only likely to occur if relatively inexpensive forms are developed.

b. Nitrification inhibitors. As discussed in Chapter 3, in recent years considerable interest has been shown in the use of chemicals that inhibit nitrification. Such chemicals generally inhibit the activity of the autotrophic NH_4^+ oxidizers, *Nitrosomonas, Nitrosospira,* and *Nitrosolobus.* Numerous diverse compounds (e.g., pyridines, pyrimidines, mercapto compounds, thiazoles, succinimates, triazine derivatives) are known to inhibit nitrification (Hauck and Koshino, 1971; Prasad *et al.,* 1971; Hauck, 1972; Huber *et al.,* 1977; Sahrawat, 1980; Slagen and Kerkhoff, 1984). Most field data have been obtained using 2-chloro-6-(trichloromethyl) pyridine (nitrapyrin) although five other compounds are currently in commercial production for general use. These are 2-amino-4-chloro-6-methyl pyrimidine (AM), sulfathiazole (ST), 2-mercaptobenzothiazole (MAST), thiourea, and dicyandiamide. A fungicide, 5-ethoxy-3-trichloromethyl-1,2,4-triadiazole (DWELL), can also be used as a nitrification inhibitor.

The effectiveness of nitrification inhibitors is subject to modification by many factors. For instance, the activity of nitrapyrin in soils is lessened by sorption onto organic matter, chemical hydrolysis, and volatilization (Goring, 1962a; Briggs, 1975; Keeney, 1980), while recovery of nitrifiers is more rapid in neutral and alkaline soils (Goring, 1962b; Hendrickson *et al.,* 1978a). Further, genera and strains of nitrifier organisms differ in their sensitivity to nitrapyrin (Belser and Schmidt, 1981) and under field conditions only a portion of the soil is treated with the chemical.

In view of the many factors influencing the effectiveness of nitrification inhibitors, it is not surprising that data on their effect on yield and quality of various crops are rather inconclusive (Hauck, 1972; Hendrickson *et al.,* 1978a; Onken, 1980; Touchton and Boswell, 1980; Keeney, 1982a; Slagen and Kerkhoff, 1984). Positive crop responses have often occurred in high rainfall areas, high rainfall years, or under irrigation when inhibitors have been applied to light-textured sandy soils that are subject to leaching (Hergert and Wiese, 1980; Onken, 1980; Papendick and Engibous, 1980). The use of nitrification inhibitors has increased N recovery by crops from both fall- (Nelson and Huber, 1980; Mouchova and Apltauer, 1983; Rodgers *et al.,* 1983; Malhi and Nyborg, 1984) and spring-applied N (Hergert and Wiese, 1980; Touchton and Boswell, 1980; Malhi and Nyborg, 1984; Sahota and Singh, 1984). Their use may also have potential under flooded rice culture, where losses of N through denitrification can be large (Prasad *et al.,* 1971). However, the performance of commercially available inhibitors under such conditions has not been outstanding (Rong-ye and Zhao-Liang, 1980; Savant and De Datta, 1982). Particularly on soils of high pH, the use of nitrification inhibitors can increase the amount of fertilizer N lost by NH_3 volatilization, hence counteracting any possible benefits from inhibition of nitrification (Liang-mo *et al.*, 1980; Rodgers, 1983).

As noted in Section II,C, nitrification inhibitors can be used in conjunction with NH_4^+ fertilizers to alleviate crop quality problems associated with accumulation of large amounts of NO_3^- or organic acids (e.g., oxalate) in plants (Sahrawat and Keeney, 1984).

c. Urease inhibitors. The use of urea as a fertilizer has increased greatly in the last decade but some problems are encountered with its use. Urea is rapidly hydrolyzed to $(NH_4)_2CO_3$ through soil urease activity and the accompanying rise in pH and accumulation of NH_4^+ can cause damage to germinating seedlings and young plants through NH_4^+ and/or NO_2^- toxicity (Engelstad and Hauck, 1974) as well as large losses of urea N by NH_3 volatilization (see Chapter 5).

An experimental approach to reducing such problems has been to find compounds that will inhibit urease hydrolysis when applied to soils in conjunction with urea fertilizers. A large number of such compounds have been found (Sahrawat, 1980; Mulvaney and Bremner, 1981). They include a vast array of compounds, the most important being quinones and polyhydric phenols, hydroxamic acid derivatives, heterocyclic mercaptans, substituted ureas, phenylureas, metallic compounds, boron-containing compounds, and organophosphorus and carbamate insecticides.

The activity of urease inhibitors varies for different compounds and is affected by several soil factors. Their effectiveness is often much greater on sands and sandy loams than on heavier-textured soils (Pugh and Waid, 1969; Bremner and Douglas, 1971, 1973; Martens and Bremner, 1984).

Some urease inhibitors such as phenylphosphorodiamidate (PPD) have given promising results in reducing gaseous losses of urea N as NH_3 in greenhouse and field research under both nonflooded and flooded soil conditions (e.g., Matzel *et al.*, 1978; Vlek *et al.*, 1980; Byrnes *et al.*, 1983).

B. Role of Symbiotic Dinitrogen Fixation

The global consumption of about 60 Tg N yr^{-1} can be compared with the estimated 44 Tg N yr^{-1} annually fixed in croplands (Burns and Hardy, 1975). The various associations between legumes and *Rhizobium* bacteria play the most important role in the supply of biologically fixed N_2. It is thought that by the end of the century the amount of fertilizer N added to croplands by man could equal the amount of N_2 fixed in all terrestrial ecosystems (Söderlund and Svensson, 1976).

Symbiotic N_2 fixation contributes to agricultural production in three major ways: (1) when legumes are used as green manures, (2) when the crop itself is a legume, and (3) when legumes are used as forage crops in permanent pastures or in cropping rotations. The amount of N_2 fixed by such crops varies considerably according to species and cultivar of legume, strain of rhizobium, effectiveness of plant–*Rhizobium* symbiosis, environmental conditions, soil fertility, and crop management (Parker *et al.*, 1977; Gibson, 1977). Estimates of the amounts fixed are often in the following ranges: pulse crops (e.g., peas, beans), 10–100 kg N ha^{-1}; soybeans, 40–150 kg N ha^{-1}; and forage legumes, 100–250 kg N ha^{-1} (LaRue and Patterson, 1981).

As already discussed in Section III,C, low rates of fertilizer N may be required, particularly in the initial growth stages, to obtain maximum N_2 fixation and yields from leguminous crops.

1. Leguminous Green Manure Crops

The type of cropping system that maximizes transfer of symbiotically fixed N to other crops is one that uses the legume as a green manure. This allows the whole legume plant to be plowed under but it is seldom profitable to grow legumes solely for their effect on soil fertility. The initial flush of mineralization from legume residues, which provides most of the N transfer, is directly related to the N concentration of the residues and to the quantity of residues. There is ample scope for green manure leguminous crops when cover crops are required for erosion control (see Chapter 5).

Leguminous plants such as vetches (*Vicia* spp.) and lupines (*Lupinus* spp.) used as green manures are useful for improving the soil fertility of newly reclaimed land. Lupines, for instance, are particularly useful in reclamation of poor sandy soils such as coastal sand dunes (Gadgil, 1971).

2. Leguminous Grain Crops

Economically, soybeans are the most important leguminous grain crop while peanuts (*Arachis hypogaea*) are second to soybeans in total world production (Döbereiner and Campelo, 1977). Both crops thrive at relatively high temperatures such as in the tropics or in regions with high summer temperatures. Peas and beans are the most important grain legumes in temperate regions (Mulder *et al.*, 1977).

Grain legumes are annual crops and their growth period is therefore relatively short. High demands are therefore made on the N_2-fixing system and apart from providing *Rhizobium* strains of maximum efficiency, soil and climatic conditions need to be optimal for maximum crop yields to be attained. Even under optimum conditions the amount of N_2 fixed is not always sufficient for optimum growth and supplemental N fertilizer is sometimes provided, particularly at sowing (see Section III,F).

Since a large proportion of the symbiotically fixed N_2 is removed during grain harvest, the contribution of grain legumes to N nutrition of succeeding crops is not always large (Graham and Chatel, 1983). For soybeans, for example, approximately 75% of their N content can be removed at harvest (Thomas and Gilliam, 1978). There is, however, convincing evidence for N transfer in a range of legume–cereal rotations (Agboola and Fayemi, 1972; Giri and De, 1975; Schroder and Hinson, 1975; Boundy, 1978; White *et al.*, 1978; Sanford and Hairston, 1984). Schroder and Hinson (1975) concluded that soybeans contributed the equivalent of about 20 kg N ha^{-1} of fertilizer N to subsequent rye crops while Rennie *et al.* (1982) estimated their contribution as up to 60 kg N ha^{-1}.

Intercropping (the growth of two or more crops simultaneously on the same field) is common practice in Third World countries (Willey, 1979a,b). The short growth period of the common grain legumes limits the N gained by companion crops grown in association (Graham and Chatel, 1983). Shading by the companion crop can reduce both yield and N_2 fixation by the legume (Francis, 1978; Wahua and Miller, 1978). However, substantial overall yield increases using legume intercrops have been recorded by a number of workers (Agboola and Fayemi, 1972; De *et al.*, 1978; Remison, 1978; Singh, 1981).

3. Leguminous Forage Crops

a. Ley farming. Forage legumes grown for cutting or grazing, either in monocultures or in mixed pasture leys, are an integral part of crop rotations in many parts of the world. In contrast to growing legumes solely for green manure, ley systems provide economic return as well as supplying legume N for use by the following crops. When legume leys are cut and much of the plant (containing a considerable proportion of legume N) is removed from the land, transfer of N to the subsequent crop is not likely to be large (Henzell and Vallis, 1977). Nonetheless, the culture of forage legumes cut for conservation is often beneficial to growth of the following crops (e.g., Hoyt and Leitch, 1983). For example, an alfalfa crop cut for conservation can appreciably reduce the N fertilizer requirement of a following grain crop (Hoyt and Henning, 1971; Tucker *et al.*, 1971; Asghari and Hanson, 1984).

In contrast to leys cut for conservation, only a small proportion of legume N is taken from grazed pastures in animal products. Most of the N in forage ingested by animals is returned to the pasture, albeit unevenly, via urine and dung (Henzell and Ross, 1973; Wilkinson and Lowrey, 1973). Thus, the contribution to the N nutrition of the subsequent crop is likely to be greater from grazed than from cut leys (Henzell and Vallis, 1977). Indeed, the use of grazed, legume-containing leys generally results in a considerable lowering of the fertilizer N requirement for following grain crops (White *et al.*, 1978; Holford, 1980; Tuohey and Robson, 1980; Watson *et al.*, 1980; Cooke, 1982; Marty *et al.*, 1982). In general, the longer the ley phase has been maintained the greater is its effect in reducing the N requirement of the following grain crop (Cooke, 1982). Holford (1980) found that a grazed lucerne ley of 3½ years duration eliminated the need for fertilizer N for wheat crops for 3 to 5 years.

b. Permanent pastures. Grass–legume pastures are used in many parts of the world because a greater total herbage yield may be obtained by growing a grass and legume in association rather than in individual swards without application of fertilizer N (Haynes, 1981). The use of legumes in pastures may also result in increased N content and digestibility and a high, well balanced mineral content of herbage, all of which are important in animal nutrition (Chestnutt and Lowe, 1970).

The grass component derives most of its N from the legumes. Under most conditions N is principally transferred to the grass component aboveground through legume leaf litter and green leaf trampled into the ground by grazing animals and through feces, and particularly urine, from animals grazing the legumes (Henzell and Vallis, 1977; Vallis, 1978). Belowground transfer can occur when roots and nodules die back following heavy defoliation or death of the legume (Vallis, 1978).

In pastures containing a significant proportion of legume, grasses frequently contain low concentrations of N (Ball *et al.*, 1978; Hoglund and Brock, 1978; Ball and Field, 1982) and the nonlegume component of such pastures is N deficient, at least during part of the growing season. On the other hand, herbaceous pasture legumes do not persist in significant proportions in association with grasses given an optimum supply of fertilizer N (Whitehead, 1970). The superior competitive ability of grasses for environmental factors such as other nutrients and light appears to be the major reason why grasses have to be restricted by seasonal N deficiency to maintain legume growth and N_2-fixing ability in mixed pastures (Henzell, 1981).

Pasture management techniques are used to maintain a balance between the legumes and grasses. Since low-growing legumes such as clovers are at a competitive disadvantage for light over the faster-growing taller grasses they are favored by heavy, close grazing or cutting (Brougham *et al.*, 1978; Gillard and Fisher, 1978; Jones and Jones, 1978; Haynes, 1981). Infrequent grazing results in a grass-dominant sward while overgrazing results in a close mat of white clover (Suckling, 1975).

VI. CONCLUSIONS

When supplies of soil water are adequate, N is most commonly the key limiting factor for crop production. Thus, considerably more N than any other element is supplied to crops as fertilizer and is removed from agricultural lands in harvested crops.

For some field crops the rate of N uptake can be extremely high (3–5 kg N ha^{-1} day^{-1}) during the rapid phase of growth. During the latter stages of development of grain crops large amounts of N, previously accumulated in vegetative plant parts, can be mobilized and translocated to developing grain. The quantities of N present in different crops at maturity varies greatly with species, cultivar, and environmental conditions and hence N requirements of crops also vary greatly. Tissue analysis (e.g., total N or NO_3^--N) can be used to identify nutritional conditions of deficiency, sufficiency, and excess in plants but time of sampling and plant part sampled must be specified since both factors can have significant effects on tissue N content.

When N is the major limiting factor, the simplest dry matter response of plants to applied N is that yield increases with increasing rates of N up to a maximum and then either stays constant or declines with further N addition. The data of an N response curve can be fitted to relatively simple mathematical models (e.g., linear, quadratic, or exponential functions) or more complex simulation models. The overall yield response to applied N is mediated by the effects of N on various yield components such as leaf number, leaf area, leaf area duration, grain number, and grain weight.

The predominant positive impact of fertilizer N on crop quality is in its enhancement of the protein content of crops. However, excess N tends to result in accumulation of NO_3^- and sometimes organic acids (e.g., oxalate) in plant parts. Both substances can be toxic to man and livestock if ingested in large amounts.

Many factors interact to influence the response of a crop to applied N. Among the most important are form, timing, and method of fertilizer application, N-supplying power of the soil (residual mineral N and potentially mineralizable N), availability of soil water, tillage method, influence of plant genotype (plant growth rate and/or capacity for uptake and metabolism of N), presence of leguminous crops, and disease and pest incidence.

Much time and effort have been devoted to developing laboratory indices of N availability. These can be divided into biological methods (aerobic and anaerobic incubations) and chemical extractants and they provide an estimate of the capacity of a soil to supply N to plants. Most indices have not shown close correlations with N uptake by crops under field conditions and none has gained wide acceptance. Given natural soil and weather variability, acceptable N recommendations can be made with more or less subjective approaches taking into account expected crop yields (and therefore N uptake) some subjective index of organic N status, previous cropping history, and residual profile NO_3^-, particularly in arid regions where summer leaching of NO_3^- is minimal.

In 1981–1982 the world consumption of fertilizer N was 60.4 Tg N, which contrasts with the estimated 44 Tg N yr^{-1} annually fixed in croplands by symbiotic N_2 fixation. The bulk of commercial N fertilizers is produced synthetically from atmospheric N_2 via the Haber–Bosch process of NH_3 synthesis. Of the fertilizer N applied to croplands approximately 30–70% is recovered by the crop so that there is ample scope for developing and implementing practices to increase efficiency of fertilizer use. Some recent developments include the commercial production of slow-release fertilizers and nitrification inhibitors and the experimental use of urease inhibitors.

The amount of N_2 fixed by leguminous crops varies considerably according to species and cultivar of legume, strain of *Rhizobium*, environmental conditions, and crop management practices. The amount of fixed N_2 transferred to subsequent crops is a function of the N concentration in the residues and the quantity of residues remaining. It is therefore greatest for green manures and least for grain crops and forage cut for conservation where a large proportion of the crop N is harvested.

REFERENCES

Agboola, A. A. (1972). The relationship between the yields of eight varieties of Nigerian maize and content of nitrogen, phosphorus and potassium in the leaf at flowering stage. *J. Agric. Sci.* **79**, 391–396.

Agboola, A. A. (1978). Influence of soil organic matter on cowpea's response to N fertilizer. *Agron. J.* **70**, 25–28.

Agboola, A. A., and Fayemi, A. A. A. (1972). Fixation and excretion of nitrogen by tropical legumes. *Agron. J.* **64**, 409–412.

Aktan, S. (1976). Nitrate nitrogen accumulation and distribution in the soil profile during a fallow grain rotation as influenced by different levels of soil profile moisture. M.S. Thesis, Oregon State University, Corvallis.

Alexander, J. T., Schmer, C. C., Orleans, L. P., and Cotton, R. H. (1954). The effect of fertilizer applications on leaf and yield of sugar beets. *Proc. Am. Soc. Sugar Beet Technol.* **8**, 370–379.

Allen, S. E., Terman, G. L., and Kennedy, H. G. (1978). Nutrient uptake by grass and leaching losses from soluble and S-coated urea and KCl. *Agron. J.* **70**, 264–268.

Allison, F. E. (1966). The fate of nitrogen to soils. *Adv. Agron.* **18,** 219–258.

Allos, H. F., and Bartholomew, W. V. (1959). Replacement of symbiotic fixation by available nitrogen. *Soil Sci.* **87,** 61–66.

Altman, D. W., McCuistion, W. L., and Kronstad, W. E. (1983). Grain protein percentage, kernel hardness, and grain yield of winter wheat with foliar applied urea. *Agron. J.* **75,** 87–91.

Anderson, E. L., Kamprath, E. J., and Moll, R. H. (1984). Nitrogen fertility affects on accumulation, remobilization, and partitioning of N and dry matter in corn genotypes differing in prolificacy. *Agron. J.* **76,** 397–404.

Anderson, F. N., Peterson, G. A., and Olson, R. A. (1972). Uptake patterns of 15N tagged nitrate by sugar beets as related to soil nitrate level and time. *J. Am. Soc. Sugar Beet Technol.* **17,** 42–48.

Arnon, I. (1975). Physiological principles of dryland crop production. *In* "Physiological Aspects of Dryland Farming" (U. S. Gupta, ed.), pp. 3–145. Ram, Delhi.

Asghari, M., and Hanson, R. G. (1984). Nitrogen, climate, and previous crop effects on corn yield and grain N. *Agron. J.* **76,** 536–542.

Aulakh, M. S., Rennie, D. A., and Paul, E. A. (1984). Gaseous nitrogen losses from soils under zero-till as compared with conventional-till management systems. *J. Environ. Qual.* **13,** 130–136.

Austin, R. B., Ford, M. A., Edrich, J. A., and Blackwell, R. D. (1977). The nitrogen economy of winter wheat. *J. Agric. Sci.* **88,** 159–167.

Baerug, R., Lyngstad, I., Selmer-Olsen, A. R., and Øien, A. (1973). Studies on soil nitrogen. I. An evaluation of laboratory methods for available nitrogen in soils from arable and ley-arable rotations. *Acta Agric. Scand.* **23,** 173–181.

Bailley-Fenech, G., Kpodar, M. P., Piquemal, M., and Latche, J. (1979). Reparttition de l'oxalate et reduction des nitrates dans les différents organes de *Fagopyrum esculentum. M. C. R. Hebd. Seances Acad. Sci., Ser. D* **288,** 327–330.

Baker, J. M., and Tucker, B. B. (1971). Effect of rates of N and P on the accumulation of NO_3^--N in wheat, oats, rye and barley on different sampling dates. *Agron. J.* **63,** 204–207.

Balasubramian, V., and Chari, A. V. (1983). Effect of irrigation scheduling on grain yield and nitrogen use efficiency of irrigated wheat at Kadawa and Bakura, northern Nigeria. *Fert. Res.* **4,** 201–210.

Balko, L. G., and Russell, W. A. (1980). Response of maize inbred lines to N fertilizer. *Agron. J.* **72,** 723–728.

Ball, P. R., and Field, T. R. O. (1982). Responses to nitrogen as affected by pasture characteristics, season and grazing management. *In* "Nitrogen Fertilizers in New Zealand Agriculture" (P. B. Lynch, ed.), pp. 65–76. N.Z. Inst. Agric. Sci., Wellington.

Ball, R., Molloy, L. F., and Ross, D. J. (1978). Influence of fertilizer nitrogen on herbage dry matter and nitrogen yields, and botanical composition of grazed grass-clover pasture. *N. Z. J. Agric. Res.* **21,** 47–55.

Bandel, V. A., Dzienia, S., Stanford, G., and Legg, J. O. (1975). N behaviour under no-till vs. conventional corn culture. I. First year results using unlabelled N fertilizer. *Agron. J.* **67,** 782–786.

Barker, A. V. (1975). Organic vs. inorganic nutrition and horticultural crop quality. *HortScience* **10,** 50–53.

Barker, A. V., and Maynard, D. N. (1972). Cation and nitrate accumulation in pea and cucumber plants as influenced by nitrogen nutrition. *J. Am. Soc. Hortic. Sci.* **97,** 27–30.

Barker, A. V., Peck, N. H., and MacDonald, G. E. (1971). Nitrate accumulation in vegetables. I. Spinach grown in upland soils. *Agron. J.* **63**, 126–129.

Barker, A. V., Maynard, D. N., and Mills, H. A. (1974). Variations in nitrate accumulation among spinach cultivars. *J. Am. Soc. Hortic. Sci.* **99**, 132–134.

Barnes, A., Greenwood, D. J., and Cleaver, T. J. (1976). A dynamic model for the effects of potassium and nitrogen fertilizers on growth and nutrient uptake of crops. *J. Agric. Sci.* **86**, 225–244.

Beachell, H. M., Khush, R. S., and Juliano, B. O. (1972). Breeding for high protein content in rice. *Rice Breed. Pap. Symp.*, 1971, pp. 419–428.

Beaton, J. D. (1978) Urea: Its popularity grows as a dry source of nitrogen. *Crops Soil* **30**, 11–14.

Beauchamp, E. G. (1977). Slow release N fertilizers applied in fall for corn. *Can. J. Soil Sci.* **57**, 487–496.

Beauchamp, E. G., Kannenberg, L. W., and Hunter, R. B. (1976). Nitrogen accumulation and translocation in corn genotypes following silking. *Agron. J.* **68**, 418–422.

Becker, F. A., and Aufhammer, W. C. (1982). Nitrogen fertilization and methods of predicting the N requirements of winter wheat in the Federal Republic of Germany. *Proc.—Fert Soc.* **211**, 33–66.

Below, F. E., Lambert, R. J., and Hageman, R. H. (1984). Foliar applications of nutrients on maize. I. Yield and N content of grain and stover. *Agron. J.* **76**, 773–777.

Belser, L. W., and Schmidt, E. L. (1981). Inhibitory effect of nitrapyrin on three genera of ammonia-oxidising nitrifiers. *Appl. Environ. Microbiol.* **41**, 819–821.

Benzian, B., and Lane, P. (1979). Some relationships between grain yield and grain protein of wheat experiments in south-east England and comparisons with such relationships elsewhere. *J. Sci. Food Agric.* **30**, 59–70.

Benzian, B., and Lane, P. (1981). Interrelationship between nitrogen concentration in grain, grain yield and added fertilizer nitrogen in wheat experiments of south-east England. *J. Sci. Food Agric.* **32**, 35–43.

Benzian, B., Darby, R. J., Lane, P., Widdowson, F. V., and Verstraeten, L. M. J. (1983). Relationship between N concentration of grain and grain yield in recent winter-wheat experiments in England and Belgium, some large yields. *J. Sci. Food Agric.* **34**, 685–695.

Bhangoo, M. S., and Albritton, D. J. (1976). Nodulating and non-nodulating Lee soybean isolines response to applied nitrogen. *Agron. J.* **68**, 642–645.

Bigeriego, M., Hauck, R. D., and Olson, R. A. (1979). Uptake, translocation and utilization of ^{15}N-depleted fertilizer in irrigated corn. *Soil Sci. Soc. Am. J.* **43**, 528–533.

Biggar, J. W. (1978). Spatial variability of nitrogen in soils. In "Nitrogen in the Environment" (D. R. Nielsen and J. G. MacDonald, eds.), Vol. 1, pp. 201–211. Academic Press, New York.

Blanc, D., Mars, S., and Otto, C. (1979). The effects of some exogenous and endogenous factors on the accumulation of nitrate ions by carrot root. *Acta Hortic.* **93**, 173–185.

Blevins, D. G., Hiatt, A. J., and Lowe, R. H. (1974). The influence of nitrate and chloride uptake on expressed cell sap pH, organic acid synthesis, and potassium accumulation in higher plants. *Plant Physiol.* **54**, 83–87.

Blevins, R. L., Thomas, G. W., and Cornelius, P. L. (1977). Influence of no-tillage and nitrogen fertilization on certain soil properties after five years of continuous corn. *Agron. J.* **69**, 383–386.

Blevins, R. L., Thomas, G. W., Smith, M. S., Frye, W. W., and Cornelius, P. L. (1983).

Changes in soil properties after 10 years non-tilled and conventionally tilled corn. *Soil Tillage Res.* **3,** 135–146.

Blue, W. G., and Eno, C. F. (1954). Distribution and retention of anhydrous ammonia in sandy soils. *Soil Sci. Soc. Am. Proc.* **12,** 157–164.

Boawn, L. C., Viets, F. G., Crawford, C. L., and Nelson, J. L. (1960). Effect of nitrogen carrier, nitrogen rate, zinc rate, and soil pH on zinc uptake by sorghum, potatoes and sugar beets. *Soil Sci.* **90,** 329–337.

Boundy, K. (1978). The exciting promise of lupins in crop rotations. *J. Agric. (Victoria, Aust.)* **76,** 8–9.

Bowerman, P., and Harris, P. B. (1974). The rate and time of application of nitrogen on continuous spring barley. *Exp. Husb.* **27,** 45–49.

Box, G. E. P., and Hunter, J. S. (1957). Multi-factor experimental designs for exploring response surfaces. *Ann. Math. Stat.* **28,** 195–241.

Boyd, D. A., Yuen, L. T. K., and Needham, P. (1976). Nitrogen requirement of cereals. 1. Response curves. *J. Agric. Sci.* **87,** 149–162.

Breimer, T. (1982). Environmental factors and cultural measures affecting the nitrate content of spinach. *Fert. Res.* **3,** 191–292.

Bremner, J. M. (1965). Nitrogen availability indexes. *In* "Methods of Soil Analysis" (C. A. Black, ed.), Vol. 2, pp. 1324–1345. Am. Soc. Agron., Madison, Wisconsin.

Bremner, J. M., and Douglas, L. A. (1971). Decomposition of urea phosphate in soils. *Soil Sci. Soc. Am. Proc.* **35,** 575–578.

Bremner, J. M., and Douglas, L. A. (1973). Effects of some urease inhibitors on urea hydrolysis in soils. *Soil Sci. Soc. Am. Proc.* **37,** 225–226.

Breteler, H., and Smit, A. L. (1974). Effect of ammonium nutrition on uptake and metabolism of nitrate in wheat. *Neth. J. Agric. Soil* **22,** 73–81.

Briggs, G. G. (1975). The behaviour of the nitrification inhibitor "N-Serve" in broadcast and incorporated applications to soil. *J. Sci. Food Agric.* **26,** 1083–1092.

Broadbent, F. E., and Carlton, A. B. (1978). Field trials with isotopically labeled nitrogen fertilizer. *In* "Nitrogen in the Environment" (D. R. Nielsen and J. G. MacDonald, eds.), Vol. 1, pp. 1–41. Academic Press, New York.

Broadbent, F. E., and Mikkelsen, D. S. (1968). Influence of placement on uptake of tagged nitrogen by rice. *Agron. J.* **60,** 674–677.

Brougham, R. W., Ball, P. R., and Williams, W. (1978). The ecology and management of white clover-based pastures. *In* "Plant Relations in Pastures" (J. R. Wilson, ed.), pp. 309–324. CSIRO, Brisbane, Australia.

Bullen, E. R., and Lessells, W. J. (1957). The effect of nitrogen on cereal yields. *J. Agric. Sci.* **49,** 319–328.

Burford, J. R., Dowdell, R. J., and Crees, R. (1981). Emission of nitrous oxide to the atmosphere from direct-drilled and ploughed clay soils. *J. Sci. Food Agric.* **32,** 219–223.

Burns, R. C., and Hardy, R. W. F. (1975). "Nitrogen Fixation in Bacteria and Higher Plants." Springer-Verlag, Berlin and New York.

Butler, G. W., Barclay, P. C., and Glenday, A. C. (1962). Genetic and environmental differences in the mineral composition of ryegrass herbage. *Plant Soil* **16,** 214–228.

Byrnes, B. H., Savant, N. K., and Craswell, E. T. (1983). Effect of a urease inhibitor phenylphosphorodiamidate on the efficiency of urea applied to rice. *Soil Sci. Soc. Am. J.* **47,** 270–274.

Campbell, C. A. (1978). Soil organic carbon, nitrogen and fertility. *In* "Soil Organic Matter" (M. Schnitzer and S. U. Kahn, eds.), pp. 173–271. Am. Elsevier, New York.

Campbell, C. A., Biederbeck, V. O., and Hinman, W. C. (1975). Relationships between nitrate in summer-fallowed surface soil and some environmental variables. *Can. J. Soil Sci.* **55**, 213–223.

Cannell, R. Q. (1982). Cereal root systems: Factors affecting their growth and function. *In* "Opportunities for Manipulation of Cereal Productivity" (A. F. Hawkins and B. Jeffcoat, eds.), pp. 118–129. ARC Letcombe Laboratory, Wantage, England.

Cantliffe, D. J. (1972a). Nitrate accumulation in vegetable crops as affected by photoperiod and light duration. *J. Am. Soc. Hortic. Sci.* **97**, 414–418.

Cantliffe, D. J. (1972b). Nitrate accumulation in spinach grown at different temperatures. *J. Am. Soc. Hortic. Sci.* **97**, 674–676.

Cantliffe, D. J., MacDonald, G. E., and Peck, N. H. (1974). Reduction of nitrate accumulation by molybdenum in spinach grown at low pH. *Commun. Soil Sci. Plant Anal.* **5**, 273–282.

Carson, P. L. (1975). Recommended nitrate-nitrogen tests. *N. D., Agric. Exp. Stn., Bull.* **449**, 13–15.

Carter, J. N., and Traveller, D. J. (1981). Effect of time and amount of nitrogen uptake on sugarbeet growth and yield. *Agron. J.* **73**, 665–671.

Carter, J. N., Jensen, M. E., and Bosma, S. M. (1974). Determining nitrogen fertilizer needs for sugar beets from residual nitrate and mineralizable nitrogen. *Agron. J.* **66**, 319–323.

Carter, J. N., Westermann, D. T., Jensen, M. E., and Bosma, S. M. (1975). Predicting nitrogen fertilizer needs for sugar beets from residual nitrate and mineralizable N. *J. Am. Soc. Sugar Beet Technol.* **18**, 232–244.

Carter, J. N., Westerman, D. T., and Jensen, M. E. (1976). Sugar beet yield and quality as affected by nitrogen level. *Agron. J.* **68**, 49–55.

Carter, M. R., and Rennie, D. A. (1984). Dynamics of soil microbial biomass N under zero and shallow tillage for spring wheat, using ^{15}N urea. *Plant Soil* **76**, 157–164.

Catchpoole, V. R. (1975). Pathways for losses of fertilizer nitrogen from a Rhodes grass pasture in southeastern Queensland. *Aust. J. Agric. Res.* **26**, 259–268.

Chandler, R. F. (1969). Plant morphology and stand geometry in relation to nitrogen. *In* "Physiological Aspects of Crop Yields" (J. D. Eastin, F. A. Haskins, C. Y. Sullivan, and C. H. M. Van Bavel, eds.), pp. 265–289. Am. Soc. Agron., Madison, Wisconsin.

Chandler, R. F. (1970). Overcoming physiological barriers to higher yields through plant breeding. *Potassium Symp.* **9**, 421–434.

Chavalier, P., and Schrader, L. E. (1977). Genotypic differences in nitrate absorption and partitioning of N among plant parts in maize. *Crop Sci.* **17**, 897–901.

Chesney, H. A. D. (1975). Fertilizer studies with groundnuts on the brown sands of Guyana. II. Effect of nitrogen, phosphorus, potassium, and gypsum and timing of phosphorus application. *Agron. J.* **67**, 10–13.

Chestnutt, D. M. B., and Lowe, J. (1970). Agronomy of white clover/grass swards. *In* "White Clover Research" (J. Lowe, ed.), Proc. Occas. Symp. B. Grassl. Soc., pp. 191–213. British Grassland Society, Grassland Research Institute, Berkshire, England.

Clark, R. B. (1983). Plant genotype differences in the uptake, translocation, accumulation, and use of mineral elements required for plant growth. *Plant Soil* **72**, 175–196.

Cooke, G. W. (1977). Waste of fertilizer. *Philos. Trans. R. Soc. London, Ser. B* **281**, 231–241.

Cooke, G. W. (1982). "Fertilizing for Maximum Yield." Granada, London.

Cornforth, I. S. (1968). The potential availability of organic nitrogen fractions in some West Indian soils. *Exp. Agric.* **4**, 193–201.

Cornforth, I. S., and Walmsley, D. (1971). Methods of measuring available nutrients in West Indian soils. *Plant Soil* **35**, 389–399.

Cox, D., and Addiscott, T. M. (1976). Sulfur-coated urea as a fertilizer for potatoes. *J. Sci. Food Agric.* **27**, 1015–1020.

Crawford, R. L., Kennedy, W. K., and Johnson, W. C. (1961). Some factors that affect nitrate accumulation in forages. *Agron. J.* **53**, 158–162.

Dahnke, W. C., and Vasey, E. H. (1973). Testing soils for nitrogen. *In* "Soil Testing and Plant Analysis" (L. M. Walsh and J. D. Beaton, eds.), pp. 97–114. Soil Sci. Soc. Am., Madison, Wisconsin.

Daigger, L. A., and Sander, D. H. (1976). Nitrogen availability to wheat as affected by depth of nitrogen placement. *Agron. J.* **68**, 524–526.

Daling, M. J., Boland, G., and Wilson, J. H. (1976). Relation between acid proteinase activity and redistribution of N during grain development in wheat. *Aust. J. Plant Physiol.* **3**, 721–730.

Dart, P. J., and Mercer, F. V. (1965). The effect of growth, temperature, level of ammonium nitrate, and light intensity on the growth and nodulation of cowpea (*Vigna sinensis* Endl. Ex Hassk.). *Aust. J. Agric Res.* **16**, 321–345.

Dart, P. J., and Wildon, D. C. (1970). Nodulation and nitrogen fixations by *Vigna sinensis* and *Vicia atropurpurea:* The influence of concentration, form, and site of application of combined nitrogen. *Aust. J. Agric. Res.* **21**, 45–56.

Darwinkel, A. (1976). Effect of sward age on nitrate accumulation in ryegrass. *Neth. J. Agric. Sci.* **24**, 266–273.

Dazzo, F. B., and Brill, W. J. (1978). Regulation by fixed nitrogen of host–symbiont recognition in the *Rhizobium*–clover symbiosis. *Plant Physiol.* **62**, 18–21.

De, R., Gupta, R. S., Singh, S. P., Pal, M., Singh, S. N., Sharma, R. N., and Kaushik, S. K. (1978). Interplanting maize, sorghum and pearl-millet with short duration grain legumes. *Indian J. Agric. Sci.* **48**, 132–140.

De Datta, S. K., Obeemea, W. N., and Jana, R. K. (1972). Protein content of rice grain as affected by nitrogen fertilizer and some triazines and substituted ureas. *Agron. J.* **64**, 785–788.

de Franca, G. E. (1981). Differences in dry-matter yield and the uptake, distribution, and use of nitrogen by sorghum genotypes. Ph.D. Thesis, University of Nebraska, Lincoln (*Diss. Abstr.* **41**, 4018B).

Deinum, B., and Sibma, B. (1980). Nitrate content of herbage in relation to nitrogen fertilization and management. *Proc. Int. Symp. Eur. Grassl. Fed. Role Nitrogen Intensive Grassl. Prod., 1980*, pp. 95–102.

de Wit, C. T., Dijkshoorn, W., and Noggle, J. C. (1963). Ionic balance and growth of plants. *Versl. Landbouwkd. Onderz.* No. 69.

Dijkshoorn, W. (1973). Organic acids and their role in ion uptake. *In* "Chemistry and Biochemistry of Herbage" (G. W. Butler and R. W. Bailey, eds.), Vol. 2, pp. 163–188. Academic Press, New York.

Dilz, K., Darwinkel, A., Boon, R., and Verstraeten, L. M. J. (1982). Intensive wheat production as related to nitrogen fertilization, crop protection and soil nitrogen: Experience in the Benelux. *Proc.—Fert. Soc.* **211**, 93–124.

Döbereiner, J., and Campelo, A. B. (1977). Importance of legumes and their contribution to tropical agriculture. *In* "A Treatise on Dinitrogen Fixation, Section IV. Agronomy and Ecology" (R. W. F. Hardy and A. H. Gibson, eds.), pp. 191–220. Wiley, New York.

Dobson, J. W., Beaty, E. R., and Fisher, C. D. (1978). Tall fescue yield, tillering and invaders as related to management. *Agron. J.* **70**, 662–666.

Dougherty, C. T., Love, B. G., and Mountier, N. S. (1978). Response surfaces of semidwarf wheat for seeding rate, and levels and times of application of nitrogen fertilizer. *N. Z. J. Agric. Res.* **21**, 655–663.

Dougherty, C. T., Love, B. G., and Mountier, N. S. (1979). Response surfaces of 'Kopara' wheat for seeding rate, and levels and times of application of nitrogen fertilizer. *N. Z. J. Agric. Res.* **22**, 47–54.

Douglas, J. T., and Goss, M. J. (1982). Stability and organic matter content of surface soil aggregates under different methods of cultivation and in grassland. *Soil Tillage Res.* **2**, 155–175.

Dowdell, R. J., and Cannell, R. Q. (1975). Effect of ploughing and direct drilling on soil nitrate content. *J. Soil Sci.* **26**, 53–61.

Dowdell, R. J., and Crees, R. (1980). The uptake of ^{15}N labelled fertilizer by winter wheat and its immobilization in a clay soil after direct drilling or ploughing. *J. Sci. Food Agric.* **31**, 992–996.

Dowdell, R. J., Crees, R., and Cannell, R. Q. (1983). A field study of effects of contrasting methods of cultivation on soil nitrate content during autumn, winter and spring. *J. Soil Sci.* **34**, 367–379.

Duffy, J., Chung, C., Boast, C., and Franklin, M. (1975). A simulation model of biophysico-chemical transformations of nitrogen in tile-drained corn belt soil. *J. Environ. Qual.* **4**, 477–486.

Eaglesham, A. R. J., Hassouna, S., and Seegers, R. (1983). Fertilizer-N effects on N_2 fixation by cowpea and soybean. *Agron. J.* **75**, 61–66.

Easson, D. L. (1984). The timing of nitrogen application for spring barley. *J. Agric. Sci.* **102**, 673–678.

Eck, H. V. (1984). Irrigated corn yield response to nitrogen and water. *Agron. J.* **76**, 421–428.

Ellen, J., and Spiertz, J. H. J. (1975). The influence of nitrogen and Benlate on leaf area duration, grain growth and pattern of N-, P- and K-uptake of winter wheat (*Triticum aestivum*). *Z. Acker- Pflanzenbau* **141**, 231–239.

Ellen, J., and Spiertz, J. H. J. (1980). Effect of rate and timing of nitrogen dressings on grain yield formation of winter wheat (*Triticum aestivum* L.). *Fert. Res.* **1**, 177–190.

Elliott, L. F., Cochran, V. L., and Papendick, R. I. (1981). Wheat residue and nitrogen placement effects on wheat growth in the greenhouse. *Soil Sci.* **131**, 48–52.

Ellis, F. B., Graham, J. P., and Christian, D. G. (1982). Interacting effects of tillage method, nitrogen fertilizer and secondary drainage on winter wheat production on a calcareous clay soil. *J. Sci. Food Agric.* **34**, 1068–1076.

Engelstad, O. P., and Hauck, R. D. (1974). Urea: Will it become the most popular nitrogen carrier? *Crops Soil* **26**, 11–14.

Eppenderfer, W. H. (1975). Effects of fertilizers on quality and nutritional value of grain protein. *Proc. Colloq. Int. Potash Inst.,* **11**, 249–263.

Evans, L. T., and Wardlaw, I. F. (1976). Aspects of the comparative physiology of grain yield in cereals. *Adv. Agron.* **28**, 301–359.

Fassett, D. W. (1973). Oxalates. *In* "Toxicants Occurring Naturally in Foods," 2nd ed, pp. 346–362. Natl. Acad. Sci., Washington, D.C.

Feigenbaum, S., Seligman, N. G., Benjamin, R. W., and Feinerman, D. (1983). Recovery of tagged fertilizer nitrogen applied to rainfed spring wheat (*Triticum aestivum* L.) subjected to severe moisture stress. *Plant Soil* **73**, 265–274.

Fenn, L. B., and Kissel, D. E. (1976). The influence of cation exchange capacity and depth of incorporation on ammonium volatilization from ammonium compounds applied to calcareous soils. *Soil Sci. Soc. Am. J.* **40**, 394–398.

Feyter, C., and Cossens, G. G. (1977). Effects of rates and methods of nitrogen application on grain yields and yield components of spring-sown wheats in South Otago, New Zealand. *N. Z. J. Exp. Agric.* **5**, 371–376.

Finney, K. F., Meyer, J. W., Smith, F. W., and Fryer, H. C. (1957). Effect of foliar spraying of Pawnee wheat with urea solutions on yield, protein content, and protein quality. *Agron. J.* **49**, 341–347.

Fjell, D. L., Paulsen, G. M., Walter, T. L., and Lawless, J. R. (1984). Relationship among nitrogen and phosphorus contents of vegetable parts and agronomic traits of normal- and high-protein wheats. *J. Plant Nutr.* **7**, 1093–1102.

Fleige, H., and Baeumer, K. (1974). Effect of zero-tillage on organic carbon and total nitrogen content and their distribution in different nitrogen fractions in loessial soils. *Agro-Ecosystems* **1**, 19–29.

Food and Agriculture Organization (1982). "FAO Fertilizer Yearbook," Vol. 32. FAO, Rome.

Fowler, J. L., and Hageman, J. H. (1978). Nitrogen fertilization of irrigated Russian-thistle forage. I. Yield and water use efficiency. *Agron. J.* **70**, 992–995.

Fox, R. H., and Piekielek, W. P. (1978a). Field testing of several nitrogen availability indexes. *Soil Sci. Soc. Am. J.* **42**, 747–750.

Fox, R. H., and Piekielek, W. P. (1978b). A rapid method for estimating the nitrogen-supplying capacity of a soil. *Soil Sci. Soc. Am. J.* **42**, 751–753.

Fox, R. H., and Piekielek, W. P. (1983). Response of corn to nitrogen fertilizer and the prediction of soil nitrogen availability with chemical tests in Pennsylvania. *Bull.— Pa., Agric. Exp. Stn.* **843**.

Fox, R. H., and Piekielek, W. P. (1984). Relationships among anaerobically mineralized nitrogen, chemical indexes, and nitrogen availability to corn. *Soil Sci. Soc. Am. J.* **48**, 1087–1090.

Franceschi, V. R., and Horner, H. T. (1980). Calcium oxalate crystals in plants. *Bot. Rev.* **46**, 361–427.

Francis, C. A. (1978). Multiple cropping potentials of beans and maize. *HortScience* **13**, 13–17.

Frederickson, J. K., Koehler, F. E., and Cheng, H. H. (1982). Availability of ^{15}N-labelled nitrogen in fertilizer and in wheat straw to wheat in tilled and no-till soil. *Soil Sci. Soc. Am. J.* **46**, 1212–1222.

Frissel, M. J., and Van Veen, J. A., eds. (1981). "Simulation of Nitrogen Behaviour of Soil–Plant Systems." Pudoc, Wageningen, The Netherlands.

Gadgil, R. L. (1971). The nutritional role of *Lupinus arboreus* in coastal sand dune forestry. 3. Nitrogen distribution in the ecosystem before tree planting. *Plant Soil* **35**, 113–126.

Gallagher, P. W., and Bartholomew, W. V. (1964). Comparison of nitrate production and other procedures in determining nitrogen availability in southeastern coastal plain soils. *Agron. J.* **56**, 179–184.

Garcia, L. R., and Hanway, J. J. (1976). Foliar fertilization of soybeans during the seed-filling period. *Agron. J.* **68**, 653–657.

Gascho, G. J., Hook, J. E., and Mitchell, G. A. (1984). Sprinkler-applied and side-dressed nitrogen for irrigated corn grown on sand. *Agron. J.* **76**, 77–81.

Gass, W. B., Peterson, G. A., Hauck, R. D., and Olson, R. A. (1971). Recovery of residual nitrogen by corn (*Zea mays* L.) from various soil depths as measured by ^{15}N tracer techniques. *Soil Sci. Soc. Am. Proc.* **35**, 290–294.

Gasser, J. K. R., and Iordanou, I. G. (1967). Effects of ammonium sulphate and calcium nitrate on the growth, yield and nitrogen uptake by barley, wheat and oats. *J. Agric. Sci.* **68**, 307–316.

Gasser, J. K. R., and Jephcott, B. M. (1964). Soil nitrogen. VIII. Some factors affecting correlations between measurements of soil-N status and crop performance. *J. Sci. Food Agric.* **15**, 422–428.

Gasser, J. K. R., and Kalembasa, S. J. (1976). Soil nitrogen. IX. The effects of leys and organic manures on the available-N in clay and sandy soils. *J. Soil Sci.* **27**, 237–249.

Geist, J. M. (1977). Nitrogen response relationships of some volcanic ash soils. *Soil Sci. Soc. Am. J.* **41**, 996–1000.

Geist, J. M., and Hazard, J. W. (1975). Total nitrogen using a sodium hydroxide index and double sampling theory. *Soil Sci. Soc. Am. Proc.* **39**, 340–343.

Geraldson, C. M., Klacan, G. R., and Lorenz, O. A. (1973). Plant analysis as an aid in fertilizing vegetable crops. *In* "Soil Testing and Plant Analysis" (L. M. Walsh and J. D. Beaton, eds.), pp. 365–379. Soil Sci. Soc. Am., Madison, Wisconsin.

Gibson, A. H. (1974). Consideration of the legume as a symbiotic association. *Proc. Indian Natl. Sci. Acad., Part B* **40**, 741–767.

Gibson, A. H. (1977). The influence of the environment and management practices on the legume–rhizobium symbiosis. *In* "A Treatise on Dinitrogen Fixation, Section IV. Agronomy and Ecology" (R. W. F. Hardy and A. H. Gibson, eds.), pp. 393–450. Wiley, New York.

Giles, J. F., Reuss, J. O., and Ludwick, A. E. (1975). Prediction of nitrogen status of sugar beets by soil analysis. *Agron. J.* **57**, 454–459.

Gillard, P., and Fisher, M. J. (1978). The ecology of Townsville stylo-based pastures in northern Australia. *In* "Plant Relations in Pastures" (J. R. Wilson, ed.), pp. 340–352. CSIRO, Brisbane, Australia.

Giri, G., and De, R. (1979). Effect of proceeding grain legumes on dryland pearl millet in NW India. *Exp. Agric.* **15**, 169–172.

Goh, K. M., and Ali, N. S. (1983). Effects of nitrogen fertilizers, calcium and water regime on the incidence of cavity spot in carrot. *Fert. Res.* **4**, 223–230.

Goh, K. M., and Kee, K. K. (1978). Effects of nitrogen and sulphur fertilization on the digestibility and chemical composition of perennial ryegrass (*Lolium perenne* L.). *Plant Soil* **50**, 161–177.

Goring, C. A. I. (1962a). Control of nitrification by 2-chloro-6-(trichloromethyl) pyridine. *Soil Sci.* **93**, 211–218.

Goring, C. A. I. (1962b). Control of nitrification of ammonium fertilizers and urea by 2-chloro-6-(trichloromethyl) pyridine. *Soil Sci.* **93**, 431–439.

Goss, M. J., Howse, K. R., and Harris, W. (1978). Effects of cultivation on soil water retention and water use by cereals in clay soils. *J. Soil Sci.* **29**, 475–488.

Graham, P. H., and Chatel, D. L. (1983). Agronomy. *In* "Nitrogen Fixation" (W. J. Broughton, ed.), Vol. 3, pp. 56–98. Oxford Univ. Press (Clarendon), London and New York.

Graham, R. D., Geytenbeck, P. E., and Radcliffe, B. C. (1983). Responses of triticale, wheat, rye and barley to nitrogen fertilizer. *Aust. J. Exp. Agric. Anim. Husb.* **23**, 73–79.

Greenwood, D. J. (1982). Nitrogen supply and crop yield: The global scene. *Plant Soil* **67**, 45–59.

Greenwood, D. J., Wood, J. T., Cleaver, T. J., and Hunt, J. (1971). A theory for fertilizer response. *J. Agric. Sci.* **77**, 511–523.

Greenwood, D. J., Cleaver, T. J., and Niendorf, K. B. (1974). Effect of weather conditions on the response of lettuce to applied fertilizers. *J. Agric. Sci.* **82**, 217–232.

Greenwood, D. J., Cleaver, T. J., Turner, M. K., Hunt, J., Niendorg, K. B., and Loquens, S. M. H. (1980). Comparison of the effects of nitrogen fertilizer on the yield, nitrogen content and quality of 21 different vegetable and agricultural crops. *J. Agric. Sci.* **95**, 471–485.

Greenwood, D. J., Draycott, A., Last, P. J., and Draycott, A. P. (1984). A concise simulation model for interpreting N-fertilizer trials. *Fert. Res.* **5**, 355–369.

Gregg, P. E. H. (1976). Field investigations into the fate of fertilizer sulphur added to pasture–soil systems. Ph.D. Thesis, University of Canterbury, New Zealand.

Halloran, G. M., and Lee, J. W. (1979). Plant nitrogen distribution in wheat cultivars. *Aust. J. Agric. Res.* **30**, 779–789.

Ham, G. E., and Caldwell, A. C. (1978). Fertilizer placement effects on soybean yield, N_2 fixation and ^{33}P uptake. *Agron. J.* **70**, 779–783.

Hanway, J. J. (1962a). Corn growth and composition in relation to soil fertility. I. Growth of different plant parts and relation between leaf weight and grain yield. *Agron. J.* **54**, 143–146.

Hanway, J. J. (1962b). Corn growth and composition in relation to soil fertility. II. Uptake of N, P and K and their distribution in different plant parts during the growing season. *Agron. J.* **54**, 217–222.

Hanway, J. J. (1962c). Corn growth and composition in relation to soil fertility. III. Percentages of N, P, and K in different plant parts in relation to stage of growth. *Agron. J.* **54**, 222–229.

Harder, H. J., Carlson, R. E., and Shaw, R. H. (1982). Corn grain yield and nutrient response to foliar fertilizers applied during grain fill. *Agron. J.* **74**, 106–110.

Harmsen, K. (1984). Nitrogen fertilizer use in rainfed agriculture. *Fert. Res.* **5**, 371–382.

Harper, F. (1983). "Principals of Arable Crop Production." Granada, London.

Hart, P. B. S., and Goh, K. M. (1980). Regression equations to monitor inorganic nitrogen changes in fallow and wheat soils. *Soil Biol. Biochem.* **12**, 147–151.

Hart, P. B. S., Goh, K. M., and Ludecke, T. E. (1979). Nitrogen mineralization in fallow and wheat soils under field and laboratory conditions. *N. Z. J. Agric. Res.* **22**, 11–25.

Hauck, R. D. (1971). Quantitative estimates of nitrogen-cycle processes: Concepts and review. *In* "Nitrogen-15 in Soil Plant Studies," pp. 65–80. IAEA, Vienna.

Hauck, R. D. (1972). Synthetic slow-release fertilizers and fertilizer amendments. *In* "Organic Chemicals in the Soil Environment" (C. A. I. Goring and J. M. Hamaker, eds.), pp. 633–690. Dekker, New York.

Hauck, R. D. (1973). Nitrogen tracers in nitrogen cycle studies—Past use and future needs. *J. Environ. Qual.* **2**, 317–327.

Hauck, R. D. (1981). Nitrogen fertilizer effects on nitrogen cycle processes. *In* "Terrestrial Nitrogen Cycles: Processes, Ecosystem Strategies and Management Impacts" (F. E. Clark and T. Rosswall, eds.), pp. 551–562. Ecological Bulletins, Stockholm.

Hauck, R. D., and Koshino, M. (1971). Slow release and amended fertilizers. *In* "Fertilizer Technology and Use" (R. A. Olson, ed.), pp. 455–484. Soil Sci. Soc. Am., Madison, Wisconsin.

Hauck, R. D., and Tanji, K. K. (1982). Nitrogen transfers and mass balances. *In* "Nitrogen in Agricultural Soils" (F. J. Stevenson, ed.), pp. 891–925. Am. Soc. Agron., Madison, Wisconsin.

Haynes, R. J. (1981). Competitive aspects of the grass–legume association. *Adv. Agron.* **33**, 227–261.

Haynes, R. J. (1985). Principles of fertilizer use for trickle irrigated crops. *Fert. Res.* **6**, 235–255.

Helyard, K. R. (1976). Nitrogen cycling and soil acidification. *J. Aust. Inst. Agric. Sci.* **42**, 217–221.

Hendrickson, L. L., Keeney, D. R., Walsh, L. M., and Liegel, E. A. (1978a). Evaluation of nitrapyrin as a means of improving nitrogen efficiency in irrigated sands. *Agron. J.* **70**, 699–703.

Hendrickson, L. L., Walsh, L. M., and Keeney, D. R. (1978b). Effectiveness of nitrapyrin in controlling nitrification of fall and spring-applied anhydrous ammonia. *Agron. J.* **70**, 704–708.

Henis, Y. (1976). Effect of mineral nutrients on soil-borne pathogens and host resistance. *Proc. Colloq. Int. Potash Inst.* **12**, 101–112.

Henzell, E. F. (1981). Forage legumes. *In* "Nitrogen Fixation" (W. J. Broughton, ed.), Vol. 1, pp. 264–289. Oxford Univ. Press (Clarendon) London and New York.

Henzell, E. F., and Ross, P. J. (1973). The nitrogen cycle of pasture ecosystems. *In* "Chemistry and Biochemistry of Herbage" (G. W. Butler and R. W. Bailey, eds.), Vol. 2, pp. 227–245. Academic Press, New York.

Henzell, E. F., and Vallis, I. (1977). Transfer of nitrogen between legumes and other crops. *In* "Biological Nitrogen Fixation in Farming Systems of the Tropics" (A. Ayanaba and P. J. Dart, eds.), pp. 78–88. Wiley, New York.

Hergert, G. W., and Wiese, R. A. (1980). Performance of nitrification inhibitors in the midwest (west). *In* "Nitrification Inhibitors—Potentials and Limitations" (J. J. Meisinger, ed.), pp. 89–105. Am. Soc. Agron., Madison, Wisconsin.

Herlihy, M. (1972). Some soil and cropping factors associated with mineralization and availability of nitrogen in field experiments. *Ir. J. Agric. Res.* **11**, 271–279.

Herron, G. M., Dreier, A. F., Flowerday, A. D., Colville, W. L., and Olson, R. A. (1971). Residual mineral N accumulation in soil and its utilization by irrigated corn. *Agron. J.* **63**, 322–327.

Hildebrandt, A. (1976). Zur Problematik der nitrosamine in der Pflanzenernährung. Ph.D. Thesis, Justus Liebig Universität, Giessen.

Hodgson, D. R., Proud, J. R., and Browne, S. (1977). Cultivation systems for spring barley with special reference to direct drilling (1971–1974). *J. Agric. Sci.* **68**, 631–644.

Hoff, J. E., and Wilcox, G. E. (1970). Accumulation of nitrate in tomato fruit and its effect on detinning. *J. Am. Soc. Hortic. Sci.* **95**, 92–94.

Hoglund, J. H., and Brock, J. L. (1978). Regulation of nitrogen fixation in a grazed pasture. *N. Z. J. Agric. Res.* **21**, 73–82.

Holford, I. C. R. (1980). Effect of duration of grazed lucerne on long-term yields and nitrogen uptake of subsequent wheat. *Aust. J. Agric. Res.* **31**, 239–250.

Holmes, D. P. (1973). Influorescence development of semidwarf and standard height wheat cultivars in different photoperiod and nitrogen treatments. *Can. J. Bot.* **51**, 941–956.

Hoyt, P. B., and Hennig, A. M. F. (1971). Effects of alfalfa and grasses on yield of subsequent wheat crops and some chemical properties of a Gray Wooded soil. *Can J. Soil Sci.* **51**, 177–183.

Hoyt, P. B., and Leitch, R. H. (1983). Effects of forage legume species on soil moisture, nitrogen, and yield of succeeding barley crops. *Can. J. Soil Sci.* **63**, 125–136.

Hsiao, T. C. (1973). Plant responses to water stress. *Annu. Rev. Plant. Physiol.* **24**, 519–570.

Hsiao, T., and Acevedo, E. (1974). Plant response to water deficits, water use efficiency and drought resistance. *Agric. Meteorol.* **14**, 59–84.

Huber, D. M., Painter, G. C., McKay, H. C., and Peterson, D. L. (1968). Effect of nitrogen fertilization on take-all of winter wheat. *Phytopathology* **58**, 1470–1472.

Huber, D. M., Warren, H. L., Nelson, D. W., and Tsai, C. Y. (1977). Nitrification inhibitors—New tools for food production. *BioScience* **27**, 523–529.

Huxley, P. A. (1980). Nitrogen nutrition of cowpea (*Vigna unguiculata*). IV. Uptake and distribution of a single dose of early applied nitrogen. *Trop. Agric. (Trinidad)* **57**, 193–202.

Ingestad, T. (1977). Nitrogen and plant growth: Maximum efficiency of nitrogen fertilizers. *Ambio* **6**, 146–151.

International Atomic Energy Agency (1970). "Fertilizer Management Practices for Maize: Results of Experiments with Isotopes," Tech. Rep. Ser. No. 121. IAEA, Vienna.

Janzen, H. H., and Bettany, J. R. (1984). Sulfur nutrition of rapeseed. I. Influence of fertilizer nitrogen and sulfur rates. *Soil Sci. Soc. Am. J.* **48**, 100–107.

Jenkinson, D. S. (1968). Chemical tests for potentially available nitrogen in soil. *J. Sci. Food Agric.* **19**, 160–168.

Jenkinson, D. S., and Ladd, J. N. (1981). Microbial biomass in soil: Measurement and turnover. *In* "Soil Biochemistry" (E. A. Paul and J. N. Ladd, eds.), Vol. 5, pp. 415–471. Dekker, New York.

Jenkinson, D. S., and Powlson, D. S. (1976). The effects of biocidal treatments on metabolism in soil. V. *Soil Biol. Biochem.* **8**, 209–213.

Jenkyn, J. F. (1976). Nitrogen and leaf diseases of spring barley. *Proc. Colloq. Int. Potash Inst.* **12**, 119–128.

Jenkyn, J. F., and Griffiths, E. (1976). Some effects of nutrition on *Rhynchosporium secalis*. *Trans. Br. Mycol. Soc.* **66**, 329–332.

Jeppson, R. G., Johnson, R. R., and Hadley, H. H. (1978). Variation in mobilization of plant nitrogen to the grain in nodulating and nonnodulating soybean genotypes. *Crop Sci.* **18**, 1058–1062.

Johnson, V. A., Mattern, P. J., and Schmidt, J. W. (1967). Nitrogen relations during spring growth in varieties of *Triticum aestivum* L. differing in grain protein content. *Crop Sci.* **7**, 664–667.

Johnson, V. A., Dreir, A. F., and Grabouski, P. H. (1973). Yield and protein responses to nitrogen fertilizer of two winter wheat varieties differing in inherent protein content of their grain. *Agron. J.* **65**, 259–263.

Jolley, V. D., and Pierre, W. H. (1977). Profile accumulation of fertilizer-derived nitrate and total nitrogen recovery in two long-term nitrogen-rate experiments with corn. *Soil. Sci. Soc. Am. J.* **41**, 373–378.

Jones, F. G. W. (1976). Pests, resistance and fertilizers. *Proc. Colloq. Int. Potash Inst.* **12**, 233–258.

Jones, J. B. (1970). Distribution of 15 elements in corn leaves. *Commun. Soil. Sci. Plant Anal.* **1**, 27–34.

Jones, J. B., and Eck, H. V. (1973). Plant analysis as an aid in fertilizing corn and grain sorghum. *In* "Soil Testing and Plant Analysis" (L. M. Walsh and J. D. Beaton, eds.), pp. 349–364. Soil Sci. Soc. Am., Madison, Wisconsin.

Jones, R. J., and Jones, R. M. (1978). The ecology of Siratro-based pastures. *In* "Plant

Relations in Pastures" (J. R. Wilson, ed.), pp. 353–367. CSIRO, Brisbane, Australia.

Jones, R. J., Seawright, A. A., and Little, D. A. (1970). Oxalate poisoning in animals grazing the tropical grass *Setaria sphacelata. J. Aust. Inst. Agric. Sci.* **36,** 41.

Jones, W. W., and Embleton, T. W. (1965). Urea foliage sprays. *Calif. Citrogr.* **50,** 334–359.

Jones, W. W., Embleton, T. W., Boswell, S. B., Goodall, G. E., and Barnhart, E. L. (1970). Nitrogen rate effects on lemon production, quality and leaf nitrogen. *J. Am. Soc. Hortic. Sci.* **95,** 46–49.

Jordan, H. V., Laird, K. D., and Ferguson, D. D. (1950). Growth rates and nutrient uptake by corn in a fertilizer-spacing experiment. *Agron. J.* **42,** 261–268.

Joshi, M. D., Bishnoi, S. R., and Singh, B. (1973). Verification of organic carbon content as an index of available nitrogen in soils. *Indian J. Agric. Res.* **7,** 123–124.

Joy, K. W. (1964). Accumulation of oxalate in tissues of sugar beet, and the effect of nitrogen supply. *Ann. Bot. (London)* [N.S.] **28,** 689–701.

Juliano, B. O. (1972). Physicochemical properties of starch and protein in relation to grain quality and nutritional value of rice. *Rice Breed., Pap. Symp., 1971,* pp. 389–405.

Jung, P. E., Peterson, L. A., and Schrader, L. E. (1972). Response of irrigated corn to time, rate and source of applied N on sandy soils. *Agron. J.* **64,** 668–670.

Jurkowska, H. (1971). Effect of dicyandiamide on the content of oxalic acid in spinach. *Agrochimica* **15,** 445–453.

Kadirgamathoiyah, S., and MacKenzie, A. F. (1970). A study of soil nitrogen organic fractions and correlations with yield response of Sudan-sorghum hybrid grass on Quebec soils. *Plant Soil* **33,** 120–128.

Kamprath, E. J., Moll, R. H., and Rodriguez, N. (1982). Effects of nitrogen fertilization and recurrent selection on performance of hybrid populations of corn. *Agron. J.* **74,** 955–958.

Karlovsky, J. (1982). The balance sheet approach to determination of phosphate maintenance requirements. *Fert. Res.* **3,** 111–125.

Keeney, D. R. (1970). Protein amino acid composition of maize grain as influenced by variety and fertility. *J. Sci. Food Agric.* **21,** 182–184.

Keeney, D. R. (1980). Factors affecting the persistence and bioactivity of nitrification inhibitors. *In* "Nitrification Inhibitors—Potentials and Limitations" (J. J. Meisinger, ed.), pp. 33–46. Am. Soc. Agron., Madison, Wisconsin.

Keeney, D. R. (1982a). Nitrogen management for maximum efficiency and minimum pollution. *In* "Nitrogen in Agricultural Soils" (F. J. Stevenson, ed.), pp. 605–647. Am. Soc. Agron., Madison, Wisconsin.

Keeney, D. R. (1982b). Nitrogen-availability indices. *In* "Methods of Soil Analysis. Part 2. Chemical and Microbiological Properties" (A. L. Page, ed.), pp. 711–733. Am. Soc. Agron., Madison, Wisconsin.

Keeney, D. R., and Bremner, J. M. (1966a). Comparison and evaluation of laboratory methods of obtaining an index of soil nitrogen availability. *Agron. J.* **58,** 498–503.

Keeney, D. R., and Bremner, J. M. (1966b). Characterization of mineralizable nitrogen in soils. *Soil Sci. Soc. Am. Proc.* **30,** 714–718.

Keeney, D. R., and Bremner, J. M. (1967). Determination and isotope-ratio analysis of different forms of nitrogen in soils. IV. Mineralizable nitrogen. *Soil Sci. Soc. Am. Proc.* **31,** 34–39.

Kiraly, Z. (1976). Plant disease resistance as influenced by biochemical effects of nutrients in fertilizers. *Proc. Colloq. Int. Potash Inst.* **12,** 33–46.

Kirkby, E. A. (1968). Influence of ammonium and nitrate nutrition on the cation–anion balance and nitrogen and carbohydrate metabolism of white mustard plants grown in dilute nutrient solutions. *Soil Sci.* **105**, 133–141.

Kirkby, E. A., and Knight, A. H. (1977). Influence of the level of nitrate nutrition on ion uptake and assimilation, organic acid accumulation, and cation–anion balance in whole tomato plants. *Plant Physiol.* **60**, 349–353.

Kitur, B. K., Smith, M. S., Blevins, R. L., and Frye, W. W. (1984). Fate of ^{15}N-depleted ammonium nitrate applied to no-tillage and conventional tillage corn. *Agron. J.* **76**, 240–242.

Knapp, E. B. (1979). Inorganic nitrogen status in the soil under no-tilled winter wheat. M.S. Thesis, Washington State University, Pullman.

Knipmeyer, J. W., Hageman, R. H., Earley, E. B., and Seif, R. D. (1962). Effect of light intensity on certain metabolites of the corn plant (*Zea mays* L.). *Crop Sci.* **2**, 1–5.

Kowal, J. J., and Barker, A. V. (1981). Growth and composition of cabbage as affected by nitrogen nutrition. *Commun. Soil Sci. Plant Anal.* **12**, 979–995.

Krüger, W. (1976). The influence of fertilizers on fungal diseases of maize. *Proc. Colloq. Int. Potash Inst.* **12**, 145–156.

Kundler, P. (1970). Utilization, fixation, and loss of fertilizer nitrogen. *Albrecht-Thaer-Arch.* **14**, 191–210.

Kutscha-Lissberg, P., and Prillinger, F. (1982). Rapid determination of EUF-extractable nitrogen and boron. *Plant Soil* **64**, 63–66.

Lal, P., Reddy, G. G., and Modi, M. S. (1978). Accumulation and redistribution of dry matter and N in triticale and wheat varieties under stress condition. *Agron. J.* **70**, 623–626.

Langer, R. H. M. (1979). The dynamics of wheat yield. *N. Z. Wheat Rev.* **14**, 32–40.

Langer, R. H. M., and Hanif, M. (1973). A study of floret development in wheat (*Triticum aestivum* L.). *Ann. Bot. (London)* [N.S.] **37**, 743–751.

Langer, R. H. M., and Liew, F. K. J. (1973). Effect of varying nitrogen supply at different stages of the reproductive phase on spikelet and grain production and on grain nitrogen of wheat. *Aust. J. Agric. Res.* **24**, 647–656.

LaRue, T. A., and Patterson, T. G. (1981). How much nitrogen do legumes fix? *Adv. Agron.* **34**, 15–38.

Last, P. J., and Draycott, A. P. (1975a). Growth and yield of sugar beet on contrasting soils in relation to nitrogen supply. I. Soil nitrogen analysis and yield. *J. Agric. Sci.* **85**, 19–26.

Last, P. J., and Draycott, A. P. (1975b). Growth and yield of sugar beet on contrasting soils in relation to nitrogen supply. II. Growth, uptake and leaching of nitrogen. *J. Agric. Sci.* **85**, 27–37.

Lathwell, D. J., Dubey, H. D., and Fox, R. H. (1972). Nitrogen supplying power of some tropical soils of Puerto Rico and methods of its evaluation. *Agron. J.* **64**, 763–766.

Lawn, R. J., and Brun, W. A. (1974a). Symbiotic nitrogen fixation in soybeans. I. Effect of photosynthetic source–sink manipulations. *Crop Sci.* **14**, 11–16.

Lawn, R. J., and Brun, W. A. (1974b). Symbiotic nitrogen fixation in soybeans. III. Effect of supplemental nitrogen and intervarietal grafting. *Crop Sci.* **14**, 22–25.

Legg, J. O., and Meisinger, J. J. (1982). Soil nitrogen budgets. *In* "Nitrogen in Agricultural Soils" (F. J. Stevenson, ed.), pp. 503–566. Am. Soc. Agron., Madison, Wisconsin.

Legg, J. O., Stanford, G., and Bennett, O. L. (1979). Utilization of labelled-N fertilizer by silage corn under conventional tillage and no-till culture. *Agron. J.* **71**, 1009–1015.

Lemaire, J. M., and Jouan, B. (1976). Fertilizers and root diseases of cereals. *Proc. Colloq. Int. Potash Inst.* **12**, 113–118.

Lewis, D. A., and Tatchell, J. A. (1978). The role of fertilizer energy in agricultural production. *Phosphore Agric.* **74**, 1–13.

Liang-mo, L., Shuang, Z., Xiu-ra, Z., and Ying-hua, P. (1980). Effect of nitrapyrin on the inhibition of nitrification in some paddy soils of China. *In* "Proceedings of Symposium on Paddy Soil," pp. 837–844. Science Press, Beijing.

Liegel, E. A., and Walsh, L. M. (1976). Evaluation of sulfur-coated urea SCU applied to irrigated potatoes and corn. *Agron. J.* **68**, 457–463.

Littel, R. C., and Mott, C. D. (1974). Computer assisted design and analysis of response surface experiments in agronomy. *Proc.—Soil Crop Sci. Soc. Fla.* **34**, 94–97.

Lorenz, O. A. (1978). Potential nitrate levels in edible plant parts. *In* "Nitrogen in the Environment" (D. R. Nielsen and J. G. MacDonald, eds.), Vol. 2, pp. 201–252. Academic Press, New York.

Lorenz, O. A., and Maynard, D. N. (1980). "Knotts Handbook for Vegetable Growers." Wiley, New York.

Ludecke, T. E. (1974). Prediction of grain yield responses in wheat to nitrogen fertilizers. *Proc. Agron. Soc. N.Z.* **4**, 27–29.

Ludwick, A. E., Reuss, J. O., and Langin, E. J. (1976). Soil nitrates following four years continuous corn and as surveyed in irrigated farm fields of central and Eastern Colorado. *J. Environ. Qual.* **5**, 82–86.

Ludwick, A. E., Soltanpour, P. N., and Reuss, J. O. (1977). Nitrate distribution and variability in irrigated fields of northeastern Colorado. *Agron. J.* **69**, 710–713.

Lyda, S. D. (1981). Alleviating pathogen stress. *In* "Modifying the Root Environment to Reduce Crop Stress" (G. F. Arkin and H. M. Taylor, eds.), pp. 195–214. Am. Soc. Agric. Eng., St. Joseph, Michigan.

Lynch, J. M. (1983). "Soil Biotechnology. Microbiological Factors in Crop Productivity." Blackwell, Oxford.

MacLean, A. A. (1964). Measurement of nitrogen-supplying power of soils by extraction with sodium bicarbonate. *Nature (London)* **203**, 1307–1308.

McMahon, M. A., and Thomas, G. W., (1976). Anion leaching in two Kentucky soils under conventional tillage and killed-sod mulch. *Agron. J.* **68**, 437–442.

McNeal, F. H., Berg, M. A., Brown, P. L., and McGuire, C. F. (1971). Productivity and quality response of five spring wheat genotypes, *Triticum aestivum* L., to nitrogen fertilizers. *Agron. J.* **63**, 908–910.

McNeal, F. H., Berg, M. A., McGuire, C. F., Stewart, V. R., and Baldridge, D. E. (1972). Grain and plant nitrogen relationships in eight spring wheat crosses, *Triticum aestivum* L. *Crop Sci.* **12**, 599–602.

Maga, J. A., Moore, F. D., and Oshima, N. (1976). Yield, nitrate levels and sensory properties of spinach as influenced by organic and mineral nitrogen fertilizer levels. *J. Sci. Food Agric.* **27**, 109–114.

Mahon, J. D., and Child, J. J. (1979). Growth response of inoculated peas (*Pisum sativum*) to combined nitrogen. *Can. J. Bot.* **57**, 1687–1693.

Malhi, S. S., and Nyborg, M. (1984). Inhibiting nitrification and increasing yield of barley by band placement of thiourea with fall-applied urea. *Plant Soil* **77**, 193–206.

Martens, D. A., and Bremner, J. M. (1984). Urea hydrolysis in soils: Factors influencing the effectiveness of phenylphosphorodiamidate as a retardant. *Soil Biol. Biochem.* **16**, 515–519.

Martin, C. (1976). Nutrition and virus deseases of plants. *Proc. Colloq. Int. Potash Inst.* **12**, 193–200.

Marty, J. R., Hilaire, A., and Cabelguenne, M. (1982). Some aspects of the role of leguminous oil crops in cultural rotations. *C. R. Seances Acad. Agric. Fr.* **68**, 1251–1261.

Marumoto, T. (1984). Mineralization of C and N from microbial biomass in paddy soil. *Plant Soil* **76**, 165–173.

Marumoto, T., Anderson, J. P. E., and Domsch, K. H. (1982). Mineralization of nutrients from soil microbial biomass. *Soil Biol. Biochem.* **14**, 469–475.

Matzel, W., Lippold, H., and Heber, R. (1978). Nitrogen uptake and balance of the fertilizers urea, urea with urease inhibitor, and ammonium nitrate applied to spring wheat at stem elongation growth stage. *In* "Isotopes and Radiation in Research on Soil–Plant Relationships," pp. 67–82. IAEA, Vienna.

Maynard, D. N. (1978). Critique of potential nitrate levels in edible plant parts. *In* "Nitrogen in the Environment" (D. R. Nielsen and J. G. MacDonald, eds.), Vol. 2, pp. 221–233. Academic Press, New York.

Maynard, D. N., and Barker, A. V. (1972). Nitrate content of vegetable crops. *HortScience* **7**, 224–226.

Maynard, D. N., and Barker, A. V. (1974). Nitrate accumulation in spinach as influenced by leaf type. *J. Am. Soc. Hortic. Sci.* **99**, 135–138.

Maynard, D. N., and Barker, A. V. (1979). Regulation of nitrate accumulation in vegetables. *Acta Hortic.* **93**, 153–161.

Maynard, D. N., and Lorenz, O. A. (1979). Controlled-release fertilizers for horticultural crops. *Hortic. Rev.* **1**, 79–123.

Maynard, D. N., Barker, A. V., Minotti, P. L., and Peck, N. H. (1976). Nitrate accumulation in vegetables. *Adv. Agron.* **28**, 71–118.

Mengel, K. (1983). Responses of various crop species and cultivars to fertilizer application. *Plant Soil* **72**, 305–319.

Michrina, B. P., Fox, R. H., and Piekielek, W. P. (1981). A comparison of laboratory and field indicators of nitrogen availability. *Commun. Soil Sci. Plant Anal.* **12**, 519–535.

Middleton, K. R., and Smith, G. A. (1978). The concept of a climax in relation to the fertilizer input of a pastoral ecosystem. *Plant Soil* **50**, 595–614.

Mikesell, M. E., and Paulsen, G. M. (1971). Nitrogen translocation and the role of individual leaves in protein accumulation in wheat grain. *Crop Sci.* **11**, 919–922.

Mikkelsen, D. A., and De Datta, S. K. (1980). Rice culture. *In* "Rice Production and Utilization" (B. S. Luh, ed.), pp. 147–234. AVI Publ. Co., Westport, Connecticut.

Mikkelsen, D. S., De Datta, S. K., and Obcemea, W. N. (1978). Ammonia volatilization losses from flooded rice soils. *Soil Sci. Soc. Am. J.* **42**, 725–730.

Miller, H. F., Kavanaugh, J., and Thomas, G. W. (1975). Time of N application and yields of corn in wet, alluvial soils. *Agron. J.* **67**, 401–404.

Miller, R. J., Rolston, D. E., Rauschkolb, R. S., and Wolfe, D. W. (1976). Drip application of nitrogen is efficient. *Calif. Agric.* **30**(11), 16–18.

Mills, H. A., and Jones, J. B. (1979). Nutrient deficiencies and toxicities in plants: Nitrogen. *J. Plant Nutr.* **1**, 101–122.

Mills, H. A., Barker, A. V., and Maynard, D. N. (1976). Nitrate accumulation in radish as affected by nitrapyrin. *Agron. J.* **68**, 13–17.

Minotti, P. L., and Stankey, D. L. (1973). Diurnal variation in the nitrate concentration of beets. *HortScience* **8**, 33–34.

Moschler, W. W., and Martens, D. C. (1975). Nitrogen, phosphorus and potassium requirement in no-tillage and conventionally tilled corn. *Soil Sci. Soc. Am. Proc.* **39**, 886–891.

Moschler, W. W., Martens, D. C., Rich, C. I., and Shear, G. M. (1972). Comparative lime effects on continuous no-tillage and conventionally tilled corn. *Agron. J.* **65,** 781–783.

Mouchova, H., and Apltauer, J. (1983). Effects of the nitrification inhibitor N-serve on the utilization of fall-applied urea by wheat. *Fert. Res.* **4,** 165–180.

Mulder, E. G., Lie, T. A., and Houwers, A. (1977). The importance of legumes under temperate conditions. *In* "A Treatise on Dinitrogen Fixation. Section IV. Agronomy and Ecology" (R. W. F. Hardy and A. H. Gibson, eds.), pp. 221–242. Wiley, New York.

Mulvaney, R. L., and Bremner, J. M. (1981). Control of urea transformations in soils. *In* "Soil Biochemistry" (E. A. Paul and J. N. Ladd, eds.), Vol. 5, pp. 153–196. Dekker, New York.

Munns, D. A. (1968). Nodulation of *Medicago sativa* in solution culture. III. Effects of nitrate on root hairs and infection. *Plant Soil* **28,** 33–47.

Munns, D. N., and Fox, R. L. (1977). Comparative lime requirements of temperate and tropical legumes. *Plant Soil* **46,** 533–548.

Munson, R. D., and Nelson, W. L. (1973). Principles and practices in plant analysis. *In* "Soil Testing and Plant Analysis" (L. M. Walsh and J. D. Beaton, eds.), pp. 201–248. Soil Sci. Soc. Am., Madison, Wisconsin.

Myers, R. J. K. (1984). A simple model for estimating the nitrogen fertilizer requirement of a cereal crop. *Fert. Res.* **5,** 95–108.

National Research Council (NRC) (1972). "Accumulation of Nitrate." Natl. Acad. Sci., Washington, D.C.

National Research Council (NRC) (1978). "Nitrates: An Environmental Assessment." Natl. Acad. Sci., Washington, D.C.

Needham, P. (1982). The role of nitrogen in wheat production: Response, interaction and production of nitrogen requirements in the UK. *Proc.—Fert. Soc.* **211,** 125–147.

Nelder, J. A. (1966). Inverse polynomials, a useful group of multi-factor response functions. *Biometrics* **22,** 128–141.

Nelson, D. W., and Huber, D. M. (1980). Performance of nitrification inhibitors in the Midwest (east). *In* "Nitrification Inhibitors—Potentials and Limitations" (J. J. Meisinger, ed.), pp. 75–88. Am. Soc. Agron., Madison, Wisconsin.

Nelson, L. E., and Selby, R. (1974). The effect of nitrogen sources and iron levels on the growth and composition of sitka spruce, and scots pine. *Plant Soil* **41,** 573–588.

Németh, K. (1979). The availability of nutrients in the soil as determined by electro-ultrafiltration (EUF). *Adv. Agron.* **31,** 155–188.

Németh, K., Makhdum, I. Q., Koch, K., and Beringer, H. (1979). Determination of categories of soil nitrogen by electro-ultrafiltration (EUF). *Plant Soil* **53,** 445–453.

Nnadi, L. A. (1975). Comparison of leaching losses of nitrogen from soluble and slow-release fertilizers. *Samaru Agric. Newsl.* **17,** 82–85.

Nommik, H. (1976). Predicting the nitrogen-supplying power of acid forest soils from data on the release of CO_2 and NH_3 on partial oxidation. *Commun. Soil Sci. Plant Anal.* **7,** 569–584.

Novoa, R., and Loomis, R. S. (1981). Nitrogen and plant production. *Plant Soil* **58,** 177–204.

Nowakowski, T. Z. (1962). Effects of nitrogen fertilizers on total nitrogen, soluble nitrogen and soluble carbohydrate contents of grass. *J. Agric. Sci.* **59,** 387–392.

Nyborg, M., and Malhi, S. S. (1979). Increasing the efficiency of fall-applied urea fertilizer by placing in big pellets or in nests. *Plant Soil* **52,** 461–465.

Nye, P. H. (1981). Changes of pH across the rhizosphere induced by roots. *Plant Soil* **61**, 7–26.

O'Connor, M. B. (1982). Nitrogen fertilizers for the production of out-of-season grass. *In* "Nitrogen Fertilizers in New Zealand Agriculture" (P. B. Lynch, ed.), pp. 65–76. N.Z. Inst. Agric. Sci., Wellington.

Øeien, A., and Selmer-Olson, A. R. (1980). A laboratory method for evaluation of available nitrogen in soil. *Acta Agric. Scand.* **30**, 149–156.

Oertli, J. J. (1975). Efficiency of nitrogen recovery from controlled-release urea under conditions of heavy leaching. *Agrochimica* **19**, 326–335.

Oertli, J. J. (1980). Controlled-release fertilizers. *Fert. Res.* **1**, 103–123.

Oghoghorie, C. G. O., and Pate, J. S. (1971). The nitrate stress syndrome of the nodulated field pea (*Pisum arvense* L.). *Plant Soil (Spec. Vol.)*, pp. 185–202.

Olday, F. C., Barker, A. V., and Maynard, D. N. (1976a). A physiological basis for different patterns of nitrate accumulation in two spinach cultivars. *J. Am. Soc. Hortic. Sci.* **101**, 217–219.

Olday, F. C., Barker, A. V., and Maynard, D. N. (1976b). A physiological basis for different patterns of nitrate accumulation in cucumber and pea. *J. Am. Hortic. Sci.* **101**, 219–221.

Olson, R. A., and Kurtz, L. T. (1982). Crop nitrogen requirements, utilization, and fertilization. *In* "Nitrogen in Agricultural Soils" (F. J. Stevenson, ed.), pp. 567–604. Am. Soc. Agron., Madison, Wisconsin.

Olson, R. A., Dreier, A. F., Thompson, C., Frank, K., and Grabouski, P. H. (1964). Using fertilizer nitrogen effectively on grain crops. *Nebr., Agric. Exp. Stn., Bull.* **479.**

Olson, R. A., Frank, K. D., Deibert, E. J., Dreier, A. F., Sander, D. H., and Johnson, V. A. (1976). Impact of residual mineral N in soil on grain protein yields of winter wheat and corn. *Agron. J.* **68**, 769–772.

Olson, R. V., and Swallow, C. W. (1984). Fate of labelled nitrogen fertilizer applied to winter wheat for five years. *Soil Sci. Soc. Am. J.* **48**, 583–586.

Onken, A. B. (1980). Performance of nitrification inhibitors in the Southwest. *In* "Nitrification Inhibitors—Potentials and Limitations" (J. J. Meisinger, ed.), pp. 119–129. Am. Soc. Agron., Madison, Wisconsin.

Onken. A. B., and Sunderman, H. D. (1972). Applied and residual nitrate-nitrogen effects on irrigated grain sorghum yield. *Soil Sci. Soc. Am. Proc.* **36**, 94–97.

Osborne, G. J. (1977). Chemical fractionation of soil nitrogen in six soils from southern New South Wales. *Aust. J. Soil Res.* **15**, 159–165.

Osborne, G. J., and Storrier, R. R. (1976). Influence of different sources of nitrogen fertilizer on the value of soil nitrogen tests. *Aust. J. Exp. Agric. Anim. Husb.* **16**, 881–886.

Osmond, C. B. (1963). Oxalates and ionic equilibria in Australian salt bushes (*Atriplex*). *Nature (London)* **198**, 503–504.

Osmond, C. B. (1967). Acid metabolism in *Atriplex*. I. Regulation of oxalate synthesis by the apparent excess cation absorption of leaf tissue. *Aust. J. Biol. Sci.* **20**, 575–587.

O'Sullivan, J., Gabelman, W. H., and Gerloff, G. G. (1974). Variation in efficiency of nitrogen utilization in tomatoes (*Lycopersicon esculentum* Mill.) grown under nitrogen stress. *J. Am. Soc. Hortic. Sci.* **99**, 543–547.

Overrein, L. M., and Moe, P. G. (1967). Factors affecting urea hydrolysis and ammonia volatilization in soil. *Soil Sci. Soc. Am. Proc.* **31**, 57–61.

Oyanedel, C., and Rodriguez, S. (1977). Estimation of N mineralization in soils. *Cienc. Invest. Agrar.* **4**, 33–44.

Pankhurst, C. E. (1978). Effects of plant nutrient supply on nodule effectiveness and *Rhizo-*

bium strain competition for nodulation of *Lotus pedunculatus*. *Plant Soil* **60**, 325–339.

Papendick, R. I., and Engibous, J. C. (1980). Performance of nitrification inhibitors in the Northwest. *In* "Nitrification Inhibitors—Potentials and Limitations" (J. J. Meisinger, ed.), pp. 107–117. Am. Soc. Agron., Madison, Wisconsin.

Parker, C. A., Trinick, M. J., and Chatel, D. L. (1977). Rhizobia as soil and rhizosphere inhabitants. *In* "A Treatise on Dinitrogen Fixation. Section IV. Agronomy and Ecology" (R. W. F. Hardy and A. H. Gibson, eds.), pp. 311–352. Wiley, New York.

Parker, M. B., and Boswell, F. C. (1980). Foliar injury, nutrient intake, and yield of soybeans as influenced by foliar fertilization. *Agron. J.* **72**, 110–113.

Parr, J. F. (1973). Chemical and biochemical considerations for maximizing the efficiency of fertilizer nitrogen. *J. Environ. Qual.* **2**, 75–84.

Parr, J. F., and Papendick, R. I. (1966). Retention of anhydrous ammonia by soil. II. Effect of ammonia concentration and soil moisture. *Soil Sci.* **10**, 109–119.

Pate, J. S., and Dart, P. J. (1961). Nodulation studies in legumes. IV. The influence of inoculum strain and time of application of ammonium nitrate on symbiotic response. *Plant Soil* **15**, 329–346.

Pearman, I., Thomas, S. M., and Thorne, G. N. (1977). Effects of nitrogen fertilizer on growth and yield of spring wheat. *Ann. Bot. (London)* [N.S.] **41**, 93–108.

Pearson, C. J., and Muirhead, W. A. (1984). Nitrogen uptake. *In* "Control of Crop Productivity" (C. J. Pearson, ed.), pp. 73–88. Academic Press, New York.

Peck, N. H., and MacDonald, G. E. (1984). Snap bean responses to nitrogen fertilization. *Agron. J.* **76**, 247–253.

Peck, N. H., Barker, A. V., MacDonald, G. E., and Shallenberger, R. S. (1971). Nitrate accumulation in vegetables. II. Table beets grown in upland soils. *Agron. J.* **63**, 130–132.

Perry, L. J., and Olson, R. A. (1975). Yield and quality of corn and grain sorghum grain and residues as influenced by fertilization. *Agron. J.* **67**, 816–818.

Pesek, J., Stanford, G., and Case, N. L. (1971). Nitrogen production and use. *In* "Fertilizer Technology and Use" (R. A. Olson, ed.), pp. 217–269. Soil Sci. Soc. Am., Madison, Wisconsin.

Pierre, W. H., Webb, J. R., and Shrader, W. D. (1971). Quantitative effects of nitrogen fertilizer on the development and downward movement of soil acidity in relation to level of fertilization and crop removal in a continuous corn cropping system. *Agron. J.* **63**, 291–297.

Pimpini, F., Venter, F., and Wunsch, A. (1970). Untersuchungen Über den Nitratgehalt in Blumenkohl. *Landwirtsch. Forsch.* **23**, 363–370.

Poole, W. D., Randall, G. W., and Ham, G. E. (1983a). Foliar fertilization of soybeans. I. Effect of fertilizer sources, rates, and frequency of application. *Agron. J.* **75**, 195–200.

Poole, W. D., Randall, G. W., and Ham, G. E. (1983b). Foliar fertilization of soybeans. II. Effect of biuret and application time of day. *Agron. J.* **75**, 201–203.

Power, J. F. (1981). Nitrogen in the cultivated ecosystem. *In* "Terrestrial Nitrogen Cycles: Processes, Ecosystem Strategies and Management Impacts" (F. E. Clark and T. Rosswall, eds.), pp. 529–546. Ecological Bulletins, Stockholm.

Power, J. F., and Legg, J. O. (1984). Nitrogen-15 recovery for five years after application of ammonium nitrate to crested wheatgrass. *Soil Sci. Soc. Am. J.* **48**, 322–326.

Powers, R. F. (1980). Mineralizable soil nitrogen as an index of nitrogen availability to forest trees. *Soil Sci. Soc. Am. J.* **44**, 1314–1320.

Prado, O., and Rodriguez, J. (1978). Nitrogen fertilizer requirement estimates of wheat. *Cienc. Invest. Agrar.* **5,** 29–40.

Prasad, R., and Rao, E. V. S. P. (1978). Nitrogen management in rice soils. *In* "Nitrogen Assimilation and Crop Productivity" (S. P. Sen, Y. P. Abrol, and S. K. Sinha, eds.), pp. 97–112. Associated Publ., New Delhi, India.

Prasad, R., Rajale, G. B., and Lakhive, B. A. (1971). Nitrification retarders and slow-release nitrogen fertilizers. *Adv. Agron.* **23,** 337–383.

Pugh, K. B., and Waid, J. S. (1969). The influence of hydroxamates on ammonia loss from various soils treated with urea. *Soil Biol. Biochem.* **1,** 207–217.

Pushman, F. M., and Bingham, J. (1976). The effect of a granular nitrogen fertilizer and a foliar spray urea on the yield and breadmaking quality of winter wheats. *J. Agric. Sci.* **87,** 281–292.

Quin, B. F., Drewitt, E. G., and Stephen, R. C. (1982). A soil incubation test for estimating wheat yields and nitrogen requirements. *Proc. Agron. Soc. N. Z.* **12,** 35–40.

Ramig, R. E., and Rhoades, H. F. (1963). Interrelationship of soil moisture level at planting time and nitrogen fertilization of winter wheat production. *Agron. J.* **55,** 123–127.

Rankin, J. D. (1978). Catalysts in ammonia production. *Proc.—Fert. Soc.* **168.**

Raven, J. A., and Smith, F. A. (1976). Nitrogen assimilation and transport in vascular land plants in relation to intracellular pH regulation. *New Phytol.* **76,** 415–431.

Reddy, K. R., and Patrick, W. H. (1978). Residual fertilizer nitrogen in a flooded rice soil. *Soil Sci. Soc. Am. J.* **42,** 316–318.

Reed, A. J., and Hageman, R. H. (1980). Relationship between nitrate uptake, flux, and reduction and the accumulation of reduced nitrogen in maize (*Zea mays* L.). I. Genotypic variation. *Plant Physiol.* **66,** 1179–1183.

Rehm, G. W., and Wiese, R. A. (1975). Effect of method of nitrogen application on corn (*Zea mays* L.) grown on irrigated sandy soils. *Soil Sci. Soc. Am. Proc.* **39,** 1217–1220.

Reinertsen, M. R., Cochran, V. L., and Morrow, L. A. (1984). Response of spring wheat to N fertilizer placement, row spacing, and wild oat herbicides in a no-till system. *Agron. J.* **76,** 753–756.

Remison, S. U. (1978). Neighbour effects between maize and cowpea at various levels of N and P. *Exp. Agric.* **14,** 205–212.

Remy, J. C., and Viaux, Ph. (1982). The use of nitrogen fertilizers in intensive wheat growing in France. *Proc.—Fert. Soc.* **211,** 67–92.

Rendiz, V. V., and Jimenez, J. (1978). Nitrogen nutrition as a regulator of biosynthesis of storage proteins in maize (*Zea mays* L.) grain. *In* "Nitrogen in the Environment" (D. R. Nielsen and J. G. MacDonald, eds.), Vol. 2, pp. 253–278. Academic Press, New York.

Rennie, R. J., Dubetz, S., Bole, J. B., and Muendel, H. H. (1982). Dinitrogen fixation measured by ^{15}N isotope dilution in two Canadian soybean cultivars. *Agron. J.* **74,** 725–730.

Rice, C. W., and Smith, M. S. (1982). Denitrification in no-till and plowed soils. *Soil Sci. Soc. Am. J.* **46,** 1168–1173.

Ris, J., Smilde, K. W., and Wijnen, G. (1981). Nitrogen fertilizer recommendations for arable crops as based on soil analysis. *Fert. Res.* **2,** 21–32.

Robinson, J. B. D. (1968a). A simple available soil nitrogen index. I. Laboratory and greenhouse studies. *J. Soil Sci.* **19,** 269–279.

Robinson, J. B. D. (1968b). A simple available soil nitrogen index. II. Field crop evaluation. *J. Soil Sci.* **19,** 280–290.

Robinson, R. G., Warnes, D. D., Nelson, W. W., Ford, J. H., and Smith, L. J. (1974). Field beans. Rates of planting, width of row, and effects of irrigation and nitrogen on yield and seed quality. *Misc. Rep.—Minn., Agric. Exp. Stn.* **124.**

Rodgers, G. A. (1983). Effect of dicyandiamide on ammonia volatilization from urea in soil. *Fert. Res.* **4,** 361–367.

Rodgers, G. A., Widdowson, F. V., Penny, A., and Hewitt, M. V. (1983). Effects of several nitrification inhibitors, when injected with aqueous urea, on yields and nitrogen recoveries of ryegrass leys. *J. Agric. Sci.* **101,** 637–656.

Rodriquez, P. M. S. (1977). Varietal differences in maize in the uptake of nitrogen and its translocation to the grain. *Diss. Abstr.* **38,** 5690-B.

Rong-ye, C., and Zhao-Liang, Z. (1980). The fate of nitrogen fertilizer in paddy soils. *In* "Proceedings of Symposium on Paddy Soil," pp. 597–602. Science Press, Beijing.

Roughan, P. G., Grattan, P., and Warrington, I. J. (1976). Effect of nitrogen source on oxalate accumulation in *Setaria sphacelata* (cv. "Kazungala"). *J. Sci. Food Agric.* **27,** 281–286.

Rush, C. M., McClung, C. M., and Lyda, S. D. (1979). Cellular effects of anhydrous ammonia on *Phymatotrichum omnivorum* sclerotia. *Phytopathology* **69,** 1044.

Russell, E. W. (1968). "The Place of Fertilizers in Food Crop Economy of Tropical Agriculture." The Fertilizer Society, London.

Russell, R. S. (1977). "Plant Root Systems: Their Function and Interaction with the Soil." McGraw-Hill, New York.

Russelle, M. P., Deibert, E. J., Hauck, R. D., Stevanovic, M., and Olson, R. A. (1981). Effects of water and nitrogen management on yield and ^{15}N-depleted fertilizer use efficiency of irrigated corn. *Soil Sci. Soc. Am. J.* **45,** 553–558.

Russelle, M. P., Hauck, R. D., and Olson, R. A. (1983). Nitrogen accumulation rates of irrigated maize. *Agron. J.* **75,** 593–598.

Sadaphal, M. N., and Das, N. P. (1966). Effect of spraying urea on winter wheat (*Triticum aestivum*). *Agron. J.* **58,** 137–141.

Sahota, T. S., and Singh, M. (1984). Relative efficiency of N fertilizers as influenced by N-serve in the potato crop. *Plant Soil* **79,** 143–152.

Sahrawat, K. L. (1980). Control of urea hydrolysis and nitrification in soil by chemicals— Prospects and problems. *Plant Soil* **57,** 335–352.

Sahrawat, K. L. (1982). Simple modification of the Walkley–Black method for simultaneous determination of organic carbon and potentially mineralizable nitrogen in tropical rice soils. *Plant Soil* **69,** 73–77.

Sahrawat, K. L. (1983). Nitrogen availability indexes for submerged rice soils. *Adv. Agron.* **36,** 415–451.

Sahrawat, K. L., and Keeney, D. R. (1984). Effects of nitrification inhibitors on chemical composition of plants: A review. *J. Plant Nutr.* **7,** 1251–1288.

Sanchez, P. A., Gavidia, O. A., Ramirez, G. E., Vergara, R., and Minguillo, F. (1973). Performance of sulfur coated urea under intermittently flooded rice culture in Peru. *Soil Sci. Soc. Am. Proc.* **37,** 789–792.

Sanford, J. O., and Hairston, J. E. (1984). Effects of N fertilization on yield, growth and extraction of water by wheat following soybeans and grain sorghum. *Agron. J.* **76,** 623–626.

Sauderbeck, D., and Timmermann, F. (1983). The efficient use of fertilizer nutrients as influenced by soil testing. Application, technique and timing. *In* "Efficient Use of Fertilizers in Agriculture," pp. 171–195. Martinus Nijhoff/Dr. W. Junk, The Hague.

Savant, N. K., and De Datta, S. K. (1982). Nitrogen transformations in wetland rice soils. *Adv. Agron.* **35,** 241–302.

Scaife, M. A. (1974). Computer simulation of nitrogen uptake and growth. *Plant Anal. Fert. Probl., Colloq., 7th,* Vol. 2, pp. 413–425.

Schenk, M. A. M., de Faria Filho, T. T., Pimentel, D. M., and Thiago, L. R. L. de S. (1982). Oxalate poisoning of dairy cows in *Setaria* pastures. *Pesqui. Agropecu. Bras.* **17,** 1403–1407.

Schnug, E., and Finck, A. (1982). Trace element mobilization by acidifying fertilizers. *In* "Plant Nutrition 1982" (A. Scaife, ed.), Vol. 2, pp. 583–587. Commonw. Agric. Bur., Slough, England.

Schroder, V. N., and Hinson, K. (1975). Soil nitrogen from soybeans [*Glycine max* (L.) Merr.]. *Proc.—Soil Crop Sci. Soc. Fla.* **34,** 101–103.

Schuphan, W., Bengtsson, B., Bosund, I., and Hylmo, B. (1967). Nitrate accumulation in spinach. *Qual. Plant Foods Hum. Nutr.* **14,** 317–330.

Shen, S. M., Pruden, G., and Jenkinson, D. S. (1984). Mineralization and immobilization of nitrogen in fumigated soil and the measurement of microbial biomass nitrogen. *Soil Biol. Biochem.* **16,** 437–444.

Siegel, O., and Vogt, G. (1975). Uber den Einfluss langsam fliessender Stickstoffquellen auf den Aufbau der Stickstoffuerbindungen in Spinat and Gerste. *Landwirtsch. Forsch.* **28,** 235–247.

Singh, N. T., Vig, A. C., Singh, R., and Chaudhury, M. R. (1979). Influence of different levels of irrigation and nitrogen on yield and nutrient uptake by wheat. *Agron. J.* **71,** 401–404.

Singh, R., and Brar, S. P. S. (1973). Correlation of different soil tests with maize response to nitrogen and phosphorus at different soil fertility levels. *J. Indian Soc. Soil Sci.* **21,** 213–217.

Singh, S. P. (1981). Studies on spatial arrangement in sorghum–legume intercropping systems. *J. Agric. Sci.* **97,** 655–661.

Slagen, J. H. G., and Kerkhoff, P. (1984). Nitrification inhibitors in agriculture and horticulture: A literature review. *Fert. Res.* **5,** 1–76.

Smika, D. E., and Grabouski, P. H. (1976). Anhydrous ammonia applications during fallow for winter wheat production. *Agron. J.* **68,** 919–922.

Smiley, R. W. (1975). Forms of nitrogen and the pH in the root zone and their importance to root infections. *In* "Biology and Control of Soil-Borne Plant Pathogens" (G. W. Bruehl, ed.), pp. 55–62. Am. Phytopathol. Soc., St. Paul, Minnesota.

Smiley, R. W., and Cooke, R. J. (1973). Relationship between take-all of wheat and rhizosphere pH in soils fertilized with ammonium vs nitrate-nitrogen. *Phytopathology* **63,** 882–890.

Smith, F. W. (1972). Potassium nutrition, ionic relations, and oxalic acid accumulation in three cultivars of *Setaria sphacelata. Aust. J. Agric. Res.* **23,** 969–980.

Smith, F. W. (1978). The effect of potassium and nitrogen on ionic relations and organic and accumulation in *Panicum maximum* var. trichoglume. *Plant Soil* **49,** 367–379.

Smith, J. A. (1966). An evaluation of nitrogen soil test methods for Ontario soils. *Can. J. Soil Sci.* **46,** 185–194.

Smith, K. A., and Howard, R. S. (1980). Field studies of nitrogen uptake using ^{15}N-tracer methods. *J. Sci. Food Agric.* **31,** 839–840.

Smith, M. W., Kenworthy, A. L., and Bedford, C. L. (1979). The response of fruit trees to injection of nitrogen through a trickle irrigation system. *J. Am. Soc. Hortic. Sci.* **194,** 311–313.

Smith, S. J., and Stanford, G. (1971). Evaluation of a chemical index of soil nitrogen availability. *Soil Sci.* **111,** 228–232.

Smith, S. J., Young, L. B., and Miller, G. E. (1977). Evaluation of soil nitrogen mineralization potential under modified field conditions. *Soil Sci. Soc. Am. J.* **41**, 74–76.

Söderlund, R., and Svensson, B. H. (1976). The global nitrogen cycle. *In* "Nitrogen, Phosphorus and Sulphur—Global Cycles" (B. H. Svensson and R. Söderlund, eds.), pp. 23–77. Ecological Bulletins, Stockholm.

Soper, R. J., Racz, G. J., and Fehr, P. I. (1971). Nitrate nitrogen in the soil as a means of predicting the fertilizer nitrogen requirements of barley. *Can. J. Soil Sci.* **51**, 45–49.

Sparrow, P. E. (1979). The comparison of five response curves for representing the relationship between annual dry matter yield of grass herbage and fertilizer nitrogen. *J. Agric. Sci.* **93**, 513–520.

Spiertz, J. H. J., and Ellen, J. (1978). Effects of nitrogen on crop development and grain growth of winter wheat in relation to assimilation and utilization of assimilates and nutrients. *Neth. J. Agric. Sci.* **25**, 210–231.

Spiertz, J. H. K., and van der Haar, H. (1978). Differences in grain growth, crop photosynthesis and distribution of assimilates between semidwarf and standard cultivars of winter wheat. *Neth. J. Agric. Sci.* **26**, 233–249.

Spratt, E. D., and Gasser, J. K. R. (1970). Effects of fertilizer nitrogen and water supply on the distribution of dry matter and nitrogen between the different parts of wheat. *Can. J. Plant Sci.* **50**, 613–625.

Stalk, S. A., and Clapp, C. E. (1980). Residual nitrogen availability from soils treated with sewage sludge in a field experiment. *J. Environ. Qual.* **9**, 505–512.

Stanford, G. (1968). Extractable organic nitrogen and nitrogen mineralization in soils. *Soil Sci.* **106**, 345–361.

Stanford, G. (1973). Rationale for optimum nitrogen fertilization in corn production. *J. Environ. Qual.* **2**, 159–166.

Stanford, G. (1977). Evaluating the nitrogen-supplying capacities of soils. *Proc. Int. Semin. Soil Environ. Fertil. Manage. Intensive Agric., 1977*, pp. 412–418.

Stanford, G. (1978). Evaluation of ammonium release by alkaline permanganate as an index of soil nitrogen availability. *Soil Sci.* **126**, 244–253.

Stanford, G. (1982). Assessment of soil nitrogen availability. *In* "Nitrogen in Agricultural Soils" (F. J. Stevenson, ed.), pp. 651–688. American Society of Agronomy, Madison, Wisconsin.

Stanford, G., and DeMar, W. H. (1969). Extraction of soil nitrogen by autoclaving in water. I. The NaOH-distillable fraction as an index of soil nitrogen availability. *Soil Sci.* **107**, 203–205.

Stanford, G., and Legg, J. O. (1968). Correlation of soil nitrogen availability indexes with nitrogen uptake by plants. *Soil Sci.* **105**, 320–326.

Stanford, G., and Smith, S. J. (1972). Nitrogen mineralization potentials of soils. *Soil Sci. Soc. Am. Proc.* **36**, 465–472.

Stanford, G., and Smith, S. J. (1976). Estimating potentially mineralizable soil nitrogen from a chemical index of soil nitrogen availability. *Soil Sci.* **122**, 71–76.

Stanford, G., and Smith, S. J. (1978). Oxidative release of potentially mineralizable soil nitrogen by acid permanganate extraction. *Soil Sci.* **126**, 210–218.

Stanford, G., Carter, J. N., and Smith, S. J. (1974). Estimates of potentially mineralizable soil nitrogen based on short-term incubations. *Soil Sci. Soc. Am. Proc.* **38**, 99–102.

Stanford, G., Carter, J. N., Westerman, D. T., and Meisinger, J. J. (1977). Residual nitrate and mineralizable soil nitrogen in relation to nitrogen uptake by irrigated sugar beets. *Agron. J.* **69**, 303–308.

Stanley, R. L., and Rhoades, F. M. (1977). Effect of time, rate, and increment of applied fertilizer on nutrient uptake and yield of corn (*Zea mays* L.). *Proc.—Soil Crop Sci. Soc. Fla.* **36,** 181–184.

Starbursk, A., and Heide, O. M. (1974). Protein content and amino-acid spectrum of finger millet (*Eleusine coracana* L. Gaertn.) as influenced by nitrogen and sulphur dioxide. *Plant Soil* **41,** 549–571.

Steele, K. W., Cooper, D. M., and Dyson, C. B. (1982). Estimating nitrogen fertilizer requirements in maize grain production, I. Determination of available soil nitrogen and prediction of grain yield increase to applied nitrogen. *N. Z. J. Agric. Res.* **25,** 199–206.

Stephen, R. C. (1982). Nitrogen fertilizers in cereal production. *In* "Nitrogen Fertilizers in New Zealand Agriculture" (P. B. Lynch, ed.), pp. 77–93. N.Z. Inst. Agric. Sci., Wellington.

Stevenson, C. K., and Baldwin, C. S. (1969). Effect of time and method of nitrogen application and source of nitrogen on the yield and nitrogen content of corn. (*Zea mays* L.). *Agron. J.* **61,** 381–384.

Sturm, H., and Effland, H. (1982). Nitrogen fertilization and its interaction with other cultural measures: Experience in the Federal Republic of Germany. *Proc.—Fert. Soc.* **211,** 5–32.

Suckling, F. E. T. (1975). Pasture management trials on unploughable hill country at Te Awa. *N. Z. J. Exp. Agric.* **3,** 351–436.

Summerfield, R. J., Dart, P. J., Huxley, P. A., Eaglesham, A. R. J., Minchin, F. R., and Day, J. M. (1977). Nitrogen nutrition of cowpea (*Vigna unguiculata*). I. Effects of applied nitrogen and symbiotic nitrogen fixation on growth and seed yield. *Exp. Agric.* **13,** 129–142.

Syverud, T. D., Walsh, L. M., Oplinger, E. S., and Kelling, K. A. (1980). Foliar fertilization of soybeans (*Glycine max.* L.). *Commun. Soil Sci. Plant Anal.* **11,** 637–651.

Tanji, K. K. (1982). Modelling of the soil nitrogen cycle. *In* "Nitrogen in Agricultural Soils" (F. J. Stevenson, ed.), pp. 721–772. Am. Soc. Agron., Madison, Wisconsin.

Temiz, K. (1976). Interaction of fertilizers with septoria leaf blotch of wheat. *Proc. Colloq. Int. Potash Inst.* **12,** 129–132.

Terman, G. L. (1977). Yields and nutrient accumulation by determinate soybeans, as affected by applied nutrients. *Agron. J.* **69,** 234–238.

Terman, G. L., Noggle, J. C., and Hunt, C. M. (1976). Nitrate nitrogen and total nitrogen concentration relationship in several plant species. *Agron. J.* **68,** 556–560.

Tesha, A. J., and Eck, P. (1983). Effect of nitrogen rate and water stress on growth and water relations of young Sweet Corn plants. *J. Am. Soc. Hortic. Sci.* **108,** 1049–1053.

Tesha, A. J., and Kumar, D. (1978). Efects of fertilizer nitrogen on drought resistance in *Coffee arabica* L. *J. Agric. Sci.* **90,** 625–631.

Thom, W. O., Miller, T. C., and Bowerman, D. H. (1981). Foliar fertilization of rice after midseason. *Agron. J.* **73,** 411–414.

Thomas, G. W., and Gilliam, J. W. (1978). Agro-ecosystems in the USA. *In* "Cycling of Mineral Nutrients in Agricultural Ecosystems" (M. J. Frissel, ed.), pp. 101–122. Elsevier, Amsterdam.

Thomas, S. M., Thorne, G. N., and Pearman, I. (1978). Effect of nitrogen on growth, yield and photorespiratory activity in spring wheat. *Ann. Bot. (London)* [N.S.] **42,** 827–837.

Tinker, P. B. H. (1978). Uptake and consumption of soil nitrogen in relation to agronomic practice. *In* "Nitrogen Assimilation of Plants" (E. J. Hewitt and C. V. Cutting, eds.), pp. 101–122. Academic Press, New York.

Tinker, P. B. H., and Widdowson, V. (1982). Maximizing wheat yields, and some causes of yield variation. *Proc.—Fert. Soc.* **211,** 149–184.

Todd, G. W. (1972). Water deficits and enzymatic activity. *In* "Water Deficits and Plant Growth" (T. T. Kozlowski, ed.), Vol. 3, pp. 177–216. Academic Press, New York.

Toews, W. H., and Soper, R. J. (1978). Effects of nitrogen source, method of placement and soil type on seedling emergence and barley crop yields. *Can. J. Soil Sci.* **58,** 311–320.

Tomar, J. S., and Soper, R. J. (1981). Fate of tagged urea in the field with different methods of N and organic matter placement. *Agron. J.* **73,** 991–995.

Touchton, J. T., and Boswell, F. C. (1980). Performance of nitrification inhibitors in the Southeast. *In* "Nitrification Inhibitors—Potentials and Limitations" (J. J. Meisinger, ed.), pp. 63–74, Am. Soc. Agron., Madison, Wisconsin.

Tucker, B. B., Cox, M. B., and Eck, H. V. (1971). Effect of rotations, tillage methods, and N fertilization on winter wheat production. *Agron. J.* **63,** 699–702.

Tuohey, C. L., and Robson, A. D. (1980). The effects of cropping after medic and non-medic pastures on total soil nitrogen and on the grain yield and nitrogen content of wheat. *Aust. J. Exp. Agric. Anim. Husb.* **20,** 220–228.

Vaidyanathan, L. V., and Leitch, M. H. (1980). Use of fertilizer and soil nitrogen by winter wheat established with and without cultivation prior to drilling. *J. Sci. Food Agric.* **31,** 852–853.

Vallis, I. (1978). Nitrogen relationships in grass/legume mixtures. *In* "Plant Relations in Pastures" (J. R. Wilson, ed.), pp. 190–201. CSIRO, Canberra, Australia.

Van Tuil, H. D. W. (1965). Organic salts in plants in relation to nutrition and growth. *Versl. Landbouwkd. Onderz.* **657.**

Van Tuil, H. D. W. (1970). The organic salt content of plants in relation to growth and nitrogen nutrition. *In* "Nitrogen Nutrition of the Plant" (E. A. Kirkby, ed.), pp. 45–60. University of Leeds, Leeds.

Vasilas, B. L., Legg, J. O., and Wolf, D. C. (1980). Foliar fertilization of soybeans: Adsorption and translocation of ^{15}N-labelled urea. *Agron. J.* **72,** 271–275.

Venter, F. (1979). Nitrate contents in carrots (*Daucus carota* L.) as influenced by fertilization. *Acta Hortic.* **93,** 163–171.

Viets, F. G. (1962). Fertilizers and the efficient use of water. *Adv. Agron.* **14,** 223–264.

Viets, F. G. (1965). The plant's need for and use of nitrogen. *In* "Soil Nitrogen" (W. V. Bartholomew and F. E. Clark, eds.), pp. 503–549. Am. Soc. Agron., Madison, Wisconsin.

Viets, F. G. (1967). Nutrient availability in relation to soil water. *In* "Irrigation of Agricultural Lands" (R. M. Hagan, R. H. Haise, and W. Edminster, eds.), pp. 458–467. Am. Soc. Agron., Madison, Wisconsin.

Viets, F. G., and Hageman, R. H. (1971). Factors affecting the accumulation of nitrate in soil, water and plants. *U.S. Dep. Agric., Agric. Handb.* **413.**

Vityakon, P. (1979). Yield, nitrate accumulation, cation–anion balance and oxalate production of spinach and beetroot as affected by rates and forms of applied nitrogen fertilizers. M. Hort. Sci. Thesis, Lincoln College.

Vlek, P. L. G., Hong, C. W., and Youngdahl, L. J. (1979). An analysis of N nutrition on yield components for the improvement of rice fertilization in Korea. *Agron. J.* **71,** 829–833.

Vlek, P. L. G., Stumpfe, J. M., and Byrnes, B. H. (1980). Urease activity and inhibition in flooded soil systems. *Fert. Res.* **1**, 191–202.

Vlek, P. L. G., Fillery, I. R. P., and Burford, J. R. (1981). Accession, transformation, and loss of nitrogen in soils of the arid region. *In* "Soil Water and Nitrogen in Mediterranean-Type Environments" (J. Monteith and C. Webb, eds.), pp. 133–175. Martinus Nijhoff/Dr. W. Junk, The Hague.

Vose, P. B., and Breese, E. L. (1964). Genetic variation in the utilization of nitrogen by ryegrass species *Lolium perenne* and *L. multiflorum. Ann. Bot. (London)* [N.S.] **28**, 253–270.

Wahua, T. A. T., and Miller, D. A. (1978). Effects of intercropping on soybean N_2-fixation and plant composition of associated sorghum and soybeans. *Agron. J.* **70**, 292–295.

Walmsley, D., and Forde, S. C. M. (1976). Further studies on the evaluation and calibration of soil analysis methods for N, P and K in the Eastern Caribbean. *Trop. Agric. (Trinidad)* **53**, 281–291.

Walters, C. L., and Walker, R. (1978). Consequences of accumulation of nitrate in plants. *In* "Nitrogen Assimilation of Plants" (E. J. Hewitt and C. V. Cutting, eds.), pp. 637–648. Academic Press, New York.

Ward, R. C., Whitney, D. A., and Westfall, D. G. (1973). Plant analysis as an aid in fertilizing small grains. *In* "Soil Testing and Plant Analysis" (L. M. Walsh and J. D. Beaton, eds.), pp. 329–348. Soil Sci. Soc. Am., Madison, Wisconsin.

Warren, H. L., Huber, D. M., Nelson, D. W., and Mann, O. W. (1975). Stalk rot incidence and yield of corn as affected by inhibiting nitrification of fall-applied ammonium. *Agron. J.* **67**, 655–660.

Watson, E. R., Lapins, P., and Barron, R. J. W. (1980). Effects of subterranean, rose and cupped clovers on soil nitrogen and on subsequent cereal crop. *Aust. J. Exp. Agric. Anim. Husb.* **20**, 354–358.

Watts, D. A., and Martin, D. L. (1981). Effects of water and nitrogen management on nitrate leaching loss from sands. *Trans. ASAE* **4**, 911–916.

Weil, R. R., and Ohlrogge, A. J. (1972). The seasonal development of the effect of interplant competition on soybean nodules. *Agron. Abstr.*, p. 59.

Welch, L. F., Boone, L. V., Chambliss, C. G., Christiansen, A. T., Mulvaney, D. L., Oldham, M. G., and Pendleton, J. W. (1973). Soybean yields with direct and residual fertilization. *Agron. J.* **65**, 547–550.

Wesley, K. W. (1979). Irrigated corn production and moisture management. *Univ. Ga., Coop. Ext. Ser., Bull.* **820.**

Westerman, R. L., Kurtz, L. T., and Hauck, R. D. (1972). Recovery of ^{15}N-labelled fertilizers in field experiments. *Soil Sci. Soc. Am. Proc.* **36**, 82–86.

Westermann, D. T., Kleinkopf, G. E., Porter, L. K., and Leggett, G. E. (1981). Nitrogen sources for bean production. *Agron. J.* **73**, 660–664.

Wetselaar, R. (1961). Nitrate distribution in tropical soils. I. Possible causes of nitrate accumulation near the soil surface after a long period. *Plant Soil* **15**, 110–120.

White, D. H., Elliot, B. R., Sharkey, M. J., and Reeves, T. G. (1978). Efficiency of land-use systems involving crops and pastures. *J. Aust. Inst. Agric. Sci.* **44**, 21–27.

White, D. J. (1977). Prospects for greater efficiency in the use of different energy sources. *Philos. Trans. Roy. Soc. London, Ser. B* **281**, 261–275.

Whitehead, D. C. (1970). "The Role of Nitrogen in Grassland Productivity," C.A.B. Pastures and Field Crops Bull. No. 48. Commonw. Agric. Bur., Slough, England.

Whitehead, D. C. (1981). An improved chemical extraction method for predicting the supply of available nitrogen. *J. Sci. Food Agric.* **32**, 359–365.

Whitehead, D. C., Barnes, R. J., and Morrison, J. (1981). An investigation of analytical procedures for predicting soil nitrogen supply to grass in the field. *J. Sci. Food Agric.* **32**, 211–218.

Widdowson, F. V., Penny, A., and Williams, R. J. B. (1961). Applying nitrogen fertilizers for spring barley. *J. Agric. Sci.* **56**, 39–47.

Wilkes, J. M., and Scarisbrick, D. H. (1974). The effect of nitrogenous fertilizer on the seed yield of the navy bean (*Phaseolus vulgaris*). *J. Agric. Sci.* **83**, 175–176.

Wilkinson, S. R., and Lowrey, R. W. (1973). Cycling of mineral nutrients in pasture ecosystems. *In* "Chemistry and Biochemistry of Herbage" (G. W. Butler and R. W. Bailey, eds.), Vol. 2, pp. 248–309. Academic Press, New York.

Willey, R. W. (1979a). Intercropping—Its importance and research needs. Part 1. Competition and yield advantages. *Field Crop Abstr.* **32**, 1–10.

Willey, R. W. (1979b). Intercropping—Its importance and research needs. Part 2. Agronomy and research approaches. *Field Crop Abstr.* **32**, 73–85.

Wilman, D. (1980). Early spring and late autumn response to applied nitrogen in four grasses. I. Yield, number of tillers and chemical composition. *J. Agric. Sci.* **94**, 425–442.

Wilman, D., and Mohamed, A. A. (1980). Early spring and late autumn response to applied nitrogen in four grasses. 2. Leaf development. *J. Agric. Sci.* **94**, 443–453.

Wilman, D., and Mohamed, A. A. (1981). Response to nitrogen application and interval between harvests in five grasses. 2. Leaf development. *Fert. Res.* **2**, 3–20.

Wilman, D., and Wright, P. T. (19830. Some effects of applied nitrogen on the growth and chemical composition of temperate grasses. *Herb. Abstr.* **53**, 387–393.

Wilman, D., Koocheki, A., and Lwoga, A. B. (1976). The effect of interval between harvests and nitrogen application on the proportion and yield of crop fractions and on the digestibility and digestible yield and nitrogen content and yield of two perennial ryegrass varieties in the second harvest year. *J. Agric. Sci.* **87**, 59–74.

Wilman, D., Droushiotis, D., Mzamane, M. N., and Shim, J. S. (1977). The effect of interval between harvests and nitrogen application on initiation, emergence and longevity of leaves, longevity of tillers and dimensions and weights of leaves and 'stems' in *Lolium. J. Agric. Sci.* **89**, 65–79.

Wilson, J. R. (1982). Environmental and nutritional factors affecting nutritional quality. *In* "Nutritional Limits to Animal Production from Pastures" (J. B. Hackler, ed.), pp. 111–131. Commw. Agric. Bur., Slough, England.

Wilson, R. K., and Flynn, A. V. (1979). Effects of fertilizer N, wilting and delayed sealing on the biochemical composition of grass silages made in laboratory silos. *Ir. J. Agric. Res.* **18**, 13–23.

Wolff, I. A., and Wasserman, A. E. (1972). Nitrates, nitrites, and nitrosamines. *Science* **177**, 15–19.

Wood, J. (1980). The mathematical expression of crop response to inputs. *Proc. Colloq. Int. Potash Inst.* **15**, 263–271.

Wright, M., and Davison, K. L. (1964). Nitrate accumulation in crops and nitrate poisoning of animals. *Adv. Agron.* **16**, 197–247.

Zamyatina, V. B. (1971). Nitrogen balance studies using ^{15}N-labelled fertilizers based on nitrogen-15 studies in the USSR. *In* "Nitrogen-15 in Soil–Plant Studies," pp. 33–45. IAEA, Vienna.

Zeiher, C., Egli, D. B., Legget, J. E., and Reicosky, D. A. (1982). Cultivar differences in N redistribution in soybeans. *Agron. J.* **74**, 375–379.

Zindler-Frank, E. (1976). Oxalate biosynthesis in relation to photosynthetic pathway and plant productivity—A survey. *Z. Pflanzenphysiol.* **80**, 1–13.

Index

no damage found 4/9/98 -NR